Treatment of Micropollutants
in Water and Wastewater

Integrated Environmental Technology Series

The *Integrated Environmental Technology Series* addresses key themes and issues in the field of environmental technology from a multidisciplinary and integrated perspective.

An integrated approach is potentially the most viable solution to the major pollution issues that face the globe in the 21st century.

World experts are brought together to contribute to each volume, presenting a comprehensive blend of fundamental principles and applied technologies for each topic. Current practices and the state-of-the-art are reviewed, new developments in analytics, science and biotechnology are presented and, crucially, the theme of each volume is presented in relation to adjacent scientific, social and economic fields to provide solutions from a truly integrated perspective.

The *Integrated Environmental Technology Series* will form an invaluable and definitive resource in this rapidly evolving discipline.

Series Editor

Dr Ir Piet Lens, Sub-department of Environmental Technology, IHE-UNESCO, P.O. Box 3015, 2601 DA Delft, The Netherlands (p.lens@unesco-ihe.org).

Published titles

Biofilms in Medicine, Industry and Environmental Biotechnology: *Characteristics, analysis and control*
Decentralised Sanitation and Reuse: *Concepts, systems and implementation*
Environmental Technologies to Treat Sulfur Pollution: *Principles and engineering*
Phosphorus in Environmental Technology: *Principles and applications*
Soil and Sediment Remediation: *Mechanisms, techniques and applications*
Water Recycling and Resource Recovery in Industries: *Analysis, technologies and implementation*
Environmental Technologies to treat Nitrogen Pollution
Bioelectrochemical Systems: *From extracellular electron transfer to biotechnological application*
Treatment of Micropollutants in Water and Wastewater

Integrated Environmental Technologies series

Treatment of Micropollutants in Water and Wastewater

Editors
Jurate Virkutyte (University of Eastern Finland, Finland),
Rajender S. Varma (USEPA, Cincinnati, OH, USA),
Veeriah Jegatheesan (James Cook University, Australia)

Publishing
London • New York

Published by   IWA Publishing
                 Alliance House
                 12 Caxton Street
                 London SW1H 0QS, UK
                 Telephone: +44 (0)20 7654 5500
                 Fax: +44 (0)20 7654 5555
                 Email: publications@iwap.co.uk
                 Web: www.iwapublishing.com

First published 2010
© 2010 IWA Publishing

Typeset in India by OKS Prepress Services.
Printed by Lightning Source.

Apart from any fair dealing for the purposes of research or private study, or criticism or review, as permitted under the UK Copyright, Designs and Patents Act (1998), no part of this publication may be reproduced, stored or transmitted in any form or by any means, without the prior permission in writing of the publisher, or, in the case of photographic reproduction, in accordance with the terms of licences issued by the Copyright Licensing Agency in the UK, or in accordance with the terms of licenses issued by the appropriate reproduction rights organization outside the UK. Enquiries concerning reproduction outside the terms stated here should be sent to IWA Publishing at the address printed above.

The publisher makes no representation, express or implied, with regard to the accuracy of the information contained in this book and cannot accept any legal responsibility or liability for errors or omissions that may be made.

Disclaimer
The information provided and the opinions given in this publication are not necessarily those of IWA Publishing and should not be acted upon without independent consideration and professional advice. IWA Publishing and the Author will not accept responsibility for any loss or damage suffered by any person acting or refraining from acting upon any material contained in this publication.

British Library Cataloguing in Publication Data
A CIP catalogue record for this book is available from the British Library

Library of Congress Cataloging-in-Publication Data
A catalog record for this book is available from the Library of Congress

ISBN 10: 1843393166
ISBN 13: 9781843393160

Contents

Preface xv

1 MICROPOLLUTANTS AND AQUATIC ENVIRONMENT 1
1.1 Introduction 1
1.2 Pesticides 2
    1.2.1 Organochlorine insecticides 3
        1.2.1.1 Fate 4
        1.2.1.2 Effects 5
    1.2.2 Organophosporous insecticides 6
        1.2.2.1 Fate 7
        1.2.2.2 Effects 8
    1.2.3 Triazine herbicides 8
        1.2.3.1 Fate 9
        1.2.3.2 Effects 10
    1.2.4 Substituted ureas 10

©2010 IWA Publishing. *Treatment of Micropollutants in Water and Wastewater*. Edited by Jurate Virkutyte, Veeriah Jegatheesan and Rajender S. Varma. ISBN: 9781843393160. Published by IWA Publishing, London, UK.

	1.2.4.1 Fate	11
	1.2.4.2 Effects	11
1.2.5	Legislation	12
1.3 Pharmaceuticals		13
1.3.1	Fate	15
1.3.2	Effects	19
1.3.3	Legislation	20
1.4 Steroid Hormones		21
1.4.1	Fate	22
1.4.2	Effects	24
1.4.3	Legislation	25
1.5 Surfactants and Personal Care Products		25
1.5.1	Fate	27
1.5.2	Effects	29
1.5.3	Legislation	30
1.6 Perfluorinated Compounds		30
1.6.1	Fate	31
1.6.2	Effects	32
1.6.3	Legislation	33
1.7 References		35

2 ANALYTICAL METHODS FOR THE IDENTIFICATION OF MICROPOLLUTANTS AND THEIR TRANSFORMATION PRODUCTS 53

2.1 Introduction 53
2.2 Theoretical Approaches to the Analytics of Micropollutants 60
    2.2.1 Computational methods to evaluate the degradation of micropollutants 60
        2.2.1.1 Frontier electron density analysis *(Frontier orbital theory)* 61
    2.2.2 Chemometrics in analysis 63
        2.2.2.1 Parafac 64
        2.2.2.2 MCR 64
        2.2.2.3 BLLS 65
        2.2.2.4 U-PLS 65
        2.2.2.5 ANN 66
2.3 Instrumental Methods 66
    2.3.1 Sample preparation 67
        2.3.1.1 Sample extraction 67

		2.3.1.2	Chromatographic separation	69
		2.3.1.3	Capillary electrophoresis (CE)	69
	2.3.2	Detection of micropollutants transformation products		71
		2.3.2.1	Mass spectrometry	72
	2.3.3	UV-Visible spectroscopy		76
	2.3.4	NMR spectroscopy		76
	2.3.5	Biological assessment of the degradation products		78
		2.3.5.1	Ecotoxicological assessment of environmental risk (toxicity)	79
		2.3.5.2	Assessment of estrogenic activity	80
		2.3.5.3	Assessment of antimicrobial activity	81
		2.3.5.4	Biosensors	81
2.4	Identification Levels of Micropollutants Transformation Products			82
2.5	Conclusions			85
2.6	References			85

3 SENSORS AND BIOSENSORS FOR ENDOCRINE DISRUPTING CHEMICALS: STATE-OF-THE-ART AND FUTURE TRENDS 93

3.1	Introduction		93
3.2	Sensors and Biosensors		94
	3.2.1	The need for alternative methods	94
	3.2.2	Electrochemical sensors	95
	3.2.3	Biosensors	96
	3.2.4	New generation immunosensors	100
3.3	Trends in Sensors and Biosensors		106
	3.3.1	Screen printed sensors and biosensors	106
	3.3.2	Nanotechnology applications	106
	3.3.3	Molecular imprinted polymer sensors	108
	3.3.4	Conducting polymers	112
3.4	Future of Sensing		114
3.5	References		115

4 NANOFILTRATION MEMBRANES AND NANOFILTERS 129

4.1	Introduction of Nanofiltration	129
4.2	Nanofiltration Membrane Materials	132
4.3	Separation and Fouling of Nanofiltration	136

viii   Treatment of Micropollutants in Water and Wastewater

4.4 Nanofiltration of Micropollutants in Water 142
4.5 References 152

5 PHYSICO-CHEMICAL TREATMENT OF MICROPOLLUTANTS: ADSORPTION AND ION EXCHANGE 165

5.1 Introduction 165
5.2 The Main Stages of Adsorption & Ion Exchange Science Development 167
5.3 Carbons in Water Treatment and Medicine 169
5.4 Zeolites (Clays) 172
5.5 Ion Exchange Resins or Ion Exchange Polymers 173
5.6 Inorganic Ion-Exchangers 176
    5.6.1 Ferrocyanides adsorbents 177
    5.6.2 Synthesis of inorganic ion exchangers 184
5.7 Biosorbents (Biomasses): Agricultural and Industrial By-Products, Microorganisms 187
5.8 Hybrid and Composite Adsorbents and Ion Exchangers 191
5.9 Comments on the Theory and Future of Adsorption and Ion-Exchange Science 192
5.10 Acknowledgement 194
5.11 References 194

6 PHYSICO-CHEMICAL TREATMENT OF MICROPOLLUTANTS: COAGULATION AND MEMBRANE PROCESSES 205

6.1 Coagulation 205
    6.1.1 Enhanced coagulation 206
        a) Effects of physical-chemical properties of micropollutants 208
        b) Choice of coagulants and dosage 210
        c) pH and alkalinity 211
    6.1.2 Coagulation-oxidation 213
6.2 Membrane Processes 215
    6.2.1 Mechanisms of solute rejection during membrane treatment 216
    6.2.2 Micropollutant removal by microfiltration 217
    6.2.3 Micropollutant removal by ultrafiltration 218
        a) Ultrafiltration alone 218
        b) Combination of ultrafiltration and powdered activated carbon 219

c) Combination of ultrafiltration and biological module
      (membrane bioreactor) 220
   6.2.4 Micropollutant removal by reverse osmosis 225
   6.2.5 Electrodialysis 227
6.3 References 233

7 BIOLOGICAL TREATMENT OF MICROPOLLUTANTS ... 239
7.1 Introduction 239
7.2 Municipal Sewage as the Source of Micropollutants 240
   7.2.1 Urine source separation and possible advantages 243
   7.2.2 Biological degradation in source separated urine 247
7.3 Biological Treatment of Micropollutants 249
   7.3.1 Analysis of Micropollutants 250
      7.3.1.1 Analytical techniques used for wastewater and
              sludge samples 251
      7.3.1.2 Endocrine disrupting effect 252
   7.3.2 Removal mechanisms of Micropollutants 254
      7.3.2.1 Sorption 254
      7.3.2.2 Abiotic degradation and volatilization 257
      7.3.2.3 Biodegradation 258
   7.3.3 Factors affecting the biological removal efficiency 259
      7.3.3.1 Compound structure 259
      7.3.3.2 Bioavailability 261
      7.3.3.3 Dissolved oxygen and pH 261
      7.3.3.4 Hydraulic and sludge retention time 263
      7.3.3.5 Organic load rate 266
      7.3.3.6 Temperature 266
   7.3.4 Biological treatment of Micropollutants in different
         processes 268
      7.3.4.1 Activated sludge systems 268
      7.3.4.2 Wetlands 272
      7.3.4.3 Membrane bioreactors 274
      7.3.4.4 Anaerobic treatment 275
      7.3.4.5 Other bioreactors 277
   7.3.5 Biological treatment of Micropollutants in sludge 278
   7.3.6 Specific microorganisms/cultures used for biodegradation
         of Micropollutants 278
   7.3.7 Formation of by-products during biodegradation 280
7.4 References 281

8 UV IRRADIATION FOR MICROPOLLUTANT REMOVAL FROM AQUEOUS SOLUTIONS IN THE PRESENCE OF H_2O_2 295

8.1 Introduction 295
8.2 Theory of UV/H_2O_2 296
    8.2.1 General 296
    8.2.2 Photolysis 298
    8.2.3 Mechanisms UV/H_2O_2 oxidation 299
    8.2.4 Ozone/UV 300
8.3 Laboratory Scale Experiments of UV/H_2O_2 300
    8.3.1 General 300
    8.3.2 Treatment of contaminated groundwater 300
    8.3.3 Drinking water applications 303
    8.3.4 Municipal waste water 305
    8.3.5 Paper and pulp industry 307
8.4 Other UV Based Techniques 307
8.5 Alternative Radiation Sources 309
8.6 Practical Issues of UV/H_2O_2 Treatment 310
8.7 Cost Estimation & Performance 313
8.8 References 316

9 HYBRID ADVANCED OXIDATION TECHNIQUES BASED ON CAVITATION FOR MICROPOLLUTANTS DEGRADATION 321

9.1 Introduction 321
9.2 Theory of Ultrasound 322
    9.2.1 Cavitation phenomena 322
    9.2.2 The general hypothesis in sonochemical processing 322
    9.2.3 Cavitation effects 323
    9.2.4 Factors affecting the efficiency of sonochemical degradation 325
        9.2.4.1 Ultrasonic frequency 325
        9.2.4.2 Input electrical power 326
        9.2.4.3 Nature of the compound and the reaction pH 326
        9.2.4.4 The reaction temperature 327
        9.2.4.5 The presence of additives 327
        9.2.4.6 Ultrasonic equipment 329
9.3 Hybrid Cavitation-Based Technologies 330

      9.3.1 US/oxidant 330
            9.3.1.1 US/H_2O_2 330
            9.3.1.2 US/O_3 331
      9.3.2 US/UV 333
      9.3.3 US/A 333
      9.3.4 US/EO 334
      9.3.5 US/MW 335
9.4 Degradation of Micropollutants 336
      9.4.1 Degradation of pharmaceuticals by hybrid techniques based on cavitation 336
      9.4.2 Degradation of organic dyes by hybrid techniques based on cavitation 340
      9.4.3 Degradation of pesticides by hybrid techniques based on cavitation 343
9.5 Scale-Up Considerations 348
9.6 Economical Aspects of Cavitation Based Treatment 350
9.7 Conclusions 353
9.8 References 353

10 ADVANCED CATALYTIC OXIDATION OF EMERGING MICROPOLLUTANTS 361

10.1 Introduction 361
10.2 Heterogeneous Catalysis 362
      10.2.1 Desirable properties of the catalyst 363
10.3 Environmental Catalysis 364
10.4 Advanced Catalytic Oxidation Processes for the Removal of Emerging Contaminants from the Aqueous Phase 365
      10.4.1 Catalytic wet peroxide oxidation processes (CWPO) 365
            10.4.1.1 Homogeneous Fenton process 365
            10.4.1.2 Heterogeneous Fenton process 368
            10.4.1.3 Heterogenized catalyst for the micropollutants removal 369
      10.4.2 Other metal catalysts in wet peroxide oxidation of micropollutants 371
      10.4.3 Catalytic ozonation of micropollutants 373
      10.4.4 Photocatalytic degradation of micropollutants 375
            10.4.4.1 Titanium dioxide catalyzed degradation of micropollutants 377

| | | 10.4.4.2 | Photo-Fenton process for the degradation of micropollutants .... | 379 |

| | | 10.4.4.3 | Other photocatalysts in the degradation of micropollutants .... | 381 |

- 10.4.5 Sonocatalytic degradation of micropollutants 383
- 10.4.6 Microwave-assisted catalytic degradation of miropollutants 388
- 10.4.7 Electrocatalytic oxidation 390
  - 10.4.7.1 Degradation of micropollutants with electrocatalytic and coupled electrocatalytic methods 392
- 10.4.8 Biocatalytic oxidation of micropollutants 395
- 10.4.9 Catalytic wet air oxidation of micropollutants 397

10.5 Advanced Nanocatalytic Oxidation of Micropollutants........ 399
10.6 Conclusions.. 414
10.7 References.. 415

11 EXISTENCE, IMPACTS, TRANSPORT AND TREATMENTS OF HERBICIDES IN GREAT BARRIER REEF CATCHMENTS IN AUSTRALIA 425

11.1 Introduction.. 425
11.2 Persistent Organic Pollutants (POPs) 426
11.3 Herbicides and Pesticides... 433
11.4 Great Barrier Reef (GBR).. 438
    11.4.1 Background 438
    11.4.2 Transport of herbicides and pesticides into the GBR 439
11.5 Persistence of Herbicides and Pesticides in the GBR Catchments and Lagoon... 442
11.6 Impact to the GBR Ecosystem due to the Persistence of Herbicides and Pesticides... 445
11.7 Removal of Herbicides by Different Water Treatment Processes.. 447
11.8 Possible Methods of Treatment of POPs Including Herbicides and Pesticides from Catchment Discharges 450
    11.8.1 Biological processes 450
    11.8.2 Adsorption processes 451
    11.8.3 Wetland processes 451
    11.8.4 Pressure driven membrane filtration processes 452
    11.8.5 Hybrid systems 453

       11.8.6 Hybrid systems – membrane bioreactors (MBR) 454
       11.8.7 Other Processes 457
11.9 Conclusions.. 457
11.10 References... 457

Index... **465**

Preface

Micropollutants – compounds which are found in the μg L^{-1} or ng L^{-1} concentration range in water, soil and wastewater are considered to be potential threats to environmental ecosystems. Over the last few years there has been a growing concern of the scientific community due the increasing concentration of micropollutants originating from a great variety of sources including pharmaceutical, chemical engineering and personal care product industries in rivers, lakes, soil and groundwater.

Once released into the environment, micropollutants are subjected to different processes such as distribution between different phases, biological and abiotic degradation. These processes contribute to their elimination and affect their bioavailability. The role of the aforementioned processes in micropollutants' fate depends on the physico-chemical properties of these compounds (polarity, water solubility, vapor pressure) and the type of the environment (natural or mechanical) where the micropollutants are present (groundwater, surface water, sediment, wastewater treatment systems, drinking-water facilities). As a result,

©2010 IWA Publishing. *Treatment of Micropollutants in Water and Wastewater*. Edited by Jurate Virkutyte, Veeriah Jegatheesan and Rajender S. Varma. ISBN: 9781843393160. Published by IWA Publishing, London, UK.

different transformation reactions can produce metabolites that often differ in their environmental behavior and ecotoxicological profile from the parent compounds.

The current concern of micropollutants in the receiving waters calls for new approaches in wastewater treatment. Unfortunately, conventional wastewater treatment plants are designed to deal with bulk substances that arrive regularly and in large quantities (primarily organic matter and the nutrients or nitrogen and phosphorus). On the contrary, micropollutants are entirely different due to their unique characteristics and behavior in the treatment plant. Thus, new measures must be taken into account to reduce or entirely eliminate these contaminants from water and wastewater.

The 10 chapters of this book have been arranged in such a way that it forms the core of Micropollutants science area: their occurrence in aquatic systems, detection and analysis utilizing the newest trends in sensors and biosensors fields, biological, physical and chemical treatment methods exploiting sole and hybrid techniques as well as presenting a case study – effect of pesticides on the one of the most precious wonders of the natural world – Great Barrier Reef (Australia). Most chapters have designed to include (i) a theoretical background, (ii) a review on the actual knowledge and (iii) cutting-edge research results. Therefore this book will be suitable for water and wastewater professionals as well for students and researchers in civil engineering, environmental chemistry, environmental engineering and process engineering fields, especially to those who wish to pinpoint the actual frontiers of science in this specific domain.

We wish to thank all authors for providing high quality manuscripts. We are indebted to Ms Maggie Smith, the Editor of IWA for having accepted our proposal to design this book. Moreover, Prof. Piet Lens, the Editor of Integrated Environmental Technology Series is greatly acknowledged for the valuable comments and recommendations regarding the layout and scientific presentation of the book.

Drs Jurate Virkutyte, Rajender S. Varma
and Veeriah Jegatheesan.
March, 2010.

Chapter 1
Micropollutants and Aquatic Environment

A. S. Stasinakis and G. Gatidou

1.1 INTRODUCTION

Micropollutants are compounds which are found in the µg L^{-1} or ng L^{-1} concentration range in the aquatic environment and are considered to be potential threats to environmental ecosystems. Different groups of compounds are included in this category such as pesticides, PCBs, PAHs, flame retardants, perfluorinated compounds, pharmaceuticals, surfactants and personal care products. Recent studies have indicated the often detection of these compounds in the aquatic environment (Kolpin *et al.*, 2002; Loos *et al.*, 2009). The way that these compounds enter the environment depends on their uses and the mode of application. The major routes seem to be agricultural and urban runoff, municipal and industrial wastewater discharge, sludge disposal and accidental spills (Ashton *et al.*, 2004; Becker *et al.*, 2008; Mompelat *et al.*, 2009).

©2010 IWA Publishing. *Treatment of Micropollutants in Water and Wastewater.* Edited by Jurate Virkutyte, Veeriah Jegatheesan and Rajender S. Varma. ISBN: 9781843393160. Published by IWA Publishing, London, UK.

Once released into the environment, micropollutants are subjected to different processes such as distribution between different phases, biological and abiotic degradation (Halling-Sørensen *et al.,* 1998; Hebberer, 2002a; Birkett and Lester, 2003; Farre *et al.,* 2008). These processes contribute to their elimination and affect their bioavailability. The role of the aforementioned processes in micropollutants' fate depends on the physico-chemical properties of these compounds (polarity, water solubility, vapor pressure) and the type of the environment (natural or mechanical) where the micropollutants are present (groundwater, surface water, sediment, wastewater treatment systems, drinking-water facilities). As a result, different transformation reactions can take place, producing metabolites that often differ in their environmental behavior and ecotoxicological profile from the parent compounds. So far, several effects of these compounds on aquatic organisms have been reported such as acute and chronic toxicity, endocrine disruption, bioaccumulation and biomagnifications (Oaks *et al.,* 2004; Fent *et al.,* 2006; Darbre and Harvey, 2008).

In the next paragraphs, data for the occurrence, fate and effects of some micropollutants' categories will be given. Pesticides have been selected due to their wide use and their well-defined toxicological effects and environmental fate. Five categories of emerging contaminants will be also presented (pharmaceuticals, steroid hormones, perfluorinated compounds, surfactants and personal care products) due to the great interest that has recently arisen for their occurrence in the environment.

1.2 PESTICIDES

The word "pesticide" precisely is referred to an agent that is used to kill an unwanted organism (Rana, 2006). According to EPA (USEPA) pesticide is called an organic compound (or mixture of compounds) which acts against pests (insect, rodent, fungus, weed etc.) by several ways like prevention, destruction, repulse or mitigation. These biologically active chemicals are often called biocides and include several classes such as herbicides, insecticides and fungicides, depending on the type of the pests that they control.

Before 1940, inorganic compounds and few natural agents originated from plants were used as pesticides (Rana, 2006). The great production of synthetic organic compounds for use in pest control was started with the discovery of DDT's insecticidal activity in 1938 and it was continued during and following the Second World War (Matthews, 2006). During the next decades, an exponential increase in production and use of synthetic pesticides was observed worldwide (Rana, 2006). Today, many different classes of pesticides are used, including among others chlorinated hydrocarbons, organophosphoric compounds, substituted ureas and triazines. Despite the positive effects from the usage of synthetic pesticides on public health and global economy, the extensive use of these compounds resulted in serious environmental contamination problems worldwide and deleterious effects on humans and ecosystems. Risks associated with pesticides usage were firstly mentioned by Rachel Carson in her book *Silent Spring*.

1.2.1 Organochlorine insecticides

Insecticides are one of the most significant types of pesticides due to the fact that they can be applied in a short time before harvesting and after crops collection (Manahan, 2004). Initially, insecticides were divided into two main groups: the organochlorines and organophosphates (Matthews, 2006). Both of these groups affect the nervous system of the organisms by inhibiting the enzyme called acetylcholinesterase (Walker *et al.*, 2006).

Organochlorine insecticides are halogenated solid organic compounds, highly lipophilic, with very low water solubility and high persistence. These properties result in remaining of residues in the environment for long time and further accumulation to several animals (Matthews, 2006; Walker *et al.*, 2006). Among the different classes of organochlorine insecticides, dichlorodiphenylethanes, chlorinated cyclodienes (or chlordanes) and hexachlorocyclohexanes are of great concern due to their potential risk to human health and environmental fate (Qiu *et al.*, 2009). The most known dichlorodiphenylethane pesticide is DDT (Figure 1.1). This compound was mainly used for vector control during the Second World War and thereafter was extensively used in agriculture (Walker *et al.*, 2006).

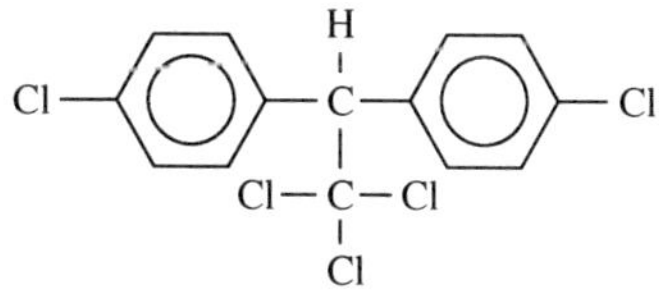

Figure 1.1 The organochlorine insecticide DDT (p'p-dichlorodophenyltrichloroethane)

Chlorinated cyclodienes, such as aldrin, dieldrin or heptachlor (Figure 1.2), were introduced in 1950s and they were used both for crop protection against pests and certain vectors of disease (e.g., tsetse fly) (Walker *et al.*, 2006).

The organochloride insecticides belonging in the group of hexachlorocyclohexanes (HCH) were introduced in the market as a crude mixture of isomers (Walker *et al.*, 2006). Between the five isomers of this mixture, only γ isomer, commonly referred as γ-HCH or lindane (Figure 1.3), found to be effective as insecticide (Manahan, 2004).

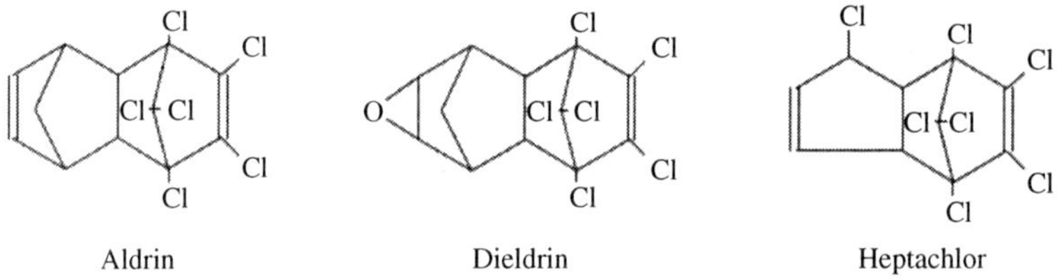

Figure 1.2 The organochloride insecticides aldrin, dieldrin and heptachlor

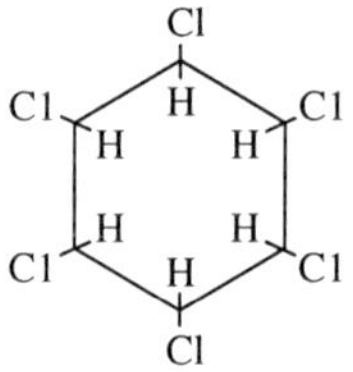

Figure 1.3 The organochloride insecticide lindane (1,2,3,4,5,6-hexachlorocyclohexane)

1.2.1.1 Fate

The widespread used organochlorine pesticides present ubiquitous persistence in different environmental media. For instance, a half-life ranging between 3 and 4 years has been estimated for dieldrin in soils, whereas a much higher half-life time (up to 15 years) has been calculated for DDT (UNEP, 2002). These half-lives indicate the the very slow elimination of compounds like DDT by most leaving organisms (IARC/WHO, 1991).

These compounds can enter the aquatic environment in several ways such as run-off from non-point sources or discharge of industrial wastewater. Despite their low water solubility, several organochlorine pesticides are detected worldwide in water column (Table 1.1).

It is entirely known that organochlorine pesticides present high affinity for lipid tissues. As a result, they can be bioaccumulated and biomagnified. According to Zhou *et al.* (2008), log bioconcentration factors (BCFs) of several organochlorine compounds vary from 2.88 to 6.28 for fish, from 3.78 to 6.17 for shrimp and from 3.13 to 5.42 for clams. Through the consumption of aquatic organisms, drinking water and agricultural crops, these compounds reach humans and are excreted to breast milk. Dahmardeh-Behrooz *et al.* (2009) reported mean concentration of total DDTs and HCHs equal to

3563 and 5742 ng g^{-1} (lipid weight), respectively, in human milk in Iran. Besides their banning, these compounds are also detected in human milk in developed countries. Kalantzi *et al.* (2004) reported concentrations as high as 220 and 40 ng g^{-1} (lipid weight) for DDTs and HCHs, respectively, in human milk from women in England. Moreover, Polder *et al.* (2003) detected even higher concentrations of DDTs (1200 ng g^{-1} lipid weight) and HCHs (320 ng g^{-1} lipid weight) after analyzing human milk in Russia.

Table 1.1 Typical water concentrations of several organochlorine (OC) insecticides

Compound(s)	Concentration (ng L^{-1})	Country	Reference
Total OCs[a]	0.01–9.83	China	Luo *et al.*, 2004
Total OCs	<LOD[b]-112	Greece	Golfinopoulos *et al.*, 2003
Total OCs	0.1–973	China	Zhou *et al.*, 2001
DDT	150–190	India	Shukla *et al.*, 2006
DDT	3.0–33.2	India	Pandit *et al.*, 2002
Lindane	680–1380	India	Shukla *et al.*, 2006
HCH	0.16–15.9	India	Pandit *et al.*, 2002

[a] Referred to the concentration range of compounds: DDTs (DDT and metabolites), aldrin, dieldrin, heptachlor and lindane.
[b] LOD = Limit of Detection.

DDTs or HCH have also been detected in other human matrices such as serum and adipose tissues. Koppen *et al.* (2002) detected concentrations of *p,p*-DDT (2.6 ng g^{-1} fat), *p,p*-DDE (871.3 ng g^{-1} fat) and γ-HCH (5.7 ng g^{-1} fat) in human serum from women in Belgium. Similarly, Botella *et al.* (2004) examined the presence of several organochlorine pesticides in human adipose tissues and blood samples in women from Spain. The detected mean concentrations of total DDTs in the two types of samples were 543.25 ng g^{-1} (human adipose tissues) and 12.10 ng mL^{-1} (blood samples), indicating either a relatively recent exposure or cumulative past exposure.

1.2.1.2 Effects

Toxicity of organochlorine pesticides depends on several parameters including the structure of each compound, the different moieties attached to initial molecule, the nature of substituents (Kaushik and Kaushik, 2007). In many cases, compounds are considered to be moderately toxic to mammals and highly toxic to aquatic organisms. For instance, DDT is considered to be moderately toxic to mammals with LD_{50} values ranging from 113 to 118 mg Kg^{-1} (body weight), while a concentration of 0.6 mg kg^{-1}, reported to cause egg shell thinning for the black duck (UNEP, 2002). Similarly, heptachlor is also considered moderately toxic to mammals but its toxicity to aquatic organisms is high. An LC_{50} value equal to 0.11 µg L^{-1} has been found for pink shrimp (UNEP, 2002). Aldrin and dieldrin are characterized as compounds with high toxicity on aquatic organisms too (Vorkamp *et al.*, 2004). Aldrin's toxicity to aquatic organisms can vary from 1–200 µg L^{-1} for aquatic insects to 2.2–53 µg L^{-1} for fish (96-h LC_{50}). On the contrary, lindane is considered moderately toxic for these organisms. According to UNEP (2002), its estimated LC_{50} values vary from 20 to 90 µg L^{-1} for invertebrates and fish (UNEP, 2002).

Additionally to acute toxicity, organochlorine insecticides are known for their endocrine disrupting effects (Luo *et al.*, 2004). According to Soto *et al.* (1995) and Chen *et al.* (1997) *p,p'*-DDE and *p,p'*-DDT interact with human ERα. Furthermore, *p,p'*-DDE has been reported to act as an antagonist for a human AR (Kelce and Wilson, 1997).

1.2.2 Organophosporous insecticides

Organophosphorous insecticides are synthetic organic compounds which contain phosphorus in their molecule and are organic esters of orthophosphoric acid, phosphonic or phosphorothioic or related acids (Manahan, 2004; Rana, 2006). These compounds were firstly produced for two uses: as insecticides and as chemical warfare gases during Second World War (Walker *et al.*, 2006). Nowadays, most of the organophospates are used as insecticides and their molecules are described with the general formula shown in Figure 1.4.

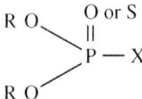

Figure 1.4 The general formula of organophosphorous insecticides. R: alkyl group, X: leaving group

Organophosphates are lipophilic compounds which present higher water solubility and lower stability comparing to organochlorine insecticides. As a result of their lower stability, these pesticides can break down easier by several physicochemical processes leading to shorter remaining times after their release in the environment (Walker *et al.*, 2006). The most commonly used organophosphorus insecticides are phosphorothionates such as methyl parathion and chlorpyrifos (Figure 1.5). These compounds have a sulphur atom (S), instead of oxygen atom (O), connected by double bond with phosphorous atom in their molecule (Manahan, 2004).

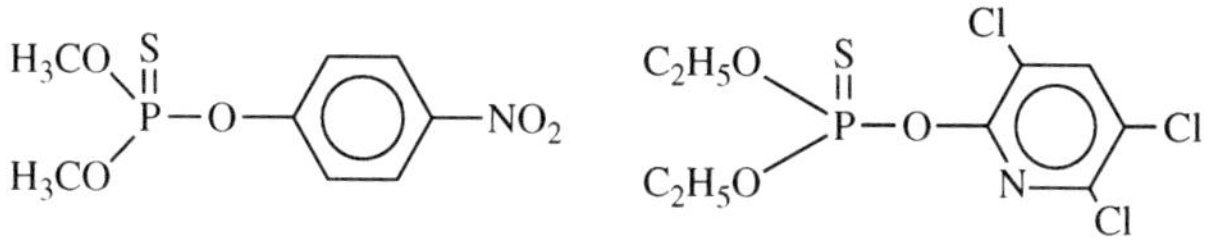

Methyl parathion   Chlorpyrifos

Figure 1.5 The organophosphorous insecticides parathion and chlorpyrifos

1.2.2.1 Fate

Beside the short life of organophosphates in the environment and their low water solubility, these compounds are detected in water column. Some indicative water concentrations of the compounds methyl parathion and chlorpyrifos are presented in Table 1.2.

Table 1.2 Typical water concentrations of the organophosphorous insecticides chlorpyrifos and methyl parathion

Compound	Concentration (ng L^{-1})	Country	Reference
Methyl parathion	<LOD-480	China	Gao et al., 2009
Methyl parathion	<LOD-41	Spain	Claver et al., 2006
Methyl parathion	13–332	Germany	Götz et al., 1998
Methyl parathion	20–270	Spain	Planas et al., 1997
Chlorpyrifos	<LOD-19.41	Italy	Carafa et al., 2007
Chlorpyrifos	<LOD-312	Spain	Claver et al., 2006

LOD: Limit of Detection.

Regarding their fate in the aquatic environment, organophosphorous insecticides can be degraded by oxidation, direct or indirect photodegradation, hydrolysis and adsorption (Pehkonen and Zhang, 2002). All these processes are responsible for the relative short lives of organophosphates in the environment. Araújo et al. (2007) investigated the photodegradation of methyl parathion under sunlight and they reported a half life time of about 5 days. Similarly, in another study, Castillo et al. (1997) investigated methyl parathion fate in natural waters and they reported half lives of 3 days (groundwater) and 4 days (estuarine and river water). Finally, Wu et al. (2006) found that chlorpyrifos is also photodegraded under sunlight and they determined a half life time of about 20 days. Biotic degradation of organophophates can also be occured in the environment. Liu *et al.* (2006) investigated the biodegradation of methyl parathion by two bacteria species *Shewanella* and *Vibrio parahaemolyticus* isolated from river sediments. According to their results, an initial concentration of 50 mg L^{-1} of the compound was almost totally disappeared during one week.

1.2.2.2 Effects

Organophosphates present varying degrees of toxicity in several organisms. For instance, methyl parathion is highly toxic to aquatic organisms and WHO has classified this compound as "extremely hazardous" for the environment, whereas chlorpyrifos is characterized as "moderately hazardous" (WHO, 2004). Furthermore, the transformation of organophosphates may result in the formation of more toxic and persistent metabolites. Dzyadevych *et al.* (2002) reported that methyl paraoxon, a photodegradation product of methyl parathion, found to be at least 10 times more toxic than the parent compound, regarding the inhibition on acetylcholinesterase activity.

Regarding the effects on humans, organophosphates are known for their neurotoxic effects. Additionally, genotoxic effects have been reported. Methyl parathion has been found to produce chromatid exchange in human lymphocytes, while interactions with the double-stranded DNA have also been reported (Rupa *et al.*, 1990; Blasiak *et al.*, 1995). Furthermore, adverse effects of these compounds in reproductive system have been reported in the literature. Salazar-Arredondo *et al.* (2008) investigated human sperm DNA damage of healthy spermatozoa by several organophosphates and their oxon metabolites. According to their results, the tested compounds found to be toxic on sperm DNA, while their metabolites were more toxic then the parent compounds.

1.2.3 Triazine herbicides

The term of herbicide is used to characterize another large group of pesticides which is used against weed control. These compounds act on contact with the plants or are translocated within the plants. According to the time of application, they can be classified to pre- or post-emergence. Furthermore, an herbicide can be a broad spectrum compound or a selective one (Matthews, 2006). Based on their chemical structure, many different groups of herbicides have been reported. Among them, triazines and substituted ureas have significant research interest due to their wide use, persistence and toxic effects.

Triazinic compounds contain three hetorocyclic nitrogen atoms in their molecule (Manahan, 2004). In case that the nitrogen and carbon atoms are interchanged in the ring structure, the herbicide is called symmetric (*s*) triazine. Otherwise, the compound is called non symmetric (*as*) triazine (Figure 1.6). The most widespread and extensively used triazinic herbicide is atrazine. However, there are other members of this group (e.g., simazine) which are also used widely (Strandberg and Scott-Fordsmand, 2002).

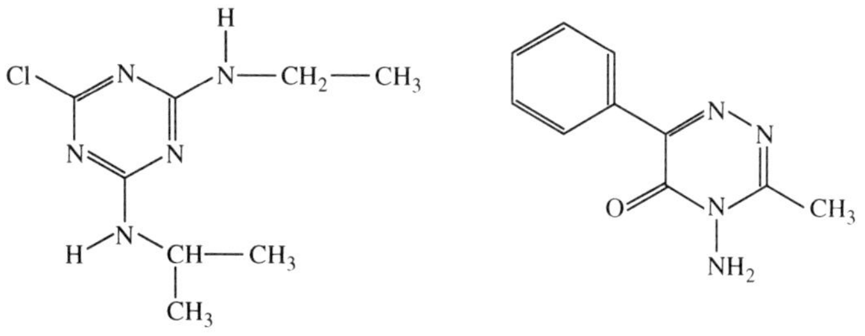

Figure 1.6 Chemical structure of a symmetric (e.g. atrazine) and non symmetric (e.g. metamitron) triazine

Triazinic compounds act as photosystem-II (PSII) inhibitors, affecting photosynthetic electron transport in chloroplasts (Corbet, 1974). Their selectivity is achieved by the inability of target weeds to metabolize and detoxify the herbicidal compound (Manahan, 2004). Triazines are solids, with low vapour pressure at room temperature and varying water solubility, ranging between 5 and 750 mg L^{-1} (Sabik *et al.*, 2000).

1.2.3.1 Fate

Triazines end up to environment via both point (e.g., industrial effluents) and diffuse sources (e.g., agriculture runoff). So far, there are several data regarding their occurrence in the aquatic environment. Some typical water concentrations for two widespread used triazinic herbicides (atrazine and simazine) are presented in Table 1.3.

Triazines are hydrolyzed quickly under acidic or alkaline pH, but at neutral pH are rather stable (Humburg *et al.*, 1989). Photo- and biodegradation of these compounds can also be occurred but s-triazines are found to be more resistant to microbes' attacks. For instance, atrazine is considered as a persistent organic pollutant, with half life ranging between 30 and 100 days (Worthing and Walker, 1987). The aforementioned biotic and abiotic transformations of triazines lead to the formation of metabolites by several mechanisms such as dehalogenation, dezalkylation and deamination (Peñuela and Barceló, 1998).

Table 1.3 Typical water concentrations of atrazine and simazine

Compound	Concentration (ng L^{-1})	Country	Reference
Atrazine	1.27–8.18	Italy	Carafa *et al.*, 2007
Atrazine	52–451	Spain	Claver *et al.*, 2006
Atrazine	<LOD-110	Australia	McMahon *et al.*, 2005
Atrazine	<LOD-3870	Greece	Albanis *et al.*, 2004
Atrazine	20–230	Greece	Lambropoulou *et al.*, 2002
Simazine	1.45–25.96	Italy	Carafa *et al.*, 2007
Simazine	49–183	Spain	Claver *et al.*, 2006
Simazine	<LOD-50	Australia	McMahon *et al.*, 2005
Simazine	<LOD-490	Greece	Albanis *et al.*, 2004

LOD: Limit of Detection.

1.2.3.2 Effects

Since triazines have been chemically designed to inhibit photosynthesis and animals lack a photosynthetic mechanism, these compounds are more toxic to plants. Acute toxicity to mammals and birds is low. For instance, atrazine has been classified by WHO (2004) as a compound unlikely to present acute hazard in normal use. Its oral LD$_{50}$ to rats is 3090 mg Kg^{-1} body weight. The compound is slightly toxic to fish and other aquatic organisms and practically nontoxic to birds (UNEP, 2002).

Despite the expected low toxicity to mammals, some triazines have been characterized as potential endocrine disruptors. Atrazine inhibits androgen-mediated development and produces estrogen-like effects in exposed organisms. Furthermore, the occurrence of atrazine in water has been considered responsible for affection of semen quality and fertility in men farmers, as well as increase of breast cancer in women (Fan *et al.*, 2007).

1.2.4 Substituted ureas

The herbicides of this group (e.g., diuron, isoproturon) are derived when hydrogen atoms in the urea molecule are substituted by several chemical groups (Figure 1.7). They have the same biochemical mode of action with triazines, inhibiting photosynthesis (Corbet, 1974).

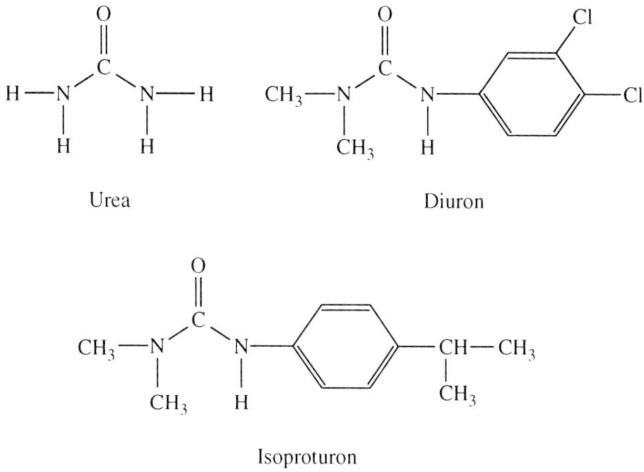

Figure 1.7 The substituted urea herbicides, diuron and isoproturon

1.2.4.1 Fate

Substituted ureas are transferred to the aquatic environment after their application in crops, via run-off. As a result, they are detected worldwide at concentrations up to few μg L^{-1} (Table 1.4).

Substituted ureas are transformed by both abiotic and biotic processes. Isoproturon, which is a hydrophophic compound, is hydrolysed both in low and high pH values. (Gangwar and Rafiquee, 2007). Salvestrini et al. (2002) reported that despite the slow hydrolysis rate of diuron in natural solutions, when this abiotic process takes place is irreversible and the only metabolite is 3,4-dichloroaniline (DCA). Phototransformation of urea herbicides can also be occurred (Shankar et al., 2008). Furthermore, substituted ureas can be subjected to biodegradation and give metabolites which may be more toxic than the parent compound. Goody et al. (2002) reported that diuron degradation leads to the formation of the toxic metabolite DCA. In a recent study, Stasinakis et al. (2009a) reported that under aerobic and anoxic conditions diuron can be biotransformed to DCA, DCPMU (1-(3,4-dichlorophenyl)-3-methylurea) and DCPU (1-3,4-dichlorophenylurea). Except of these metabolites, a significant part of Diuron seems to be mineralized or/and biotransformed to other unknown compounds.

1.2.4.2 Effects

Similarly to triazines, substituted ureas are expected to be highly toxic in photosynthetic organisms due to their biochemical mode of action. Gatidou and Thomaidis (2007) investigated the toxic effects of diuron on the photosynthetic microorganism *Dunaniella teriolecta* and they estimated an EC_{50} value (after 96h of exposure) of about 6 μg L^{-1}. Fernandez-Alba et al. (2002) estimated an EC_{50} value equal to 3.2 μg L^{-1} for seagrass. On the other hand, significant lower toxicity has been reported for crustaceans (8.6 mg L^{-1}, 48 h EC_{50}) and fish (74 mg L^{-1}, 7 day LC_{50}) (Fernandez-Alba et al., 2002).

Table 1.4 Typical water concentrations of the substituted ureas herbicides diuron and isoproturon

Compound	Concentration (ng L^{-1})	Country	Reference
Diuron	<LOD-366	UK	Gatidou et al., 2007
Diuron	7.64–40.78	Italy	Carafa et al., 2007
Diuron	<LOD-105	Spain	Claver et al., 2006
Diuron	30–560	Greece	Gatidou et al., 2005
Diuron	<LOD-80	Australia	McMahon et al., 2005
Diuron	<LOD-3054	Japan	Okamura et al., 2003
Isoproturon	<LOD-92	China	Müller et al., 2008
Isoproturon	0.22–32.08	Italy	Carafa et al., 2007
Isoproturon	LOD<30	Spain	Claver et al., 2006

LOD: Limit of Detection.

In mammals, substituted ureas pose slight toxicity. An oral LD_{50} of diuron in rats equal to 3.4 g Kg^{-1} or a dermal LD_{50} greater than 2 g Kg^{-1} are indicative of its low toxicity in mammals (Giacomazzi and Cochet, 2004). According to WHO (2004), some compounds like isoproturon are classified as slightly hazardous and others such as linuron or diuron as compounds unlikely to present acute hazard in normal use. Regarding ureas' metabolites, DCA has been found to be highly toxic and has been classified as a secondary poisonous substance (Giacomazzi and Cochet, 2004).

1.2.5 Legislation

The excessive usage of synthetic pesticides resulting in contamination problems and harmful effects on humans and ecosystems led several countries to take action concerning their presence in the aquatic environment. European Union with Directive 98/83/EC (EU, 1998) set maximum allowable concentrations in drinking water for individual pesticides (0.1 μg L^{-1}) and for total pesticides (0.5 μg L^{-1}). Furthermore, with the Decision No 2455/2001/EC European

Community established the list of priority substances in the field of water policy, amending Directive 2000/60/EC. Pesticides such as atrazine, simazine, diuron, isoproturon, heptachlor, aldrin, dieldrin, lindane, chlorpyrifos are considered as priority compounds (EU, 2001). Additionally, some European countries have set environmental quality standards (EQSs) for priority compounds. Italy established EQSs for some pesticides in water ranging from 1 ng L^{-1} (chlorpyrifos) to 50 ng L^{-1} (atrazine) (Carafa *et al.*, 2007).

UNEP Governing Council decided in 1997 the immediate international action for the reduction and/or elimination of the emissions and discharges of 12 persistent organic pollutants (POPs) due to their persistent, toxicity and bioaccumulation. This decicion led to the adoption of Stockholm Convention in 2001. Among the different compounds which compose the also known "dirty dozen" are the organochlorine pesticides: aldrin, DDT, dieldrin, heptachlor (UNEP, 2003). According to Master List of Actions Report of UNEP (UNEP, 2003), the above organochlorine pesticides have been banned in most countries worldwide. Furthermore, an International Code of Conduct has been promoting since 1985 by Food and Agriculture Organization of the United Nations. This Code sets standards for governments, pesticide industry and pesticide users (FAO, 2003).

1.3 PHARMACEUTICALS

Pharmaceutically active compounds (pharmaceuticals) are complex molecules with molecular weights ranging from 200 to 500/1000 Da, which are developed and used due to their specific biological activity (Kummerer, 2009). A great number of pharmaceutical compounds (more than 4000 compounds in Europe) are discharged to the environment after human and veterinary usage (Mompelat *et al.*, 2009). In contrast to other micropollutants, that their concentrations will be decreased in the future due to the existed laws and regulations, the use of pharmaceuticals is expected to be increased due to their beneficial health effects.

The first study on human drugs' occurrence in environmental samples appeared in the late 1970s (Hignite and Azarnoff, 1977). The research regarding the effects of these compounds in the environment started in 1990s, when it was discovered that some of these compounds interfere with ecosystems at concentration levels of a few micrograms per liter (Halling-Sørensen *et al.*, 1998). In parallel, during that decade the first optimized analytical methods were developed for the determination of low concentrations of pharmaceuticals in environmental samples (Hirsch *et al.*, 1996; Ternes *et al.*, 1998).

Among pharmaceuticals, non-steroidal anti-inflammatory drugs (NSAIDs), anticonvulsants, lipid regulators and antibiotics are often detected into the aquatic environment and they are considered as a potential group of environmental contaminants. NSAIDs are drugs with analgesic, antipyretic and anti-inflammatory effects. Typical representatives of this category are ibuprofen (IBF, $C_{13}H_{18}O_2$) and diclofenac (DCF, $C_{14}H_{11}C_{12}NO_2$) (Figure 1.8). Anticonvulsants are used in the treatment of epileptic seizures, with carbamazepine (CBZ, $C_{15}H_{12}N_2O$) being the compound which is often reported in relevant papers. Almost 1000 tons CBZ are estimated to be consumed worldwide (Zhang *et al.*, 2008). Lipid regulators such as gemfibrozil (GEM, $C_{15}H_{22}O_3$) are used to lower lipid levels. Antibiotics are characterized by the great variety of substances such as penicillins, tetracyclines, sulfonamides and fluoroquinolones. In the literature there are available data for all these categories. In this text, data will be given for erythromycin and trimethoprim. Erythromycin ($C_{37}H_{67}NO_{13}$) is a macrolide antibiotic used as human and veterinary medicine, as well as in aquacultures. Trimethoprim (TMP, $C_{14}H_{18}N_4O_3$) is mainly used for treatment of urinary tract infections (Figure 1.8).

Pharmaceuticals are not completely metabolized by the human/animal body. As a result, they are excreted via urine and faeces as unchanged parent compound and as metabolites or conjugates (Heberer, 2002a). Excretion rates are significantly depending on the compound and the mode of application (oral, dermal). Regarding CBZ, after oral administration, almost 28% of the parent compound is discharged through the faeces to the environment, while the rest is absorbed and metabolized by the liver (Zhang *et al.*, 2008). The metabolites of CBZ are excreted with urine. Among them, the most important seem to be 10,11-dihydro-10,11-expoxycarbamazepine (CBZ-epoxide) and trans-10,11-dihydro-10,11-dihydroxycarbamazepine (CBZ-diol) (Reith *et al.*, 2000). Regarding DCF, almost 65% of its oral dosage is excreted through urine (Zhang *et al.*, 2008). The main DCF metabolites detected in urine are 4′-hydroxy-diclofenac (4′-OH-DFC) and 4′,5-dihydroxy-diclofenac (4′-5-diOH-DFC) (Schneider and Degan, 1981). IBF is extensively metabolized in the liver to 2-[4-(2-hydroxy-2-methylpropyl)phenyl]-propionic acid (hydroxyl-IBF) and 2-[4-(carboxypropyl)phenyl]-propionic acid (carboxy-IBF) (Winker *et al.*, 2008). Regarding TMP, almost 80% of the parent compound is excreted, while its main metabolites are 1,3-oxides and 3′,4-hydroxy derivatives (Kasprzyk-Hordern *et al.*, 2007). Only 5% of erythromycin is excreted unchanged, while its major metabolite is Erythromycin-H_2O (Kasprzyk-Hordern *et al.*, 2007). GEM is metabolized by the liver to four main metabolites and almost 70% of the initial compound is excreted as the glucuronide conjugate in the urine (Zimetbaun *et al.*, 1991).

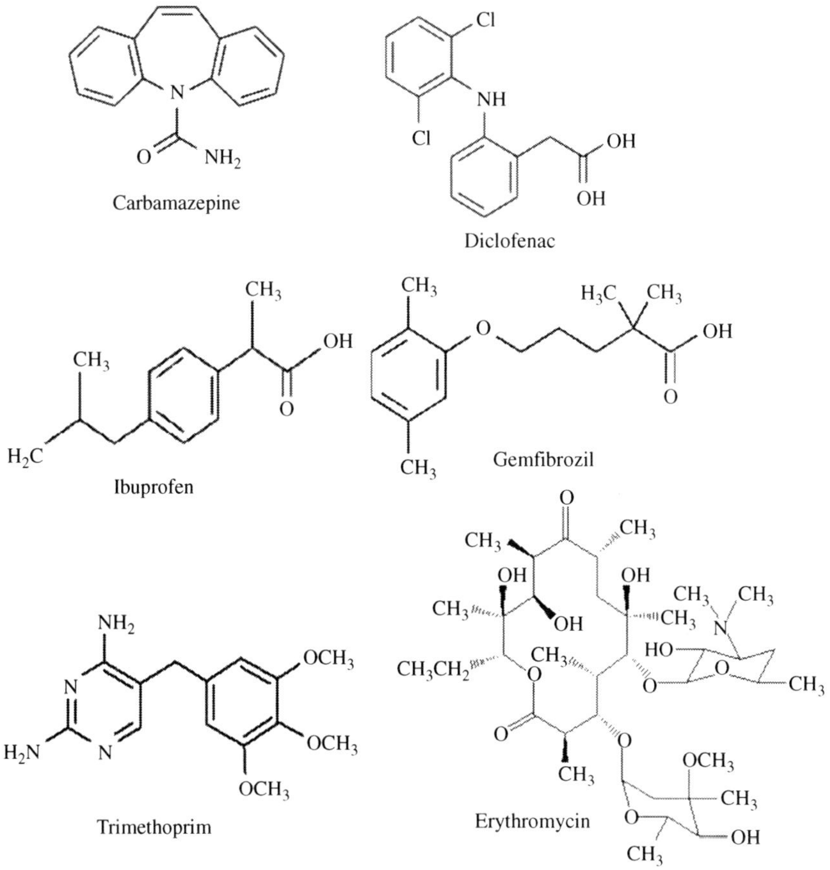

Figure 1.8 Chemical structures of selected pharmaceuticals

1.3.1 Fate

Municipal wastewater is the main way for the introduction of human pharmaceuticals and their metabolites in the environment. Moreover, hospital wastewater, wastewater from production industries and landfill leachates may contain significant concentrations of these compounds (Bound et al., 2006; Gomez et al., 2007). Regarding veterinary drugs, they can directly (e.g., use in aquacultures) or indirectly (e.g., manure application and runoff) be released to the environment (Sarmah et al., 2006).

In the literature there are several data regarding removal efficiency of pharmaceuticals in WWTSs. Removal rates are variable between different WWTSs and different compounds (Fent *et al.*, 2006), indicating that they depend on the chemical properties of the compound as well as the treatment process applied. The main mechanisms affecting pharmaceuticals' removal in WWTSs are sorption and biodegradation. NSAIDs and GEM occur as ions at neutral pH and they have little tendency of adsorption to the suspended solids in WWTSs. As a result, they remain in the dissolved phase and they disposed to the environment via the treated wastewater (Fent *et al.*, 2006). On the other hand, basic pharmaceuticals (e.g., fluoroquinolone antibiotics) can be adsorbed to suspended solids and accumulated to the sludge. Regarding the role of biodegradation in WWTSs, this seems to be significant for some compounds (e.g., DCF), while it is of minor importance for others (e.g., CBZ) (Metcalfe *et al.*, 2003; Kreuzinger *et al.*, 2004).

Due to the partial removal of pharmaceuticals during WWTSs, significant concentrations of these compounds are often detected in effluent wastewater, while lower concentrations are detected in surface water and groundwater (Segura *et al.*, 2009; Table 1.5). A recent survey in European river waters revealed that CBZ, DCF, IBF and GEM were detected in 95%, 83%, 62% and 35% of collected samples respectively (Loos *et al.*, 2009). Beside the fact that a significant part of pharmaceuticals are excreted from human/animal bodies as metabolites, so far, in most research papers concentrations of the parent compounds are reported, while there are limited data for the concentrations of their metabolites in the aquatic environment. In a previous study, IBF and its major metabolites hydroxyl- and carboxy-IBF were determined in wastewater and seawater (Weiger *et al.*, 2004). According to the results, hydroxyl-IBF was the major component in treated wastewater (concentrations ranging between 210 to 1130 ng L^{-1}), whereas carboxy-IBF was dominant in seawater samples (concentrations up to 7 ng L^{-1}). The elevated concentrations of hydroxyl-IBF in treated wastewater seem to be due to its excretion from human body, as well as to its formation in activated sludge process (Zwiener *et al.*, 2002). In another study, CBZ and five metabolites were detected in wastewater samples (Miao and Metcalfe, 2003). Among them, 10,11-dihydro-10,11-dihydroxycarbamazepine was detected at much higher concentrations than the parent compound. Moreover, in a recent study, Leclercq *et al.* (2009) reported the presence of six metabolites of CBZ in wastewater samples. Among them, 10,11-dihydro-10,11-trans-dihydroxycarbamazepine was detected at a higher concentration than the parent compound.

Table 1.5 Concentrations of pharmaceuticals in water samples

Substance	Concentration (ng L^{-1})	Country	Reference
Surface Water			
Trimethoprim	<LOD-183	UK	Kasprzyk-Hordern et al., 2008
Erythromycin-H$_2$O	<LOD-351	UK	Kasprzyk-Hordern et al., 2008
IBF	<LOD-100	UK	Kasprzyk-Hordern et al., 2008
DCF	<LOD-261	UK	Kasprzyk-Hordern et al., 2008
CBZ	<LOD-684	UK	Kasprzyk-Hordern et al., 2008
IBF	<LOD-5044	UK	Ashton et al., 2004
DCF	<LOD-568	UK	Ashton et al., 2004
Erythromycin	<LOD-1022	UK	Ashton et al., 2004
Trimethoprim	<LOD-42	UK	Ashton et al., 2004
Groundwater			
DCF	<LOD-380	Germany	Heberer, 2002b
IBF	<LOD-200	Germany	Heberer, 2002b
GEM	<LOD-340	Germany	Heberer, 2002b
CBZ	<LOD-2.4	USA	Standley et al., 2008
IBF	<LOD-19	USA	Standley et al., 2008
Trimethoprim	1.4–11	USA	Standley et al., 2008
Treated Wastewater			
IBF	20–1820	Europe	Andreozzi et al., 2003
DCF	<LOD-5450	Europe	Andreozzi et al., 2003
CBZ	300–1200	Europe	Andreozzi et al., 2003
Trimethoprim	20–130	Europe	Andreozzi et al., 2003
IBF	780–48240	Spain	Santos et al., 2007
CBZ	<LOD-1290	Spain	Santos et al., 2007
DCF	<LOD	Spain	Santos et al., 2007

(*continued*)

Table 1.5 (*continued*)

Substance	Concentration (ng L^{-1})	Country	Reference
Erythromycin	<LOD-1842	UK	Ashton *et al.*, 2004
Trimethophim	<LOD-1288	UK	Ashton *et al.*, 2004
IBF	240–28000	Spain	Gomez *et al.*, 2007
DCF	140–2200	Spain	Gomez *et al.*, 2007
CBZ	110–230	Spain	Gomez *et al.*, 2007

LOD: Limit of Detection.

Regarding the fate of these compounds to the aquatic environment, they can be adsorbed on suspended solids, colloids and dissolved organic matter or/and undergo biotic, chemical and physico-chemical transformations (Yamamoto *et al.*, 2009). Data on the sorption of pharmaceuticals in sediment and soil have been reported in several studies in the literature (Tolls, 2001; Figueroa *et al.*, 2004; Drillia *et al.*, 2005; Kim and Carlson, 2007). In most of those studies, higher sorption coefficients of pharmaceuticals than those predicted from octanol–water partitioning coefficients (log K_{ow}) were found, suggesting that mechanisms other than hydrophobic partitioning play a significant role in sorption of these compounds (Tolls, 2001). In cases that treated wastewater or sludge are reused for agricultural purposes, highly mobile pharmaceuticals can contaminate groundwater, whereas strongly sorbing compounds can accumulate in the top soil layer (Thiele-Bruhn, 2003). Sorption of pharmaceuticals to soils is affected by the solution chemistry, the type of mineral and organic sorbents and the concentration of dissolved organic matter (DOM) in reused wastewater (Nelson *et al.*, 2007; Blackwell *et al.*, 2007). Experiments with NSAIDs showed that CBZ and DCF can be classified as slow-mobile compounds in soil layers which are rich in organic matter, while their mobility increases significantly in soils which are poor in organic matter (Chefetz *et al.*, 2007).

Pharmaceuticals photodegradation depends on several factors such as the intensity of solar irradiation, the concentration of nitrates, DOM and bicarbonates (Lam and Mabury, 2004). The role of process in pharmaceuticals' fate varies significantly between different compounds of this category (Lam *et al.*, 2004; Benotti and Brownawell, 2009). For instance, phototransformation seems to be the major mechanism of DCF removal in surface water (Buser *et al.*, 1998; Andreozzi *et al.*, 2003). A half-life lower than 1 h has been reported under natural sunlight, while the initial product of DCF photodegradation was 8-chlorocarbazole-1-acetic acid, which is photodegraded even faster than the parent compound (Poiser et al., 2001). In other experiments, Lin and Reinhard (2005) calculated a half-life of GEM equal to 15 hours for river water. On the other hand, CBZ and IBF are photodegraded with much slower rate under sunlight irradiation (Yamamoto et al., 2009). Regarding CBZ, a half-life of 115 hours has been calculated, while 10,11-epoxycarbamazepine was its major phototransformation product (Lam and Mabury, 2005).

So far, most biodegradation studies with pharmaceuticals have focused on their removal during wastewater treatment (Joss et al., 2005; Radjenovic et al., 2009). On the other hand, there are limited data for their biodegradation in the aquatic environment. Lam et al. (2004) conducted experiments in a microcosm with several pharmaceuticals and found that photolysis was more important mechanism than biodegradation for CBZ and trimethoprim. Biodegradation experiments with river water showed that IBF and CBZ were relatively stable against microbes (Yamamoto et al., 2009). Half lives of 450–480 h^{-1} and 3000–5600 h^{-1} were calculated for IBF and CBZ, respectively (Yamamoto et al., 2009). In another study, IBF was biodegraded in a river biofilm reactor and its main metabolites were hydroxyl–IBF and carboxy–IBF (Winkler et al., 2001). Experiments with river sediment showed that under aerobic conditions DCF can be biodegraded and its major metabolite was p-benzoquinone imine of 5-hydroxydiclofenac (Groning et al., 2007). Experiments with CBZ and trimethoprim showed that half-lives higher than 40 days were calculated for the biodegradation of these compounds in seawater (Benotti and Brownawell, 2009).

1.3.2 Effects

Experiments with single compounds have shown that acute toxicity of most pharmaceuticals on aquatic organisms seems unlikely for environmental relevant concentrations (Choi et al., 2008; Zhang et al., 2008). Acute effects have been observed at much higher concentrations (100–1000 times) than those usually determined in the aquatic environment (Farre et al., 2008). However, it should be mentioned that pharmaceuticals are usually occurred in the environment as mixtures. Based on the above, several studies have shown that their toxicity to non-target organisms may be occurring at environmentally relevant concentrations due to combined and synergistic effects (Pomati et al., 2008; Quinn et al., 2009). Specifically, toxicity experiments with a mixture of NSAIDs showed that mixture toxicity was found at concentrations at which the single compound showed no or only little effects (Cleuvers, 2004). Moreover, ecotoxicity tests with antibiotics showed that combined toxicity of two antibiotics can lead to either synergistic, antagonistic, or additive effects (Christensen et al., 2006).

On the other hand, chronic toxicity of pharmaceuticals is a matter of great interest, as several aquatic species are exposed to these compounds for their entire life cycle. Beside the above, so far, fewer data are available regarding the long-term effects of pharmaceuticals to aquatic organisms. Schwaiger et al. (2004) studied DCF possible effects in rainbow trout after prolonged exposure and they reported histopathological changes of kidney and liver when fish was exposed to 5 µg L^{-1} DCF for a period of 28 days. In another study, Triebskorn et al. (2004) reported that the lowest observed effect concentration (LOEC) for cytological alterations in liver, kidney and gills of rainbow trout was 1 µg L^{-1} DCF.

Some of the pharmaceuticals seem to bioconcentrate and transport through food chain in other species. Mimeault et al. (2005) investigated the uptake of GEM in goldfish and reported that exposure to environmental levels of GEM results to bioconcentration of this compound in plasma. Schwaiger et al. (2004) reported that DCF is bioconcentrated mainly in liver and kidney of rainbow trout. Brown et al. (2007) reported the bioaccumulation of DCF, IBF and GEM in fish blood of rainbow trout. Several studies have related the presence of DCF residues with decline of vultures' population in India and Pakistan (Oaks et al., 2004; Schultz et al., 2004). Other toxicity effects which have been reported in the literature, include estrogenic activity (Isidori et al., 2009), as well as mutagenic and genotoxic potential of GEM (Isidori et al., 2007). Finally, the release of antibiotics as well as their metabolites into the environment increases the risk of developing bacterial resistance to antibiotics in aquatic ecosystems (Costanzo et al., 2005; Thomas et al., 2005).

1.3.3 Legislation

Despite the great amounts of pharmaceuticals released to the environment, regulations for ecological risk assessment are largely missing. In USA, environmental assessments of veterinary pharmaceuticals are required by the U.S. Food and Drug Administration (FDA) since 1980 (Boxall et al., 2003). Regarding human pharmaceuticals, an environmental assessment report should be provided in cases that the expected concentration of the active ingredient of the pharmaceutical in the aquatic environment is expected to be equal to or higher than 1 µg L^{-1} (FDA-CDER, 1998). In European Union, the first requirement for ecotoxicity testing was established in 1995 for veterinary pharmaceuticals, according to the European Union Directive 92/18/EEC and the corresponding "Note for Guidance" (EMEA, 1998). During the last decade,

European Commission published Directive 2001/83/EC amended by Directive 2004/27/EC (for human pharmaceuticals) and Directive 2001/82/EC amended by 2004/28/EC (for veterinary pharmaceuticals), indicating that authorization for pharmaceuticals must be accompanied by environmental risk assessment.

There are rare cases where limit values have been set for the presence of pharmaceuticals in the aquatic environment. In such a case, California set water quality standard (19 ng L^{-1}) for lindane (a compound used as pharmaceutical in treatment of head lice) in drinking water sources.

1.4 STEROID HORMONES

Steroid hormones are a group of compounds controlling endocrine and immune system. The major classes of natural hormones are estrogens (e.g., estradiol, estrone, estriol), androgens (e.g., progesterone, androstenedione), progestagents (e.g., progesterone) and corticoids (e.g., cortisol). Several synthetic hormones such as ethinylestradiol, mestranol, dexamethanose have also been produced apart from the aforementioned endogenous hormones.

Among these compounds, estrone (E1), 17β-estradiol (E2), estriol (E3) and ethinylestradiol (EE2) (Figure 1.9) have received more scientific attention since they consider to be the most important contributors to estrogenicity of treated wastewaters and surface waters (Rodgers-Grey *et al.*, 2000). These compounds end up in the environment through wastewater effluents, untreated discharges, runoff of manure and sewage sludge reuse. Aquaculture is another important source of estrogens in the environment (Fent *et al.*, 2006). Fish food additives containing hormones are directly added into the water. Therefore, these compounds can end up in the aquatic environment due to overfeeding or loss of appetite of fish (a phenomenon normally observed in sick organisms).

Steroid hormones are excreted by humans (Länge *et al.*, 2002). Several studies have shown that the gender, the pregnancy or menopause can differentiate the excretion rates of these compounds. For instance, E1 found to have an excretion rate of about 11, 5 and 1194 μg d^{-1} for premenopausal, postmenopausal and pregnant women, respectively, indicating that pregnant women may contribute in a large extent to the total amount of natural estrogens excreted by humans. For men, the excretion rate of E1 estimated to be 3.9 μg d^{-1} (Liu *et al.*, 2009). Natural estrogens in urine are mainly excreted in sulfate or glucuronide conjugates. However, free estrogens have also been detected in feces. Glucuronides were reported to easily change to their free estrogens, whereas sulfates were more resistant to biotransformation (D'Ascenzo *et al.*, 2003).

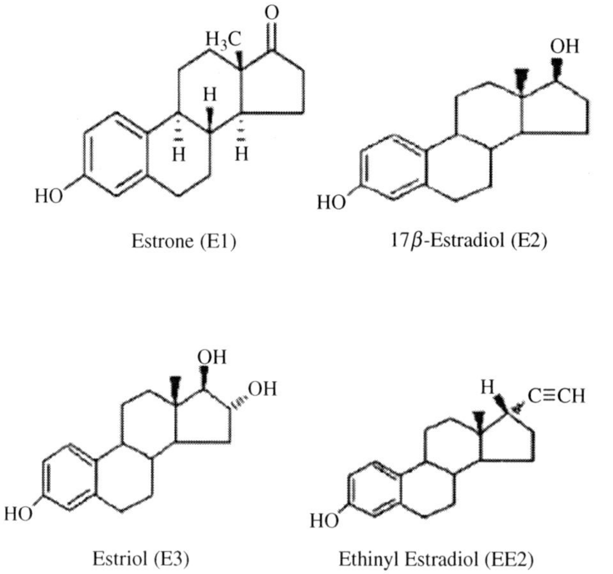

Figure 1.9 Molecular formula of estrogens: estrone (E1), 17β-estradiol (E2), estriol (E3) and ethinyl estradiol (EE2)

1.4.1 Fate

Estrogenic compounds have been detected in WWTSs and surface waters. A survey of the US Geological Service indicated that these compounds are often detected in water bodies. Specifically, they were detected at percentages varied from 6% to 21% of the total number of analyzed samples. The median concentrations found to be between 0.03 and 0.16 µg L^{-1} (Kolpin *et al.*, 2002). Other authors have also reported the presence of steroids in surface and drinking water (Table 1.6). Detection of steroid hormones in drinking water at concentration levels similar to those found in surface waters indicate that these compounds do not totally being removed during water treatment (Ning *et al.*, 2007).

Wastewater seems to be the major transport route of these compounds in the environment. Servos *et al.* (2005) detected considerable effluent estrogenicity, possibly due to hormonally active intermediates of estrogens formed either due to degradation during wastewater treatment or by cleavage of estrogen conjugates (Ning *et al.*, 2007). Some indicative concentrations of E1, E2 and EE2 compounds in treated wastewater are given in Table 1.7.

Table 1.6 Occurrence of steroid hormones in water column

Compound	Concentration (ng L^{-1})	Country	Reference
Estrone (E1)	DW: 0.70	Germany	Kuch and Ballschmiter, 2001
	SW: 1.5–12	Italy	Lagana et al., 2004
	SW: 1.4–1.8	France	Cargouet et al., 2004
17β-estradiol (E2)	SW: 0.60 DW: 0.70	Germany	Kuch and Ballschmiter, 2001
	SW: 2–5	Italy	Lagana et al., 2004
	SW: 1.7–2.1	France	Cargouet et al., 2004
	SW: <LOD	USA	Vanderford et al., 2003
17a-ethinylestradiol (EE2)	SW: 0.80 DW: 0.35	Germany	Kuch and Ballschmiter, 2001
	SW: n.d.-1	Italy	Lagana et al., 2004
	SW: 1.3–1.4	France	Cargouet et al., 2004
	SW: 3.6–14	Nevada	Vanderford et al., 2003
Estriol (E3)	SW: 2–6	Italy	Lagana et al., 2004
	SW: 1.8–2.2	France	Cargouet et al., 2004

SW: surface water; DW: drinking water; LOD: Limit of Detection.

Table 1.7 Occurrence of steroid hormones in wastewater

Compound	Concentration (ng L^{-1})	Country	Reference
Estrone (E1)	1–100	Canada	Servos et al., 2005
	4–7	France	Cargouet et al., 2004
17β-estradiol (E2)	1–15	Canada	Servos et al., 2005
	5–9	France	Cargouet et al., 2004
17-ethinylestradiol (EE2)	3–5	France	Cargouet et al., 2004

Regarding the fate of steroid hormones in the environment, these compounds are highly hydrophobic, low volatile and present low polarities with octanol–water partition coefficients (K_{ow}) ranging between 10^3 and 10^5. Due to the above, they are significantly sorbed on the suspended solids and sediments. Lai et al. (2000) investigated the distribution of several estrogens between water column and sediments and they reported that they can be rapidly removed from water phase to sediments. Additionally, Jürgens *et al.* (1999) indicated that estrogens ended up to bed sediment at percentage up to 92% during the first 24 h.

Biodegradation of estrogens can be occurred in natural environments. Jürgens *et al.* (2002) studied the behaviour of E2 and EE2 in surface waters. The authors concluded that microorganisms capable to transform E2 to E1 were present in river water. As a result, half-lives ranging from 0.2 to 9 days (at 20°C) were calculated. E1 was further degraded at similar rates. On the contrary, EE2 found to be more resistant to biodegradation but susceptible to photodegradation. Interconversion between E1 and E2 occurs, favoring E1 (Birkett and Lester, 2003). This explains the highest E1 concentrations in aquatic environment in many cases.

Furthermore, bioaccumulation of these compounds by several organisms has been reported. Larsson *et al.* (1999) established bioconcentration factors (BCFs) for E1, E2 and EE2 on juvenile rainbow trout. BCF values ranged between 104 and 106. Lai *et al.* (2002) investigated the possible uptake and accumulation of steroid compounds by the alga *Chlorella vulgaris*. They calculated a BCF of 27 for E1, within 48 h under both light and dark conditions. Gomes *et al.* (2004) studied the bioaccumulation of E1 in *Daphnia magna* and found that uptake of the compound via aqueous phase occurred within the first 16 h and a BCF value equal to 228 was estimated. According to the same study, feeding of *Daphnia magna* with algae contaminated with E1, resulted in a partitioning factor of 24 for the crustacean. This fact was an indication of possible biomagnification of E1 via food.

1.4.2 Effects

Steroid hormones are mainly known for their estrogenic action. These compounds present higher endocrine disrupting activity compared with other chemicals (Christiansen *et al.*, 1998). They cause effects such as feminization of male fish and induction of vitellogenesis. Even at concentrations close to detection limits, steroids can cause deleterious effects. For instance, concentrations of 17β-ethinylestradiol (EE2) at level of 0.1 ng L^{-1}, induce the expression of vitellogenin in fish (Purdum *et al.*, 1994). Additionally, affection of sex differentiation (Van Aerle *et al.*, 2002) and fecundity of organisms have been reported. At concentration of 4 ng L^{-1} of EE2, the development of normal secondary sexual characteristics on male fathead minnows was prevented (Länge *et al.*, 2001). Few ng L^{-1} of estradiols also led to induction of vitellogenin in juvenile rainbow trout (Thorpe *et al.*, 2001).

Steroids can cause adverse effects not only in fish but also in other organisms such as amphibians, reptiles or invertebrate. For instance, an oral dosage of EE2 in the range of 0.005–0.09 mg Kg^{-1} d^{-1} can produce carcinogenic effects in female mice (Seibert, 1996). According to Palmer and Palmer (1995), 1 µg g^{-1} of E2 can induce vitellogenesis both in frogs and turtles during a week. Furthermore, inhibition of barnacle settlement due to E2 has been reported (Billinghurst et al., 1998). Effects on plants have also been reported in the literature. Shore et al. (1992) indicated that irrigation of alfalfa with water containing estrogens such as E1 and E2, affected the growth of plants.

Humans are also affected by steroids. EE2 has been linked with prostate cancer development (Hess-Wilson and Knudsen, 2006). E2 has also been related with diseases like breast cancer and endometriosis (Dizerega et al., 1980; Thomas, 1984).

1.4.3 Legislation

Control of steroid hormones is a difficult issue due to the fact that natural estrogens cannot be banned or replaced by other compounds. Despite the difficulties, several countries have activated towards this direction. European Union has banned the use of hormones as growth promoters in food-production animals according to the Directive 88/146/EEC. Similarly, the use of compounds such as progesterone, testosterone, estradiol, zeranol and trenbolone acetate for animal food production has been regulated by the US Food and Drug Administration (FDA) and by FAO/WHO.

1.5 SURFACTANTS AND PERSONAL CARE PRODUCTS

Surfactants are a group of synthetic organic compounds consisting of a polar head group and a nonpolar hydrocarbon tail. They are widely used in detergents, textiles, polymer, paper industries and their major classes are anionic (e.g., linear alkylbenzene sulphonates), cationic (e.g., quaternary ammonium compounds) and non ionic surfactants (e.g., alkylphenol ethoxylates) (Ying et al., 2005). Among them, alkylphenol ethoxylates (APEs) constitute a large portion of the surfactant market (production equal to 500.000 t in 1997) (Renner et al., 2007). Significant scientific attention has been given during the last decade to nonylphenol ethoxylates (NPE) which represent almost 80% of the worldwide production of APEs (Brook et al., 2005). The microbial breakdown of NPEs results to the formation of nonylphenol (NP, $C_{15}H_{24}O$) (Figure 1.10) which is much more toxic than the parent compounds and induces estrogenic effects in several aquatic organisms (Birkett and Lester, 2003; Soares et al., 2008).

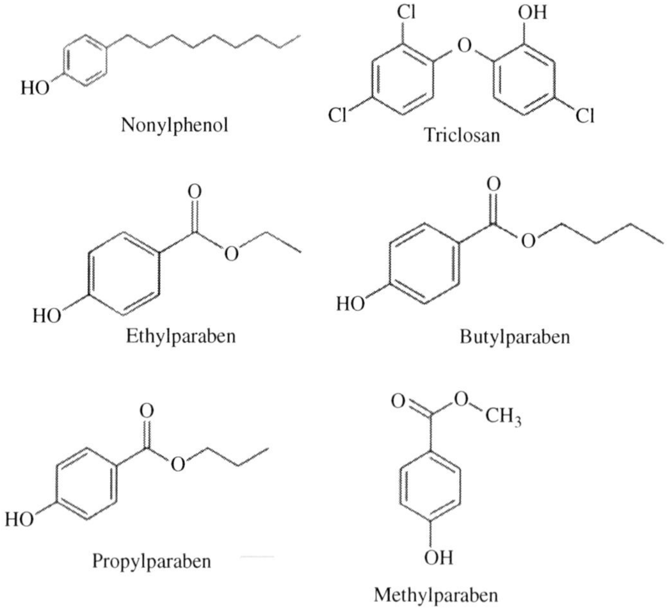

Figure 1.10 Chemical structures of selected surfactants and personal care products

Personal care products include products used for beautification and personal hygiene (skin care products, soaps, shampoos, dental care). These products contain significant concentrations of synthetic organic chemicals such as antimicrobial disinfectants (e.g., triclosan, triclocarban), preservatives (e.g., methylparaben, $C_8H_8O_3$; ethylparaben, $C_9H_{10}O_3$; butylparaben, $C_{11}H_{14}O_3$; propylparaben, $C_{10}H_{12}O_3$) and sunscreen agents (e.g., benzophenone-3, octyl methoxycinnamate) which are introduced to the aquatic environment during regular use (Ternes *et al.*, 2003; Kunz and Fent, 2006). Among them, triclosan (TCS, $C_{12}H_7Cl_3O_2$) and parabens (Figure 1.10) seem to be compounds of significant research and practical interest due to their wide use and toxicological properties (Kolpin *et al.*, 2002). Parabens are used in more than 22000 cosmetic products (Andersen, 2008), while approximately 350 t of TCS are produced annually in Europe for commercial applications (Singer *et al.*, 2002).

1.5.1 Fate

The major source of all these compounds in the environment is the discharge of wastewater. So far, there are several studies investigating their elimination in WWTSs. Regarding NP, contradictory results have been reported for its removal, ranging from minus 9% (Stasinakis *et al.*, 2008) to 98% (Planas *et al.*, 2002; Gonzalez *et al.*, 2007; Jonkers *et al.*, 2009). These differences are due to the fact that NP can be formed during activated sludge process from the biotransformation of NPEs (Ahel *et al.*, 1994). On the other hand, removal efficiency of TCS seem to be more consistent and exceed 90% in most published papers (Heidler and Halden, 2006; Stasinakis *et al.*, 2008). Regarding parabens, in a recent study it was reported that they are almost totally removed during wastewater treatment (Jonkers *et al.*, 2009). The main mechanisms, affecting the removal of these compounds from the dissolved phase of wastewater, are adsorption on the suspended solids and biotransformation to unknown metabolites (Ahel *et al.*, 1994; Heidler and Halden, 2007; Stasinakis *et al.*, 2007; Stasinakis *et al.*, 2008; Stasinakis *et al.*, 2009b). The partial elimination of some of the aforementioned substances (e.g., nonylphenol) during wastewater treatment or/and the disposal of untreated wastewater in the environment result to frequent detection of these compounds in surface waters (Kolpin *et al.*, 2002). Trace concentrations of NP have also been detected in drinking water (Petrovic *et al.*, 2003). In Table 1.8, a few recent data are given concerning the concentrations of these compounds in treated wastewater and surface water.

NP is a hydrophobic compound (log K_{ow} equal to 4.48) with low solubility in water, therefore it partitions mainly to organic matter (John *et al.*, 2000). In natural waters, NP can be photodegraded with a half-life of 10 to 15 hours (Ahel *et al.*, 1994). NP biodegradation is affected by several factors as the existence of aerobic and anaerobic conditions, the type of microorganisms used and their acclimatization on this compound. According to Lalah *et al.* (2003), NP partitions significantly into sediments, while it is resistant to biodegradation in river water and sediment. Other studies have shown that NP can be biodegraded with a slow rate under aerobic conditions in river sediment (half lives ranging between 14 to 99 days) (Yuan *et al.*, 2004) or under anaerobic conditions in mangrove sediments (half lives ranging between 53 to 87 days) (Chang *et al.*, 2009). On the other hand, experiments using river water – sediment and groundwater – aquifer material showed that NP can be rapidly degraded under aerobic and anaerobic conditions (half lives ranging between 0.4 to 1.1 days) due to biotic and abiotic factors (Sarmah and Northcott, 2008). In a recent study, it has been shown that NP biodegradation is differentiated for the different isomers which compose this chemical (Gabriel *et al.*, 2008).

Table 1.8 Concentrations of surfactants and personal care products in water samples

Substance	Concentration (ng L^{-1})	Country	Reference
Surface Water			
NP	<29–195	Switzerland	Jonkers et al. (2009)
Methylparaben	3.1–17	Switzerland	Jonkers et al. (2009)
Ethylparaben	<LOD-1.6	Switzerland	Jonkers et al. (2009)
Propylparaben	<LOD-5.8	Switzerland	Jonkers et al. (2009)
Butylparaben	<LOD-2.8	Switzerland	Jonkers et al. (2009)
NP	36–33231	China	Peng et al. (2008)
Methylparaben	<LOD-1062	China	Peng et al. (2008)
Propylparaben	<LOD-2142	China	Peng et al. (2008)
TCS	35–1023	China	Peng et al. (2008)
NP	0.1–7300	China	Shao et al. (2005)
Treated Wastewater			
NP	<LOD-281	Switzerland	Jonkers et al. (2009)
Methylparaben	4.6–423	Switzerland	Jonkers et al. (2009)
Ethylparaben	<LOD-17	Switzerland	Jonkers et al. (2009)
Propylparaben	<LOD-28	Switzerland	Jonkers et al. (2009)
Butylparaben	<LOD-12	Switzerland	Jonkers et al. (2009)
TCS	<LOD-6880	Greece	Stasinakis et al. (2008)
TCS	80–400	Spain	Gomez et al. (2007)

LOD: Limit of Detection

TCS is slightly soluble in water, hydrolytically stable and relatively non-volatile (Mc Avoy *et al.*, 2002). Due to the hydrophobicity of its protonated form (log K_{ow} equal to 5.4), it can be sorded to the suspended solids (Singer *et al.*, 2002). TCS is subjected to photolytic transformation in surface waters (Tixier *et al.*, 2002). Photodegradation experiments with natural sunlight showed that its elimination was followed by formation of the more toxic metabolite 2,7/2,8-dibenzodichloro-p-dioxin (Mezcua *et al.*, 2004). In another study, Latch *et al.* (2005) reported that during TCS photodegradation, 2,8-dichlorodibenzo-p-dioxin and 2,4-dichlorophenol are produced. Despite the fact that there are a few data reporting TCS biodegradation in activated sludge process (Federle *et al.*, 2002; Stasinakis *et al.*, 2007) and soil (Ying *et al.*, 2007), there is a lack of data for its biodegradation potential in surface waters.

Regarding parabens, so far there is a lack of data on their fate in the aquatic environment. In a recent study investigating photodegradation and biodegradation of butylparaben, it was reported that this compound is highly stable against sunlight, while it is biodegraded in riverine water (Yamamoto et al., 2007).

1.5.2 Effects

NP is considered as an endocrine disruptor (Birkett and Lester, 2003). A great number of data are available in the literature regarding its effects on aquatic organisms, while several review papers have been published (Staples et al., 2004; Vazquez – Duhalt et al., 2005; Soares et al., 2008). These papers indicate that the effects of NP are very diverse and they are depended on the test organism (species, stage of development) and the characteristics of the environment. In a recent study, Hirano et al. (2009) reported that environmentally relevant concentrations of NP can disrupt growth of the mysid crustacean *Americamysis bahia*. Moreover, exposure of Atlantic salmon smolts to 10 μg L^{-1} NP for 21 days caused direct and delayed mortalities (Lerner et al., 2007). In another study, Schubert et al. (2008) reported that a mixture of NP, E1 and E2 at concentration levels of a few ng L^{-1} would not adversely affect reproductive capability of brown trout. Due to the fact that NP is composed of several isomers, recent studies have been focused on correlating isomer molecular structure and endocrine activity (Gabriel et al., 2008; Preuss et al., 2009). Bioaccumulation of NP has been observed in algae, fish and aquatic birds (Ahel et al., 1993; Hu et al., 2007).

TCS toxicity has been investigated in algae, invertebrates, and fish (Orvos et al., 2002; Dussault et al., 2008). Recent studies have also shown that this compound may act as an endocrine disruptor (Veldhoen et al., 2006; Kumar et al., 2009). According to Veldhoen et al. (2006), environmentally relevant TCS concentrations altered thyroid hormone receptor mRNA expression in the American bullfrog. TCS is rapidly accumulated in algae and freshwater snails (Coogan and La Point, 2008). Moreover, it has been reported the bioaccumulation of TCS and its biotransformation product, methyl-TCS, in fish (Balmer et al., 2004). Recently, TCS was detected in the plasma of dolphins at concentrations ranging between 0.025 to 0.27 ng g^{-1} wet weight, indicating its bioaccumulation in marine mammals (Fair et al., 2009).

The presence of parabens in human urine and their ability to penetrate human skin have been demonstrated by several authors (Ye et al., 2006; El Hussein et al., 2007; Darbre and Harvey, 2008). These compounds seem to have low acute toxicity (Andersen, 2008), while they show little tendency to bioaccumulate in aquatic organisms (Alslev *et al.*, 2005). Recently, suspicions have been raised concerning their potential for causing endocrine disrupting effects in rainbow trout (propyl and butyl parabens) (Bjerregaard *et al.*, 2003; Alslev *et al.*, 2005) and medaka (propyl paraben) (Inui *et al.*, 2003).

1.5.3 Legislation

Due to the fact that surfactants and personal care products are emerging contaminants, there are limited regulations for their concentrations in the environment. Among the studied compounds, regulations have been established only for nonylphenols. Specifically, NP and its ethoxylates have been listed as priority substances in the Water Framework Directive (EU, 2001) and most of their uses are currently regulated (EU, 2003). EPA prepared a guideline setting quality criteria for NP in freshwater and saltwater (Brooke and Thursby, 2005). Moreover, Canada has set stringent water quality guidelines for NP and NPEs (Enironment Canada, 2001; 2002). European Union in an attempt to set some limit values for trace organic contaminants in sludge, proposed in a Working Document a limit value of 50 $\mu g\ g^{-1}$ dry weight for NPEs (sum of NP, NP1EO, NP2EO) (EU, 2000). However, at the moment, only few countries such as Switzerland and Denmark have legislation about the concentrations of NPEs in sewage sludge (JRC, 2001).

1.6 PERFLUORINATED COMPOUNDS

Perfluorinated compounds (PFCs) have been produced since 1950. They are used in a great number of industrial and consumer applications due to their physico-chemical characteristics such as thermal and chemical stability, surface active properties and low surface free energy (Lehmler, 2005). The bond C–F is particularly strong and as a result these compounds are resistant to various modes of degradation, such as oxidation, reduction and reaction with acids and bases (Kissa, 2001). Among PFCs, the most commonly studied substances are perfluorinated sulfonates (PFAS) and perfluorinated carboxylates (PFCA). These molecules consist of one perfluorinated carbon chain and one sulfonic (PFAS) or carboxylic group (PFCA). Among these, several compounds have been detected in the environment such as perfluorononanoic acid (PFNA, $C_9HF_{17}O_2$), perfluorodecanoic acid (PFDA, $C_{10}HF_{19}O_2$), perfluoroundecanoic acid (PFUnA, $C_{11}HF_{21}O_2$), perfluorododecanoic acid (PFDoA, $C_{12}HF_{23}O_2$) and perfluorooctanesulfonamide (PFOSA, $C_8H_2F_{17}NO_2S$).

However, perfluorooctane sulfonate (PFOS, $C_8HF_{17}O_3S$) and perfluorooctanoate (PFOA, $C_8HF_{15}O_2$) seem to be of greatest concern (Figure 1.11), due to their extended uses, the concentration levels detected, their behavior and toxicity. The production of PFOS was almost 3500 metric tons in 2000 (Lau et al., 2007). Due to the fact that 3M Company, the major manufacturer of PFOS, phased out production in 2002, the global production of this chemical decreased to 175 metric tons by 2003 (3M Company, 2003). Regarding PFOA, its production was estimated to be almost 500 metric tons in 2000, while it was increased to 1200 metric tons by 2004 (Lau et al., 2007). PFOS and its precursors are used in many applications such as food packing materials, surfactants in diverse cleaning agents, cosmetics, fire-fighting foams, electronic and photographic devices, (OECD, 2002 OECD (Organization for Economic Co-operation and Development), 2002. Co-operation on existing chemicals. Hazard assessment of perfluorooctane sulfonate (PFOS) and its salts. ENV/JM/RD(2002)17/FINAL, Paris.Kissa 2001). PFOA is mainly used during the production of certain fluoropolymers such as polytetrafluoroethylene (PTFE) and to a lesser extent in other industrial applications (OECD, 2005).

$CF_3(CF_2)_nCOO^-$

Perfluorocarboxylate (PFCA)

$CF_3(CF_2)_nSO_3^-$

Perfluoroalkyl sulfonate (PFAS)

Figure 1.11 Chemical structures of PFCA and PFAS

1.6.1 Fate

PFCs are commercially synthesized by electrochemical fluorination or telomerization (Fromme et al., 2009). However, PFCs can also be formed in the environment from biotic and abiotic transformation of commercially synthesized precursors. For instance, it has been reported that perfluorooctane sulfonamides can be biotransformed to PFOS (Tomy et al., 2004), while fluorotelomer alcohols (FTOH) can be subsequently transformed into PFOA (Wang et al., 2005).

PFCs have been detected in potable water, surface water, groundwater and wastewater worldwide (Table 1.9). Moreover, they have been detected in remote areas, reflecting the widespread global pollution for these compounds (Giesy and Kannan, 2001; Houde *et al.*, 2006). So far, a few data have been reported for the fate of PFOA and PFOS in the environment and the mechanisms affecting their distribution and transport. PFOA and PFOS are considered stable compounds due to their resistance in abiotic and biotic degradation (Giesy and Kannan, 2002). Yamashita *et al.* (2005) reported that there is a long range transport of these compounds by oceanic currents, while other authors proposed the atmospheric transport of volatile precursor chemicals and their transformation to PFOS and PFOA as an explanation for the presence of these anthropogenic chemicals in remote regions (Young *et al.*, 2007).

In a recent study investigating the occurrence of organic micropollutants in European rivers, PFOA and PFOS were detected in 97% and 94% of samples, respectively (Loos *et al.*, 2009). The discharge of municipal wastewater seems to be one of the major routes introducing these compounds into the aquatic environment (Sinclair and Kannan, 2006; Becker *et al.*, 2008). In some papers, the concentrations of these compounds in effluent wastewater are higher that those detected in influent wastewater, indicating possible biodegradation of their precursors during biological treatment processes (Sinclair and Kannan, 2006; Murakami *et al.*, 2009).

1.6.2 Effects

The toxicity of PFOS and PFOA has been studied extensively. Hepatotoxicity, immunotoxicity, developmental toxicity, hormonal effects and a carcinogenic potency are the effects of main concern (Lau *et al.*, 2004; Lau *et al.*, 2007). Animal studies have shown that these compounds are mainly distributed to serum, liver and kidney (Seacat *et al.*, 2002; Hundley *et al.*, 2006). The elimination half-lives of PFOA and PFOS are significantly differentiated for different species or different gender of the same species, ranging from few hours to 30 days (Kemper, 2003; Butenhoff *et al.*, 2004).

PFCs have also been detected in human blood and tissue samples from occupationally and non-occupationally exposed humans throughout the world (Kannan *et al.*, 2004; Calafat *et al.*, 2006; Olsen *et al.*, 2007). Food intake and drinking water consumption seems to be the major contemporary exposure pathway for the background population (Vestergren and Cousins, 2009). According to a recent study, PFOA concentrations in human blood range between 2 and 8 µg L^{-1} for background exposed population in industrialized countries (Vestergren and Cousins, 2009). Epidemiologic data related to PFC exposure are limited and they have been mainly performed on PFC production plant workers. In general, serum fluorochemical levels have not been associated with adverse health effects (Lau *et al.*, 2007). Studies in retirees from PFC production facilities showed a mean elimination half-life of 3.8 years and 5.4 years for PFOA and PFOS, respectively (Olsen *et al.*, 2007). In a recent review paper, Fromme *et al.* (2009) describe the different pathways which are responsible for human exposure to PFCs (exposure via inhalation from outdoor and indoor air, oral exposure via food and water consumption). Regarding the acute toxicity of these compounds in aquatic organisms, it seems that at concentration levels which are similar to those detected in the environment, no acute toxicity of PFOS or PFOA has been observed (Sanderson *et al.*, 2004; Li, 2009).

The persistence of PFCs in the environment and their potential to accumulate and biomagnificate in the food chain is a matter of significant toxicological concern. PFCs have been detected in serum or plasma of wildlife worldwide (Keller *et al.*, 2005; Tao *et al.*, 2006). Determination of PFOS concentrations in water and organisms at various trophic levels (Kannan *et al.*, 2005) showed that its concentrations in benthic invertebrates were 1000-fold greater than those in surrounding water. Moreover, a biomagnification factor (BMF) equal to 20 was calculated for bald eagles (Kannan *et al.*, 2005). Bioconcentration experiments with different PFCs (Martin *et al.*, 2003) showed that bioconcentration factors (BCFs) increased with increasing length of the perfluoroalkyl chain, while carboxylates had lower BFCs than sulfonates of equal perfluoroalkyl chain length. As a result, BCF values equal to 27 ± 9.7, 4300 ± 570 and 40000 ± 4500 (L/Kg) were calculated in blood of rainbow trout for PFOA, PFOS and PFDoA, respectively (Martin *et al.*, 2003).

1.6.3 Legislation

From a regulatory point of view, PFOS has been classified as very persistent, very bioaccumulative and toxic compound, fulfilling the criteria for being considered as a persistent organic pollutant under the Stockholm Convention (EU, 2006). In May 2009, PFOS was added to Annex B of Stockholm Convention on Persistent Organic Pollutants. Canada banned the importation and use of several long chain perfluorinated carboxylic acids due to their effects on human and environment (Canadian Government Department of the Environment, 2008). In 2002, EPA issued a Significant New Use Rule regulating the import and production of several perfluorooctanyl-based chemicals (EPA, 2002). Finally, in 2006, EPA established the 2010/15 PFOA Stewardship Program. Targets of this Program are the reduction of global facility emissions and product content of PFOA and related chemicals by 95 percent up to 2010 and the elimination of emissions and product content of these compounds up to 2015.

Table 1.9 Concentrations of PFCs in water samples

Substance	Concentration (ng L^{-1})	Country	Reference
Surface Water			
PFOA	0.6–15.9	Italy	Loos et al., 2007
PFOS	<LOD-38.5	Italy	Loos et al., 2007
PFNA	0.2–16.2	Italy	Loos et al., 2007
PFDA	<LOD-10.8	Italy	Loos et al., 2007
PFUnA	0.1–38.0	Italy	Loos et al., 2007
PFDoA	<LOD-14.1	Italy	Loos et al., 2007
PFOS	9.3–56	USA	Plumlee et al., 2008
PFOA	<LOD-36	USA	Plumlee et al., 2008
PFDA	<LOD-19	USA	Plumlee et al., 2008
PFOA	0.8–14	Germany	Becker et al., 2008
PFOS	<LOD-15	Germany	Becker et al., 2008
Groundwater			
PFOS	19–192	USA	Plumlee et al., 2008
PFOA	<LOD-28	USA	Plumlee et al., 2008
PFDA	<LOD-19	USA	Plumlee et al., 2008
Potable Water			
PFOA	1.0–2.9	Italy	Loos et al., 2007
PFOS	6.2–9.7	Italy	Loos et al., 2007
PFNA	0.3–0.7	Italy	Loos et al., 2007
PFDA	0.1–0.3	Italy	Loos et al., 2007
PFUnA	0.1–0.4	Italy	Loos et al., 2007
PFDoA	0.1–2.8	Italy	Loos et al., 2007
PFOS	<LOD-22	Japan	Takagi et al., 2008
PFOA	2.3–84	Japan	Takagi et al., 2008
Treated Municipal Wastewater			
PFOS	3–68	USA	Sinclair and Kannan, 2006
PFOA	58–1050	USA	Sinclair and Kannan, 2006
PFNA	<LOD-376	USA	Sinclair and Kannan, 2006
PFDA	<LOD-47	USA	Sinclair and Kannan, 2006
PFOS	20–190	USA	Plumlee et al., 2008
PFOA	12–190	USA	Plumlee et al., 2008

(*continued*)

Table 1.9 (*continued*)

Substance	Concentration (ng L^{-1})	Country	Reference
PFDA	<LOD-11	USA	Plumlee et al., 2008
PFNA	<LOD-32	USA	Plumlee et al., 2008
PFOA	8.7–250	Germany	Becker et al., 2008
PFOS	2.4–195	Germany	Becker et al., 2008

LOD: Limit of Detection

1.7 REFERENCES

3M Company (2003) Environmental and health assessment of perfluorooctanesulfonate and its salts. US EPA Administrative Record. AR-226–1486.

Ahel, M., Giger, W. and Koch, M. (1994) Behaviour of alkylphenol polyethoxylate surfactants in the aquatic environment – I. Occurrence and transformation in sewage treatment. *Water Res.* **28**, 1131–1142.

Ahel, M., Mc Evoy, J. and Giger, W. (1993) Bioaccumulation of the lipophilic metabolites of nonionic surfactants in freshwater organisms. *Environ. Pollut.* **79**, 243–248.

Ahel, M., Scully, F. E., Hoigne, J. and Giger, W. (1994) Photochemical degradation of nonylphenol and nonylphenol polyethoxylates in natural-waters. *Chemosphere* **28**, 1361–1368.

Albanis, T. A., Hela, D. G., Lambropoulou, D. A. and Sakkas, V. A. (2004) Gas chromatographicmass spectrometric methodology using solid phase microextraction for the multiresidue determination of pesticides in surface waters (N.W. Greece). *Int. J. Environ. An. Ch.* **84**, 1079–1092.

Alslev, B., Korsgaard, B. and Bjerregaard, P. (2005) Estrogenicity of butylparaben in rainbow trout Oncorhynchus mykiss exposed via food and water. *Aquat. Toxicol.* **72**, 295–304.

Andersen, F. A. (2008) Final amended report on the safety assessment of methylparaben, ethylparaben, propylparaben, isopropylparaben, butylparaben, isobutylparaben, and benzylparaben as used in cosmetic products. *Int. J. Toxicol.* **27**, 1–82.

Andreozzi, R., Raffaele, M. and Paxéus, N. (2003) Pharmaceuticals in STP effluents and their solar photodegradation in aquatic environment. *Chemosphere* **50**, 1319–1330.

Araújo, T. M., Campos, M. N. N. and Canela, M. C. (2007) Studying the photochemical fate of methyl parathion in natural waters under tropical conditions. *Int. J. Environ. An. Ch.* **87**, 937–947.

Ashton, D., Hilton, M. and Thomas, K. V. (2004) Investigating the environmental transport of human pharmaceuticals to streams in the United Kingdom. *Sci. Total Environ.* **333**, 167–184.

Balmer, M. E., Poiger, T., Droz, C., Romanin, K., Bergqvist, P. A. and Mueller J. F. (2004) Occurrence of methyl Triclosan, a transforation product of the bactericide triclosan, in fish from various lakes in Switzerland. *Environ Sci. Technol.* **38**, 390–395.

Becker, A. M., Gerstmann, S. and Frank, H. (2008) Perfluoroctane surfactants in waste waters, the major source of river pollution. *Chemosphere* **72**, 115–121.
Bennoti, M. J. and Brownawell, B. J. (2009) Microbial degradation of pharmaceuticals in estuarine and coastal seawater. *Environ. Pollut.* **157**, 994–1002.
Billinghurst, Z., Clare, A. S., Fileman, T., McEvoy, J., Readman, J. and Depledge, M. H. (1998) Inhibition of barnacle settlement by the environmental oestrogen 4-nonylphenol and the natural oestrogen 17 beta oestradiol. *Mar. Pollut. Bull.* **36**, 833–839.
Birkett, J. W. and Lester, J. N. (2003) Endocrine disrupters in wastewater and sludge treatment processes. CRC Press LLC, Florida.
Bjerregaard, P., Andersen, D. N., Pedersen, K. L., Pedersen, S. N. and Korsgaard, B. (2003) Estrogenic effect of propylparaben (propylhydroxybenzoate) in rainbow trout Oncorhynchus mykiss after exposure via food and water. *Comp. Biochem. Phys. Part C.* **136**, 309–317.
Blackwell, P. A., Kay, P. and Boxall, A. B. A. (2007) The dissipation and transport of veterinary antibiotics in a sandy loam soil. *Chemosphere* **67**, 292–299.
Blasiak, J., Kleinwachter, V., Walter, Z. and Zaludova, R. (1995) Interaction of organophosphorus insecticide methyl parathion with calf thymus DNA and a synthetic DNA duplex. *Z Naturforsch C.* **50**, 820–823.
Botella, B., Crespo, J., Rivas, A., Cerrillo, S., Olea-Serrano, M.-F. and Olea, N. (2004) Exposure of women to organochlorine pesticides in Southern Spain. *Environ. Res.* **96**, 34–40.
Bound, J. P., Kitsou, K. and Voulvoulis, N. (2006) Household disposal of pharmaceuticals and perception of risk to the environment. *Environ. Toxicol. Phar.* **21**, 301–307.
Boxall, A. B., Kolpin, D. W., Halling-Sorensen, B. and Tolls, J. (2003) Are veterinary medicines causing environmental risks? *Environ. Sci. Technol.* **37**, 286–294.
Brook, D., Crookes, M., Johnson, I., Mitchell, R. and Watts, C. (2005) National Centre for Ecotoxicology and Hazardous Substances, Environmental Agency, Bristol U.K. 2005. Prioritasation of alkylphenols for environmental risk assessment.
Brooke, L. and Thursby, G. (2005) Ambient aquatic life water quality criteria for nonylphenol. Washington DC, USA: Report for the United States EPA, Office of Water, Office of Science and Technology.
Brown J. N., Paxeus N. and Forlin L. (2007) Variations in bioconcentration of human pharmaceuticals from sewage effluents into fish blood plasma. *Environ. Toxicol. Phar.* **24**, 267–274.
Buser, H.-R., Poiger, T. and Müller, M. D. (1998) Occurrence and fate of the pharmaceutical drug diclofenac in surface waters: rapid photodegradation in a lake. *Environ. Sci. Technol.* **33**, 3449–3456.
Butenhoff, J. L., Kennedy, G. L., Hindliter, P. M., Lieder, P. H., Hansen, K. J., Gorman, G. S., Noker, P. E. and Thomford, P. J. (2004) Pharmacokinetics of perfluorooctanoate in Cynomolgus monkeys. *Toxicol. Sci.* **82**, 394–406.
Calafat, A. M., Kuklenyik, Z., Caudill, S. P., Reidy, J. A. and Needham, L. L. (2006) Perfluorochemicals in pooled serum samples from the United States residents in 2001 and 2002. *Environ. Sci. Technol.* **40**, 2128–2134.
Canadian Government Department of the Environment (2008). Perfluorooctanesulfonate and its salts and certain other compounds regulations. Canada Gazette, Part II **142**, 322–1325.

Carafa, R., Wollgast, J., Canuti, E., Ligthart, J., Dueri, S., Hanke, G., Eisenreich, S. J., Viaroli, P. and Zaldívar, J. M. (2007) Seasonal variations of selected herbicides and related metabolites in water, sediment,seaweed and clams in the Sacca di Goro coastal lagoon (Northern Adriatic). *Chemosphere* **69**, 1625–1637.

Cargouet, M., Perdiz, D., Mouatassim-Souali, A., Tamisier-Karolak, S. and Levi, Y. (2004) Assessment of River Contamination by Estrogenic Compounds in Paris Area (France). *Sci. Total Environ.* **324**, 55–66.

Castillo, M., Domingues, R., Alpendurada, M. F. and Barceló, D. (1997) Persistence of selected pesticides and their phenolic transformation products in natural waters using off-line liquid solid extraction followed by liquid chromatographic techniques. *Anal. Chim. Acta* **353**, 133–142.

Chang, B. V., Lu, Z. J. and Yuan, S. Y. (2009) Anaerobic degradation of nonylphenol in subtropical mangrove sediments. *J. Hazard. Mater.* **165**, 162–167.

Chen, C. W., Hurd, C., Vorojeikina, D. P., Arnold, S. F. and Notides, A. C. (1997) Transcriptional activation of the human estrogen receptor by DDT isomers and metabolites in yeast and MCF-7 cells. *Biochem. Pharmacol.* **53**, 1161–1172.

Choi, K., Kim, Y., Jung, J., Kim, M. H., Kim, C. S., Kim, N. H. and Park, J. (2008) Occurrences and ecological risk of roxithromycin, trimethoprim and chloramphenicol in the Han river, Korea. *Environ. Toxicol. Chem.* **27**, 711–719.

Christensen, A. M., Ingerslev, F. and Baun, A. (2006) Ecotoxicity of mixtures of antibiotics used in aquacultures. *Environ. Toxicol. Chem.* **25**, 2208–2215.

Christiansen, T., Korsgaard, B. and Jespersen, A. (1998) Effects of nonylphenol and 17β-oestradiol on vitellogenin synthesis, testicular structure and cytology in male eelpout Zoarces viviparous. *J. Exp. Biol.* **201**, 179–192.

Claver, A., Ormad, P., Rodríguez, L. and Ovelleiro, J.-L. (2006) Study of the presence of pesticides in surface waters in the Ebro river basin (Spain). *Chemosphere* **64**, 1437–1443.

Cleuvers, M. (2004) Mixture toxicity of the anti-inflammatory drugs diclofenac, ibuprofen, naproxen, and acetylsalicylic acid. *Ecotox. Environ. Safe.* **59**, 309–315.

Coogan, M. A. and La Point, T. W. (2008) Snail bioaccumulation of triclocarban, triclosan, and methyltriclosan in a North Texas, USA, stream affected by wastewater treatment plant runoff. *Environ. Toxicol. Chem.* **27**, 1788–1793.

Corbet, J. R. (1974) The Biochemical mode of action of pesticides, Academic Press, London.

Costanzo, S. D., Murby, J. and Bates, J. (2005) Ecosystem response to antibiotics entering the aquatic environment. *Mar. Pollut. Bull.* **51**, 218–223.

D'Ascenzo, G., Corcia, A. D., Mancini, A. G. R., Mastropasqua, R., Nazzari, M. and Samperi, R. (2003) Fate of natural estrogen conjugates in municipal sewage transport and treatment facilities. *Sci. Total Environ.* **302**, 199–209.

Dahmardeh-Behrooz, R., Esmaili Sari, A., Bahramifar, N. and Ghasempouri, S. M. (2009) Organochlorine pesticide and polychlorinated biphenyl residues in human milk from the Southern Coast of Caspian Sea, Iran. *Chemosphere* **74**, 931–937.

Darbre, P. D. and Harvey, P. W. (2008) Paraben esters: Review of recent studies of endocrine toxicity, absorption, esterase and human exposure, and discussion of potential human health risks. *J. Appl. Toxicol.* **28**, 561–578.

Dizerega, G. S., Barber, D. L. and Hodgen, G. D. (1980) Endometriosis: role of ovarian steroids in initiation, maintenance, and suppression. *Fertil. Steril.* **33**, 649–653.

Drillia, P., Stamatelatou, K. and Lyberatos, G. (2005) Fate and mobility of pharmaceuticals in solid matrices. *Chemosphere* **60**, 1034–1044.

Dussault, E. B., Balakrishnan, V. K., Sverko, E., Solomon, K. R. and Sibley, P. K. (2008) Toxicity of human pharmaceuticals and personal care products to benthic invertebrates. *Environ. Toxicol. Chem.* **27**, 425–432.

Dzyadevych, S. V., Soldatkin, A. P. and Chovelon, J.-M. (2002) Assessment of the toxicity of methyl parathion and its photodegradation products in water samples using conductometric enzyme biosensors. *Anal. Chim. Acta* **459**, 33–41.

El Hussein, S., Muret, P., Berard, M., Makki, S. and Humbert, P. (2007) Assessment of principal parabens used in cosmetics after their passage through human epidermis-dermis layer (ex-vivo study). *Exp. Dermatol.* **16**, 830–836.

EMEA (1998) Note for guidance: environmental risk assessment for veterinary medicinal products other than GMO-containing and immunological products, EMEA, London (EMEA/CVMP/055/96).

Environment Canada (2001). Nonylphenol and its ethoxylates: Priority substances list assessment report. Report no. EN40-215-/57E.

Environment Canada (2002). Canadian environmental quality guidelines for nonylphenol and its ethoxylates (water, sediment and soil). Scientific Supporting Document. Ecosystem Helath: Sciencebased solutions report No 1–3. National Guidelines and Standard Office, Environmental Quality Branch, Environment Canada, Ottawa.

EPA (2002). Rules and regulations. United States Federal Register 67, pp. 72854–72867.

EU (1988) Council Directive 88/146/EEC for prohibiting the use in livestock farming of certain substances having a hormonal action.

EU (1998) Council Directive on the Quality of Water Intended for Human Consumption, 98/83/CE.

EU (2000) Working document on sludge, Third Draft, European Union, Brussels, Belgium, April 27, 2000.

EU (2001) European Union, Decision No 2455/2001/EC of the European Parliament and of the council of 20 November 2001 establishing the list of priority substances in the field of water policy and amending directive 2000/60/EC, *Off. J.* L331 (15/12/2001).

EU (2003) Directive 2003/53/EC, Amending for the 26th time the Council directive 76/769/EEC relating to restrictions on the marketing and use of certain dangerous substances and preparations (nonylphenol, nonylphenol ethoxylate and cement), Luxembourg, Luxembourg: European Parliament and the Council of the European Union.

EU (2006) Directive 2006/122/ECOF of the European Parliament and of the Council of 12 December 2006. Official Journal of the European Union, L/372/32–34, 27.12.2006.

EU (European Union), 2004. Directive 2004/27/EC, Amending directive 2001/83/EC on the community code relating to medicinal products for human use. Official Journal of the European Union, L/136/34–57, 30.04.2004.

EU (European Union), 2004. Directive 2004/28/EC, Amending directive 2001/82/EC on the community code relating to veterinary medicinal products. Official Journal of the European Union, L/136/58–84, 30.04.2004.

Fair, P. A., Lee, H. B., Adams, J., Darling, C., Pacepavicius, G., Alaee, M., Bossart, G. D., Henry, N. and Muir, D. (2009) Occurrence of triclosan in plasma of wild Atlantic bottlenose dolphins (Tursiops truncatus) and in their environment. *Environ. Pollut.* **157**, 2248–2254.

Fan, W., Yanase, T., Morinaga, H., Gondo, S., Okabe, T., Nomura, M., Hayes, T. B., Takayanagi, R. and Nawata, H. (2007) Herbicide atrazine activates SF-1 by direct affinity and concomitant co-activators recruitments to induce aromatase expression via promoter II. *Biochem. Bioph. Res. Co.* **355**, 1012–1018.

Farre, M., Perez, S., Kantiani, L. and Barcelo, D. (2008) Fate and toxicity of emerging pollutants, their metabolites and transformation products. *TrAC Trend Anal. Chem.*, **27**, 991–1007.

FDA-CDER (1998) Guidance for Industry-Environmental Assessment of Human Drugs and Biologics Applications, Revision 1, FDA Center for Drug Evaluation and Research, Rockville.

Federle, T. W., Kaiser, S. K. and Nuck, B. A. (2002) Fate and effects of triclosan in activated sludge. *Environ. Toxicol. Chem.* **21**, 1330–1337.

Fent, K., Weston, A. A. and Caminada, D. (2006) Ecotoxicology of human pharmaceuticals. *Aquat. Toxicol.* **76**, 122–159.

Fernadez-Alba, A. R., Hernando, M. D., Piedra, L. and Chisti, Y. (2002) Toxicity evaluation of single and mixed antifouling biocides measured with acute toxicity bioassays. *Anal. Chim. Acta.* **456**, 303–312.

Figueroa, R. A., Leonard, A. and Mackay, A. N. (2004) Modeling tetracycline antibiotics sorption to clays. *Environ. Sci. Technol.* **38**, 476–483.

Fromme, H., Tittlemier, S. A., Volkel, W., Wilhelm, M. and Twardella, D. (2009) Perfluorinated compounds – Exposure assessment for the general population in western countries. *Int. J. Hyg. Envir. Heal.*, **212**, 239–270.

Gabriel, F. L. P., Routledge, E. J., Heidlberger, A., Rentsch, D., Guenther, K., Giger, W., Sumpter, J. P. and Kohler, H. P. E. (2008) Isomer-specific degradation and endocrine disrupting activity of nonylphenols. *Environ. Sci. Technol.* **42**, 6399–6408.

Gangwar, S. K. and Rafiquee, M. Z. A. (2007) Kinetics of the acid hydrolysis of isoproturon in the absence and presence of sodium lauryl sulfate micelles. *Colloid Polym. Sci.* **285**, 587–592.

Gao, J., Liu, L., Liu, X., Zhou, H., Lu, J., Huang, S. and Wang, Z. (2009) The occurrence and spatial distribution of organophosphorous pesticides in Chinese surface water. *B. Environ. Contam. Tox.* **82**, 223–229.

Gatidou, G., Kotrikla, A., Thomaidis, N. S. and Lekkas, T. D. (2004) Determination of the antifouling booster biocides irgarol 1051 and diuron and their metabolites in seawater by high performance liquid chromatography–diode array detector. *Anal. Chim. Acta* **528**, 89–99.

Gatidou, G. and Thomaidis, N. S. (2007) Evaluation of single and joint toxic effects of two antifouling biocides, their main metabolites and copper using phytoplankton bioassays. *Aquat. Toxicol.* **85**, 184–191.

Gatidou, G., Thomaidis, N. S. and Zhou, J. L. (2007) Fate of Irgarol 1051, diuron and their main metabolites in two UK marine systems after restrictions in antifouling paints. *Environ. Int.* **33**, 70–77.

Giacomazzi, S. and Cochet, N. (2004) Environmental impact of diuron transformation: a review. *Chemosphere* **56**, 1021–1032.

Giesy, J. P. and Kannan, K. (2001) Global distribution of perfluoroctane sulfonate in wildlife. *Environ. Sci. Technol.* **35**, 1339–1342.

Giesy, J. P. and Kannan, K. (2002) Perfluorochemicals in the environment. *Environ. Sci. Technol.* **36**, 147–152.

Golfinopoulos, S.K, Nikolaou, A. D., Kostopoulou, M. N., Xilourgidis, N. K., Vagi, M. C. and Lekkas, D. T. (2003) Organochlorine pesticides in the surface waters of Northern Greece. *Chemosphere* **50**, 507–516.

Gomes, R. L., Deacon, H. E., Lai, K. M., Birkett, J. W., Scrimshaw, M. D. and Lester, J. N. (2004) An assessment of the bioaccumulation of estrone in daphnia magna. *Environ. Toxicol. Chem.* **23**, 105–108.

Gómez, M. J., Martínez Bueno, M. J., Lacorte, S., Fernández-Alba, A. R. and Agüera, A. (2007) Pilot survey monitoring pharmaceuticals and related compounds in a sewage treatment plant located on the Mediterranean coast. *Chemosphere* **66**, 993–1002.

Gonzalez, S., Petrovic, M. and Barcelo, D. (2007) Removal of a broad range of surfactants from municipal wastewater – Comparison between membrane bioreactor and conventional activated sludge treatment. *Chemosphere* **67**, 335–343.

Gooddy, D. C., Chilton, P. J. and Harrison, I. (2002) A field study to assess the degradation and transport of diuron and its metabolites in a calcareous soil. *Sci. Total Environ.* **297**, 67–83.

Götz, R., Bauer, O. H., Friesel, P. and Roch, K. (1998) Organic trace compounds in the water of the River Elbe near Hamburg part II. *Chemosphere* **36**, 2103–2118.

Government of Canada, 2006. Perfluorooctane sulfonate and its salts and certain other compounds regulations. *Can. Gazette Part. I* **140**, 4265–4284.

Groning, J., Held, C., Garten, C., Claussnitzer, U., Kaschabek, S. R. and Schlomann, M. (2007) Transformation of diclofenac by the indigenous microflora of river sediments and identification of a major intermediate. *Chemosphere* **69**, 509–516.

Halling-Sørensen, B., Nors Nielsen, S., Lanzky, P. F., Ingerslev, F., Holten Lützhoft, H. C. and Jørgensen, S. E. (1998) Occurrence, fate and effects of pharmaceutical substances in the environment – a review. *Chemosphere* **36**, 357–393.

Heberer, T. (2002a) Occurrence, fate, and removal of pharmaceutical residues in the aquatic environment: a review of recent research data. *Toxicol. Lett.* **131**, 5–17.

Heberer, T. (2002b) Tracking persistent pharmaceutical residues from municipal sewage to drinking water. *J. Hydrol.* **266**, 175–189.

Heidler, J. and Halden, R. U. (2007) Mass balance of triclosan removal during conventional sewage treatment. *Chemosphere* **66**, 362–369.

Hess-Wilson, J. K. and Knudsen, K. E. (2006) Endocrine disrupting compounds and prostate cancer. *Cancer Lett.* **241**, 1–12.

Hignite, C. and Azarnoff, D. L. (1977) Drugs and drug metabolites as environmental contaminants: Chlorophenoxyisobutyrate and salicylic acid in sewage water effluent. *Life Sci.* **20**, 337–341.

Hirano, M., Ishibashi, H., Kim, J. W., Matsumura, N. and Arizono, K. (2009) Effects of environmentally relevant concentrations of nonylphenol on growth and 20-hydroxyecdysone levels in mysid crustacean, Americamysis bahia. *Comp. Biochem. Phys. Part C: Toxicol. Pharm.* **149**, 368–373.

Hirsch, R., Ternes, T. A., Haberer, K. and Kratz, K. L. (1996) Determination of betablockers and β-sympathomimetics in the aquatic environment. *Vom Wasser* **87**, 263–274.

Houde, M., Martin, J. W., Letcher, R. J., Solomon, K. R. and Muir, D. C. G. (2006) Biological monitoring of polyfluoroalkyl substances: A review. *Environ. Sci. Technol.* **40**, 3463–3473.

Hu, J., Jin, F., Wan, Y., Yang, M., An, L., An, W. and Tao, S. (2005) Trophodynamic behavior of 4-nonylphenol and nonylphenol polyethoxylate in a marine aquatic food web from Bohai Bay, North China: Comparison to DDTs. Environ. Sci. Technol. **39**, 4801–4807.

Humburg, N. E., Colby, S. R. and Hill, E. R. (1989) Herbicide Handbook of the Weed Science Society of America (sixth ed.), Weed Science Society of America, Champaign, IL.

Hundley, S. G., Sarrif, A. M. and Kennedy, G. L. (2006) Absorption, distribution, and excretion of ammonium perfluorooctanoate (APFO) after oral administration to various species. Drug Chem. Toxicol. **29**, 137–145.

IARC/WHO (1991) Occupational exposures in insecticide application, and some pesticides, Lyon7 IARC, vol. 53.

Inui, M., Adachi, T., Takenaka, S., Inui, H., Nakazawa, M., Ueda, M., Watanabe, H., Mori, C., Iguchi, T. and Miyatake, K. (2003) Effect of UV screens and preservatives on vitellogenin and choriogenin production in male medaka (Oryzias latipes). Toxicol. **194**, 43–50.

Isidori, M., Bellotta, M., Cangiano, M. and Parella, A. (2009) Estrogenic activity of pharmaceuticals in the aquatic environment. Environ. Int. **35**, 826–829.

Isidori, M., Nardelli, A., Pascarella, L., Rubino, M. and Parella, A. (2007) Toxic and genotoxic impact of fibrates and their photoproducts on non-target organisms. Environ. Int. **33**, 635–641.

John, D. M., House, W. A. and White, G. F. (2000) Environmental fate of nonylphenol ethoxylates: differential adsorption of homologs to components of river sediment. Environ. Toxicol. Chem. **19**, 293–300.

Jonkers, N., Kohler, H. P., Dammshauser, A. and Giger, W. (2009) Mass flows of endocrine disruptors in the Glatt River during varying weather conditions. Environ. Pollut. **157**, 714–723.

Joss, A., Keller, E., Alder, A. C., Gobel, A., McArdell, C. S., Ternes, T. and Siegrist, H. (2005) Removal of pharmaceuticals and fragrances in biological wastewater treatment. Water Res. **39**, 3139–3152.

JRC (Joint Research Center) (2001), Organic Contaminants in sewage sludge for agricultural use (http://ec.europa.eu/environment/ waste/sludge/pdf/ organics_in_ ludge.pdf, retrieved 05.11.2009).

Jürgens, M. D., Williams, R. J. and Johnson, A. C. (1999) R&D Technical Report P161, Environment Agency, Bristol, UK.

Jürgens, M. D., Holthaus, K. I. E., Johnson, A. C., Smith, J. J. L., Hetheridge, M. and Williams, R. J. (2002) The potential for estradiol and ethinylestradiol degradation in English rivers. Environ. Toxicol. Chem. **21**, 480–488.

Kalantzi, O. L., Martin, F. L., Thomas, G. O., Alcock, R. E., Tang, H. R., Drury, S. C., Carmichael, P. L., Nicholson, J. K. and Jones, K. C. (2004) Different levels of polybrominated diphenyl ethers (PBDEs) and chlorinated compounds in breast milk from two U.K. regions. Environ. Health Persp. **112**, 1085–1091.

Kannan, K., Corsolini, S., Falandysz, J., Fillmann, G., Kumar, K. S., Loganathan, B. G., Mohd, M. A., Olivero, J., Van Wouwe, N., Yang, J. H. and Aldoust, K. M. (2004) Perfluorooctanesulfonate and related fluorochemicals in human blood from several countries. Environ. Sci. Technol. **38**, 4489–4495.

Kannan, K., Tao, L., Sinclair, E., Pastva, S. D., Jude, D. J. and Giesy, J. R. (2005) Perfluorinated compounds in aquatic organisms at various trophic levels in a Great Lakes food chain. *Arch. Environ. Cont. Toxicol.* **48**, 559–566.

Kasprzyk-Hordern, B., Dinsdale, R. M. and Guwy, A. J. (2007) Multi-residue method for the determination of basic/neutral pharmaceuticals and illicit drugs in surface water by solid-phase extraction and ultra performance liquid chromatography–positive electrospray ionisation tandem mass spectrometry. *J. Chromatogr. A* **1161**, 132–145.

Kasprzyk-Hordern, B., Dinsdale, R. M. and Guwy, A. J. (2008) The occurrence of pharmaceuticals, personal care products, endocrine disruptors and illicit drugs in surface water in South Wales, UK. *Water Res.* **42**, 3498–3518.

Kaushik, P. and Kaushik, G (2007) An assessment of structure and toxicity correlation in organochlorine pesticides. *J. Hazard. Mater.* **143**, 102–111.

Kelce, W. R. and Wilson, E. M. (1997) Environmental antiandrogens: developmental effects, molecular mechanisms, and clinical implications. *J. Mol. Med.* **75**, 198–207.

Keller, J. M., Kannan, K., Taniyasu, S., Yamashita, N., Day, R. D., Arendt, M. D., Segars, A. L. and Kucklick, J. R. (2005) Perfluorinated compounds in the plasma of loggerhead and Kemp's ridley sea turtles from the southeastern coast of the United States. *Environ. Sci. Technol.* **39**, 9101–9108.

Kemper, R. A. (2003) Perfluorooctanoic acid: Toxicokinetics in the rat. DuPont Haskell Laboratories, Project No. DuPont-7473. US EPA Administrative Record, AR-226-1499.

Kim, S.-C. and Carlson, K. (2007) Temporal and spatial trends in the occurrence of human and veterinary antibiotics in aqueous and river sediment matrices. *Environ. Sci. Technol.* **41**, 50–57.

Kissa, E. (2001) Fluorinated Surfactants and Repellents (second ed), Marcel Dekker, Inc., New York, NY, USA.

Kolpin, D. W., Furlong, E. T., Meyer, M. T., Thurman, E. M., Zaugg, S. D., Barber, L. B. and Buxton, H. T. (2002) Pharmaceuticals, hormones and other organic wastewater contaminants in U.S. Streams, 1999−2000: A National Reconnaissance. *Environ. Sci. Technol.* **36**, 1202–1211.

Koppen, G., Covaci, A., Van Cleuvenbergen, R., Schepens, P., Winneke, G., Nelen, V., van Larebeke, N., Vlietinck, R. and Schoeters, G. (2008) Persistent organochlorine pollutants in human serum of 50–65 years old women in the Flanders Environmental and Health Study (FLEHS). Part 1: concentrations and regional differences. *Chemosphere* **48**, 811–825.

Kreuzinger, N., Clara, M., Strenn, B. and Kroiss, H. (2004) Relevance of the sludge retention time (SRT) as design criteria for wastewater treatment plants for the removal of endocrine disruptors and pharmaceuticals from wastewater. *Water Sci. Technol.* **50**, 149–156.

Kuch, H.M. and Ballschmiter, K. (2001) Determination of endocrine disrupting phenolic compounds and estrogens in surface and drinking water by HRGC-(NCI)-MS in the picogram per liter range. *Environ. Sci. Technol.* **35**, 3201–3206.

Kumar, V., Chakraborty, A., Kural, M. R. and Poy, P. (2009) Alteration of testicular steroidogenesis and histopathology of reproductive system in male rats treated with triclosan. *Reprod. Toxicol.* **27**, 177–185.

Kummerer, K. (2009) The presence of pharmaceuticals in the environment due to human use – present knowledge and future challenges. *J. Environ. Manage.* **90**, 2354–2366.

Kunz, P. Y. and Fent, K. (2006) Estrogenic activity of UV filter mixtures. *Toxicol. Appl. Pharm.* **217**, 86–99.

Lagana, A., Bacaloni, A., De Leva, I., Faberi, A., Fago, G. and Marino, A. (2004) Analytical methodologies for determining the occurrence of endocrine disrupting chemicals in sewage treatment plants and natural waters. *Anal. Chim. Acta* **501**, 79–88.

Lai, K. M., Scrimshaw, M. D. and Lester, J. N. (2002) Biotransformation and bioconcentration of steroid estrogens by *Chlorella vulgaris*. *App. Environ. Microb.* **68**, 859–864.

Lai, K. M., Johnson, K. L., Scrimshaw, M. D. and Lester, J. N. (2000) Binding of waterborne steroid estrogens to solid phases in river and estuarine systems. *Environ. Sci. Technol.* **34**, 3890–3894.

Lalah, J. D., Schramm, K. W., Henkelmann, B., Lenoir, D., Behechti, A., Gunther, K. and Kettrup, A. (2003) The dissipation, distribution and fate of a branched ^{14}C-nonylphenol isomer in lake water/sediment systems. *Environ. Pollut.* **122**, 195–203.

Lam, M. and Mabury, S. A. (2005) Photodegradation of the pharmaceuticals atorvastatin, carbamazepine, levofloxacin, and sulfamethoxazole in natural waters. *Aquat. Sci.* **67**, 177–188.

Lam, M. W., Young, C. J., Brain, R. A., Johnson, D. J., Hanson, M. A., Wilson, C. J., Richards, S. M., Solomon, K. R. and Mabury, S. A. (2004) Aquatic persistence of eight pharmaceuticals in a microcosm study. *Environ. Toxicol. Chem.* **23**, 1431–1440.

Lambropoulou, D. A., Sakkas, V. A., Hela, D. G. and Albanis, T. A. (2002) Application of solid phase microextraction (SPME) in monitoring of priority pesticides in Kalamas River (N.W. Greece). *J. Chromatogr. A* **963**, 107–116.

Lange, I. G., Daxenberger, A., Schiffer, B., Witters, H., Ibarreta, D. and Meyer, H. H. (2002) Sex hormones originating from different livestock production systems: fate and potential disrupting activity in the environment. *Anal. Chim. Acta* **473**, 27–37.

Länge, R., Hutchinson, T. H., Croudace, C. P. and Siegmund, F. (2001) Effects of the synthetic estrogen 17 alpha-ethinylestradiol on the life-cycle of the fathead minnow (Pimephales promelas). *Environ. Toxicol. Chem.* **20**, 1216–1227.

Larsson, D. G. J., Adolfsson-Erici, M., Parkkonen, J., Pettersson, M., Berg, A. H., Olsson, P. E. and Förlin, L. (1999) Ethinyloestradiol: an undesired fish contraceptive? *Aquat. Toxicol.* **45**, 91–97.

Latch, D. E., Packer, J. L., Stender, B. L., VanOverbeke, J., Arnold, W. A. and McNeill, K. (2005) Aqueous photochemistry of triclosan: formation of 2,4-dichlorophenol, 2,8-dichlorodibenzo-p-dioxin and oligomerization products. *Environ. Toxicol. Chem.* **24**, 517–525.

Lau, C., Anitole, K., Hodes, C., Lai, D., Pfahles-Hutchens, A. and Seed, J. (2007) Perfluoroalkyl acids: A review of monitoring and toxicological findings. *Toxicol. Sci.* **99**, 366–394.

Lau, C., Butenhoff, J. L. and Rogers, J. M. (2004) The developmental toxicity of perfluoroalkyl acids and their derivatives. *Toxicol. Appl. Pharmacol.* **198**, 231–241.

Leclercq, M., Mathieu, O.,Gomez, E., Casellas, C., Fenet, H. and Hillaire-Buys, D. (2009) Presence and fate of carbamazepine, oxcarbazepine, and seven of their metabolites at wastewater treatment plants. *Arch. Environ. Cont. Toxicol.* **56**, 408–415.

Lehmler, H. J. (2005) Synthesis of environmentally relevant fluorinated surfactants – a review. *Chemosphere* **58**, 1471–1496.

Lerner, D. T., Bjornsson, B. T. and Mccormick, S. D. (2007) Larval exposure to 4-nonylphenol and 17β-estradiol affects physiological and behavioral development of seawater adaptation in Atlantic salmon smolts. *Environ. Sci. Technol.* **41**, 4479–4485.

Li, M. H. (2009) Toxicity of perfluorooctane sulfonate and perfluorooctanoic acid to plants and aquatic invertebrates. *Environ. Toxicol.* **24**, 95–101.

Lin, A. Y. and Reinhard, M. (2005) Photodegradation of common environmental pharmaceuticals and estrogens in river water. *Environ. Toxicol. Chem.* **24**, 1303–1309.

Liu, J., Wang, L., Zheng, L., Wang, X. and Lee, F. S. C. (2006) Analysis of bacteria degradation products of methyl parathion by liquid chromatography/electrospray time-of-flight mass spectrometry and gas chromatography/mass spectrometry. *J. Chromatogr. A* **29**, 180–187.

Liu, Z.-H., Kanjo, Y. and Mizutani, S. (2009) Urinary excretion rates of natural estrogens and androgens from humans, and their occurrence and fate in the environment: A review. *Sci. Total Environ.* **407**, 4975–4985.

Löffler, D., Römbke, J., Meller, M. and Ternes, T. A. (2005) Environmental fate of pharmaceuticals in water/sediment systems. *Environ. Sci. Technol.* **39**, 5209–5218.

Loos, R., Gawlik, B. M., Locoro, G., Rimaviciute, E., Contini, S. and Bidoglio, G. (2009) EU-wide survey of polar organic persistent pollutants in European river waters. *Environ. Pollut.* **157**, 561–568.

Loos, R., Wollgast, J., Huber, T. and Hanke, G. (2007) Polar herbicides, pharmaceutical products, perfluorooctanesulfonate (PFOS), perfluorooctanoate (PFOA), and non-ylphenol and its carboxylates and ethoxylates in surface and tap waters around Lake Maggiore inn Northern Italy. *Anal. Bioanal. Chem.* **387**, 1469–1478.

Lopez-Avila, V. and Hites, R. A. (1980) Organic compounds in an industrial wastewater. Their transport into sediment. *Environ. Sci. Technol.* **14**, 1382–1390.

Luo, X., Mai, B., Yang, Q., Fu, J., Sheng, G. and Wang, Z., (2004) Polycyclic aromatic hydrocarbons (PAHs) and organochlorine pesticides in water columns from the Pearl River and the Macao harbor in the Pearl River Delta in South China. *Mar. Pollut. Bull.* **48**, 1102–1115.

Manahan, S. E. (2004) Environmental Chemistry, CRC Press, New York, USA.

Martin, J. W., Mabury, S. A., Solomon, K. R. and Muir, D. C. G. (2003) Bioconcentration and tissue distribution of perfluorinated acids in rainbow trout (*Oncorhynchus Mykiss*). *Environ. Toxicol. Chem.* 196–204.

Matthews, G. (2006) Pesticides: health, safety and the environment, Blackwell Publishing, Oxford, UK.

Mc Avoy, D. C., Schatowitz, B., Jacob, M., Hauk, A. and Eckhoff, W. S. (2002) Measurement of triclosan in wastewater treatment systems. *Environ. Toxicol. Chem.* **21**, 1323–1329.

McMahon, K., Bengtson–Nash, S., Eaglesham, G, Müller, J. F., Duke, N. C. and Winderlich, S. (2005) Herbicide contamination and the potential impact to seagrass meadows in Hervey Bay, Queensland, Australia. *Mar. Pollut. Bull.* **51**, 325–334.

Metcalfe, C. D., Koenig, B. G., Bennie, D. T., Servos, M., Ternes, T. A. and Hirsch, R. (2003) Occurrence of neutral and acidic drugs in the effluents of Canadian sewage treatment plants. *Environ. Toxicol. Chem.* **22**, 2872–2880.

Mezcua, M., Gomez, M. J., Ferrer, I., Aguera, A., Hernando, M. D. and Fernandez-Alba, A. R. (2004) Evidence of 2,7/2,8-dibenzodichloro-p-dioxin as a photodegradation product of triclosan in water and wastewater samples. *Anal. Chim. Acta* **524**, 241–247.

Miao, X. S. and Metcalf, C. D. (2003) Determination of carbamazepine and its metabolites in aqueous samples using Liquid Chromatography–Electrospray Tandem Mass Spectrometry. *Anal. Chem.* **75**, 3731–3738.

Mimeault, C., Woodhouse, A. J., Miao, X. S., Metcalfe, C. D., Moon, T. W. and Trudeau, V. L. (2005) The human lipid regulator, gemfibrozil bioconcentrates and reduces testoterone in the goldfish, Carassius auratus. *Aquat. Toxicol.* **73**, 44–54.

Mompelat, S., Le Bot, B. Thomas, O. (2009) Occurrence and fate of pharmaceutical products and by-products, from resource to drinking water. *Environ. Int.* **35**, 803–814.

Müller, B., Berg, M., Yao, Z.-P., Zhang, X.-F., Wang, D. and Pfluger, A. (2008) How polluted is the Yangtze river? Water quality downstream from the Three Gorges Dam. *Sci. Total Environ.* **402**, 232–247.

Murakami, M., Shinohara, H. and Takada, H. (2009) Evaluation of wastewater and street runoff as sources of perfluorinated surfactants (PFSs). *Chemosphere* **74**, 487–493.

Nelson, S. D., Letey, J., Farmer, W. J. and Ben-Hur, M. (1998) Facilitated transport of nanpropamide by dissolved organic matter in sewage sludge-amended soil. *J. Environ. Qual.* **27**, 1194–2000.

Ning, B., Graham, N., Zhang, Y., Nakonechny, M. and El-Din, M. G. (2007) Degradation of Endocrine Disrupting Chemicals by Ozone/AOPs. *Ozone-Sci. Eng.* **29**, 153–176.

Oaks, J. L., Gilbert, M., Virani, M. Z., Watson, R. T., Meteyer, C. U., Rideout, B. A., Shivaprasad, H. L., Ahmed, S., Chaudhry, M. J. I., Arshad, M., Mahmood, S., Ali, A. and Khan, A. A. (2004) Diclofenac residues as the cause of vulture population decline in Pakistan. *Nature* **427**, 630–633.

OECD (Organization for Economic Co-operation and Development), 2005. Results of survey on production and use of PFOS, PFAS and PFOA, related substances and products/mixtures containing these substances. ENV/JM/MONO(2005)1, Paris.

Okamura, H., Aoyama, I., Ono, Y. and Nishida, T. (2003) Antifouling herbicides in the coastal waters of western Japan, *Mar. Pollut. Bull.* **47**, 59–67.

Olsen, G. W., Burris, J. M., Ehresman, D. J., Froehlich, J. W., Seacat, A. M., Butenhoff, J. L. and Zobel, L. R. (2007) Half-life of serum elimination of perfluorooctanesulfonate, perfluorohexanesulfonate, and perfluorooctanoate in retired fluorochemical production workers. *Environ. Health Perspect.* **115**, 1298–1305.

Orvos, D. R., Versteeg, D. J., Inauen, J., Capdevielle, M., Rodethenstein, A. and Cunningham, V. (2002) Aquatic toxicity of triclosan. *Environ. Toxicol. Chem.* **21**, 1338–1349.

Palmer, B. D. and Palmer, S. K. (1995) Vitellogenin induction by xenobiotic estrogens in the red-eared turtle and African clawed frog. Environ. *Health Perspect.* **103**, 19–25.

Pandit, G. G., Mohan Rao, A. M., Jha, S. K., Krishnamoorthy, T. M., Kale, S. P., Raghu, K. and Murthy, N. B. K. (2001) Monitoring of organochlorine pesticide residues in the Indian marine environment. *Chemosphere* **44**, 301–305.

Pehkonen, S.O. and Zhang, Q. (2002) The degradation of organophosphorus pesticides in natural waters: a critical review. *Crit. Rev. Env. Sci. Tec.* **32**, 17–72.

Peng, X., Yu, Y., Tang, C., Tan, J. Xuang, Q. and Wang, Z. (2008) Occurrence of steroid estrogens, endocrine-disrupting phenols, and acid pharmaceutical residues in urban riverine water of the Pearl River Delta, South China. *Sci. Total Environ.* **397**, 158–166.

Peñuela, G. A. and Barceló, D. (1998) Photosensitized degradation of organic pollutants in water: processes and analytical applications. *Trend. Anal. Chem.* **17**, 605–612.

Petrovic, M., Diaz, A., Ventura, F. and Barcelo, D. (2003) Occurrence and removal of estrogenic short-chain ethoxy nonylphenolic compounds and their halogenated derivatives during drinking water production. *Environ. Sci. Technol.* **37**, 4442–4448.

Planas, C., Caixach, J., Santos, F. J. and Rivera, J. (1997) Occurrence of pesticides in Spanish surface waters. Analysis by high resolution gas chromatography coupled to mass spectrometry. *Chemosphere* **34**, 2393–2406.

Planas, C., Guadayol, J. M., Droguet, M., Escalas, A., Rivera, J. and Caixach, J. (2002) Degradation of polyethoxylated nonylphenols in a sewage treatment plant. Quantitative analysis by isotopic dilution-HRGC/MS. *Water Res.* **36**, 982–988.

Plumlee, M. H., Larabee, J. and Reinhard, M. (2008) Perfluorochemicals in water reuse. *Chemosphere* **72**, 1541–1547.

Poiger, T., Buser, H.-R. and Müller, M. D. (2001) Photodegradation of the pharmaceutical drug diclofenac in a lake: pathway, field measurements, and mathematical modeling. *Environ. Toxicol. Chem.* **20**, 256–263.

Polder, A., Odland, J. O., Tkachev, A., Foreid, S., Savinova, T. N. and Skaare, J. U. (2003) Geographical variation of chlorinated pesticides, toxaphenes and PCBs in human milk from sub-arctic and arctic locations in Russia. *Sci. Total Environ.* **306**, 79–195.

Pomati, F., Orlandi, C., Clerici, M., Luciani, F. and Zuccato, E. (2008) Effects and interactions in an environmentally relevant mixture of pharmaceuticals. *Toxicol. Sci.* **102**, 129–137.

Preuss, T. G., Gurer-Orham, H., Meerman, J. and Ratte, H. T. (2009) Some nonylphenol isomers show antiestrogenic potency in the MVLN cell assay. Toxicology in Vitro (in press, doi: 10.1016/j.tiv.2009.08.017).

Purdum, C. E., Hardiman, P. A., Bye, V. J., Eno, N. C., Tyler ,C. R. and Sumpter, J. P. (1994) Oestrogenic effects of effluent from sewage treatment works. *Chem. Ecol.* **8**, 275–285.

Qiu, Y.-W., Zhang, G., Guo, L.-L., Cheng, H.-R., Wang, W.-X., Li, X.-D. and Wai, W. H. (2009) Current status and historical trends of organochlorine pesticides in the ecosystem of Deep Bay, South China. *Estuar. Coast. Shelf S.* **85**, 265–272.

Quinn, B., Gagné, F. and Blaise, C. (2009) Evaluation of the acute, chronic and teratogenic effects of a mixture of eleven pharmaceuticals on the cnidarian, Hydra attenuate. *Sci. Total Environ.* **407**, 1072–1079.

Radjenovic, J., Petrovic, M. and Barcelo, D. (2009) Fate and distribution of pharmaceuticals in wastewater and sewage sludge of the conventional activated sludge (CAS) and advanced membrane bioreactor (MBR) treatment. *Water Res.* **43**, 831–841.

Rana, S. V. S. (2006) Environmental Pollution, Health and Toxicology, Alpha Science International Lts., Oxford, UK.

Reith, D. M., Appleton, D. B., Hooper, W. and Eadie, M. J. (2000) The effect of body size on the metabolic clearance of carbamazepine. *Biopharm. Drug Dispos.* **21**, 103–111.

Renner, R. (1997) European bans on surfactant trigger transatlantic debate. *Environ. Sci. Technol.* **31**, 316–320.

Rodgers-Gray, T. P., Jobling, S., Morris, S., Kelly, C., Kirby, S., Janbakhsh, A., Harries, J. E., Waldock, M. J., Sumpter, J. P. and Tyler, C. R. (2000) Long-term temporal

changes in the estrogenic composition of treated sewage effluent and its biological effects on fish. *Environ Sci Technol.* **34**, 1521–1528.
Rupa, D. S., Reddy, P. P. and Reddi, O. S. (1990) Cytogeneticity of quinalphos and methyl parathion in human peripheral lymphocytes. *Hum. Exp. Toxicol.* **9**, 385–387.
Sabik, H., Jeannot, R. and Rondeaua, B. (2000) Multiresidue methods using solid-phase extraction techniques for monitoring priority pesticides, including triazines and degradation products, in ground and surface waters. *J. Chromatogr. A*, **885**, 217–236.
Salazar-Arredondo, E., Solís-Heredia, M. de, J., Rojas-García, E., Hernández-Ochoa, I. and Betzabet Quintanilla-Vega, B. (2008) Sperm chromatin alteration and DNA damage by methyl-parathion, chlorpyrifos and diazinon and their oxon metabolites in human spermatozoa. *Reprod. Toxicol.* **25**, 455–460.
Salvestrini, S., Di Cerbo, P. and Capasso, S. (2002) Kinetics of the chemical degradation of diuron. *Chemosphere* **48**, 69–73.
Sanderson, H., Boudreau, T. M., Mabury, S. A. and Solomon, K. R. (2004) Effects of perfluorooctane sulfonate and perfluorooctanoic acid on the zooplanktonic community. *Ecotox. Environ. Safe.* **58**, 68–76.
Santos, L., Aparicio, I. and Alonso, E. (2007) Occurrence and risk assessment of pharmaceutically active compounds in wastewater treatment plants. A case study: Seville city (Spain). *Environ. Int.* **33**, 596–601.
Sarmah, A. K., Meyer, M. T. and Boxall, A. B. A. (2006) A global perspective on the use, sales, exposure pathways, occurrence, fate and effects of veterinary antibiotics (Vas) in the environment. *Chemosphere* **65**, 725–759.
Sarmah, A. K. and Northcott, G. L. (2008) Laboratory degradation studies of four endocrine disruptors in two environmental media. *Environ. Toxicol. Chem.* **27**, 819–827.
Scheytt, T., Mersmann, P., Lindstädt, R. and Heberer, T. (2005) Determination of pharmaceutically active substances carbamazepine, diclofenac, and ibuprofen, in sandy sediments. *Chemosphere* **60**, 245–253.
Schneider, W. and Degen, P. H. (1981) Simultaneous determination of diclofenac sodium and its hydroxy metabolites by capillary column gas chromatography with electron-capture detection. *J. Chromatogr.* **217**, 263–271.
Schubert, S., Peter, A., Burki, R., Schonenberger, R., Suter, M. J. F. and Segner, H., Burkhardt-Holm P. (2008) Sensitivity of brown trout reproduction to long-term estrogenic exposure. *Aquat. Toxicol.* **90**, 65–72.
Schwaiger, J., Ferling, H., Mallow, U., Wintermayr, H. and Negele, R. D. (2004) Toxic effects of the non-steroidal anti-inflammatory drug diclofenac. Part I: Histopathological alterations and bioaccumulation in rainbow trout. *Aquat. Toxicol.* **68**, 141–150.
Seacat, A. M., Thomford, P. J., Hansen, K. J., Olsen, G. W., Case, M. T. and Butenhoff, J. L. (2002) Subchronic toxicity studies on perfluorooctanesulfonate potassium salt in cynomolgus monkeys. *Toxicol. Sci.* **68**, 249–264.
Segura, P. A., Francois, M., Cagnon, C. and Sauve, S. (2009) Review of the occurrence of anti-infectives in contaminated wastewaters and natural and drinking waters. *Environ. Health Persp.* **117**, 675–684.
Seibert, B. (1996) Data from animal experiments and epidemiology data on tumorigenicity of estradiol valerate and ethinyl estradiol, cited in: Endocrinically Active Chemcials in the Environment, UBA TEXTE 3/96, Berlin, pp. 88–95.

Servos, M. R., Bennie, D. T., Burnison, B. K., Jurkovic, A., McInnis, R., Neheli, T., Schnell, A, Seto, P., Smyth, S. A. and Ternes, T. A. (2005) Distribution of estrogens, 17b-estradiol and estrone, in Canadian municipal wastewater treatment plants. *Sci. Total Environ.* **336**, 155–170.

Shankar, M. V., Nélieu, S., Kerhoas, L. and Einhorn, J. (2008) Natural sunlight $NO^{3\ -}/NO^{2-}$-induced photo-degradation of phenylurea herbicides in water. *Chemosphere* **71**, 1461–1468.

Shao, B., Hu, J., Yang, M., An, W. and Tao, S. (2005) Nonylphenol and nonylphenol ethoxylates in river water, drinking water, and fish Tissues in the area of Chongqing, China. *Arch. Environ. Cont. Toxicol.* **48**, 467–473.

Shore, L. S., Kapulnik, Y., Ben-Dov, B., Fridman, Y., Wininger, S. and Shemesh, M. (1992) Effects of estrone and 17 β estradiol on vegetative growth of Medicago sativa, Physiol. *Plant.* **84**, 217–222.

Shukla, G., Kumar, A., Bhanti, M., Joseph, P. E. and Taneja, A. (2006) Organochlorine pesticide contamination of ground water in the city of Hyderabad. *Environ. Int.* **32**, 244–247.

Shultz, S., Baral, H. S., Charman, S., Cunningham, A. A., Das, D., Ghalsasi, G. R., Goudar, M., Green, R. E., Jones, A., Nighot, P., Pain, D. J. and Prakash, V. (2004) Diclofenac poisoning is widespread in declining vulture populations across the Indian subcontinent. *Proc. Roy. Soc. London B* **271**, S458–S460.

Sinclair, E. and Kannan, K. (2006) Mass loading and fate of perfluoroalkyl surfactants in wastewater treatment plants. *Environ. Sci. Technol.* **40**, 1408–1414.

Singer, H., Muller, S., Tixier, C. and Pillonel, L. (2002) Triclosan: Occurrence and fate of a widely used biocide in the aquatic environment: Field measurements in wastewater treatment plants, surface waters, and lake sediments. *Environ. Sci. Technol.* **36**, 4998–5004.

Soares, A., Guieysse, B., Jefferson, B., Cartmell, E. and Lester, J. N. (2008) Nonylphenol in the environment: A critical review on occurrence, fate, toxicity and treatment in wastewaters. *Environ. Int.* **34**, 1033–1049.

Soto, A. M., Sonnenschein, C., Chung, K. L., Fernandez, M. F., Olea, N. and Serrano, F. O. (1995) The E-SCREEN assay as a tool to identify estrogens: an update on estrogenic environmental pollutants. *Environ. Health Persp.* **103**, 113–122.

Standley, L. J., Rudel, R. A., Swartz, C. H., Attfield, K. R., Christian, J., Erickson, M. and Brody, J. G. (2008) Wastewater-contaminated groundwater as a source of endogenous hormones and pharmaceuticals to surface water ecosystems. *Environ. Tox. Chem.* **27**, 2457–2468.

Staples, C., Mihaich E., Carbone J., Woodburn K., Klecka G. (2004) A weight of evidence analysis of the chronic ecotoxicity of nonylphenol ethoxylates, nonylphenol ether carboxylates, and nonylphenol. *Hum. Ecol. Risk Assess.* **10**, 999–1017.

Stasinakis, A. S., Gatidou, G., Mamais, D., Thomaidis, N. S. and Lekkas, T. D. (2008) Occurrence and fate of endocrine disrupters in Greek sewage treatment plants. *Water Res.* **42**, 1796–1804.

Stasinakis, A. S., Kordoutis, C., I., Tsiouma, V. C., Gatidou, G. and Thomaidis, N. S. (2009b) Removal of selected endocrine disrupters in activated sludge systems: Effect of sludge retention time on their sorption and biodegradation. Bioresource Technol. (in press).

Stasinakis, A. S., Petalas, A. V., Mamais, D., Thomaidis, N. S., Gatidou, G. and Lekkas, T. D. (2007) Investigation of triclosan fate and toxicity in continuous-flow activated sludge systems. *Chemosphere*, **68**, 375–381.

Stasinaksi, A. S., Kotsifa, S., Gatidou, G., Mamais, D. (2009a) Diuron biodegradation in activated sludge batch reactors under aerobic and anoxic conditions. *Water Res.* **43**, 1471–1479.

Strandberg, M. T. and Scott-Fordsmand, J.-J. (2002) Field effects of simazine at lower trophic levels–a review. *Sci. Total Environ.* **296**, 117–137.

Takagi, S., Adachi, F., Miyano, K., Koizumi, Y., Tanaka, H., Mimura, M., Watanabe, I., Tanabe, S. and Kannan, K. (2008) Perfluorooctanesulfonate and perfluorooctanoate in raw and treated tap water from Osaka, Japan. *Chemosphere* **72**, 1409–1412.

Tao, L., Kannan, K., Kajiwara, N., Costa, M., Fillman, G., Takahashi, S. and Tanabe, S. (2006) Perfluorooctanesulfonate and related flurochemicals in albatrosses, elephant seals, penguins, and polar skuas from the Southern Ocean. *Environ. Sci. Technol.* **40**, 7642–7648.

Ternes, T. A., Hirsch, R., Mueller, J. and Haberer, K. (1998) Methods for the determination of neutral drugs as well as betablockers and β_2-sympathomimetics in aqueous matrices using GC/MS and LC/MS/MS. *Fresen. J. Anal. Chem.* **362**, 329–340.

Ternes, T. A., Knacker, T. and Oehlmann, J. (2003) Persconal care products in the aquatic environment – A group of substances which has been neglected to date. Umweltwissenschaften und Schadstoff-Forschung, **15**, 169–180.

Thiele-Bruhn, S. (2003) Pharmaceutical antibiotic compounds in soils – a review. *J. Plant. Nutr. Soil Sci.* **166**, 145–167.

Thomas, D. B. (1984), Do hormones cause breast cancer? *Cancer* **53**, 595–604.

Thomas, L., Russell, A. D. and Maillard, J. Y. (2005) Antimicrobial activity of chlorhexidine diacetate and benzalkonium chloride against Pseudomonas aeruginosa and its response to biocide residues. *J. Appl. Microb.* **98**, 533–543.

Thorpe, K. L., Hutchinson, T. H., Hetheridge, M. J., Scholze, M., Sumpter, J. P. and Tyler, C. R. (2001) Assessing the biological potency of binary mixtures of environmental estrogens using vitellogenin induction in juvenile rainbow trout (Oncorhynchus mykiss). *Environ. Sci. Technol.* **35**, 2476–2481.

Thorpe, K. L., Cummings, R. I., Hutchinson, T. H., Scholze, M., Brighty, G., Sumpter, J. P. Tyler, C. R. (2003) Relative potencies and combination effects of steroidal estrogens in fish. *Environ Sci Technol* **37**, 1142–1149.

Tixier, C., Singer, H. P., Canonica, S. and Muller, S. R. (2002) Phototransformation of triclosan in surface waters: a relevant elimination process for this widely used biocide – Laboratory studies, field measurements, and modeling. *Environ. Sci. Technol.* **36**, 3482–3489.

Tolls, J. (2001) Sorption of veterinary pharmaceuticals in soils: a review. *Environ. Sci. Technol.* **17**, 3397–3406.

Tomy, G. T., Tittlemier, S. A., Palace, V. P., Budakowski, W. R., Brarkevelt, E., Brinkworth, L. and Friesen, K. (2004) Biotransformation of N-ethyl perfluorooctanesulfonamide by rainbow trout (Onchorhynchus mykiss) liver microsomes. *Environ. Sci. Technol.* **38**, 758–762.

Triebskorn, R., Casper, H., Heyd, A., Eikemper, R., Köhler, H.-R. and Schwaiger, J. (2004) Toxic effects of the non-steroidal anti-inflammatory drug diclofenac. Part II:

Cytological effects in liver, kidney, gills and intestine of rainbow trout (Oncorhynchus mykiss). *Aquat. Toxicol.* **68**, 151–166.
U.S. Environmental Protection Agency: http://www.epa.gov/opp00001/about/ (retrieved 20.10.2009).
U.S. EPA (U.S. Environmental Protection Agency) (2000) Water quality standards: establishment of numeric criteria for priority toxic pollutants for the State of California; final rule. *Fed. Reg.* **65**, 31681–31719.
UNEP (2002) Sub-Saharan Africa Regional Report: Regionally Based Assessment of Persistent Toxic Substances.
UNEP (2003) Stockholm Convention: Master List of Actions: on the reduction and/or elimination of the releases of persistent organic pollutants (Fifth ed.), United Nations Environmental Programme, Geneva, Switzerland.
Van Aerle, R., Rounds, N., Hutchinson, T. H., Maddix, S. and Tyler, C. R. (2002) Window of sensitivity for the estrogenic effects of ethinylestradiol in early life-stages of fathead minnow. *Ecotoxicology* **11**, 423–434.
Vanderford, B. J., Pearson, R. A., Rexing, D. J. and Snyder, S.A. (2003) Analysis of endocrine disruptors, pharmaceuticals and personal care products in water using Liquid Chromatography/Tandem Mass Spectrometry. *Anal. Chem.* **75**, 6265–6274.
Vazquez-Duhalt, R., Marquez-Rocha, F., Ponce, E., Licea, A. F. and Viana, M. T. (2005) Nonylphenol, an integrated vision of a pollutant. Scientific review. *Appl. Ecol. Environ. Res.* **4**, 1–25.
Veldhoen, N., Skirrow, R. C., Osachoff, H., Wigmore, H., Clapson, D. J., Gunderson, M. P., Van Aggelen, G. and Helbing, C. C. (2006) The bactericidal agent triclosan modulates thyroid hormone-associated gene expression and disrupts postembryonic anuran development. *Aquat. Toxicol.* **80**, 217–227.
Vestergren, R. and Cousins, I. T. (2009) Tracking the pathways of human exposure to perfluorocarboxylates. *Environ. Sci. Technol.* **43**, 5565–5575.
Vorkamp, K., Riget, F., Glasius, M., Pécseli, M., Lebeufand, M. and Muir, D. (2004) Chlorobenzenes, chlorinated pesticides, coplanar chlorobiphenyls and other organochlorine compounds in Greenland biota. *Sci. Total Environ.* **331**, 157–175.
Walker, C. H., Hopkin, S. P., Silby, R. M. and Peakall, D. B. (2006) Principles of Ecotoxicology, CRC Press, New York, USA.
Wang, N., Stostek, B., Folsom, P. W., Sulecki, L. M., Capka, V., Buck, R. C., Berti, W. R. and Gannon, J. T. (2005) Aerobic biotransformation of 14C-labeled 8-2 telomer B alcohol by activated sludge from domestic sewage treatment plant. *Environ. Sci. Technol.* **39**, 531–538.
Weiger, S., Berger, U., Jensen, E., Kallenborn, R., Thoresen, H. and Huhnerfuss, H. (2004) Determination of selected pharmaceuticals and caffeine in sewage and seawater from Tromsø/Norway with emphasis on ibuprofen and its metabolites. *Chemosphere*, **56**, 583–592.
WHO (World Health Organization) (2004) The WHO recommended classification of pesticides by hazard and guidelines to classification. WHO, Geneva.
Winker, M., Faika, D., Gulyas, H. and Otterpohl, R. (2008) A comparison of human pharmaceutical concentrations in raw municipal wastewater and yellowwater. *Sci. Total Environ.* **399**, 96–104.
Winkler, M., Lawrence, J. R. and Neu, T. R. (2001) Selective degradation of ibuprofen and clofibric acid in two model river biofilm systems. *Water Res.* **35**, 3197–3205.

Worthing, C. R. and Walker, S. B. (1987) The Pesticide Manual: a World Compendium (eight ed.), British Crop Protection Council, London.
Wu, X., Hua, R., Tang, F., Li, X., Cao, H. and Yue, Y. (2006) Photochemical degradation of chlorpyrifos in water. *Chinese J. App. Ecol.* **17**, 1301–1304.
Yamamoto, H., Nakamura, Y., Moriguchi, S., Nakamura, Y., Honda, Y., Tamura, I., Hirata, Y., Hayashi, A. and Sekizawa, J. (2009) Persistence and partitioning of eight selected pharmaceuticals in the aquatic environment: Laboratory photolysis, biodegradation, and sorption experiments. *Water Res.* **43**, 351–362.
Yamamoto, H., Watanabe, M., Katsuki, S., Nakamura, Y., Moriguchi, S., Nakamura, Y. and Sekizawa, J. (2007) Preliminary ecological risk assessment of butylparaben and benzylparaben-2. Fate and partitioning in aquatic environments. *Environ. Sci. : An Int. J. Environ. Phys. Toxicol.* **14**, 97–105.
Yamashita, N., Kannan, K., Taniyasu, S., Horii, Y., Petrick, G. and Gamo, T. (2005) A global survey of perfluorinated acids in oceans. *Mar. Pollut. Bull.* **51**, 658–668.
Ye, X., Bishop, A. M., Reidy, J. A., Needham, L. L. and Calafat, A. M. (2006) Parabens as urinary biomarkers of exposure in humans. *Environ. Health Persp.* **114**, 1843–1846.
Ying, G. G. (2006) Fate, behavior and effects of surfactants and their degradation products in the environment. *Environ. Int.* **32**, 417–431.
Ying, G. G., Yu, X. Y. and Kookana, R. S. (2007) Biological degradation of triclocarban and triclosan in a soil under aerobic and anaerobic conditions and comparison with environmental fate modeling. *Environ. Pollut.* **150**, 300–305.
Young, C. J., Furdui, V. I., Franklin, J., Koerner, R. M., Muir, D. C. G. and Mabury, S. A. (2007) Perfluorinated acids in arctic snow: New evidence for atmospheric formation. *Environ. Sci. Technol.* **41**, 3455–3461.
Yuan, S. Y., Yu, C. H. and Chang, B. V. (2004) Biodegradation of nonylphenol in river sediment. *Environ. Pollut.* **127**, 425–430.
Zhang, Y., Geiben, S. U. and Gal, C. (2008) Carbamazepine and diclofenac: Removal in wastewater treatment plants and occurrence in water bodies. *Chemosphere* **73**, 1151–1161.
Zhou, J. L., Maskaoui, K., Qiu, Y. W., Hong, H. S. and Wang, Z. D. (2001) Polychlorinated biphenyl congeners and organochlorine insecticides in the water column and sediments of Daya bay, China. *Environ. Pollut.* **113**, 373–384.
Zhou, R., Zhu, L., Chen, Y. and Kong, Q. (2008) Concentrations and characteristics of organochlorine pesticides in aquatic biota from Qiantang River in China. *Environ. Pollut.* **151**, 190–199.
Zimetbaum, P., Frishman, W. H. and Kahn, S. (1991) Effects of gemfibrozil and other fibric acid derivates on blood lipids and lipoproteins. *J. Clin. Pharm.* **31**, 25–37.
Zwiener, C., Seeger, S., Glauner, T. and Frimmel, F. H. (2002) Metabolites from the biodegradation of pharmaceutical residues of ibuprofen in biofilm reactors and batch experiments. *Anal. Bioanal. Chem.* **372**, 569–575.

Chapter 2
Analytical methods for the identification of Micropollutants and their transformation products

Ekaterina V. Rokhina and Jurate Virkutyte

2.1 INTRODUCTION

Micropollutants (the harmful substances detected in micro g range in the environment) are the non-regulated contaminants, which may be potential candidates for the future regulation depending on their health effects and monitoring data regarding their occurrence. The great majority of micropollutants are presented in Table 2.1. These recently emerged substances are mostly pesticides, pharmaceuticals, personal care products (PPCPs), industrial chemicals, gasoline additives and etc.

©2010 IWA Publishing. *Treatment of Micropollutants in Water and Wastewater.* Edited by Jurate Virkutyte, Veeriah Jegatheesan and Rajender S. Varma. ISBN: 9781843393160. Published by IWA Publishing, London, UK.

Table 2.1 Classes of emerging contaminants

Group of emerging compounds (Category/subcategories)	Examples	Review of analytical method for the specific class of compounds
Antifoaming agents	Surfinol-104	(Richardson, 2009)
Antioxidants	2,6-Di-tert-butylphenol 4-tert-Butylphenol BHA BHQ BHT	(Henry and Yonker, 2006) (Bakker and Qin, 2006)
Artificial sweeteners	Sucrose	(Giger, 2009)
Complexing agents	DTPA EDTA NTA Oxadixyl TAED	(Richardson et al., 2007)
Biocides	Triclosan Methyltriclosan Chlorophene	(Richardson, 2000)
Detergents Aromatic sulphonates		(Richardson, 2009)
Linear alkylbenzene sulfonates (LAS)	C10-C14-LAS C12-LAS	
Ethoxylates/carboxylates of octyl/nonyl phenols	4-Nonylphenol di-ethoxylate 4-Octylphenol di-ethoxylate	
Disinfection by-products	Bromate Cyanoformaldehyde Decabromodiphenyl ethane Hexabromocyclododecane (HBCD) NDMA	(Richardson, 2003) (Richardson et al., 2007)
Flame retardants Brominated flame retardants	Tetrabromo bisphenol A (TBBPA) Decabromodiphenyl ethane	(Hyotylainen and Hartonen, 2002)
Polybrominated diphenylethers	Technical Decabromodiphenyl ether	

(*continued*)

Table 2.1 (*continued*)

Group of emerging compounds (Category/subcategories)	Examples	Review of analytical method for the specific class of compounds
	Technical Octabromodiphenyl ether Technical Pentabromodiphenyl ether	
Organophosphates	Tri-(dichlorisopropyl)-phosphate Triethylphosphate Tri-n-butylphosphate	(Reemtsma *et al.*, 2008)
Chlorinated paraffin	Long chain PCAs (IPCAs, C>17) Technical PCA products	
Fragrances	Acetylcedrene Benzylacetate Camphor g-Methylionone	(Raynie, 2004)
Gasoline additives	Dialkyl ethers, Methyl-t-butyl ether (MTBE)	(Richardson, 2009)
Industrial chemicals	TCEP Triphenyl phosphine oxide	
Personal care products		(Hao *et al.*, 2007) (Kot-Wasik *et al.*, 2007)
Sun-screen agents	Benzophenone 4-Methylbenzylidene camphor Octocrylene Oxybenzone	
Insect repellents	N,N-diethyl-m-toluamide (DEET) Bayrepel	
Antiseptics Carriers	Triclosan, Chlorophene Octamethylcyclotrasiloxane (D4) Decamethylcyclopentasiloxane (D5)	

(*continued*)

Table 2.1 (*continued*)

Group of emerging compounds (Category/subcategories)	Examples	Review of analytical method for the specific class of compounds
	Octamethyltrisiloxane (MDM) Decamethyltetrasiloxane (MD2M) Dodecamethylpenta-siloxane (MD3M)	
Parabens (hydroxybenzoic acid esters)	Methyl-paraben Ethyl-paraben Propyl-paraben Isobutyl-paraben	
Pesticides		(Gascon *et al.*, 1997) (Kralj *et al.*, 2007) (Hernandez *et al.*, 2008) (Liu *et al.*, 2008) (Raman Suri *et al.*, 2009)
Polar pesticides and their degradation products	Amitrole Bentazone Chlorpyrifos 2,4 D Diazinon Prometon Secbumeton Terbutryn	
Other pesticides	Cypermethrin Deltamethrin Permethrin	
New pesticides	Sulfonyl urea	
Degradation products of pesticides	Desisopropylatrazine Desethylatrazine	
Plasticizers		(Chang *et al.*, 2009)
Phthalates	Benzylbutylphthalate (BBP) Diethylphthalate (DEP) Dimetylphthalate (DMP)	
Other	Bisphenol A Triphenyl phosphate	
Benzophenone derivatives	2,4-Dihydroxybenzo-phenone	

(*continued*)

Table 2.1 (*continued*)

Group of emerging compounds (Category/subcategories)	Examples	Review of analytical method for the specific class of compounds
Pharmaceuticals		(De Witte *et al.*, 2009a) (Fatta *et al.*, 2007) (Gros *et al.*, 2008) (Pavlovic *et al.*, 2007) (Radjenovic *et al.*, 2007; Radjenovic *et al.*, 2009) (Kosjek and Heath, 2008)
Antibacterial	Sulfonamides Ampicillin Ciprofloxacin Sulfamerazine	(Hernandez *et al.*, 2007) (Garcia-Galan *et al.*, 2008)
Analgesics, anti-inflammatory drugs	Codein, ibuprofene acetaminophen, acetylsalicilyc acid, diclofenac, fenoprofen	(Macia *et al.*, 2007)
Antidepressant	Tetracycline Citalopram Escitalopram Sertraline	
Antidiabetic	Glyburide Metformin	
β-blockers	Atenolol Betaxolol Carazolol Metoprolol Propranolol Sotalol	(Hernando *et al.*, 2007)
Blood viscosity agents Bronchodilators	Pentoxifylline Albuterol Albuterol sulfate Clenbuterol	
Diuretic	Caffeine Furosemide Hydrochlorothiazide	
Lipid regulators	Bezafibrate Clofibric acid Etofibrate Fenofibrate Fenofibric acid Gemfibrozil	

(*continued*)

Table 2.1 (*continued*)

Group of emerging compounds (Category/subcategories)	Examples	Review of analytical method for the specific class of compounds
Sedatives, hypnotics	Acecarbromal Allobarbital Amobarbital Butalbital Hexobarbital	
Psychiatric drugs	Amitryptiline Doxepine Diazepam Imapramine Nordiazepam	
X-ray contrast agents	Diatrizoate Iohexol Iomeprol Iopamidol Iopromide	
Steroid and hormones	17-alpha-Estradiol 17-alpha-Ethinylestradiol 17-beta-Estradiol Cholesterol Diethylstilbestrol Estriol Estrone Mestranol	(Lopez de Alda *et al.*, 2003; Gabet *et al.*, 2007) (Miege *et al.*, 2009) (Streck, 2009)
Surfactants and surfactant metabolites	Alkylphenol ethoxylates 4-nonylphnol 4-octylphenol alkylphenol carboxylates	(Lopez de Alda *et al.*, 2003)
Anticorrosives		(Richardson, 2009)
Textile dyes Acid Reactive Direct		(Poiger *et al.*, 2000) (Pinheiro *et al.*, 2004) (Oliveira *et al.*, 2007) (Kucharska and Grabka, 2010) (Petroviciu *et al.*, 2010)

Several new candidates were introduced by Richardson (2009) and Giger (2009) such as sucralose (artificial sweetener also known as Splenda), antimony (leachate from polyethylene terephthalate (PET) plastic bottles), siloxanes, and musks (fragrance additives found in perfumes, lotions, sunscreens, deodorants, and laundry detergents). Importantly, these groups of contaminants do not need to persist in the environment to cause negative effects since their high transformation/removal rates can be compensated by their continuous introduction into the water environment with domestic and industrial wastewater effluents.

Moreover, occurrence, risk assessment and ecotoxicological data are not available for most of these emerging contaminants, so it is difficult to predict what health effects they may have on humans and aquatic organisms. After the discharge into the environment, emerging pollutants are subjected to biodegradation, chemical and photochemical degradation that highly contribute to their elimination. Depending on the compartment in which the synthetic chemicals are present in the environment (e.g., groundwater, surface water and/or sediments) or in the technosphere (e.g., wastewater treatment plants (WWTPs) and drinking-water facilities), different transformations can take place, sometimes producing products that differ in their environmental behavior and ecotoxicological profile (Farre *et al.*, 2008). Unfortunately, these degradation products can exhibit even greater toxicity than parent compounds (e.g., the major biodegradation product of nonylphenol ethoxylates – nonylphenol, is much more persistent than the parent compound and can mimic estrogenic properties).

Analytical methodology available for various groups of emerging contaminants is still lacking and, although methods exist to analyze each of those compounds individually, the key issue is to develop multi-residue methods in which different compound classes can be determined by a single short analysis. Therefore, constant development in the techniques and improvement of currently existing methods is mainly focused in the field of analysis of micropollutants and their transformation products. Better instrument design and a fuller understanding of the mechanics of analytical processes would enable steady improvements to be made in sensitivity, precision, and accuracy. The ultimate aim is the development of a non-destructive method, which not only saves time but leaves the sample unchanged for further examination or processing (Fifield and Kealey, 2000).

Analysis of organic pollutants in wastewater is a complex process, basically due to the range of physico-chemical and toxicological properties of compounds included in the same group, e.g., pharmaceuticals. Taking into account public concerns on environmental issues, analytical studies and the consequent use of toxic reagents and solvents have increased to a point at which they became unsustainable to continue without an environmentally friendly perspective. The strategy to develop clean and environmentally benign material and techniques can be termed as "Sustainable Analytical Procedures" (SAPs) in the frame of Green Analytical Chemistry (GAC). GAC was initiated as a search for practical alternatives to the off-line treatment of wastes and residues in order to replace polluting methodologies with the clean ones (Armenta *et al.*, 2008). In general, GAC methods are used to replace toxic reagents, to miniaturize and to automate methods, dramatically reduce the amounts of reagents consumed and wastes generated, thus reducing or even avoiding potential side effects of currently available and future analytical methods.

The focus of the Chapter is on the general trends and the most used techniques in the adequate monitoring and analytical instrumentation to investigate the fate and the behavior of emerging pollutants in wastewater treatment plants and receiving waters. Several recent reviews providing the detailed insight on the analytics of a specific group of micropollutants are given in Table 2.1. Also, Susan D. Richardson published a biennial review covering the recent developments in water analysis for all classes of emerging environmental contaminants over the period of 2007–2008 taking into account 250 most significant references.

2.2 THEORETICAL APPROACHES TO THE ANALYTICS OF MICROPOLLUTANTS

Analytical chemistry of micropollutans usually includes samples that are far from simple, containing numerous components to be analyzed simultaneously or a few target analytes in the presence of many chemical interferents. In such cases, theoretical approaches, e.g., computational chemistry methods and mathematical tools coupled with sophisticated instrumentation are used to assist the analysis.

2.2.1 Computational methods to evaluate the degradation of micropollutants

Computational chemistry is a branch of chemistry that uses principles of computer science to assist in solving chemical problems. The results of theoretical chemistry, incorporated into the efficient computer programs are utilized to calculate the structures and properties of molecules and solids (e.g., the expected positions of the constituent atoms, absolute and relative (interaction) energies, electronic charge distributions, etc.). Computational chemists often attempt to solve the non-relativistic Schrödinger equation, with relativistic corrections:

$$\hat{H}\Psi = E\Psi \qquad (2.1)$$

where $\hat{H}$ is the Hamiltonian operator, Ψ is a wave function, and E is the energy. Solutions to Schrödinger's equation are able to describe atomic and subatomic systems, atoms and electrons.

Application of computational chemical methods allows rapid progress in estimating reaction pathways or products from theoretically obtained information on the reaction position or bond conditions. Recent advances in molecular modeling can predict the fate of a huge variety of micropollutants in the presence of different oxidizing agents. Efficient improvement in a predictive molecular-level modeling methods can also enhance the understanding of the fate of new compounds in the water environment.

Computational chemistry methods range from highly accurate to a very approximate methods. *Ab initio* (lat. "from the beginning") methods are directly derived from theoretical principles without using experimental data. Other methods are called empirical or semi-empirical because they include experimental results obtained from the acceptable models of atoms or related molecules, and can approximate or omit various selected elements of the underlying theory. However, to avoid potential errors that occur when one or some elements are omitted, the method is parameterized.

In the recent years, the impact of Density Functional Theory (DFT), one of a recently developed *ab-initio* methods in quantum chemistry, has increased enormously due to less computational requirements however providing the same accuracy of calculations as other computationally intensive methods. DFT is based on the Hohenberg–Kohn theorem that uses electron density instead of the more complex wave function to determine all atomic and molecular properties.

Several examples of computational calculations used to study micropollutants degradation are listed in Table 2.2.

2.2.1.1 Frontier electron density analysis (Frontier orbital theory)

The determination of specific sites of interaction between the two chemical species is of fundamental importance to establish the mechanism of the reaction and also to design the desired products. A number of DFT-based reactivity descriptors, such as the electronic chemical potential, hardness, softness and Fukui function have been derived to determine the specific sites of interaction between the pollutant and an oxidant (e.g., hydroxyl radical). These parameters are associated with the response of the electron density of a system to a change in number of electrons (N) or external potential [$v(r)$].

Table 2.2 Computational methods used to study advanced (catalytic) oxidation of micropollutants

Compound	AOP	Calculation method (basis set/level)	Software	Identification of products	Ref
Alizarin red	TiO$_2$ photocatalysis	single determinant Hartree–Fock level with optimization on AM1 level	MOPAC 6.0	GC MS	(Liu et al., 2000)
PAHs in ethanol	Fenton process	PM3	MOPAC 6.0	GC MS with HPLC-UV	(Lee et al., 2001)
17 β-estradiol (E2)	TiO$_2$ photocatalysis	STO-3G/unrestricted Hartree–Fock (UHF)	Gaussian 98	HPLC +NIST	(Ohko et al., 2002)
Pyridaben	TiO$_2$ photocatalysis	PM3	Hyperchem, version 5.0.	GC MS	(Zhu et al., 2004)
Polychlorinated dibenzo-p-dioxins (PCDDs),	Fe(II)/H$_2$O$_2$/ UV	AM1	MOPAC 6.0	GC MS	(Katsumata et al., 2006)
Imazapyr	TiO$_2$ photocatalysis	NR	MOPAC 6.0	HPLC-UV, LC/ESI-MS	(Carrier et al., 2006)
Alanine	chlorination	AM2	MOPAC 6.0	HPLG,GC-ESI-MS-SCAN (NIST 147)	(Chu et al., 2009)
Levoflaxine	ozone, peroxone	B3LYP/6-31 +G(d,p),BMK/B3LYP	GAUSSIAN 03	GC MS	(De Witte et al., 2009b)
Dinitronaphthalenes	TiO$_2$ photocatalysis	B3LYP/6-31G*	GAUSSIAN 03	GC MS	(Bekbolet et al., 2009)

Fukui function $f(r)$ is the most important local DFT descriptor. Compared to the frontier orbitals HOMO (highest occupied molecular orbital) and LUMO (lowest unoccupied molecular orbital), the Fukui function contains more detailed information, taking orbital relaxation effects into account (De Witte et al., 2009a). It is based on the Frontier Molecular Orbital (FMO) theory and can be defined as the mixed second derivative of the energy of the molecule with respect to N and $[v(r)]$. Physically, Fukui function reflects how sensitive the chemical potential of the system is to an external perturbation at a particular point. It shows the change in the electron density driven by a change in the number of electrons in its frontier valence region. The Fukui function is the reactivity index for orbital-controlled reactions, where larger Fukui function indicates higher reactivity. Fukui functions per atom i in a molecule can be defined as following:

$$f_i^- = [q_i(N) - q_i(N-1)] \quad (2.2)$$

$$f_i^o = \frac{[f_i^+ + f_i^-]}{2} \quad (2.3)$$

where q_i is the electron population of atom i in the molecule. f_i^- is used when the system undergoes an electrophilic, whereas $f_o\,i$ is valid when the system undergoes a radical attack.

More detailed information about the computational methods and reactivity based descriptors can be found in some recently published books (Parr and Weitao, 1994; Young, 2001).

2.2.2 Chemometrics in analysis

Critical aspect in quantitative analysis is the occurrence of matrix effects (i.e., suppression or, less frequently, enhancement of the analyte signal), which may lead to a significant difference in the response of an analyte in a sample as compared to a pure standard solution. The nature and the amount of co-eluting matrix compounds may vary between the samples in such a way that matrix effects in a series of samples can be difficult to predict. Despite a considerable development in various methods, a challenge in applying sophisticated instrumentation and mathematical tools to develop analytical figure of merits to detect and quantify the target analytes at trace levels remains unsolved. Spectroscopic data can be mathematically modeled by using two different approaches: i) by using hard-modelling-based methods, which require a reaction model to be postulated, or ii) by using soft modelling- based methods, which do not require to know the kinetic model linked to the reactions studied (Escandar et al., 2007).

The standard methods for the second-order approaches to data analysis are parallel factor analysis (PARAFAC), and multivariate curve resolution alternating least squares (MCR), as well as the most recently developed bilinear least-squares (BLLS), unfolded partial least squares/residual bilinearization (U-PLS/RBL), and artificial neural networks followed by the residual bilinearization (ANN/RBL) (Galera et al., 2007). These methods not only simplify the data by reducing dimensionality but also provide visual representation of data as well.

2.2.2.1 Parafac

Multi-way data are characterized by several sets of categorical variables that are measured in a crossed fashion (e.g., any kind of spectrum measured chromatographically for several samples). Determination of such variables would give rise to three-way data; i.e., data that can be arranged in a cube instead of a matrix as in currently available standard multivariate data sets. Assuming trilinearity, PARAFAC decomposes a three-way array X, obtained by joining second-order data from the calibration and test samples:

$$\underline{X} = \sum_{n+1}^{N} a_n \otimes b_n \otimes c_n + \underline{E} \qquad (2.4)$$

where $\otimes$ indicates the Kronecker product, N is the total number of responsive components and E is an appropriately dimensioned residual error term. The column vectors a_n, b_n and c_n are usually collected into the score matrix A and the loading matrices B and C.

2.2.2.2 MCR

The goal of multivariate curve resolution (MCR) is to mathematically decompose an instrumental response for a mixture into the pure contributions of each component involved in the system studied. This method is capable of dealing with data sets deviating from trilinearity. Data from the spectroscopic monitoring of a chemical process can be arranged in a data matrix D (r×c), the r rows of which are the number of spectra recorded throughout the process and the c columns of which are the instrumental responses measured at each wavelength (Garrido et al., 2008). The MCR decomposition of matrix D is carried out as following:

$$D = CS^T + E \qquad (2.5)$$

where C ($r \times n$) is the matrix describing how the contribution of the spectroscopically active species n involved in the process changes in the different r rows of the data matrix (concentration profiles). This equation can be solved by either non-iterative methods (e.g., window factor analysis (WFA), sub-window factor analysis (SFA), heuristic evolving latent projections (HELP), orthogonal projection resolution (OPR), parallel vector analysis (PVA)) or iterative approaches (iterative target transformation factor analysis (ITTFA), resolving factor analysis (RFA), multivariate curve resolution-alternating least squares (MCR-ALS). Iterative methods are more frequently used because of their flexibility (assumptions of a model are not required), and ability to handle different kinds of data structures and chemical problems. Moreover, they are able to integrate external information into the resolution process.

2.2.2.3 BLLS

The classic method of least squares is applied to approximate solutions of over determined systems, i.e., systems of equations in which there are more equations than unknowns. This method is usually performed in statistical contexts, particularly regression analysis. Least squares can be interpreted as a method of fitting data.

BLLS starts with a calibration step using the calibration set, in which approximations to pure analyte matrices Sn at unit concentration are found by direct least squares. To estimate the pure analyte matrices Sn, the calibration data matrices $X_{c;i}$ are first vectorized and grouped into a JK×I matrix VX. Then, a procedure analogous to classical least-squares is performed:

$$V_S = V_X \times Y^{T+} \quad (2.6)$$

where Y is an I · N_c matrix collecting the nominal calibration concentrations, N_c is the number of calibrated analytes, and VS (size JK · N_c) contains the vectorized Sn matrices.

2.2.2.4 U-PLS

In the U-PLS method, concentration of the target contaminant is first used in the calibration step (without including data for the unknown sample). The I calibration data matrices Xc i are vectorized and a usual U-PLS model is calibrated with these data and the vector of calibration concentrations. This provides a set of loadings P and weight loadings W (both of size JK×A, where A is the number of latent factors), as well as the regression coefficients v (size A×1). The parameter A can be selected by techniques such as leave-one-out cross-validation (Galera et al., 2007). If no unexpected interferences occur in the test sample, v can be employed to estimate the analyte concentration:

$$y_u = t_u^T \times v \quad (2.7)$$

where t_u is the test sample score, obtained by the projection of the (unfolded) data for the test sample X_u onto the space of the A latent factors.

2.2.2.5 ANN

ANN is a computational model simulating the structure and/or functional aspects of biological neural networks. It consists of an interconnected group of artificial neurons and processes information using a connectionist approach to computation. In most cases the ANN is an adaptive system that changes its structure based on external or internal information that flows through the network during the learning phase. In general, neural networks are non-linear statistical data modeling tools. They are utilized to model complex relationships between the inputs and outputs or to find patterns in available data. Typically, an ANN comprises three layers: input, hidden, and output. Usually, the number of input neurons equals A, where, A principal components (PCs) $\ll$ JK (the number of channels in each dimension). The value of A is estimated by computing the % of variance explained by the PCs of the unfolded training data matrix (size I×JK, I is the number of training samples), and selecting the first A PCs which explain more than a certain % (i.e., 99%) of the total variance (Galera *et al.*, 2007).

The training/prediction scheme works properly for the test samples, which have a composition that is representative of the training set. However, when unexpected constituents occur in the test sample, its scores will not be suitable for the analyte prediction using the currently trained ANN. In this case, it is necessary to resort to a technique that marks the new sample as an outlier, indicating that further actions are necessary before ANN prediction, and then isolates the contribution of the unexpected component from that of the calibrated analytes, in order to recalculate appropriate scores for the test sample (Galera *et al.*, 2007).

2.3 INSTRUMENTAL METHODS

The authentic characteristics of micropollutants, such as their occurrence at trace concentration levels, the presence of extremely diverse groups, and etc. make procedures of the detection and analysis quite challenging. Traditionally, structure proof involves several steps: purification, functional group identification, and establishment of atom and group connectivity. The modern instrumental methods have the ability to run reactions, purify products, and determine structures on the nanogram scales with the subsequent increase in the rate at which structural information can be obtained. This has resulted in an exponential growth of chemical knowledge and is directly responsible for the explosion of information being continually published in the chemical literature.

2.3.1 Sample preparation

Many techniques for separating and concentrating the species of interest have been devised in the past several decades. Such techniques are aimed at exploiting differences in physico-chemical properties of various components present in a mixture and can be divided into extraction and chromatographic separation methods.

2.3.1.1 Sample extraction

Generally, the pretreatment or extraction step plays an important role in determining the overall level of analytical performances in practice. The development of new methods for sample preparation, which can dramatically reduce the required amounts of the reagent and organic solvents, also improves the characteristics of other methodologies, which cannot be directly applied to the samples as electrochemical and chromatographic methods can (Armenta *et al.*, 2008). Current trends in sample handling focus on the development of faster, safer, and more environment friendly extraction techniques (Rubio and Perez-Bendito, 2009; Tobiszewski *et al.*, 2009). In order to select the most appropriate and efficient extraction method, several points should be taken into account: i) the solubility of the compound and ii) the difference between concentrations of the parent compound and its degradation products. Moreover, degradation products to be separated by, for instance, gas chromatography (GC) are generally dissolved in organic solvent instead of water, thus only small volumes can be injected to obtain the results. Therefore, sample concentration procedures, which provide a solvent change are needed (De Witte *et al.*, 2009a).

Solid-phase extraction (SPE) is one of the leading techniques for the extraction of pollutants from the aquatic systems (Kostopoulou and Nikolaou, 2008). One of the most important parameters in the application of a SPE method is the selection of an appropriate solid sorbent to the target analyte as well as the use of solvents for washing and the subsequent elution of the target contaminant. The sorbent can be packed into small tubes or cartridges, such as a small liquid chromatographic column or it is also available in the shape of discs with a filtration apparatus (Chang *et al.*, 2009). No losses of the reaction intermediates should occur during sample loading while 100% recovery is desired in the elution step (De Witte *et al.*, 2009a). Due to the wide range of polarities of micropollutants, non-selective sorbents such as silica (C18) or polymer-based resins are most commonly used. Recent developments in this field are mainly related to the use of new sorbent materials such as molecularly imprinted polymers (MIPs). Depending on the type of cartridges, the recovery rates of an identical analyte may vary from 10 to 90%. Moreover, Gatidou and co-workers (2007) compared several types of cartridges for the extraction of nonylphenol (NP), nonylphenol monoethoxylate (NPIEO), triclosan (TCS), bisphenol A (BPA), and nonylphenol diethoxylate (NP2EO) from aqueous solution. The analytical results of the target compound were compared for different types of the SPE cartridge: C18, EnviChrom-P, Isolute NV+, and Oasis HLB. Among them, only C18 demonstrated sufficient recoveries for the majority of the compounds, whereas the rest of studied cartridges were selectively efficient (Gatidou et al., 2007).

Nowadays, enhanced solvent-extraction techniques are widely accepted for the extraction of pollutants from environmental solid matrixes, however, serious limitations that relates to the cleanup step still remains unsolved. The latest trends for the purification of collected extracts involve the use of some fast and/ or simplified procedures to treat the liquid samples, e.g., solid-phase (SPME) and liquid phase (LPME) microextraction. In the past few years, matrix solid-phase dispersion (MSPD) became very popular in the extraction of organic pollutants from a variety of solid environmental samples and headspace microextraction procedures (e.g., HS-SPME) and gradually developed into the preferred tools for the extraction of volatile and semivolatile compounds (Rubio and Perez-Bendito, 2009). Alternative approaches to the sample preparation process are extraction techniques based on exploitation of microwave energy (i.e., microwave-assisted extraction (MAE)) and micelle creation (i.e., micelle-mediated extraction (MME), where cloud-point extraction (CPE) is of special importance. In literature, MAE is usually reserved for the liquid –liquid (LLE) and solid–liquid (SLE) extraction, which are assisted by the microwave energy (Madej, 2009).

The SPME method primarily involves the adsorption of an analyte onto the surface of a sorbent-coated silica fiber. The advantage of SPME is the direct injection of the adsorbed analyte into the chromatograph. The sensitivity of the SPME method can be assured, since the polymeric stationary phases used in SPME have a high affinity for organic molecules (Chang et al., 2009). The practical application of SPME has various aspects. Direct and headspace SPME can be differentiated, depending on where the coated fiber is placed for the extraction. Occasionally, a vigorously selected derivatization reagent (e.g., BSA, BSTFA, MSTFA, TMCS, HMDS) can be added to enhance the performance of the selected analytical method. In addition, operational factors, such as sample stirring speeds as well as the extraction time and temperature required to reach the equilibrium between the aqueous and stationary phases on the SPME fiber should be adjusted. If all operational parameters are fixed, the analysis can be automated with the reduced usage of solvents. Moreover, the fiber can be reused several times and recycled with a proper care (Chang *et al.*, 2009).

2.3.1.2 Chromatographic separation

The selection of the chromatographic separation technique is based on the polarity and thermal stability of the target compound. The most performed indispensable chromatography separation types are GC (gas chromatography) and LC (liquid chromatography). Recent advances in separation techniques of special relevance to environmental analysis of micropollutants and their degradation products have been promoted by the use of thin layer chromatography (TLC), ultraperformance liquid chromatography (UPLC), hydrophilic interaction chromatography (HILIC), fast gas chromatography (FGC), comprehensive two-dimensional gas chromatography (GC×GC), and capillary electrophoresis-based microchips (μCE) (Rubio and Perez-Bendito, 2009). LC is probably the most versatile separation method, as it allows separation of compounds of a wide range of polarity with little effort, compared to GC-MS, in sample preparation. Semi-polar compounds can be separated using reverse-phase columns, while hydrophilic columns can facilitate the quantification of target polar compounds (Chang *et al.*, 2009). The fastest growing chromatography trend still remains in the use of ultra-performance liquid chromatography (UPLC). UPLC is a recently developed LC technique that uses small diameter particles (typically 1.7μm) in the stationary phase and short columns, which allow higher pressures and, ultimately, narrower LC peaks (5–10 s wide) (Richardson 2009). UPLC has such advantages as providing narrow peaks, improved chromatographic separations, and extremely short analysis times, often to 10 min or less.

2.3.1.3 Capillary electrophoresis (CE)

The physical mechanisms of capillary electrophoresis (CE) are similar to LC. It is also a water based separation technique, which is well suited to the analysis of inherently hydrophilic compounds. The main difference from LC, which is based on the partition of the solutes between the mobile phase and the stationary phase, is that CE is derived from the differences in charge-to-mass ratio of the molecules. Therefore, a totally different selectivity is expected for the analytes, thus providing a separation method complementary to LC (Pico *et al.*, 2003).

All three separation techniques including their advantages and disadvantages are summarized in Table 2.3.

Table 2.3 Comparative analysis of separation techniques (Adapted from Pico et al. (2003))

Technique	Advantages	Disadvantages	Solutions
GC	High resolving power and ability to resolve individual analytes High sensitivity and selectivity Existence of mass-spectrum libraries for screening unknown samples	Inadequate for polar, thermo-labile and low-volatility compounds High consumption of expensive, high-purity gases	Derivatisation
LC	Application to virtually any organic solute regardless of its volatility or thermal stability Both mobile and stationary phase compositions are variables Capable of automation and miniaturisation (microchip technology)	Insufficient separation efficiency and selectivity Large amounts of expensive and toxic organic solvent used as mobile phase	Development of more efficient and selective columns materials (immuno-sorbents, MIPs and restricted access materials)
CE	High separation efficiency Small consumption of expensive reagents and toxic solvents Capable of automation and miniaturisation (microchip technology)	Inadequate limits of detection Lack of selective detectors	Sample enrichment (SPE, stacking) Increase detection path-length Development of coupling methods to combine CE with highly selective detectors

2.3.2 Detection of micropollutants transformation products

Identification of micropollutants transformation products is a challenge that resides in obtaining high-quality data suitable for the detection from the available analytical technologies. These analytical technologies (liquid chromatography (LC), mass spectrometry (MS), fragmentation pattern analysis (MS2), ultraviolet/visible spectroscopy (UV/Vis), nuclear magnetic resonance (NMR) and experimental validation by standard compounds) are summarized in Figure 2.1. In addition, information already available in literature and databases (NIST, Wiley, etc.) would also be an important identification tool to determine the transformation products of target micropollutans.

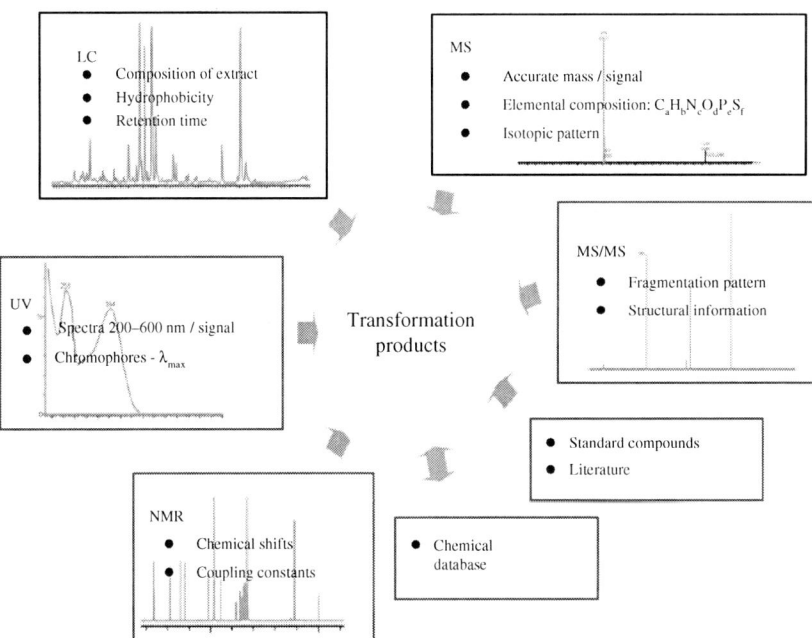

Figure 2.1 The most common methods to identify the transformation products (Adapted from Moco *et al.* (2007)

However, identification of transformation products is quite a challenging task due to several problems, associated with the identification of unknown structures. Usually, they present in low concentrations in the samples and standard material for structure elucidation is seldom available (De Witte *et al.*, 2009a). The scheme of comprehensive structure elucidation of micropollutants transformation products can include several detection methods (e.g., HPLC/ NMR, UV/MS, etc.) (Kormos et al., 2009).

2.3.2.1 Mass spectrometry

Currently, the most prevailing approach designed to analyze emerging pollutants incorporates a mass-based analysis process. Generally speaking, the mass-based methods employing mass spectrometry (MS) show relatively low detection limits as compared to other methods. MS is a spectrometric method that allows the detection of mass-to-charge species that can point to the molecular mass (MM) of the detected compound.

Mass spectrometry involves absorption of energy at particular frequencies. During the MS determinations, a molecule is purposely broken into pieces, these pieces are identified by the mass, and subsequently the original structure is then inferred from these pieces. Mass spectrometers can be divided into three fundamental parts: i) the ionization source, ii) the analyzer, and iii) the detector (Scheme 2.1). Different mass analyzers have different features, including the m/z range that can be covered, the mass accuracy, and the achievable resolution as well as the compatibility of different analyzers with different ionization methods. GC-MS instruments make use of the hard-ionization method, electron-impact (EI) ionization, while LC-MS mostly uses soft-ionization sources (e.g., atmospheric pressure ionization (API) (e.g., electrospray ionization (ESI)) and atmospheric pressure chemical ionization (APCI)) (Moco et al., 2007). In LC-MS applications, the mass detection of a molecule by MS, is conditioned by the capacity of the analyte to ionize while being part of a complex mixture and, therefore, micropollutants unable to ionize, cannot be detected.

2.3.2.1.1 Recent advances in MS based techniques. Various configurations of mass spectrometers used for MS applications, in terms of ion acceleration and mass detection, ion-production interfaces and ion fragmentation capabilities are currently available. Liquid and gas chromatography, in combination with mass spectrometry (LC-MS, GC/MS) and tandem MS (LC-MS2, GC/MS^2), continue to be the predominant techniques for the identification and quantification of organic pollutants, and their transformation products, in the environment. New methods reported include the utilization of UPLC/MS/MS, LC/TOF-MS with accurate mass determination, and online-SPE/LC/MS^2. The recent developments in UPLC allowed to vastly improve chromatographic resolution (as compared to the conventional LC), run times (often less than 10 min), and to minimize the matrix effects. Recent advances in capillary electrophoresis-mass spectrometry (CE-MS) have rendered this technique more competitive in the analysis of environmental samples; however, the inherently small (trace) amounts of pollutants present in the environment, and the high complexity of environmental sample matrixes, have placed strong demands on the detection capabilities of CE (Rubio and Perez-Bendito, 2009).

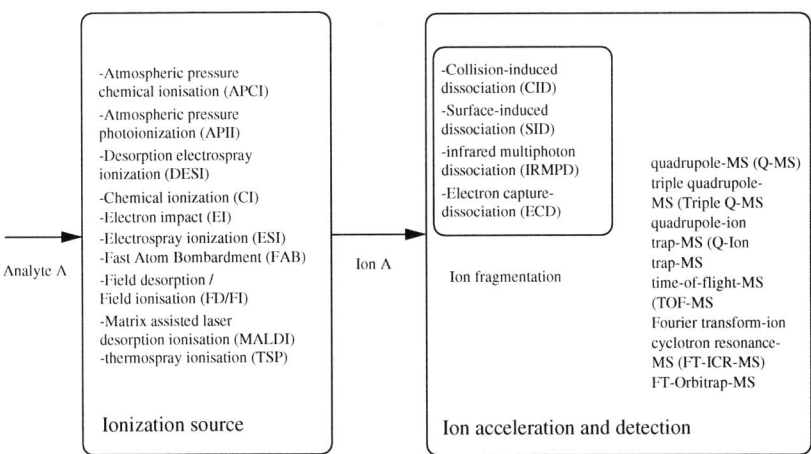

Scheme 2.1 Simplified configuration of MS (Adapted from Moco et al. (2007)

Also, there is a tremendous increase in the use of time-of-flight (TOF)-mass spectrometry (MS) and quadruple (Q)-TOF-MS for the structural elucidation and compound confirmation. As a rule, (Q)-MS instruments have a mass resolving power that is 4 times less than that of a time-of-flight (TOF)-MS, while a Fourier transform (FT)-ion cyclotron (ICR)-MS can reach a resolving power of higher than 1,000,000 (i.e., 400 times greater than a Q-MS) (Richardson, 2009). The use of hybrid QTOF, instead of single TOF, offers more possibilities in screening and identification of target micropollutants, so it is feasible to access the MSE acquisition mode, where low and high collision-energy full-scan acquisitions can be performed simultaneously, resulting in valuable fragmentation information for use in elucidation and confirmation of the unknown compounds (Ibanez et al., 2008). TOF-MS and Q-TOF-MS provide an increased resolution capability (typically 10 000–12 000 resolution), which allows the precise empirical formula assignments for the unknowns and also provides extra confidence for positive identifications in the quantitative work (Richardson, 2009). Especially, TOF-MS and Q-TOF-MS are effective means in the detection of pharmaceuticals, endocrine disrupting compounds (EDCs), and pesticide degradation products.

A higher mass accuracy facilitates a finer distinction between closely related mass-to-charge signals, so the quality and the quantity of assignments of mass signals to metabolites can be much improved by using high-resolution and ultra-high-resolution accurate mass spectrometers (Moco et al., 2007).

In addition, liquid chromatography (LC)/electrospray ionization (ESI)- and atmospheric pressure chemical ionization (APCI)-MS methods continue to dominate the new methods developed for the detection of emerging contaminants, and the use of multiple reaction monitoring (MRM) with MS/MS has become commonplace for the quantitative environmental analysis. Also, atmospheric pressure photoionization (APPI) is used with LC/MS to provide the improved ionization for more non-polar compounds, such as polybrominated diphenyl ethers (PBDEs). Furthermore, FT-MS instruments, such as the cyclotron (FT-ICR-MS) and the Orbitrap type (FT-Orbitrap-MS), enable measurements at a higher mass accuracy in a wider dynamic range. FT-ICR-MS has the highest mass-resolving power so far reported for any mass spectrometer ($>1,000,000$) and a mass accuracy is generally within 1 ppm (Hogenboom et al., 2009). The recently developed FT-Orbitrap-MS has a relatively modest performance compared to the FT-ICR-MS (maximum resolving power $>100,000$ and 2 ppm of mass accuracy with an internal standard), but it is a high-speed, high ion-transmission instrument, due to shorter accumulation times (Moco et al., 2007). Hyphenation of UPLC and TOF-MS is an efficient, advanced approach for rapid screening of non-target organic micropollutants in water. Therefore, UPLC provides a fast chromatographic run with improved resolution, minimizing the co-elution of components. Then again, the huge amount of information provided by TOF-MS, together with the measurement of accurate mass for the most representative ions, facilitates confident identification of non-target compounds in samples (Ibanez et al., 2008).

In general, the chromatographic parameters (such as temperature, pH, column, flow rate, eluents, gradient), injection parameters, sample properties, MS and MS^2 parameters (calibration and instrumental parameters, such as capillary voltage and lens orientation), and the remaining parameters related to the configuration of the system can affect the performance of analyses. An adequate configuration should be adopted, taking into account the aim of the analyses and the current limitations of the instruments.

2.3.2.1.2 Data interpretation. High resolution instruments provide very high mass accuracies, and the range of possibilities for chemical formula is limited, especially for lower m/z values and therefore, it is more proficient to determine the correct molecular formula than low resolution instruments. However, the number of possible chemical structures increases with increasing molecular mass values.

Important aspect that must be taken into account when determining the chemical structure of the target compound is the algorithm used for the calculation. There are more possible mathematical combinations of elements that fit certain molecular masses than the number of chemical formulas that exist chemically. The chemical rules (e.g., the octet rule) dictate certain limitations on chemical bonding derived from the electronic distribution of the participating atoms present in molecules (e.g., nitrogen rule). Also, the number of rings and double bonds can be calculated from the number of C, H atoms in a molecule.

The practical approach to narrow the choice of chemical structures is the isotope labeling. Isotopically labeled standards (deuterated or ^{13}C, ^{15}N-labeled) allow a more accurate quantification of a target compounds in a variety of sample matrixes. This is especially relevant for wastewater and biological samples, where matrix effects can be substantial (Richardson, 2008). For the most small organic molecules, the intensity of the second isotopic signal, corresponding to the ^{13}C signal, can indicate the number of carbons that the molecular ion contains (natural abundance of ^{13}C is 1.11%) (Moco et al., 2007). Kind and Fiehn (2006) stated that this strategy can remove more than 95% of false positives and can even outperform an analysis of accurate mass alone using a (as yet non-existent) mass spectrometer capable of 0.1 ppm mass accuracy. Several works of isotope labeling in the analysis of transformation products have been reported (Vogna et al., 2002; McDowell et al., 2005).

Another possibility to accurately determine a target compound is isolating one ion and performing MS^2 to the successively obtained fragments. It is extremely informative for tracking functional groups and connectivity of fragments for elucidating the structures of degradation products.

When a separation method is coupled to the mass spectrometer, retention time is a parameter that can give information about the polarity of the metabolite. Nowadays, retention-time variation is relatively low in stabilized (LC or GC)-MS setups, thus allowing a direct comparison of chromatograms and the construction of databases (Moco et al., 2007).

The ability to assign the reaction intermediates using MS resides in the possibility of combining different features of the MS analysis (accurate mass, fragmentation pattern, and isotopic pattern) with additional experimental parameters (e.g., retention time) and its confirmation with standard compounds. Experimental spectra can be compared against a home-made mass-spectral library (either empirical or theoretical), hence automating an efficient screening for many different compounds. Home made libraries need to contain a large number of contaminants, and exact masses need to be included in the databases for the correct candidate assignment. When a match is unsatisfactory, the deconvoluted MS spectra can be used to investigate the identity of unknown compounds not present in the library (Ibanez et al., 2008).

2.3.3 UV-Visible spectroscopy

UV-Vis spectroscopy was one of the earliest techniques to be combined with other techniques when chromatographic separation was incorporated with UV-Vis detection. Although it is common to use fixed wavelength detection, rapid digital scanning can permit essentially continuous spectral scanning of chromatographic effluents.

Absorption of radiation in the visible and ultraviolet regions of the electromagnetic spectrum results in electronic transitions between molecular orbitals. The energy changes are relatively large, corresponding to about 105 J mol^{-1}, which corresponds to a wavelength range of 200–800 nm or a wave number range of 12 000–50 000 cm^{-1} (Fifield and Kealey, 2000). The use of UV-Vis for the determination of the degradation products usually includes the measurement of absorbance spectra of the sample or individual degradation product after the chromatographic separation. The appearance and disappearance of the absorbance maxima is directly related to the presence of the particular functional groups. Unsaturated groups, known as chromophores, are responsible for the absorption mainly in the near UV and visible regions and are of most value for the identification purposes and for quantitative analysis. The positions and intensities of the absorption bands are sensitive to the substituents, which are close to the chromophore, as well as to conjugation with other chromophores, and to the solvent effects. UV-Vis generally requires quite low concentrations of the analyte (absorbances less than 2). However, the presence of the same chromophore in different compounds will result in the same UV-Vis spectra. Therefore, despite the use of UV-Vis is useful for the estimation of the degradation products, more advanced analytical methods are needed in order to identify possible reaction intermediates (De Witte et al., 2009a).

New developments in this technique include commercial versions of UV-vis imaging systems, attenuated total reflection (ATR) spectroscopy, FT-UV spectroscopy optimization, and various applications of fiber optic based and high-resolution techniques (Richardson, 2009).

2.3.4 NMR spectroscopy

Nuclear magnetic resonance (NMR) spectroscopy has become an important technique for the determination of chemical and physical properties of a wide array of organic compounds. Such determination is possible due to the absorption of energy at particular frequencies by atomic nuclei when they are placed in a magnetic field. Nuclei, which have finite magnetic moments and spin quantum numbers of I = ½ such as ^1H, ^{11}B, ^{13}C, ^{15}N, ^{17}O, ^{19}F, and ^{31}P are the most useful and common in NMR measurements. Since hydrogen and carbon are the most common nuclei found in organic compounds, the ability to probe these nuclei by NMR is invaluable for the organic structure determination. Unfortunately, as the nuclear transition energy is much lower (typically of the order of 104) than the electronic transition, NMR is relatively un-sensitive in comparison to other techniques, such as UV/Vis spectroscopy (Wishart, 2008). Furthermore, the signal-to-noise ratio (S/N) in NMR depends on many parameters (e.g., magnetic field strength of the instrument (B_0), concentration of the sample, acquisition time (NS), and the measurement temperature).

However, NMR is perhaps the most selective analytical technique currently available, being able to provide unambiguous information about a target molecule. NMR can elucidate chemical structures, and can provide highly specific evidence for the identification of a molecule (Moco et al., 2007). Moreover, NMR is a quantitative technique, as the number of nuclear spins is directly related to the intensity of the signal.

NMR can be directly applied for the structure elucidation. ^1H is the most used nucleus for NMR measurements due to its very high natural abundance (99.9816–99.9974%) and good NMR properties. In fact, several NMR techniques are available for a great variety of purposes, e.g., one-dimensional 1H NMR, two-dimensional (2D)-NMR, homonuclear ^1H-2D spectra (e.g., COSY, TOCSY and NOESY) and heteronuclear 2D spectra that can be acquired for detecting direct ^1H-^{13}C bonds by HMQC or, over a longer range, HMBC. Nowadays, the detection limit in a 14.1 Tesla (600 MHz for 1H NMR) instrument, is in the microgram (^1H-^{13}C NMR) or even sub-microgram region (^1H NMR). The sensitivity of NMR has been improving over the years, increasing the suitability of this technique for various analytical applications (Cardoza et al., 2003).

However, the direct measurements by NMR face several difficulties: i) the abundance of protons an the reaction media (usually water) prevents the attribution of chemical shifts to the individual proton, and ii) the extraction of the target compound is essential prior to NMR analysis.

There are different configurations for coupling chromatography to NMR (Exarchou et al., 2003; Exarchou et al., 2005). More recently the on-line coupling of LC to SPE and subsequently to NMR became available and improved some of the existing analytical barriers of the previous modes. In this configuration, the chromatographic peaks are trapped in SPE cartridges and can be concentrated up to several times by multi-trapping into the same cartridge (Moco et al., 2007).

Recently, the separation of 2-phenyl-1,4-benzopyrone and phenolic acids present in Greek oregano extract was accomplished by LC-UV-SPE-NMR-MS (Exarchou *et al.*, 2003). The compounds were separated by LC, trapped in SPE cartridges and subsequently eluted for NMR and MS acquisition. The work demonstrated that two related compounds co-eluting in the LC (and therefore trapped into the same cartridge) could be readily distinguished by MS and NMR (Exarchou *et al.*, 2003). This method is applicable for the analysis of rare or/and unknown compounds in complex mixtures, since it allows separation, concentration and NMR acquisition of the analyte within a single system. Moreover, concurrent interpretation of UV, MS, and NMR data obtained for every trapped peak led to the unequivocal assignment of their structure.

The identification of micropollutant degradation products can be aided by various methods, such as MS or NMR, but often the full chemical description of a molecule is only achieved by integrating the compound information taken from different sources (Moco *et al.*, 2007). Advanced analytical instrumentation employ MS to determine the MS fragmentation pathways (i.e., cleaved moieties) whereas ^{1}H and ^{13}C NMR is utilized for structural confirmation (Kormos *et al.*, 2009). Therefore, the combination of MS with NMR is one of the most powerful strategies for the identification of an unknown molecule. To combine the advantages of NMR spectroscopy and MS spectrometry, the extended hyphenation of LC-NMR-MS has been used to identify compounds (Exarchou *et al.*, 2003). The most efficient way to seize the advantages of both technologies, is to use them in parallel or, if possible, on-line. However, due to the complex analytical set-up, it is still most common to undertake analyses by LC-MS and LC-(SPE)-NMR separately (Moco *et al.*, 2007). In summary, high information content of NMR experiments makes this an attractive technique for the analysis of contaminant-transformation processes, especially when coupled with a separation method. Because of the complementary nature of the results provided by each technique, the combination of NMR and MS/MS analysis is an especially powerful approach for elucidating the structure of new transformation products (Cardoza *et al.*, 2003).

2.3.5 Biological assessment of the degradation products

Biological assessment of the micropollutant degradation products complements chemical analysis in terms of information provided on toxicity, estrogenicity and antibacterial activity of the discharged wastewater. The use of biological organisms is vital in the measurement and the evaluation of the potential impact of these contaminants (Wadhia and Thompson, 2007).

2.3.5.1 Ecotoxicological assessment of environmental risk (toxicity)

An environmental toxicant can be defined as a substance that, in a given concentration and chemical form, challenges the organisms (bioindicators) of the ecosystem and causes adverse or toxic effects (Lidman, 2005) The structure of micropollutant molecule will account for its physico-chemical characteristics and properties and subsequently, its toxicity. Molecules with similar structure have similar properties, and thus the potential for the similar toxicity (Hamblen *et al.*, 2003). Therefore, Veith *et al.* (1988) constructed a model using a K-nearest neighbor pattern recognition technique based on eight molecular topological parameters as molecular descriptors. To describe the interrelatedness of chemical structures. K-nearest neighbors explained over 90% of the observed variability in the 8 descriptors (Veith *et al.*, 1988).

Environmental samples can be ecotoxicologically tested using any level of biological organization from molecular to whole organisms and populations, communities or assemblages of organisms. However, the assessment of toxicity will be reliable only if several specific aspects will be considered during the analysis with standardized test protocols: i) environmental samples taken for testing with bioassays need to be representative, and the collection, storage and preparation procedures must not result in change in toxicity of the sample; ii) a measure of test variability needs to be ascertained; and iii) the effect of confounding variables (e.g., pH, dissolved solids, and Eh) also needs to be established.

Various ecotoxicity tests have been developed to characterize the toxicity of individual chemicals. The adoption of these tests for the toxicity evaluation of environmental samples led to contaminated media tests and/or bioassays. A bioassay can be defined as a biological assay performed to measure the effects of a substance on a living organism, so bioassays can provide qualitative or quantitative evaluation of a selected system. In the latter case, the assessment often involves an estimation of the concentration or the potency of a substance by measuring selected biological responses. A toxicity test can be considered a bioassay that allows measurement of damage (Wadhia and Thompson, 2007).

Recently, the benefits of using rapid, sensitive, reproducible and cost-effective bacterial assays such as bioluminescent bacteria (BLB) were acknowledged. One of the most recognized bioluminescent bacteria is *Vibrio fisheri* (NRRL B-11177) that emit light as a result of their normal metabolism (about 10% of its metabolic energy), and the intensity of the light emitted is an indication of their metabolic activity (Gu *et al.*, 2002). When these bacteria are exposed to a toxic substance, the light emission is reduced, thus providing a measure of the acute toxicity of a sample. Its advantages, over alternatives, are simplicity of operation, speed and low cost. Moreover, this is a standardized organism that has been studied very well, and a number of different systems are commercially available (Farre *et al.*, 2007).

The need to bioanalyze a large number of environmental samples in a relatively short period of time led to the development and the increased significance of fast, miniaturized tests for toxicity, so-called microbiotests (also known as alternative tests and second-generation tests) (Wolska *et al.*, 2007). Microbiotests have many undoubted advantages, such as: i) relatively low cost per analysis, ii) operation with small sample volumes, iii) no requirement to culture test organisms (organisms are stored in cryptobiotic form, i.e., rotifers as cysts, crustaceans as resting eggs, and algae as cells immobilized on specific medium), iv) possibility of working with several samples at once, v) short response time, vi) repeatability and reproducibility of data, and vii) tests can be performed under laboratory conditions. Furthermore, the microbiotests can be used by personnel without any special training and previous experience of working with bioindicators. Microbiotests are also widely utilized in commercially available systems such as ToxAlert 10 and ToxAlert 100 (Merck); LUMIStox (Dr. Bruno Lange); ToxTracer (Skalar); Biotox and The BioToxTM Flash, Toxkit (Aboatox); Microtox and microtox SOLO, DeltaTox Analyzer (AZUR Environmental); ToxScreen (CheckLight).

2.3.5.2 Assessment of estrogenic activity

Routledge and Sumpter (1997) anticipated that both, the position and branching of the substituent of alkylsubstituted phenolic compounds affected their estrogenic potency. In particular, optimal activity was associated with a tertiary-branched alkyl moiety of 6–8 carbon atoms substituted in the 4-position to the hydroxy group (Routledge and Sumpter, 1997). Thus, theoretically, the estrogenic activity of every compound is related to its chemical structure, e.g., relative position of the hydroxyl phenolic group in the ring due to its high affinity with the estrogen receptor, which is responsible for its estrogenic potential (Streck, 2009).

Biological analysis methods vastly improved over the last decade, especially with advances in the development of bioassays for the estrogenicity assessment. Bioassays (e.g., YES, MELN tests) have the advantage to measure the estrogenic effect related to hormones and other estrogenic disruptors present in the samples, so they can be better adapted to screen estrogenic disruption in aquatic environments exposed to urban and industrial sources of contamination. However, the possible inhibition effect from a mixture of pollutants needs to be taken into account by performing chromatographic fractionation of samples and biological testing of the isolated fractions individually (estrogenic potency). The tests of estrogenicity assessment are summarized in Table 2.4.

Analytical methods for the identification of Micropollutants 81

Table 2.4 Tests for assessment of the estrogenic activity in transformation products of estrogenic compounds

Bioassay	Basics of the method	Reference
The yeast estrogen screen (YES) test	recombinant yeast cultures expressing human estrogen receptor	(Nelson et al., 2007) (Salste et al., 2007)
E-screen test	proliferation of human breast-cancer cells MCF-7 under estrogenic control	(Korner et al., 2000)
ER-CALUX	human breast-cancer cells T47D stably transfected with luciferase reporter gene	(Murk et al., 2002)
MELN	human breast-cancer cells MCF-7 stably transfected with luciferase reporter gene	(Pillon et al., 2005)

2.3.5.3 Assessment of antimicrobial activity

There are many concerns about the widespread distribution of antimicrobial agents and preservatives (e.g., triclozan,) in the environment, which lead to the development of cross-resistance to antibiotics, the adverse effects on ecological health, and the formation of more toxic pollutants under different conditions. For instance, triclosan has been incorporated into a broad array of personal care products (e.g., hand disinfecting soaps, medical skin creams, dental products, deodorants, toothpastes), consumer products (e.g., fabrics, plastic kitchenware, sport footwear), and cleaners or disinfectants in hospitals or households thus its transformation may lead to adverse effects on various environmental systems (Lange et al., 2006; Roh et al., 2009).

The classic measure of antimicrobial potency is the minimal inhibitory concentration (MIC) for a particular agent with a given pathogen. For instance, by relating the MIC to concentrations of the selected pathogen in human tissues, it has been possible to develop empirical relationships between exposure, potency and patient response (Drlica, 2001). Reduction of microbial activity can be estimated by the cell growth measurement in aqueous solution or in agar plates. Tests on *Pseudomonas putida*, *Escherichia coli*, etc. usually indicate whether the degradation process of a target contaminant is sufficient for the reduction of its antimicrobial activity.

2.3.5.4 Biosensors

Biosensor can be defined as an integrated receptor-transducer device, which is capable of providing selective quantitative or semi-quantitative analytical information using a biological recognition element. In biosensors, a biological unit (e.g., an enzyme or an antibody), which is typically immobilized on the surface of the transducer (electrode), interacts with the analyte (which contains a target compound) and causes a change in a measurable property within the local environment near the transducer surface, thus converting a (bio)chemical process into a measurable electronic signal. The main factors that influence the biosensor response are the mass-transport kinetics of analytes and products, as well as loading of the sensing molecule. Biosensors are usually categorized by the transduction element (e.g., electrochemical, optical, piezoelectrical or thermal) or the biorecognition principle (e.g., enzymatic, immunoaffinity recognition, whole-cell sensor or DNA). Biosensors offer simplified, sensitive, rapid and reagentless real-time measurements for a wide range of biomedical and industrial applications.

Recently, enzymatic biosensor (*tyrosinase*) was applied for the detection of BPA (Carralero *et al.*, 2007), 17-β-estradiol (Notsu *et al.*, 2002), PAH (Fahnrich *et al.*, 2003) and surfactants (Taranova *et al.*, 2004). Optical biosensors with high sensitivity (up to ng range) were successfully used for the monitoring of pesticides, EDCs, and surfactants (Skladal, 1999; Tschmelak *et al.*, 2006).

Biosensor technology is a rapidly expanding field of research that has been transformed over the past two decades through new discoveries related to the novel material technologies, means of signal transduction, and powerful computer software to control devices (Farre *et al.*, 2009). More detailed information on biosensors is available in Chapter 3.

2.4 IDENTIFICATION LEVELS OF MICROPOLLUTANTS TRANSFORMATION PRODUCTS

Recent developments in analytical methodologies led to the variations in the scientific evidence that greatly support the identification of micropolutants transformation products. Therefore, so called metabolomics approaches include targeted analyses that relates to residue and a target contaminant analysis as well as untargeted analyses in which the goal is to measure as many compounds as possible. In the latter case, identification is not the primary step in data processing. The analytical information obtained from the profiles is transformed to coordinates on the basis of mass, retention and amplitude.

The GC/MS spectra are readily available for many widely studied compounds, however, for the most of micropollutants, especially taking into account the constantly growing number of the potential emerging compounds, such databases do not exist. Thus, De Witte *et al.* (2009a) proposed the utilization of the methodology created by the Chemical Analysis Working Group (CAWG) as a basis for the identification of pharmaceutical degradation products. According to the CAWG methodology, four levels of identification can be distinguished for previously characterized, identified and reported compounds. Level 1 requires a minimum of two independent data compared with a standard compound, analyzed under identical experimental conditions. Level 2 identification is associated with putatively annotated compounds (e.g., without chemical reference standards, based upon physicochemical properties and/or spectral similarity with public/commercial spectral libraries) (Sumner *et al.*, 2007). In case of the characterization based upon characteristic physicochemical properties of a chemical class of compounds, or by spectral similarity to known compounds of a chemical class, the identification level is assigned to level 3. And, finally, level 4 transformation products are labeled as unknown compounds – although unidentified or unclassified these metabolites can still be differentiated and quantified based upon available spectral data.

De Witte *et al.* (2009a) also proposed four approaches as a novel identification scheme based on CAWG identification levels for metabolites in biological organisms and the notations of Doll and Frimmel. Therefore, four approaches based on the subsequent identification method engaged with available analytical tools are depicted in Figure 2.4.

The proposed scheme is applicable not only to the pharmaceutical degradation products but also to the broader range of micropollutants (e.g., pesticides, etc.). Every approach is connected to the relevant level of identification. For instance, analysis of a standard compound in approach A, next to the analysis of the sample or the parent molecule/analogous products in approach B can strengthen identification. Approach C proposes the involvement of MS spectra bases, whereas the approach D involves only the sample analysis. Approach D is also divided into two levels because the difference has to be made between the identification based on the extended MS fragmentation (tentative identification) or/and on the molecular weight or UV spectrum determination (indicative determination). Level 1 and level 3 are not commonly possible, whereas level 2, 4 and 5 represent the classification levels more likely using the current state-of-art technologies (De Witte *et al.*, 2009a).

84   Treatment of Micropollutants in Water and Wastewater

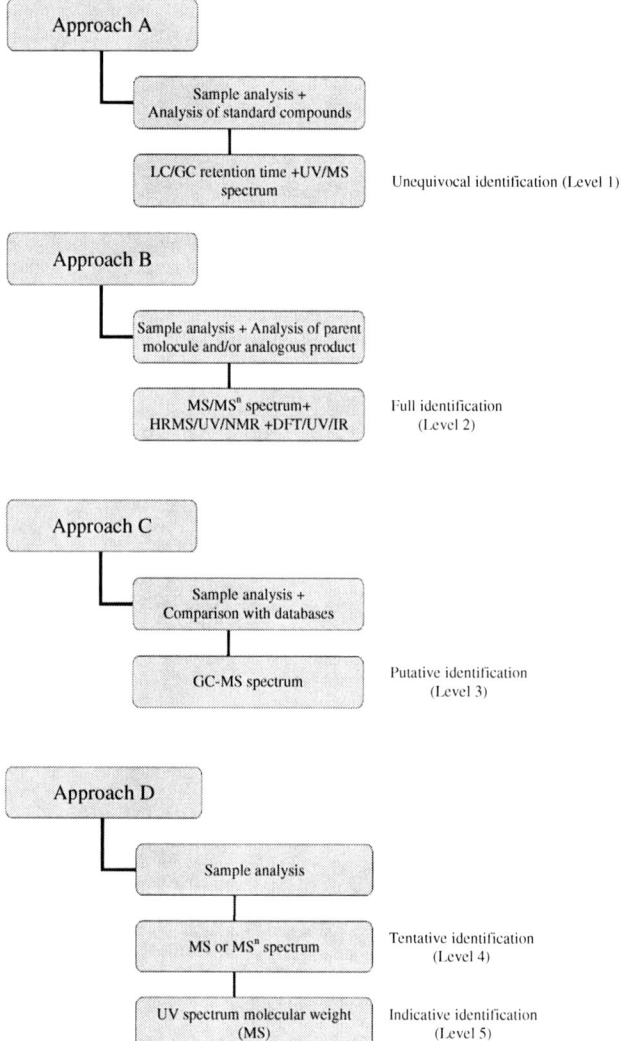

Figure 2.4 Identification levels with their required analytical tools for the identification of transformation products. Adapted form De Witte *et al.* (2009a)

2.5 CONCLUSIONS

The elucidation of the environmental fate of micropollutants is essential due to the potential formation of the persistent metabolites that may cause adverse effect on the environment. The combination of chemical and biological analyses is necessary to evaluate the potential effects posed by the degradation and intermediate products of micropollutants. The developments in analytical instrumental techniques such as the increase in chromatographic resolution, detection sensitivity and selectivity enables the extraction and the detection of target compounds even at nano g concentrations levels from extremely complex matrices. In addition, it is vitally important to develop appropriate tools to identify and quantify unknown degradation products. More sophisticated reactivity-structure linkages that use molecular orbital energy modeling to predict tendencies for neutral micropollutant molecule oxidation by the oxidant (e.g., hydroxyl radical) are currently under development. Generally, the utilization of the standard compounds and databases is recommended, however, the lack of studies for quite many of the micropollutants transformation products triggers the development of the identification methodologies.

2.6 REFERENCES

Armenta, S., Garrigues, S. and de la Guardia, M. (2008) Green Analytical Chemistry. *TrAC Trend. Anal. Chem.* **27**, 497–511.

Bakker, E. and Qin, Y. (2006) Electrochemical Sensors. *Anal. Chem.* **78**, 3965–3984.

Bekbolet, M., Çınar, Z., Kılıç, M., Uyguner, C. S., Minero, C. and Pelizzetti, E. (2009) Photocatalytic oxidation of dinitronaphthalenes: Theory and experiment. *Chemosphere* **75**, 1008–1014.

Cardoza, L. A., Almeida, V. K., Carr, A., Larive, C. K. and Graham, D. W. (2003) Separations coupled with NMR detection. *TrAC Trend. Anal. Chem.* **22**, 766 775.

Carralero, V., Gonzalez-Cortez, A., Yanez-Sedeno, P. and Pingarron, J. M. (2007) Nanostructured progesterone immunosensor using a tyrosinase–colloidal gold–graphite–Teflon biosensor as amperometric transducer. *Anal. Chim. Acta* **596**, 86–91.

Carrier, M., Perol, N., Herrmann, J.-M., Bordes, C., Horikoshi, S., Paisse, J. O., Baudot, R. and Guillard, C. (2006) Kinetics and reactional pathway of Imazapyr photocatalytic degradation Influence of pH and metallic ions. *Appl. Catal. B-Environ.* **65**, 11–20.

Chang, H.-S., Choo, K.-H., Lee, B. and Choi, S.-J. (2009) The methods of identification, analysis, and removal of endocrine disrupting compounds (EDCs) in water. *J. Hazard. Mater.* **172**, 1–12.

Chu, W.-H., Gao, N.-Y., Deng, Y. and Dong, B.-Z. (2009) Formation of chloroform during chlorination of alanine in drinking water. *Chemosphere* **77**, 1346–1351.

De Witte, B., Langenhove, H. V., Demeestere, K. and Dewulf, J. (2009a) Advanced oxidation of pharmaceuticals: Chemical analysis and biological assessment of degradation products. *Crit. Rev. Environ. Sci. Technol.* In Press, Accepted Manuscript.

De Witte, B., Langenhove, H. V., Hemelsoet, K., Demeestere, K., Wispelaere, P. D., Van Speybroeck, V. and Dewulf, J. (2009b) Levofloxacin ozonation in water: Rate determining process parameters and reaction pathway elucidation. *Chemosphere* **76**, 683–689.

Drlica, K. (2001) Antibiotic resistance: Can we beat the bugs? *Drug Discov. Today* **6**, 714–715.

Escandar, G. M., Faber, N. K. M., Goicoechea, H. C., de la Peña, A. M., Olivieri, A. C. and Poppi, R. J. (2007) Second- and third-order multivariate calibration: Data, algorithms and applications. *TrAC Trend. Anal. Chem.* **27**, 752–765.

Exarchou, V., Godejohann, M., Beek, T. A. V., Gerothanassis, I. P. and Vervoort†, J. (2003) LC-UV-Solid-Phase Extraction-NMR-MS combined with a cryogenic flow probe and its application to the identification of compounds present in Greek Oregano. *Anal. Chem.* **75**, 6288–6294.

Exarchou, V., Krucker, M., Beek, T. A. V., Vervoort, J., Gerothanassis, I. P. and Albert, K. (2005) LC-NMR coupling technology: Recent advancements and applications in natural product analysis. *Magn. Reson. Chem.* **43**, 681–687.

Fahnrich, K. A., Pravda, M. and Guilbault, G. G. (2003) Disposable amperometric immunosensor for the detection of polycyclic aromatic hydrocarbons (PAHs) using screen-printed electrodes. *Biosens. Bioelectron.* **18**, 73–82.

Farre, M., Martinez, E. and Barcelo, D. (2007) Validation of interlaboratory studies on toxicity in water samples. *TrAC Trend. Anal. Chem.* **26**, 283–292.

Farre, M., Kantiani, L., Perez, S. and Barcelo, D. (2009) Sensors and biosensors in support of EU Directives. *TrAC Trend. Anal. Chem.* **28**, 170–185.

Farre, M. I., Perez, S., Kantiani, L. and Barcelo, D. (2008) Fate and toxicity of emerging pollutants, their metabolites and transformation products in the aquatic environment. *TrAC Trend. Anal. Chem.* **27**, 991–1007.

Fatta, D., Achilleos, A., Nikolaou, A. and Meric, S. (2007) Analytical methods for tracing pharmaceutical residues in water and wastewater. *TrAC Trend. Anal. Chem.* **26**, 515–533.

Fifield, F. W. and Kealey, D. (2000) *Principles and Practice of Analytical Chemistry.* Cambridge, Blackwell Science Ltd.

Gabet, V., Miege, C., Bados, P. and Coquery, M. (2007) Analysis of estrogens in environmental matrices. *TrAC Trend. Anal. Chem.* **26**, 1113–1131.

Galera, M. M., García, M. D. G. and Goicoechea, H. C. (2007) The application to wastewaters of chemometric approaches to handling problems of highly complex matrices. *TrAC Trend. Anal. Chem.* **26**, 1032–1042.

Garcia-Galan, M. J., Silvia Diaz-Cruz, M. and Barcelo, D. (2008) Identification and determination of metabolites and degradation products of sulfonamide antibiotics. *TrAC Trend. Anal. Chem.* **27**, 1008–1022.

Garrido, M., Rius, F. and Larrechi, M. (2008) Multivariate curve resolution–alternating least squares (MCR-ALS) applied to spectroscopic data from monitoring chemical reactions processes. *Anal. Bioanal. Chem.* **390**, 2059–2066.

Gascon, J., Oubica, A. and Barcely, D. (1997) Detection of endocrine-disrupting pesticides by enzyme-linked immunosorbent assay (ELISA): Application to atrazine. *TrAC Trend. Anal. Chem.* **16**, 554–562.

Gatidou, G., Thomaidis, N. S., Stasinakis, A. S. and Lekkas, T. D. (2007) Simultaneous determination of the endocrine disrupting compounds nonylphenol, nonylphenol ethoxylates, triclosan and bisphenol A in wastewater and sewage sludge by gas chromatography-mass spectrometry. *J. Chromatogr. A* **1138**, 32–41.

Giger, W. (2009) Hydrophilic and amphiphilic water pollutants: Using advanced analytical methods for classic and emerging contaminants. *Anal. Bioanal. Chem.* **393**, 37–44.

Gros, M., Petrovic, M. and Barcelo, D. (2008) Tracing Pharmaceutical Residues of Different Therapeutic Classes in Environmental Waters by Using Liquid Chromatography/Quadrupole-Linear Ion Trap Mass Spectrometry and Automated Library Searching. *Anal. Chem.* **81**, 898–912.

Gu, M. B., Min, J. and Kim, E. J. (2002) Toxicity monitoring and classification of endocrine disrupting chemicals (EDCs) using recombinant bioluminescent bacteria. *Chemosphere* **46**, 289–294.

Hamblen, E. L., Cronin, M. T. D. and Schultz, T. W. (2003) Estrogenicity and acute toxicity of selected anilines using a recombinant yeast assay. *Chemosphere* **52**, 1173–1181.

Hao, C., Zhao, X. and Yang, P. (2007) GC-MS and HPLC-MS analysis of bioactive pharmaceuticals and personal-care products in environmental matrices. *TrAC Trend. Anal. Chem.* **26**, 569–580.

Henry, M. C. and Yonker, C. R. (2006) Supercritical Fluid Chromatography, Pressurized Liquid Extraction, and Supercritical Fluid Extraction. *Anal. Chem.* **78**, 3909–3916.

Hernandez, F., Sancho, J. V., Ibanez, M. and Grimalt, S. (2008) Investigation of pesticide metabolites in food and water by LC-TOF-MS. *TrAC Trend. Anal. Chem.* **27**, 862–872.

Hernandez, F., Sancho, J. V., Ibanez, M. and Guerrero, C. (2007) Antibiotic residue determination in environmental waters by LC-MS. *TrAC Trend. Anal. Chem.* **26**, 466–485.

Hernando, M. D., Gómez, M. J., Agüera, A. and Fernández-Alba, A. R. (2007) LC-MS analysis of basic pharmaceuticals (beta-blockers and anti-ulcer agents) in wastewater and surface water. *TrAC Trend. Anal. Chem.* **26**, 581–594.

Hogenboom, A. C., van Leerdam, J. A. and de Voogt, P. (2009) Accurate mass screening and identification of emerging contaminants in environmental samples by liquid chromatography-hybrid linear ion trap Orbitrap mass spectrometry. *J. Chromatogr. A* **1216**, 510–519.

Hyotylainen, T. and Hartonen, K. (2002) Determination of brominated flame retardants in environmental samples. *TrAC Trend. Anal. Chem.* **21**, 13–30.

Ibanez, M., Sancho, J. V., Hernandez, F., McMillan, D. and Rao, R. (2008) Rapid non-target screening of organic pollutants in water by ultraperformance liquid chromatography coupled to time-of-light mass spectrometry. *TrAC Trend. Anal. Chem.* **27**, 481–489.

Katsumata, H., Kaneco, S., Suzuki, T., Ohta, K. and Yobiko, Y. (2006) Degradation of polychlorinated dibenzo-p-dioxins in aqueous solution by Fe(II)/H2O2/UV system. *Chemosphere* **63**, 592–599.

Kormos, J. L., Schulz, M., Wagner, M. and Ternes, T. A. (2009) Multistep Approach for the Structural Identification of Biotransformation Products of Iodinated X-ray Contrast Media by Liquid Chromatography/Hybrid Triple Quadrupole Linear Ion

Trap Mass Spectrometry and 1H and 13C Nuclear Magnetic Resonance. *Anal. Chem.* **81**, 9216–9224.

Korner, W., Bolz, U., Sussmuth, W., Hiller, G., Schuller, W., Hanf, V. and Hagenmaier, H. (2000) Input/output balance of estrogenic active compounds in a major municipal sewage plant in Germany. *Chemosphere* **40**, 1131–1142.

Kosjek, T. and Heath, E. (2008) Applications of mass spectrometry to identifying pharmaceutical transformation products in water treatment. *TrAC Trend. Anal. Chem.* **27**, 807–820.

Kostopoulou, M. and Nikolaou, A. (2008) Analytical problems and the need for sample preparation in the determination of pharmaceuticals and their metabolites in aqueous environmental matrices. *TrAC Trend. Anal. Chem.* **27**, 1023–1035.

Kot-Wasik, A., Debska, J. and Namiesnik, J. (2007) Analytical techniques in studies of the environmental fate of pharmaceuticals and personal-care products. *TrAC Trend. Anal. Chem.* **26**, 557–568.

Kralj, M. B., Trebse, P. and Franko, M. (2007) Applications of bioanalytical techniques in evaluating advanced oxidation processes in pesticide degradation. *TrAC Trend. Anal. Chem.* **26**, 1020–1031.

Kucharska, M. and Grabka, J. (2010) A review of chromatographic methods for determination of synthetic food dyes. *Talanta* **80**, 1045–1051.

Lange, F., Cornelissen, S., Kubac, D., Sein, M. M., von Sonntag, J., Hannich, C. B., Golloch, A., Heipieper, H. J., Möder, M. and von Sonntag, C. (2006) Degradation of macrolide antibiotics by ozone: A mechanistic case study with clarithromycin. *Chemosphere* **65**, 17–23.

Lee, B.-D., Iso, M. and Hosomi, M. (2001) Prediction of Fenton oxidation positions in polycyclic aromatic hydrocarbons by Frontier electron density. *Chemosphere* **42**, 431–435.

Lidman, U. (2005). The nature and chemistry of toxicants. Environmental Toxicity Testing. Thompson, K. C., Wadhia, K. and Loibner, A. P. Oxford, UK, Blackwell Publishing.

Liu, G., Li, X., Zhao, J., Horikoshi, S. and Hidaka, H. (2000) Photooxidation mechanism of dye alizarin red in TiO2 dispersions under visible illumination: An experimental and theoretical examination. *J. Mol. Catal. A-Chem.* **153**, 221–229.

Liu, S., Yuan, L., Yue, X., Zheng, Z. and Tang, Z. (2008) Recent Advances in Nanosensors for Organophosphate Pesticide Detection. *Adv. Powder Tech.* **19**, 419–441.

Lopez de Alda, M. J., Díaz-Cruz, S., Petrovic, M. and Barceló, D. (2003) Liquid chromatography-(tandem) mass spectrometry of selected emerging pollutants (steroid sex hormones, drugs and alkylphenolic surfactants) in the aquatic environment. *J. Chromatogr. A* **1000**, 503–526.

Maciá, A., Borrull, F., Calull, M. and Aguilar, C. (2007) Capillary electrophoresis for the analysis of non-steroidal anti-inflammatory drugs. *TrAC Trend. Anal. Chem.* **26**, 133–153.

Madej, K. (2009) Microwave-assisted and cloud-point extraction in determination of drugs and other bioactive compounds. *TrAC Trend. Anal. Chem.* **28**, 436–446.

McDowell, D. C., Huber, M. M., Wagner, M., Gunten, U. V. and Ternes, T. A. (2005) Ozonation of carbamazepine in drinking water: Identification and kinetic study of major oxidation products. *Environ. Sci. Technol.* **39**, 8014–8022.

Miege, C., Bados, P., Brosse, C. and Coquery, M. (2009) Method validation for the analysis of estrogens (including conjugated compounds) in aqueous matrices. *TrAC Trend. Anal. Chem.* **28**, 237–244.

Moco, S., Vervoort, J., Moco, S., Bino, R. J., De Vos, R. C. H. and Bino, R. (2007) Metabolomics technologies and metabolite identification. *TrAC Trend. Anal. Chem.* **26**, 855–866.

Murk, A. J., Legler, J., Lipzig, M. M. v., Meerman, J. H., Belfroid, A. C., Spenkelink, A., Burg, B. V. D. and G.B. Rijs, D. V. (2002) Detection of estrogenic potency in wastewater and surface water with three in vitro bioassays. *Environ. Toxicol. Chem.* **21**, 16–21.

Nelson, J., Bishay, F., Roodselaar, A. v., Ikonomou, M. and Law, F. C. P. (2007) The use of in vitro bioassays to quantify endocrine disrupting chemicals in municipal wastewater treatment plant effluents. *Sci. Total Environ.* **374**, 80–90.

Notsu, H., Tatsuma, T. and Fujishima, A. (2002) Tyrosinase-modified boron-doped diamond electrodes for the determination of phenol derivatives. *J. Electroanal. Chem.* **523**, 86–92.

Ohko, Y., Iuchi, K.-i., Niwa, C., Tatsuma, T., Nakashima, T., Iguchi, T., Kubota, Y. and Fujishima, A. (2002) 17β-Estradiol degradation by TiO2 photocatalysis as a means of reducing estrogenic activity. *Environ. Sci. Technol.* **36**, 4175–4181.

Oliveira, D. P., Carneiro, P. A., Sakagami, M. K., Zanoni, M. V. B. and Umbuzeiro, G. A. (2007) Chemical characterization of a dye processing plant effluent—Identification of the mutagenic components. *Mutat. Res.-Gen.Tox. En.* **626**, 135–142.

Parr, R. G. and Weitao, Y. (1994) *Density-Functional Theory of Atoms and Molecules.* Oxford, USA, Oxford Univercity Press.

Pavlovic, D. M., Babic, S., Horvat, A. J. M. and Kastelan-Macan, M. (2007) Sample preparation in analysis of pharmaceuticals. *TrAC Trend. Anal. Chem.* **26**, 1062–1075.

Petroviciu, I., Albu, F. and Medvedovici, A. (2010) LC/MS and LC/MS/MS based protocol for identification of dyes in historic textiles. *Microchemical. J.* In Press, Accepted Manuscript.

Pico, Y., Rodriguez, R. and Manes, J. (2003) Capillary electrophoresis for the determination of pesticide residues. *TrAC Trend. Anal. Chem.* **22**, 133–151.

Pillon, A., Boussioux, A. M., Escande, A., Ait-Aissa, S., E. Gomez, Fenet, H., Ruff, M., Moras, D., Vignon, F., Duchesne, M. J., Casellas, C., Nicolas, J. C. and Balaguer, P. (2005) Binding of estrogenic compounds to recombinant estrogen receptor-alpha: Application to environmental analysis. *Environ. Health Perspect.* **113**, 278–284.

Pinheiro, H. M., Touraud, E. and Thomas, O. (2004) Aromatic amines from azo dye reduction: Status review with emphasis on direct UV spectrophotometric detection in textile industry wastewaters. *Dyes Pigments* **61**, 121–139.

Poiger, T., Richardson, S. D. and Baughman, G. L. (2000) Identification of reactive dyes in spent dyebaths and wastewater by capillary electrophoresis-mass spectrometry. *J. Chromatogr. A* **886**, 271–282.

Radjenovic, J., Petrovic, M. and Barceló, D. (2009) Complementary mass spectrometry and bioassays for evaluating pharmaceutical-transformation products in treatment of drinking water and wastewater. *TrAC Trend. Anal. Chem.* **28**, 562–580.

Radjenovic, J., Petrovic, M., Barceló, D. and Petrovic, M. (2007) Advanced mass spectrometric methods applied to the study of fate and removal of pharmaceuticals in wastewater treatment. *TrAC Trend. Anal. Chem.* **26**, 1132–1144.

Raman Suri, C., Boro, R., Nangia, Y., Gandhi, S., Sharma, P., Wangoo, N., Rajesh, K. and Shekhawat, G. S. (2009) Immunoanalytical techniques for analyzing pesticides in the environment. *TrAC Trend. Anal. Chem.* **28**, 29–39.

Raynie, D. E. (2004) Modern Extraction Techniques. *Anal. Chem.* **76**, 4659–4664.

Reemtsma, T., Quintana, J. B., Rodil, R., Garclía-López, M. and Rodríguez, I. (2008) Organophosphorus flame retardants and plasticizers in water and air I. Occurrence and fate. *TrAC Trend. Anal. Chem.* **27**, 727–737.

Richardson, S. D. (2000) Environmental Mass Spectrometry. *Anal. Chem.* **72**, 4477–4496.

Richardson, S. D. (2003) Disinfection by-products and other emerging contaminants in drinking water. *TrAC Trend. Anal. Chem.* **22**, 666–684.

Richardson, S. D. (2008) Environmental Mass Spectrometry: Emerging Contaminants and Current Issues. *Anal. Chem.* **80**, 4373–4402.

Richardson, S. D. (2009) Water Analysis: Emerging Contaminants and Current Issues. *Anal. Chem.* **81**, 4645–4677.

Richardson, S. D., Plewa, M. J., Wagner, E. D., Schoeny, R. and DeMarini, D. M. (2007) Occurrence, genotoxicity, and carcinogenicity of regulated and emerging disinfection by-products in drinking water: A review and roadmap for research. *Mutat. Res.-Rev. Mutat.* **636**, 178–242.

Roh, H., Subramanya, N., Zhao, F., Yu, C.-P., Sandt, J. and Chu, K.-H. (2009) Biodegradation potential of wastewater micropollutants by ammonia-oxidizing bacteria. *Chemosphere* **77**, 1084–1089.

Routledge, E. J. and Sumpter, J. P. (1997) Structural features of alkyl-phenolic chemicals associated with estrogenic activity. *J. Biol. Chem.* **272**, 3280–3288.

Rubio, S. and Perez-Bendito, D. (2009) Recent Advances in Environmental Analysis. *Anal. Chem.* **81**, 4601–4622.

Salste, L., Leskinen, P., Virta, M. and Kronberg, L. (2007) Determination of estrogens and estrogenic activity in wastewater effluent by chemical analysis and the bioluminescent yeast assay. *Sci. Total Environ.* **3**, 343–351.

Skladal, P. (1999) Effect of methanol on the interaction of monoclonal antibody with free and immobilized atrazine studied using the resonant mirror-based biosensor. *Biosens. Bioelectron.*, 257–263.

Streck, G. (2009) Chemical and biological analysis of estrogenic, progestagenic and androgenic steroids in the environment. *TrAC Trend. Anal. Chem.* **28**, 635–652.

Sumner, L., Amberg, A., Barrett, D., Beale, M., Beger, R., Daykin, C., Fan, T., Fiehn, O., Goodacre, R., Griffin, J., Hankemeier, T., Hardy, N., Harnly, J., Higashi, R., Kopka, J., Lane, A., Lindon, J., Marriott, P., Nicholls, A., Reily, M., Thaden, J. and Viant, M. (2007) Proposed minimum reporting standards for chemical analysis. *Metabolomics* **3**, 211–221.

Taranova, L. A., Fesay, A. P., Ivashchenko, G. V., Reshetilov, A. N., Winther-Nielsen, M. and Emneus, J. (2004) Comamonas testosteroni Strain TI as a Potential Base for a Microbial Sensor Detecting Surfactants *Appl. Biochem. Microbiol.* **40**, 404–408.

Tobiszewski, M., Mechlinska, A., Zygmunt, B. and Namiesnik, J. (2009) Green analytical chemistry in sample preparation for determination of trace organic pollutants. *TrAC Trend. Anal. Chem.* **28**, 943–951.

Tschmelak, J., Kumpf, M., Kappel, N., Proll, G. and Gauglitz, G. (2006) Total internal reflectance fluorescence (TIRF) biosensor for environmental monitoring of testosterone with commercially available immunochemistry: Antibody characterization, assay development and real sample measurements. *Talanta* **69**, 343.

Veith, G. D., Greenwood, B., Hunter, R. S., Niemi, G. J. and Regal, R. R. (1988) On the intrinsic dimensionality of chemical structure space. *Chemosphere* **17**, 1617–1630.

Vogna, D., Marotta, R., Napolitano, A. and d'Ischia, M. (2002) Advanced oxidation chemistry of paracetamol. UV/H2O2-induced hydroxylation/degradation pathways and 15N-aided inventory of nitrogenous breakdown products. *J. Org. Chem.* **67**, 6143–6151.

Wadhia, K. and Thompson, K. C. (2007) Low-cost ecotoxicity testing of environmental samples using microbiotests for potential implementation of the Water Framework Directive. *TrAC Trend. Anal. Chem.* **26**, 300–307.

Wishart, D. S. (2008) Quantitative metabolomics using NMR. *TrAC Trend. Anal. Chem.* **27**, 228–237.

Wolska, L., Sagajdakow, A., Kuczynska, A. and Namiesnik, J. (2007) Application of ecotoxicological studies in integrated environmental monitoring: Possibilities and problems. *TrAC Trend. Anal. Chem.* **26**, 332–344.

Young, D. C. (2001) *COMPUTATIONAL CHEMISTRY:A Practical Guide for Applying Techniques to Real-World Problems*, John Wiley & Sons, Inc.

Zhu, X., Feng, X., Yuan, C., Cao, X. and Li, J. (2004) Photocatalytic degradation of pesticide pyridaben in suspension of TiO2: Identification of intermediates and degradation pathways. *J. Mol. Catal. A-Chem.* **214**, 293–300.

Chapter 3
Sensors and biosensors for endocrine disrupting chemicals: State-of-the-art and future trends

Achintya N. Bezbaruah and Harjyoti Kalita

3.1 INTRODUCTION

Endocrine systems control hormones and activity-related hormones in many living organisms including mammals, birds, and fish. The endocrine system consists of various glands located throughout the body, hormones produced by the glands, and receptors in various organs and tissues that recognize and respond to the hormones (USEPA, 2010a). There are some chemicals and compounds that cause interferences in the endocrine system and these substances are known as endocrine disrupting chemicals (EDCs). Wikipedia states that EDCs or "endocrine disruptors are exogenous substances that act like hormones in the endocrine system and disrupt the physiologic function of

©2010 IWA Publishing. *Treatment of Micropollutants in Water and Wastewater.* Edited by Jurate Virkutyte, Veeriah Jegatheesan and Rajender S. Varma. ISBN: 9781843393160. Published by IWA Publishing, London, UK.

endogenous hormones. They are sometimes also referred to as hormonally active agents." EDCs can be man-made or natural. These compounds are found in plants (phytochemicals), grains, fruits and vegetables, and fungus. Alkyl-phenols found in detergents, bisphenol A used in PVC products, dioxins, various drugs, synthetic estrogens found in birth control pills, heavy metals (Pb, Hg, Cd), pesticides, pasticizers, and phenolic products are all examples of EDCs from a long list that is rapidly getting longer. It is suspected that EDCs could be harmful to living organisms, therefore, there is a concerted effort to detect and treat EDCs before they can cause harm to the ecosystem components.

In this chapter we are discussing some of the EDC sensors and biosensors which have been developed over the last few years. The first part of the chapter is dedicated to EDC sensors and biosensors. We then include other sensors while discussing trends in sensors and biosensors keeping in mind that the technology used for the other sensors can be very well adapted for the fabrication of EDC sensors. The purpose of this chapter is to offer an opportunity to the readers to have a feel of the enormous possibilities that sensor and biosensor technologies hold for detecting and quantifying micro-pollutants in the environment. The chapter is based on a number of original and review papers which are cited throughout the chapter.

3.2 SENSORS AND BIOSENSORS

3.2.1 The need for alternative methods

The most widely used methods for the determination of various EDCs are high-performance liquid chromatography (HPLC), liquid chromatography coupled with electrochemical detection (LC-ED), liquid chromatography coupled with mass spectrometry (LC-MS), capillary electrophoresis (CE), gas chromatography (GC), and gas chromatography coupled with mass spectrometry (GC–MS) (Nakata *et al.*, 2005; Petrovic *et al.*, 2005; Liu *et al.*, 2006a; Vieno *et al.*, 2006; Wen *et al.*, 2006; Gatidou *et al.*, 2007; Comerton *et al.*, 2009; Mottaleb *et al.*, 2009). These methods offer excellent selectivity and detection limits, however, they are not suitable for rapid processing of multiple samples and real-time detection. They involve highly trained operators, time-consuming detection processes, and complex pre-treatment steps. The instruments are sophisticated and expensive. Further, the methods are unsuitable for field studies and in-situ monitoring of samples (Rahman *et al.*, 2007; Rodrigues *et al.*, 2007; Huertas-Perez and Garcia-Campana, 2008; Saraji and Esteki, 2008; Blăzkova *et al.*, 2009; Le Blanc *et al.*, 2009; Suri *et al.*, 2009; Yin *et al.*, 2009). EDCs can also be detected using immunochemical techniques like enzyme-linked immunosorbent assays (ELISA) (Marchesini et al., 2005, 2007; Rodriguez-Monaz et al., 2005; Kim et al., 2007), however, these immunotechniques are less advantageous than chromatographic techniques because the stability of the biological materials used in assays is lower, and the assays involve complicated multistage steps that may involve expensive equipment. The specific antibodies or particular proteins must be obtained by recombinant techniques for assay fabrication (Le Blanc et al., 2009; Yin et al., 2009). ELISA-based methods are difficult to use by non-specialized laboratories and in the field. They involve labor intensive operations like repeated incubation and washing, and enzyme reaction for final signal generation (Blăzkova et al., 2009). Further, ELISA-based methods are specific for a single compound or, at the best, its structurally related compounds. They can not be used for multi-analyte detection and quantification. Unfortunately EDCs are structurally diverse, and there is a continuous introduction of new EDCs into the environment due to market driven evolution of chemicals (Marchesini et al., 2007).

There is a need for new, simple analytical techniques for EDCs with reliable and fast responses. High cost is a major hindrance for the introduction of new tools and equipment into existing laboratories. Lower capital, operation, and maintenance costs will make such equipment very attractive. The equipment should be simple to operate, less time consuming, have high sensitivity, and capable of real-time detection. Sensors of various kinds can be the alternative for expensive analytical methods (Yin et al., 2009). The high number and structural diversity of EDCs calls for the urgent development of sensors for monitoring activities (or measuring effects of the EDCs) rather than only the concentration of a single, or a set of, compounds (Le Blanc et al., 2009).

3.2.2 Electrochemical sensors

Electrochemical sensors are cheap, simple to fabricate, and reusable. They have high stability and sensitivity. They can potentially be used for other species with the necessary modifications (Kamyabi and Aghajanloo, 2008; Yin et al., 2009). Many phenolic compounds are successfully detected using electrochemical sensors as most sensors are oxidized at readily accessible potentials (Liu et al., 2005a). Being able to decrease the redox potential needed for the electro-chemical reaction makes the sensor more adaptable and sensitive to other EDCs. Chemically modified carbon paste electrodes have been prepared by Yin et al. (2009) for the detection of bisphenol A (BPA). Cobalt phthalocyanine modifier has been used in electrodes to help decrease the redox potential. Increased sensitivity and selectivity have been achieved for BPA in an aqueous medium. The detection limit was 1.0×10^{-8} M (Yin et al., 2009).

3.2.3 Biosensors

While chemicals and electrochemical strategies for determining contaminants are robust, they don't give us a complete picture of the ecological risks involved and impacts observed. Such information can be obtained only after proper interpretation by experts. However, combining both biological responses and chemical analyses may give us a better picture of the situation. We should be able to get results for the identification of toxic hotspots, toxic chemical characterization, and estimation of ecological risks of the contaminants at relevant spatial scales. Such assessments call for rapid, inexpensive screening to characterize the extent of the contamination (Brack *et al.,* 2007; Farré *et al.,* 2007; Blasco and Picó, 2009; Fernandez *et al.,* 2009; USEPA, 2010b). Further, the use of biological tools will help in the quantification of an EDC or any other pollutant in terms of its eco-effects (Marchesini *et al.,* 2007). Different biological tools including biosensors have been extensively used in recent years. Biomonitoring is becoming an essential component in effective environmental monitoring (Grote *et al.,* 2005; Rodriguez-Mozaz *et al.,* 2005; Barcelo and Petrovic, 2006; Gonzalez-Doncel *et al.,* 2006; González-Martinez *et al.,* 2007; Tudorache and Bala, 2007; Blasco and Picó, 2009).

A biosensor is defined by the International Union of Pure and Applied Chemistry (IUPAC) as "a device that uses specific biochemical reactions mediated by isolated enzymes, immunosystems, tissues, organelles or whole cells to detect chemical compounds usually by electrical, thermal or optical signals." Organelles include both mitochondria and chloroplasts (where photosynthesis takes place). Biosensors offer a number of advantages over conventional analytical techniques including portability, miniaturization, and on-site monitoring. They are also capable of measuring pollutants in complex matrices and with minimal sample preparation. Even though biosensors can't yet measure analytes as accurately as conventional analytical methods they are very good tools for routine testing and screening (Rodriguez-Mozaz *et al.,* 2006a). Rodriguez-Mozaz *et al.* (2006a) have carried out an extensive review of biosensors for environmental analysis and monitoring. Their review covers biosensors for the measurement of pesticides, hormones, PCBs, dioxins, bisphenol A, antibiotics, phenols, and EDC effects. The monitoring process using conventional analytical methods involves the collection of water samples followed by laboratory-based instrumental analysis, and such analyses only provide snapshots of the situation at the sampling site and time rather than more realistic information on spatio-temporal variations in water characteristics (Allan *et al.,* 2006; Rodriguez-Mozaz *et al.,* 2006a). Biosensors can be useful in situations when continuous and spatial data are needed. Biosensors have high specificity and sensitivity. Further, a biosensor can not only determine chemicals of concern but can record their biological effects (toxicity, cytotoxicity, genotoxicity or endocrinedisrupting effects). Often, information on biological effects is more relevant than the chemical composition. A biosensor can provide an assessment of both the total and bioavailable/bioaccesible contaminants. However, the majority of the biosensor systems developed is still in the lab tables or in prototype stages and needs to be validated before mass production and use (Rodriguez-Mozaz *et al.*, 2006a; Farré *et al.*, 2009a)

Rodriguez-Mozaz *et al.* (2006a) further discussed specific biosensors developed for certain contaminants. Organophosphorous hydrolase (OPH) can be combined with optical or amperometric transducers to measure absorbance or oxidation reduction currents generated by hydrolysis byproduct of many pesticides (e.g., paraoxon, parathion) and chemical warfare agents (e.g., sarin and soman). OPH [or Phosphotriesterase (PTE)] enzyme can hydrolyze organophosphate pesticides to release p-nitrophenol which is electroactive and chromophoric and can, thus, be measured with an OPH biosensor (Rodriguez-Mozaz *et al.*, 2006a).

EDCs bind to a hormone receptor site or a transport protein and express their biological effects. This can interpreted as (1) mimicking or antagonizing the effects of the endogenous hormone; or (2) disrupting the synthesis or metabolism of endogenous hormones or hormone receptors. It is possible to use the same receptors or transport proteins targeted by the EDCs as their bio-recognition elements. A method like this allows us to monitor the endocrine disrupting potency of single or multiple chemicals in a sample based on their bio-effect(s) on the receptor (Marchesini *et al.*, 2007). For example, human estrogen receptor α (ERα) group is capable of interacting with a large variety of chemicals (e.g., phytoestrogens, xenoestrogens, pesticides) that cause estrogenic effects *in-vivo*. The receptor family offers a variety of opportunities for use in tailor-made applications for EDCs. These include interaction between the ligand-binding domain (LBD) and its ligands or peptides derived from co-activator or co-repressor proteins (Fechner *et al.*, 2009), and the interaction between DNA-binding domains and certain DNA sequences (estrogen response elements) (Asano *et al.*, 2004; Le Blanc *et al.*, 2009). EDCs are chemicals that are able to interfere with interactions between ERα and these domains. Le Blanc *et al.* (2009) used this knowledge of the effects of EDCs on ERα receptors. They labeled ERα in an assay to determine the impact of EDCs. As compared to conventional methods, the new assays can determine the total effect on the receptor instead of concentrations of single compounds. The signal obtained is the response of the organism which is exposed to EDCs. The detection limit was reported to be 0.139 nM of estradiol equivalents. While standard analytical techniques are designed to find only known compounds, the results of this assay incorporate all known and unknown EDCs (and possibly other compounds). These data are difficult to compare and validate. While validation will be a necessary step in the near future, the assay can be used now to monitor changes in the estrogenicity of environmental samples over time. Sanchez-Acevedo *et al.* (2009) recently reported the detection of picomolar concentrations of bisphenol A (BPA) in water using a carbon nanotube field-effect transistor (CNTFET). The CNTFET is functionalized with ERα where ERα serves as the recognition layer for the sensor. The sensor uses the molecular recognition principles. Single-walled carbon nanotubes (SWCNTs) have been used as transducers and ERα is adsorbed onto their surface. A blocking agent has been used in order to avoid non-specific adsorption on the SWCNT surface. BPA has been detected up to 2.19×10^{-12} M in aqueous solution in 2 minutes. Fluoranthene, pentacloronitrobenzene, and malathion present in the water didn't produce any interferences. Such a biosensor can be useful in a label-free platform for detecting other analytes by using an appropriate nuclear receptor (Sanchez-Acevedo *et al.*, 2009).

Marchesini *et al.* (2006) reported the use of a plasmon resonance (SPR)-based label-free biosensor manufactured by an US manufacturer. They used this in combination with a ready-to-use biosensor chip to screen bio-effect related molecules and predicted possible SPR biosensor uses for EDC bio-effect monitoring. While it is possible to use such biosensors for EDC detection, the exorbitant price of commercially available systems and the lack of portability for in-situ analysis are the major drawbacks of SPR-based biosensors. These are the major challenges that need to be overcome for SPR-based biosensors to be popular (Marchesini *et al.*, 2007). SPR-based sensors have been used for dioxins, polychlorinated biphenyl and atrazine (Farré *et al.*, 2009b) and the sensor needed 15 min for a single sample measurement. A portable SPR immunosensor for organophosphate pesticide chlorpyrifos (detection limit of 45–64 ng/L) as well as single and multi-analyte SPR assays for the simultaneous detection of cholinesterase-inhibiting pesticides have been reported (Mauriz *et al.*, 2006a,b; Farré *et al.*, 2009b). These sensors were made re-usable through the formation of alkanethiol self-assembled monolayers.

Bacterial and other cells are also used in sensors known as whole-cell sensors. During ongoing research to detect the estrogenic properties of commonly used chemicals, products, and their ingredients, researchers have developed many different live animal, whole cell, and *in-vitro* binding assays. ER-positive breast cancer cell lines show increased proliferation due to estrogenic activity. Many *in-vitro* assays can be used to detect estrogens. Hormone responsive reporter assays in human breast cancer cells and rat fibroblast cells are examples of this. However, they typically need complex equipment and reagents, and are highly sensitive to interferences (Gawrys et al., 2009). Gawrys et al. (2009) have developed a simple detection system in which the ligand-binding domain of the estrogen receptor β (ERβ) has been incorporated into a larger allosteric reporter protein in *E. coli* cells. The reporter protein expresses itself by creating a hormone-dependent growth phenotype in thymidylate synthase deficient *E. coli* strains. If a knockout media is used then there will be a marked change in growth in the presence of various test compounds that can be detected by a simple measure of the turbidity. Estrogenic behavior in compounds found in consumer products was tested using this technique. The allosteric biosensor *E. coli* strain was used to evaluate estrogenicity of a variety of compounds and complex mixtures used in common consumer products. Perfumes, hand and body washes, deodorants, essential oils and herbal supplements were included in the samples tested with 17 β-estradiol and two thyroid hormones (as controls). The system offered an additional advantage of detecting cytotoxicity of various compounds to the sensor strains. Cytotoxicity detection was based on the loss of viability of the cells in the presence of the test compound under nonselective conditions (Skretas and Wood 2005; Gawrys et al., 2009).

A review by Farré et al. (2009) covered the developments in whole-cell biosensors. Amperometric biosensors based on genetically-engineered *Moraxella sp.* and *Pseudomonas putida* JS444 with surface-expressed OPH were used for the detection of organophosphorous pesticides (Lei et al., 2005, 2007). The sensors measured up to 277 ng/L of fenitrothion. Liu et al. (2007) used horizontally aligned SWCNTs to fabricate biosensors for the real-time detection of organophosphate. SWCNT surface immobilized OPH triggers enzymatic hydrolysis of pesticides (e.g., paraoxon). The hydrolysis causes a detectable change in the conductance of the SWCNTs which is correlated to the organophosphorous pesticide concentration. Glass electrodes have been modified with genetically-engineered *E. coli* and organophosphorous pesticides degrading bacteria *Flavobacteium sp.* (Mulchandani et al., 1998a, b; Berlein et al., 2002).

The photosynthesis reaction mechanism (photosystem II or PS II) has also been used in biosensors (Giardi and Pace, 2005; Campàs et al., 2008). The PSII-based biosensors can recognize analytes such as triazines, phenylurea, diazines, and phenolic compounds. In PSII, light is first absorbed by chlorophyll–protein complexes. The photochemically active reaction centre chlorophyll (P680) then becomes excited and donates electrons to the primary pheophytin acceptor. This charge separation is stabilized by the transfer of an electron to quinone Q_A and subsequently to Q_B. Q_A, a firmly bound plastoquinone molecule is located in the D2 subunit while Q_B is a mobile plastoquinone located in the D1 subunit of PSII. Q_A and Q_B are binding pockets. Many herbicides can bind reversibly to the "herbicide-binding niche" which is the D1 subunit of PSII within its Q_B binding pocket. Once bound to the niche, the herbicides displace the plastoquinone Q_B and inhibit natural electron transfer. Once the electron flow is stopped, oxygen evolution also stops and the fluorescence properties of PSII change (Giardi and Pace, 2006; Chaplen et al., 2007; Campàs et al., 2008). While this is an exciting way of detecting herbicides, PSII based herbicide recognition is not very dependable as heavy metals may interfere (Chaplen et al., 2007). Such interferences limit the use of PSII based biosensors (Giardi et al., 2009). There are about 65 amino acids in a herbicide binding site. Giardi et al. (2009) hypothesized that modifying only one amino acid within the Q_B binding pocket would change photosynthetic activity and herbicide-binding characteristics considerably. Also, depending on the position and type of amino acid substitution, different herbicides will show affinity for the site. They used unicellular green algae *Chlamydomonas reinhardtii* strains and modified the Q_B pocket. The mutant algae cells were then used to fabricate a re-usable and portable optical biosensor with enhanced sensitivity toward different herbicide (e.g., atrazine, diuron, linuron). The detection limits ranged from 0.9×10^{-11} to 3.0×10^{-9} M (Giardi et al., 2009).

3.2.4 New generation immunosensors

While conventional ELISA is considered inadequate for many contaminants, new generation of immunosensors are becoming increasingly popular. Electrochemical immunosensors can be used for real-time in-situ monitoring of EDCs like BPA. Immunosensors are widely used for the detection of an analyte where an enzyme is labeled with a specific antigen. Enzyme labeling is a time consuming and complicated procedure. However, label-free electrochemical immunosensors represent a very attractive technique to detect EDCs by monitoring changes in electronic properties due to immunocomplex formation on the electrode surface (Rahman et al., 2007). Rahman et al. (2007) have fabricated a label-free impedimetric immunosensor for the direct detection of BPA. They prepared antigens through the conjugation of BHPVA with bovine serum albumin and then produced a specific polyclonal antibody. A covalent immobilization technique was used during sensor fabrication to attach a polyclonal antibody onto a carboxylic acid group which was functionalized on nanoparticle-based conducting polymer (Rahman et al., 2005) coated onto a glassy carbon electrode. Silver-Silver Chloride (Ag/AgCl) and platinum (Pt) wires were used as reference and counter electrodes, respectively. The detection limit was determined to be 0.3 µg BPA/L (Rahman et al., 2007).

Suri et al. (2009) provide an elaborate discussion of immunoanalytical techniques for pesticide analysis. Immunochemical techniques offer great potential for developing inexpensive, reliable sensors for effective field monitoring of many toxic molecules. Such a sensor can be based on the specificity of the antibody–antigen (A_b–A_g) reaction. Specific antibodies can be produced against pesticide molecules. As compared to other sensors, immunosensors can provide quantitative results with similar or even greater sensitivity, accuracy, and precision. Immunosensor data are comparable to standard chemical-based methods as well. Immunosensors are important tools because they complement existing analytical methods and provide low-cost confirmatory tests for many compounds, including pharmaceuticals and pesticides. A detectable signal is obtained from binding interactions between immobilized biomolecules (A_b or A_g) and analytes (A_g or A_b) of interest. Immobilization typically happens on the transducer surface. The sensor is based on the molecular recognition characteristics and, hence, the high selectivity of an A_b can be achieved (Farré et al., 2007; Suri et al., 2009). A_b binds reversibly with a specific A_g in a solution to form an immunocomplex (A_b – A_g) (Suri et al., 2009):

$$A_b + A_g \frac{K_a}{K_d} \leftrightarrow A_b - A_g \quad (3.1)$$

where K_a = rate constants for association and K_d = rate constants for dissociation. The equilibrium constant (or the affinity constant) of the reaction is:

$$K = \frac{K_a}{K_d} = \frac{[A_b - A_g]}{[A_b][A_g]} \quad (3.2)$$

An immunocomplex typically has a low K_d value (in the range 10^{-6}–10^{-12}) and displays a high K value (~104). The equilibrium kinetics in solution suggest rapid association and dissociation while the direction of equilibrium depends on the overall affinity (Suri et al., 2009). Immunosensors were initially used for clinical diagnostics. The development and applications of immunosensors for environmental pollutants (e.g., pesticides) are relatively new. The reasons for the time lag include difficulty in finding antibodies against pesticides (pesticides being low in molecular weight) (Suri et al., 2009). A_b affinity and specificity primarily determines the analytical capability of an immunosensor and, hence, the development of antibodies represents a key step in the sensor development (Farré et al., 2007). Now, that antibodies can be produced against low-molecular mass pesticides, immunosensors are expected to become cost-effective devices for the on-site monitoring of pesticides. However, many challenges remain. One of these challenges was the development of pesticide species-specific immunosensors. Pesticides are usually nonimmunogenic and it is, therefore, crucial to synthesize a suitable hapten molecule which can be coupled with a carrier protein to make a stable carrier-hapten complex. The carrier-hapten conjugate should mimic the structure of small pesticide molecules such that a suitable A_b for a particular immunoassay for the specific target molecule can be generated (Suri et al., 2009).

With recent developments, immunoassays have been used for the measurement of both single and multiple analytes. Organic pollutants like pesticides, polychlorinated biphenyls (PCBs), and surfactants can be rapidly and efficiently determined through immunochemistry (Farré et al., 2007). Farré et al. (2009b) have discussed electrochemical immunosensors for environmental analysis (atrazine determination) which have used recombinant single-chain A_b (scA_b) fragments (Grennan et al., 2003). Automated optical immunosensors have been used to detect many organic pollutants including estrone, progesterone, and testosterone in water samples. The detection limits were reported to be sub-ng/L (Taranova et al., 2004). Labeled immunosensors have been used to detect hormones, enzymes, virus, tumor antigens, and bacterial antigens at concentrations around 10^{-12} to 10^{-9} mol/L (Campàs et al., 2008; Wang et al., 2008; Wang and Lin, 2008; Bojorge Ramírez et al., 2009; He et al., 2009)

Farré et al. (2007) list a number of limitations of the immunosensing approach including

(1) lengthy preparation time for immuno-reagents;
(2) lack of specificity as well as cross-reactivity;
(3) lack or limited response towards some groups of pollutants (e.g., perfluorinated compounds);
(4) poor stability under different thermal and pH conditions; and
(5) short life-times of biological components.

Farré et al. (2007) also enumerate key aspects that need to be addressed in future immunoassays which include

(1) development of more stable biological components;
(2) fabrication of more robust assays;
(3) assurance of better repeatability between different batches of production when disposable elements are involved; and
(4) integration of new technologies coupled to biosensors (e.g., the polymerase chain reaction).

González-Martinez et al. (2007) expect that immunosensors should be usable when

- a high number of samples need to be screened;
- on-line control is necessary;
- analysis is to be carried out in the field;
- different analytes need to be determined in a sample by different methods;
- data should be presented within minutes or in real time;
- samples need to be analyzed directly with none or hardly any pretreatment; and
- traditional methods do not work properly.

González-Martinez et al. (2007) described features of an ideal immunosensor as (1) very high sensitivity even to measure contaminants of interest in very diluted solutions; (2) high selectivity for compounds of interest without or with the minimum cross-reactivity problem; (3) applicable to the whole family of related compounds with generic immunoreagents; (4) high rapidity or speed without compromising sensitivity; (5) re-usable such that the device can work for a very long time (or large number of samples) without major maintenance; (6) capable of multiparameter determination (5–10 contaminants simultaneously); (7) versatile such that the sensor can be used for new analytes provided that the appropriate reagents are available; and (8) robust such that a sensor can be used under different conditions. Bojorge Ramírez et al. (2009) agree that there are a number of challenges to overcome for the mass production of immunosensors for a wide variety of compounds of interest. Antibodies to be used *in-vivo* need protein stability and antigen affinity. The industrial-scale mass production of antibodies is not yet possible without further advances in biochemical engineering technologies (Bojorge Ramírez et al., 2009).

Fluorophores are now preferred over enzymes in immunosensors because they are more stable in solution than enzymes, also, the assay is shortened as the signal is displayed immediately (González-Martinez et al., 2007). Wikipedia defines a fluorophore as a functional group of a molecule which becomes fluorescent under appropriate environmental conditions. Fluorophores absorb the energy of a specific wavelength and re-emit it at a different, but specific, wavelength. The quantity and wavelength of the emitted energy depend on both the fluorophore and the chemical environment to which the fluorophore is exposed (Joseph, 2006). Eu(III) chelate-dyed nanoparticles have been used as an antibody label in a fluoroimmunosensor for atrazine. The sensitivity (IC50) was reported to be ~1 μg L^{-1} for this immunosensor (Cummins et al., 2006).

Nanotechnology has made significant inroads into the immunosensor area too. Magnetic nanoparticles were functionalized with specific antibodies (A_b) and used in immunomagnetic electrochemical sensors (Andreescu et al., 2009). The use of A_b-coated magnetic nanoparticles eliminates or at least reduces the need for regeneration of the sensing surface. Quantification of the formed immunocomplex is done through enzyme labeling. Quantification can also be achieved via electrochemical detection of the reaction products after the complex is exposed to the enzymatic substrates, or through fluorescent labeling (Andreescu et al., 2009). Different environmental contaminants were detected in this way. The contaminants include PCBs (Centi et al., 2005), 2,4-dichlorophenoxy acetic acid based herbicide (also known as 2,4-D), and atrazine (Helali et al., 2006; Zacco et al., 2006). Arochlor 1248 (a PCB) detection limits of 0.4 ng/mL using screen-printed electrode strips have been achieved as well as atrazine detection limits of 0.027 nmol L^{-1} using anti-atrazine-specific antibody (Zacco et al., 2006). Andreescu et al. (2009) reported in their review paper that paraoxon was measured at a low 12 μg/L level and a linear range within 24–1920 μg/L was achieved with an electrochemical immunosensor based on A_b-labeled gold nanoparticles on a glassy carbon electrode. Polymeric nanoparticles (e.g., 2-methacryloyloxyethyl phosphorylcholine and polysterene) coated with anti-bisphenol A_b were used in a piezoelectric immunosensor for bisphenol A and an eight-fold sensitivity increase was achieved (Park et al., 2006). In a reported work, Blăzkova et al. (2009) developed a simple and rapid immunochromatographic assay for a sensitive yet inexpensive monitoring of methiocarb in surface water using a binding inhibition format on a membrane strip. In the assay, the detection reagent consisted of anti-methiocarb A_b and colloidal carbon-labeled secondary A_b. They used carbon nanoparticles to bind proteins noncovalently without changing their bioactivity. A detection limit of 0.5 ng/L was reported. The assay results (recovery 90–106%) were in a good agreement with those of ELISA (recovery 91–117%). The strips were stable for at least 2 months without any change in performance. The developed immunochromatographic assay has potential for on-site screening of environmental contaminants.

A few examples of EDC sensors and biosensors are given in Table 3.1.

Table 3.1 Examples of EDC sensors and biosensors (after Rodriguez-Mozaz et al., 2006; Farré et al., 2009a, 2007)

Analyte	Transduction method	Limit of detection	Reference
Carbamates	Potentiometric	15–25 µM	Ivanov et al., 2000
Dimethyl and diethyl dithiocarbamates	Amperometric detection	20 µM	Pita et al., 1997
Bisphenol A	Potentiometric immunosensor	0.6 ng/mL	Mita et al., 2007
Fenitrothion and ethyl p-nitrophenol	Organophosphates	4 µg/L	Rajasekar et al., 2000
Progesterone	Amperometric detection	0.43 ng/mL	Carralero et al., 2007
Parathion	Amperometric detection	10 ng/mL	Sacks et al., 2000
2,4-dichlorophenoxy acetic acid	Amperometric immunosensor	0.1 µg/L	Wilmer et al., 1997
Atrazine	Electrochemical amperometric	0.03 nmol/L	Zacco et al., 2006
Chlorsulfuron	Electrochemical and amperometric	0.01 ng/mL	Dzantiev et al., 2004
Estrogens	Total internal reflection fluorescence	0.05–0.15 ng/mL	Rodriguez-Mozaz et al., 2006b
Trifluralin	Optical wave light spectroscopy	0.03 pg/mL	Székács et al., 2003
Sulphamethoxazole	Piezoelectric	0.15 ng/mL	Melikhova et al., 2006
Isoproturon	Total internal reflection fluorescence	0.01–0.14 µg/L	Blăzkova et al., 2006
Dioxins	Quartz crystal microbalances	15 ng/L	Kurosawa et al., 2005
Paraoxon and carbofuran	Electrochemical (amperometric)	0.2 µg/L	Bachmann and Schmid, 1999
Phenols	Electrochemical	0.8 µg/L	Nistor et al., 2002
Chlorophenols	Optical chemiluminescence	1.4–1975 µg/L	Degiuli and Blum, 2000
Nonylphenol	Electrochemical	10 µg/L	Evtugyn et al., 2006

3.3 TRENDS IN SENSORS AND BIOSENSORS

3.3.1 Screen printed sensors and biosensors

There have been efforts to develop and apply environmentally friendly analytical procedures for contaminant detection. Increasing concerns about the impact of chemical waste generated during conventional analytical procedures have accelerated the search for alternatives. "Green" analytical chemistry is especially relevant to the use of instruments in the field and in decentralized laboratories where treatments for toxic and hazardous wastes are not available. A range of environmentally friendly electrochemical sensors for water monitoring is available and a number of them are now in advanced prototype stages. Conventional electrochemical cells are now being replaced with screen printed electrodes (SPEs) connected to miniaturized potentiostats. SPEs are finding uses in major analytical laboratory equipment and as hand-held field devices. A number of SPEs are commercially available and it is possible to manufacture them in the laboratory for research applications since screen-printing technology is getting cheaper and more easily available (Rico *et al.*, 2009). Farré *et al.* (2009b) reviewed many recent papers and covered many topics including SPEs. Electrochemical DNA and protein sensors based on the catalytic activity of hydrazine have been developed as SPEs (Shiddiky *et al.*, 2008). Enzyme-based high sensitivity biosensors for 2, 4-D, atrazine, and ziram have also been reported (Kim *et al.*, 2008). Gold nanoparticles have been used on tyrosinase electrode for the measurement of pesticides in water (Kim *et al.*, 2008). Organophosphorous and carbamate pesticides (e.g., monocrotophos, malathion, metasystox, and lannate) were measured electrochemically using SPEs containing immobilized acetylcholine esterase (AChE) (Dutta *et al.*, 2008). The measured concentration ranged from 0 to 10 µg/L (Farré *et al.*, 2009b).

3.3.2 Nanotechnology applications

Nanomaterials are natural or engineered materials which have at least one dimension at the nanometer scale (≤ 100 nm). Nanomaterials possess completely new and enhanced properties as compared to the parent bulk materials. Examples of advanced nanomaterials include metallic, metal oxide, polymeric, semiconductor and ceramic nanoparticles, nanowires, nanotubes, quantum dots, nanorods, and composites of these materials. The unique properties of these materials are attributed to the extremely high surface area per unit weight, and their mechanical, electrical, optical, and catalytic properties. These properties offer a wide range of opportunities for the detection of environmental contaminants and toxins in addition to their remediation (Zhang, 2003; Li *et al.*, 2006;

Jimenez-Cadena *et al.,* 2007; Pillay and Ozoemena, 2007; Vaseashta *et al.,* 2007; Khan and Dhayal, 2008; Thompson and Bezbaruah, 2008; Bezbaruah *et al.,* 2009a,b). Nanotechnology incorporation into sensors (Trojanowicz, 2006; Ambrosi *et al.,* 2008; Gomez *et al.,* 2008; Guo and Dong, 2008; Kerman *et al.,* 2008; Wang and Lin, 2008; Algar *et al.,* 2009), miniaturization of electronics, and advancements in wireless communication technology have shown the emerging trend towards environmental sensor networks that continuously and remotely monitor environmental parameters (Huang *et al.,* 2001; Burda *et al.,* 2005; Liu *et al.,* 2005b; Jun *et al.,* 2006; Blasco and Picó, 2009; Zhang *et al.,* 2009).

Major research efforts are being targeted towards the development and application of nanomaterials in sensing (He and Toh, 2006; González-Martinez *et al.,* 2007). Nanomaterials are used in designing novel sensing systems and enhancing their performance (Farré *et al.,* 2009a). Satisfactory electrical communication between the active site of the enzyme and the electrode surface is a major challenge in amperometric enzyme electrodes. Aligned CNTs have been reported to improve electrical communication in such electrodes (Farré *et al.,* 2009a). Andreescu *et al.* (2009) have discussed environmental monitoring possibilities using nanomaterials. The paper cites a number of examples where nanotechnology has been successfully used in sensors. The use of nanotechnology in sensors and sensor hardware has resulted in the development of a number of miniaturized, rapid, ultrasensitive, and inexpensive methods for in-situ and field-based environmental monitoring. While these methods are not perfect and do not necessarily meet the general expectations, they are the harbingers of things to come. Nanoscale materials have been used for the construction of gas sensors (Gouma *et al.,* 2006; Jimenez-Cadena *et al.,* 2007; Milson *et al.,* 2007; Pillay and Ozoemena, 2007; Vaseashta *et al.,* 2007) enzyme sensors, immunosensors, and genosensors, for direct wiring of enzymes at electrode surfaces and for signal amplification (Liu and Lin, 2007; Pumera *et al.,* 2007). Metal oxide nanoparticles, which have excellent catalytic properties, are used for the construction of enzymeless electrochemical sensors (Hrbac *et al.,* 1997; Yao *et al.,* 2006; Hermanek *et al.,* 2007; Salimi *et al.,* 2007). Magnetic iron oxide nanostructures have potential for providing control for electrochemical processes (Wang *et al.,* 2005, 2006). Attachment of biological recognition elements to nanomaterial surfaces have led to the development of various catalytic and affinity biosensors (Andreescu *et al.,* 2009).

Costa-Fernández *et al.* (2006) have published a review on the application of quantum dots (QDs) as nanoprobes in sensing and biosensing. Andreescu *et al.* (2009) also discussed the use of carbon nanotubes (CNTs) and QDs in sensing. The surface chemistry, high surface area, and electronic properties make the use of CNTs ideal for chemical and biochemical sensing. CNTs have the ability to enhance the binding of biomolecules and increase electrocatalytic activities.

With CNTs, the detection of several analytes (e.g., pesticides) is possible at a low-applied potential. There is no need to use electronic mediators, and, hence, interferences are reduced. Electrochemical biosensors were constructed by immobilizing enzymes like acetylcholinesterase (AChE) and organophosphate hydrolyse (OPH) (Deo *et al.*, 2005) within CNTs/hybrid composites (Arribas *et al.*, 2005, 2007; Sha *et al.*, 2006; Rivas *et al.*, 2007). A number of strategies for CNT functionalization and application in sensing and biosensing have been reported (Andreescu *et al.*, 2005, 2008). These sensors have superior sensitivity compared to macroscale material-based ones. CNT-based sensors have been used for the monitoring of organophosphorus pesticides (Deo *et al.*, 2005; Joshi *et al.*, 2005), phenolic compounds (Sha *et al.*, 2006), and herbicides (Arribas *et al.*, 2005, 2007). Gold, platinum, copper, and some other nanoparticles have been incorporated onto CNTs/polymeric composites to further enhance their characteristics (Andreescu *et al.*, 2009). Further, signal amplification has been achieved using gold nanoparticle in biosensors and in a variety of colorimetric and fluorescence assays (Andreescu *et al.*, 2009). QDs have been used as sensing probes for small metal ions (Costa-Fernández *et al.*, 2006; Somers *et al.*, 2007), pesticides (Ji *et al.*, 2005), phenols (Yuan *et al.*, 2008), and nitroaromatic explosives (Goldman *et al.*, 2005). QD-enzyme conjugates respond to enzyme substrates and inhibitors (Ji *et al.*, 2005), and antibodies (Goldman *et al.*, 2005). The conjugates approach has been used to fabricate QDs (Abad *et al.*, 2005; Ji *et al.*, 2005) for the detection of pesticides. Photoluminescence intensity of the QD bioconjugates changes in the presence of the analyte (e.g., pesticide paraoxon) and the changes in signal have been quantified and correlated to analyte concentrations (Ji *et al.*, 2005).

3.3.3 Molecular imprinted polymer sensors

A conventional biosensor selectively recognizes analytes and binds them into the specific binding layers provided. This binding creates different events such as optical, mass, thermal and electrochemical changes that produce their corresponding signals (Eggins, 2002). Lots of progress has been made on biological recognition, but there are still some complex compounds that can't be detected accurately by biosensors. Such compounds include antibodies and enzymes (Sellergren, 2001; Yan and Ramstrom, 2005). Molecularly imprinted polymers (MIPs) have drawn considerable attention in recent years. In the environmental area, they are being used for remediation and sensing. MIPs are synthesized using template (target) molecules which are cross-linked into a monomer. The cross linkers are specific to the template. The target-monomer complex is then polymerized and the template molecules are removed to leave the polymer matrix with "holes" specific to the target molecules (Haupt and

Mosbach, 2000; Widstrand *et al.*, 2006). The MIP synthesis process is illustrated in Figure 3.1. The holes capture the target molecules from a sample even if they are present in small amounts. MIP materials have high recognition affinity to the target molecule. The holes are so specific that they allow only the target molecules to enter them and reject all others. MIPs are robust, cost effective, and easy to design. There are a number of advantages of MIP materials: (1) they are small in size; (2) they have increased number of accessible complementary cavities for the target molecule; (3) they have enhanced surface catalytic activity; and (4) they can establish equilibration with the target molecule very quickly due to the limited diffusional length (Nakao and Kaeriyama, 1989; Lu *et al.*, 1999). MIPs have been used in conjunction with optical and electrochemical techniques to detect amino acids, enzymes, antibodies, pesticides, proteins, and vitamins.

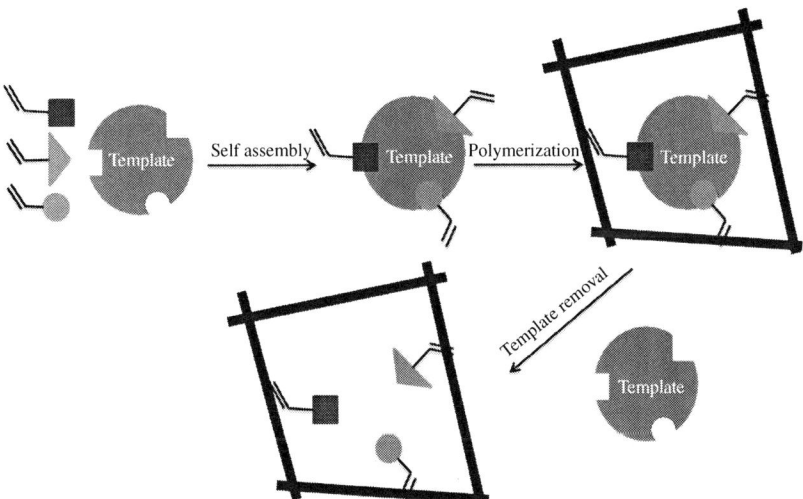

Figure 3.1 Schematic representation of molecular imprinted polymer (after Shelke *et al.*, 2008)

The efficiency of the MIP sensors depends on the interaction of the template molecule and the complementary functional monomer group (Whitcombe and Vulfson, 2001). Noncovalent and covalent interactions are observed between them. The noncovalent interactions that hold the template molecule and the functional monomer group together include hydrogen bonding, hydrophobic interaction, Van der Waals forces, and dipole-dipole interactions (Holthoff and Bright, 2007). However, if a functional group has strong covalent interactions, nonconvalent interactions are suppressed (Graham *et al.*, 2002). Reversible covalent interactions can also bind the template molecule with the functional monomer group. Wulff *et al.* (1977) first introduced the concept of covalent interactions between the functional group and the template molecule and how the template molecule can be released by cleaving the bond. These types of interactions are favored if the functional monomer has diol, aldehyde, or amine. A MIP sensor selectively binds the analyte molecules and produces a transduction scheme to detect the analyte (Figure 3.2) (Lange *et al.*, 2008). A few examples of MIP-based sensors are listed in Table 3.2. The list includes the template molecules, transduction method, and detection limits for various analytes. Both organic and inorganic materials are used to synthesize the MIPs used in the sensors. Holthoff and Bright (2007) have discussed the use of polystyrene, polyacrylate, and inorganic polysiloxane to synthesize MIP-based sensors.

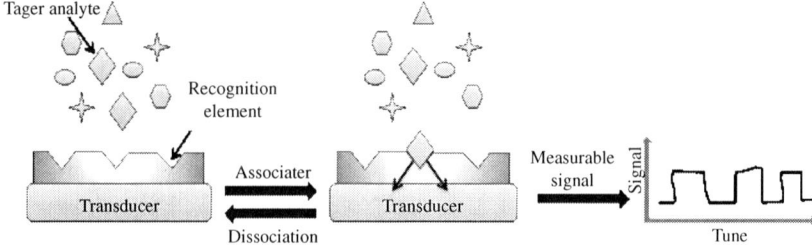

Figure 3.2 Schematic Representation of an MIP-based biosensor and its response profile (after Holthoff and Bright 2007)

Conventional MIP-based sensors are typically designed for individual analytes. The use of nanotechnology in molecular imprinting holds a lot of promise to overcome this limitation. Micro- and nano-sensors are of interest to scientists and engineers as they can be used as arrays to analyze different molecules at the same time (Alexander *et al.*, 2006). MIP materials can be appropriately patterned on a chip surface and fabricated for multi-analyte sensing applications by interfacing with transducers. Various patterning techniques that are used to make these MIP-based micro- and nano-sensors include photolithography, soft-lithography, and microspotting. UV mask lithography has been used to make the micro- and nano-MIP sensors where MIP layer is applied onto a metallic electrode and then cured by UV irradiation (Huang *et al.*, 2004). For example, Pt electrode is used in acrylic molecularly imprinted photoresist (Du *et al.*, 2008; Gomez-Caballero *et al.*, 2008), and Au and Pt electrodes are used to fabricate albuterol (a bronchodilator) MIP microsensor (Huang *et al.*, 2007).

Table 3.2 Examples of MIP-based sensors (after Holthoff and Bright 2007; Navarro-Villoslada *et al.*, 2007; Zhang *et al.*, 2008)

Target analyte	Template	Transduction method	Detection limit	Reference
Atrazine	Atrazine	Electrochemical	0.5 µM	Prasad *et al.*, 2007
Cytidine	Cytidine	Electrochemical	Not reported	Whitcombe *et al.*, 1995
Glutathione	Glutathione	Electrochemical	1.25 µM	Yang *et al.*, 2005
L-Histidine	LHistidine	Electrochemical	25 nM	Zhang *et al.*, 2005
Parathion	Parathion	Electrochemical	1 nM	Li *et al.*, 2005
L-Tryptophan	L-Tryptophan	Optical	Not reported	Liao *et al.*, 1999
Adrenaline	Adrenaline	Optical	5 µM	Matsui *et al.*, 2004
1,10-Phenan-throline	1,10-Phenan-throline	Optical	Not reported	Lin and Yamada, 2001
9-Ethyladenine	9-Ethyladenine	Optical	Not reported	Matsui *et al.*, 2000
9-Anthrol	9,10-Anthra-cenediol	Optical	0.3 µM	Shughart *et al.*, 2006
2,4-Dichloro-phenoxy-acetic acid	2,4-Dichloro-phenoxy-acetic acid	Optical	Not reported	Leung *et al.*, 2001
Penicillin G	Penicillin G	Optical	1 ppm	Zhang *et al.*, 2008
Zearalenone	Zearalenone	Optical	25 µM	Navarro-Villoslada *et al.*, 2007

Microcontact printing is one of the emerging technologies for the production of patterned microstructures (Quist *et al.*, 2005; Lin *et al.*, 2006). MIP micropatterns are created using this microcontact printing method. MIP microstructures can be synthesized using the poly(dimethyl-siloxiane) (PDMS) stamp technique (Yan and Kapua, 2001). However, incompatibility of the PDMS stamps to some organic solvents limits their applications (Vandevelde *et al.*, 2007). MIP based sensor for theophylline (a methylxanthine drug, also known as dimethylxanthine) has been synthesized using this technique and excellent selectivity for the template molecule was achieved. Similar results were reported for structurally similar caffeine (Voicu *et al.*, 2007). Micro-stereolithography technique is also used to synthesize MIP based sensors with 9-ethyl adenine as the template (Conrad *et al.*, 2003).

3.3.4 Conducting polymers

Conducting polymers have found increased applications in various industries. Some of the main classes of conducting polymers that are available for various applications include polyacetylene, polyaniline (PANI), polypyrrole (PPY), polythiophene (PTH), poly(paraphenylene), poly(paraphenylenevinylene), polyfluorene, polycarbazole, and polyindole (PI). Conducting polymers exhibit intrinsic conductivity when the conjugated backbone of the polymer is oxidized or reduced (Bredas, 1995). Apart from its conductivity, the change of electronic band in the conducting polymer affects the optical properties in the UV-visible and near IR region. The changes in conductivity and optical properties make them candidates for use as optical sensors. Chemical and electrochemical methods are used to inject charge (doping) into conducting polymers (Wallace 2003). The electrochemical method is preferred as it is easy to adjust the doping level by controlling the electrical potential.

Conducting polymers have been effectively used to detect metal ions. Polyindole and polycarbazole provide selective responses towards Cu(II) ions (Prakash *et al.*, 2002) and poly-3-octylthiophene (P3OTH) gives Nernstian responses to Ag(I) ions (Vazquez *et al.*, 2005). Extraction and stripping voltammetry method has also been used to detect Pb(II) and Hg(II) with conductive polymers (Heitzmann *et al.*, 2007). Further, ion selectivity of the sensors can be improved by introducing specific ligands (Migdalski *et al.*, 2003; Zanganeh and Amini, 2007; Mousavi *et al.*, 2008), ionophores (Cortina-Puig *et al.*, 2007) and monomers (Seol *et al.*, 2004; Heitzmann *et al.*, 2007) to the polymer backbone.

Organic molecules have an affinity towards the conducting polymer backbone, side group, and to the immobilized receptor group. This affinity is exploited to design conducting polymer sensors for organic molecules. Both biological and synthetic receptors can be used to selectively bind the organic molecule. Dopamine, ascorbic acid, and chlorpromazine sensing has been done by introducing γ-cyclodextrin receptor to poly(3-methylthiophene) (P3MTH) (Bouchta *et al.*, 2005) and β-cyclodextrin to PPY (Izaoumen *et al.*, 2005). A film of PANI and poly(3-aminophenylboronic acid) is used for the detection of saccharides by optical method (Pringsheim *et al.*, 1999). Syntheses of a variety of chemosensitive PANI and PPY conductive polymers to detect dicarboxylates, amino acids and ascorbic acid have also been reported (Volf *et al.*, 2002).

The electropolymerization method is also used to synthesize MIPs for the preparation of conductive chemosensitive film (Gomez-Caballero *et al.*, 2005; Yu *et al.*, 2005; Liu *et al.*, 2006). The electropolymerization method can control the thickness of the polymer film and this technique is compatible with the combinatorial and high-throughput approach (Potyrailo and Mirsky, 2008). Conducting MIP film of PANI has been synthesized to detect ATP, ADP, and AMP (Sreenivasan, 2007).

Synthetic and biological receptors can be used to manipulate the sensitivity of a conducting polymer for different analytes (Adhikari and Majunder, 2004; Ahuja et al., 2007). Some conducting polymers that have been modified with various receptors are listed in Table 3.3. To immobilize the receptor, it is bonded to the polymer matrix through covalent or noncovalent interaction. Physical adsorption (Lopéz et al., 2006), the Langmuir-Blodgett technique (Sharma et al., 2004), layer-by layer deposition technique (Portnov et al., 2006), and mechanical embedding method (Kan et al., 2004) are used to bind the receptor to the matrix through noncovalent bonding. Gerard et al. (2002) have discussed the advantages and limitations of these techniques.

Table 3.3 Conducting polymer-based sensors and biosensors (after Lange et al., 2008)

Analyte	Receptor	Polymer	Transduction method	Reference
Uric acid	Uricase	PANI	Optical, Amperometric	Arora et al., 2007; Kan et al., 2004
H_2O_2	Horseradish peroxidase	PANI/ polyethylene terephthalate	Optical	Caramori and Fernandes, 2004; Borole et al., 2005; Fernandes et al., 2005
Glucose	Glucose oxidase	3-methylthio-phene/thio-phene-3-acetic acid copolymer	Amperometric	Kuwahara et al., 2005
Phenol	Tyrosinase	Polyethylene-dioxythiophene	Amperometric	Vedrine et al., 2003
Organipho-sphate pesticide	Acetylcholine-sterase	PANI	Amperometric	Law and Higson, 2005
Cholesterol	Cholesterol esterase/ Cholesterol oxidase	PPY, PANI	Amperometric	Singh et al., 2004; Singh et al., 2006
Glycoproteins	Boronic acid	Poly(aniline boronic acid)	Optical	Liu et al., 2006

3.4 FUTURE OF SENSING

Sensors and biosensors have a number of disadvantages compared to standard chemical monitoring methods, however, they fulfill a number of requirements of current and emerging environmental pollution monitoring that chemical methods fail to address. Ongoing developments in material technology, computer technology, and microelectronics are expected to help sensor developers to overcome many of these problems. It is expected that progress in the development of tools and strategies to identify, record, store, and transmit parameter data will help in expanding the scope of the use of sensors on a broader scale (Blasco and Picó, 2009).

Additionally, the next generation of environmental sensors should operate as stand-alone outside the laboratory environment and with remote controls. New devices based on microelectronics and related (bio)-micro-electro-mechanical systems (MEMS) and (bio)-nano-electro-mechanical systems (NEMS) are expected to provide technological solutions. Miniaturized sensing devices, microfluidic delivery systems, and multiple sensors on one chip are needed. High reliability, potential for mass production, low cost of production, and low energy consumption are also expected and some progress has already been achieved in these areas (Farré *et al.*, 2007).

The recent developments in communication technology have not yet been fully exploited in the sensor area. New technologies like Bluetooth, WiFi and radio-frequency identification (RFID) can definitely be utilized to provide a network of distributed electronic devices in even very remote places. A wireless sensor network comprising spatially-distributed sensors or biosensors to monitor environmental conditions will contribute enormously towards continuous environmental monitoring especially in environments that are currently difficult to monitor such as coastal areas and open seas (Farré *et al.*, 2009a). Blasco and Picó (2009) expect that such a network can: (1) provide appropriate feedback during characterization or remediation of contaminated sites; (2) offer rapid warning in the case of sudden contamination; and (3) minimize the huge labor and analytical costs, as well as errors and delays, inherent to laboratory-based analyses. The laboratory-on-a-chip (LOC) is another concept that is going to impact future sensor technology. LOC involves microfabrication to achieve miniaturization and/or minimization of components of the analytical process (sample preparation, hardware, reaction time and detection) (Farré *et al.*, 2007). It has been suggested that nanoscale and ultra-miniaturized sensors could dominate the production lines in the next generation of biotechnology-based industries (Farré *et al.*, 2007).

3.5 REFERENCES

Abad, J. M., Mertens, S. F. L., Pita, M. Fernandez, V. F. and Schiffrin, D. J. (2005) Functionalization of Thioctic acid-capped gold nanoparticles for specific immobilization of histidine-tagged proteins. *J. Am. Chem. Soc.* **127**(15), 5689–5694.

Adhikari, B. Majumdar, S. (2004) Polymers in sensor applications. *Prog. Polym. Sci.* **29**(7), 699–766.

Ahuja, T., Mir, I. A., Kumar, D. and Rajesh (2007) Biomolecular immobilization on conducting polymers for biosensing applications. *Biomaterials* **28**(5), 791–895.

Alexander, C., Andersson, H. S., Andersson, L. I., Ansell, R. J., Kirsch, N., Nicholls, I. A., O'Mahony J. and Whitcombe, M. J. (2006) Molecular imprinting science and technology: A survey of the literature for the years up to and including 2003. *J. Mol. Recognit.* **19**(2), 106–180.

Algar, W. R., Massey, M. and Krull, U. J. (2009) The application of quantum dots, gold nanoparticles and molecular switches to optical nucleic-acid diagnostics. *Trends Anal. Chem.* **28**(3), 292–306.

Allan, I. J., Vrana, B., Greenwood, R., Mills, D. W., Roig, B. and Gonzalez, C. (2006) A "toolbox" for biological and chemical monitoring requirements for the European Union's Water Framework Directive. *Talanta* **69**(2), 302–322.

Ambrosi, A., Merkori, A. and de la Escosura-Muniz, A. (2008) Electrochemical analysis with nanoparticle-based biosystems. *Trends Anal. Chem.* **27**(7), 568–584.

Arora, K., Sumana, G., Saxena, V., Gupta, R. K., Gupta, S. K., Yakhmi, J. V., Pandey, M. K., Chand, S. and Malhotra, B. D. (2007) Improved performance of polyaniline-uricase biosensor. *Acta* **594**(1), 17–23.

Asano, K., Ono, A., Hashimoto, S., Inoue, T. and Kanno, J. (2004) Screening of endocrine disrupting chemicals using a surface plasmon resonance sensor. *Anal. Sci.* **20**(4), 611–616.

Andreescu, D., Andreescu, S. and Sadik, O. A. (2005) New materials for biosensors, biochips and molecular bioelectronics. In: *Biosensors and Modern Biospecific Analytical Techniques*, (ed. Gorton, L.), Elsevier, Amsterdam, pp. 285–327.

Andreescu, S., Njagi, J., Ispas, C. and Ravalli. M. T. (2009) JEM spotlight: Applications of advanced nanomaterials for environmental monitoring. *J. Environ. Monitor.* **11**(1), 27–40.

Andreescu, S., Njagi, J. Ispas, C. (2008) Nanostructured materials for enzyme immobilization and biosensors. In: *The New Frontiers of organic and composite nanotechnology*, (eds Erokhin, V., Ram, M. K. and Yavuz, O.), Elsevier, Amsterdam.

Arribas, A. S., Bermejo, E., Chicharro, M., Zapardiel, A., Luque, G. L., Ferreyra, N. F. amd Rivas, G. A. (2007) Analytical applications of glassy carbon electrodes modified with multi-wall carbon nanotubes dispersed in polyethylenimine as detectors in flow systems. *Anal. Chim. Acta* **596**(2), 183–194.

Arribas, A. S., Vazquez, T., Wang, J., Mulchandani, A. and Chen, W. (2005) Electrochemical and optical bioassays of nerve agents based on the organophosphorus-hydrolase mediated growth of cupric ferrocyanide nanoparticles. *Electrochem. Commun.* **7**(12), 1371–1374.

Bachmann, T. T. and Schmid, R. D. (1999) A disposable multielectrode biosensor for rapid simultaneous detection of the insecticides paraoxon and carbofuran at high resolution. *Anal. Chim. Acta* **401**(1–2), 95–103.

Barcelo, D. and Petrovic, M. (2006) New concepts in chemical and biological monitoring of priority and emerging pollutants in water. *Anal. Bioanal. Chem.* **385**(6), 983–984.

Berlein, S., Spener, F. and Zaborosch, C. (2002) Microbial and cytoplasmic membrane-based potentiometric biosensors for direct determination of organophosphorus insecticides. *Appl. Microbiol. Biotechnol.* **54**(5), 652–658.

Bezbaruah, A. N., Krajangpan, S., Chisholm, B. J., Khan, E. and Elorza Bermudez, J. J. (2009a) Entrapment of iron nanoparticles in calcium alginate beads for groundwater remediation applications, *J. Hazard. Mater.* **166**(2–3), 1339–1343.

Bezbaruah, A. N., Thompson, J. M. and Chisholm, B. J. (2009b) Remediation of alachlor and atrazine contaminated water with zero-valent iron nanoparticle, *J. Environ. Sci. Heal. B* **44**(6), 1–7.

Blasco, C. and Picó, Y. (2009) Prospects for combining chemical and biological methods for integrated environmental assessment. *Trends Anal. Chem.* **28**(6), 745–757.

Blăzkova, M., Mickova-Holubova, B., Rauch, P. and Fukal. L. (2009) Immunochromatographic colloidal carbon-based assay for detection of methiocarb in surface water. *Biosensors and Bioelectronics* **25**(4), 753–758.

Blăzkova, M., Karamonova, L., Greifova, M., Fukal, L., Hoza, I., Rauch, P. and Wyatt, G. (2006) Development of a rapid, simple paddle-style dipstick dye immunoassay specific for Listeria monocytogenes. *Eur. Food Res. Technol.* **223**(6), 821–827.

Bojorge Ramirez, N., Salgado, A. M. and Valdman, B. (2009) The evolution and development of immunosensors for health and environmental monitoring: Problems and perspectives, *Braz. J. Chem. Eng.* **26**(2), 227–249.

Borole, D. D., Kapadi, U.R, Mahulikar, P. P. and Hundiwale, D. G. (2005) Glucose oxidase electrodes of a terpolymer poly(aniline-co-o-anisidine-co-o-toluidine) as biosensors. *Eur. Polym. J.* **41**(9), 2183–2188.

Bouchta, D., Izaoumen, N., Zejli, H., Kaoutit, M. E. and Temsamani, K. R. (2005) A novel electrochemical synthesis of poly-3-methylthiophene-gamma-cyclodextrin film – Application for the analysis of chlorpromazine and some neurotransmitters. *Biosens. Bioelectron.* **20**(11), 2228–2235.

Brack, W., Klamer, H. J. C., de Ada, M. L. and Barcelo, D. (2007) Effect-directed analysis of key toxicants in European river basins – A review. *Environ. Sci. Pollut. Res.* **14**(1), 30–38.

Bredas, J. L. and Street, G. B. (1985) Polarons, bipolarons, and solitons in conducting polymers. *Acc. Chem. Res.* **18**(10), 309–315.

Burda, C. Chen, X., Narayanan, R. and El-Sayed, M. A. (2005) Chemistry and Properties of Nanocrystals of Different Shapes. *Chem. Rev.* **105**(4), 1025–11-2.

Campàs, M., Carpentier, R. and Rouillon, R. (2008) Plant tissue-and photosynthesis-based biosensors. *Biotechnol. Adv.* **26**(4), 370–378.

Caramori, S. S. and Fernandes, K. F. (2004) Covalent immobilisation of horseradish peroxidase onto poly (ethylene terephthalate)-poly (aniline) composite. *Process Biochem.* **39**(7), 883–888.

Carralero, V., Gonzalez-Cortes, A., Yanez-Sedeno, P. and Pingarron, J. M. (2007) Nanostructured progesterone immunosensor using a tyrosinase-colloidal gold-graphite-Teflon biosensor as amperometric transducer. *Anal. Chim. Acta* **596**(1), 86–91.

Centi, S., Laschi, S., Franek, M. and Mascini, M. (2005) A disposable immunomagnetic electrochemical sensor based on functionalised magnetic beads and carbon-based screen-printed electrodes (SPCEs) for the detection of polychlorinated biphenyls (PCBs). *Anal. Chim. Acta*, **538**(1–2), 205–212.

Chaplen, F. W. R., Vissvesvaran, G., Henry, E. C. and Jovanovic, G. N. (2007) Improvement of bioactive compound classification through integration of orthogonal cell-based biosensing methods. *Sensors*, **7**(1), 38–51.

Comerton, A. M., Andrews, R. C. and Bagley, D. M. (2009) Practical overview of analytical methods for endocrine-disrupting compounds, pharmaceuticals and personal care products in water and wastewater. *Philosophical Transactions of Royal Society A*, **367**(1904), 3923–3939.

Conrad, P. G., Nishimura, P. T., Aherne, D., Schwartz, B. J., Wu, D., Fang, N., Zhang, X., Roberts, M. J. and Shea, K. J. (2003) Functional molecularly imprinted polymer microstructures fabricated using microstereolithographic techniques. *Adv. Mater.* **15**(18), 1541–1514.

Cortina-Puig, M., Munoz-Berbel, X., del Valle, M., Munoz, F. J. and Alonso-Lomillo, M. A. (2007) Characterization of an ion-selective polypyrrole coating and application to the joint determination of potassium, sodium and ammonium by electrochemical impedance spectroscopy and partial least squares method. *Anal. Chim. Acta* **597**(2), 231–237.

Costa-Fernández, J. M., Pereiro, R. and Sanz-Medel, A. (2006) The use of luminescent quantum dots for optical sensing.*Trends. Anal. Chem.* **25**(3), 207–218.

Cummins, C. M., Koivunen. M. E., Stephanian, A., Gee, S. J., Hammock, B. D. and Kennedy I. M. (2006) Application of europium(III) chelate-dyed nanoparticle labels in a competitive atrazine fluoroimmunoassay on an ITO waveguide. *Biosens Bioelectron.* **21**(7), 1077–1085.

Degiuli, A. and Blum, L. J. (2000) Flow injection chemiluminescence detection of chlorophenols with a fiber optic biosensor. *J. Med. Biochem.* **4**(1), 32–42.

Deo, R. P., Wang, J., Block, I., Mulchandani, A., Joshi, K. A., Trojanowicz, M., Scholz, F., Chen, W. and Lin, Y. (2005) Determination of organophosphate pesticides at a carbon nanotube/organophosphorus hydrolase electrochemical biosensor. *Anal. Chim. Acta.* **530**(2), 185–189.

Du, D., Chen, S., Cai, J., Tao, Y., Tu, H. and Zhang, A. (2008) Recognition of dimethoate carried by bi-layer electrodeposition of silver nanoparticles and imprinted poly-*o*-phenylenediamine. *Electrochim. Acta* **53**(22), 6589- 6595.

Dutta, K., Bhattacharyay, D., Mukherjee, A., Setford, S. J., Turner, A. P. F. and Sarkar, P. (2008) Detection of pesticide by polymeric enzyme electrodes. *Ecotoxicol. Environ. Safety* **69**(3), 556–561.

Dzantiev, B. B., Yazynena, E. V., Zherdev, A. V., Plekhanova, Y. V., Reshetilov, A. N., Chang, S.-C. and McNeil, C. J. (2004) Determination of the herbicide chlorsulfuron by amperometric sensor based on separation-free bienzyme immunoassay. *Sens. Actuators. B* **98**(2–3), 254–261.

Eggins, B. R. (ed.), (2002) *Chemical Sensors and Biosensors*, John Wiley & Sons, Chichester, UK.

Evtugyn, G. A., Eremin, S. A., Shaljamova, R. P., Ismagilova, A. R. and Budnikov, H. C. (2006) Amperometric immunosensor for nonylphenol determination based on peroxidase indicating reaction. *Biosens. Bioelectron.* **22**(1), 56–62

Le Blanc, F. A., Albrecht, C., Bonn, T., Fechner, P., Proll, G., Pröll, F., Carlquist, M. and Gauglitz, G. (2009) A novel analytical tool for quantification of estrogenicity in river water based on fluorescence labelled estrogen receptor α. *Anal. Bioanal. Chem.* **395**(6), 1769–1776.

Farré, M., Kantiani, L., Perez, S. and Barcelo. D. (2009a) Sensors and biosensors in support of EU Directives, *Trends Anal. Chem.* **28**(2), 170–185.

Farré, M. Gajda-Schrantz, K. Kantiani, L. and Barcelo, D. (2009b) Ecotoxicity and analysis of nanomaterials in the aquatic environment. *Anal. Bioanal. Chem.* **393**(1), 81–95.

Farré, M., Kantiani, L. and Barcelo, D. (2007) Advances in immunochemical technologies for analysis of organic pollutants in the environment, *Trends Anal. Chem.* **26**(11), 1100–1112.

Fechner, P., Proell, F., Carlquist, M. and Proll, G. (2009) An advanced biosensor for the prediction of estrogenic effects of endocrinedisrupting chemicals on the estrogen receptor alpha. *Anal. Bioanal. Chem.* **393**(6–7), 1579–1585.

Fernandes, K. F., Lima, C. S., Lopes, F. M. and Collins, C. H. (2005) Hydrogen peroxide detection system consisting of chemically immobilised peroxidase and spectrometer. *Process Biochem.* **40**(11), 3441–3445.

Fernandez, M. P., Noguerol, T. N., Lacorte, S., Buchanan, I. and Pina, B. (2009) Toxicity identification fractionation of environmental estrogens in waste water and sludge using gas and liquid chromatography coupled to mass spectrometry and recombinant yeast assay. *Anal. Bioanal. Chem.* **394**(3), 957–968.

Gatidou, G., Thomaidis, N., Stasinakis, A. and Lekkas, T. (2007) Simultaneous determination of the endocrine disrupting compounds nonylphenol, nonylphenol ethoxylates, triclosan and bisphenol A in wastewater and sewage sludge by gas chromatography-mass spectrometry. *J. Chromatogr. A* **1138**(1–2), 32–41.

Gawrys, M. D., Hartman, I. Landweber, L. F. and Wood. D. W. (2009) Use of engineered *Escherichia coli* cells to detect estrogenicity in everyday consumer products, *J. Chem. Technol. Biotechnol.* **84**(12), 1834–1840.

Gerard, M., Chaubey, A. and Malhotra, B. D. (2002) Application of conducting polymers to biosensors. *Biosens. Bioelectron.* **17**(5), 345–359.

Giardi, M. T., Scognamiglio, V., Rea, G., Rodio, G., Antonacci, A., Lambreva, M., Pezzotti, G. and Johanningmeier U. (2009) Optical biosensors for environmental monitoring based on computational and biotechnological tools for engineering the photosynthetic D1 protein of *Chlamydomonas reinhardtii*. *Biosensors and Bioelectronics* **25**(2), 294–300.

Giardi, M. T. and Pace, E. (2006) Photosystem II-Based Biosensors for the Detection of Photosynthetic Herbicides, Maria Teresa Giardi and Emanuela Pace. In: *Biotechnological Applications of Photosynthetic Proteins: Biochips, Biosensors and Biodevices* (eds. Giardi, M. T. and Piletska, E.), Landes Bioscience, Springer Publishers, Georgetown, TX, USA, pp. 147–154.

Giardi, M. T. and Pace, E. (2005) Photosynthetic proteins for technological applications. *Trends Biotechnol.* **23**(5), 257–263.

Goldman, E. R., Meditnz, I. L., Whitley, J. L., Hayhurst, A., Clapp, A. R., Uyenda, H. T., Deschamps, J. R., Lessman, M. E. and Mattoussi, H. (2005) A hybrid quantum dot-antibody fragment fluorescence resonance energy transfer-based TNT sensor. *J. Am. Chem. Soc.* **127**(18), 6744–6751.

Gomez-Caballero, A., Goicolea, M. A. and Barrio, R. J. (2005) Paracetamol voltammetric microsensors based on electrocopolymerized-molecularly imprinted film modified carbon fiber microelectrodes. *Analyst* **130**(7), 1012–1018.

Gomez-Caballero, A., Unceta, N., Aranzazu Goicolea, M. and Barrio, R. J. (2008) Evaluation of the selective detection of 4,6-dinitro-*o*-cresol by a molecularly imprinted polymer based microsensor electrosynthesized in a semiorganic media. *Sens. Actuators B* **130**(2), 713–722.

Gomez, M. J., Fernandez-Romero, J. M. and Guilar-Caballos, M. P. (2008) Nanostructures as analytical tools in bioassays. *Trends Anal. Chem.* **27**(5), 394–406.

Gonzalez-Doncel, M., Ortiz, J., Izquierdo, J. J., Martın, B., Sanchez, P. and Tarazona, J. V. (2006) Statistical evaluation of chronic toxicity data on aquatic organisms for the hazard identification: The chemicals toxicity distribution approach. *Chemosphere* **63**(5), 835–844.

González-Martinez M. A., Puchades R. and Maquieira, A. (2007) Optical immunosensors for environmental monitoring: How far have we come? *Anal. Bioanal. Chem.* **387**(1), 205–218.

Gouma, P. I., Prasad, A. K. Iyer, K. K. (2006) Selective nanoprobes for 'signalling gases'. *Nanotechnol.* **17**(4), S48–S53.

Graham, A. L., Carlson, C. A. and Edmiston, P. L. (2002) Development and characterization of molecularly imprinted Sol-Gel materials for the selective detection of DDT. *Anal. Chem.* **74**(2), 458–467.

Grennan, K., Strachan, G., Porter, A. J. Killard, A. J. and Smyth, M. R. (2003) Atrazine analysis using an amperometric immunosensor based on single-chain antibody fragments and regeneration-free multi-calibrant measurement. *Anal. Chem. Acta* **500**(1–2), 287–298.

Grote, M., Brack, W., Walter, H. A. and Altenburger, R. (2005) Confirmation of cause-effect relationships using effect-directed analysis for complex environmental samples. *Environ. Toxicol. Chem.* **24**(6), 1420–1427.

Guo, S. and Dong, S. (2008) Biomolecule-nanoparticle hybrids for electrochemical biosensors. *Trends Anal. Chem.* **28**(1), 96–109.

Haupt, K. and Mosbach, K. (2000) Molecularly imprinted polymers and their use in biomimetic sensors. *Chem. Rev.* **100**(7), 2495 2504.

He, L. and Toh C. S. (2006) Recent advances in anal chem-a material approach. *Anal.Chem. Acta* **556**(1), 1–15.

He, P., Wang, Z., Zhang, L. and Yang, W. (2009) Development of a label-free electrochemical immunosensor based on carbon nanotube for rapid determination of clenbuterol. *Food Chem.* **112**(3), 707–714.

Heitzmann, M., Bucher, C., Moutet, J. C., Pereira, E., Rivas, B. L., Royal, G. and Saint-Aman, E. (2007) Complexation of poly (pyrrole-EDTA like) film modified electrodes: Application to metal cations electroanalysis. *Electrochim. Acta* **52**(9), 3082–3087.

Helali, S., Martelet, C., Abdelghani, A., Maaref, M. A. and Jaffrezic-Renault, N. (2006) A disposable immunomagnetic electrochemical sensor based on functionalized magnetic beads on gold surface for the detection of atrazine. *Electrochim. Acta,* **51**(24), 5182–5186.

Hermanek, M., Zboril, R., Medrik, I., Pechousek, J. and Gregor, C. (2007) Catalytic efficiency of iron(III) oxides in decomposition of hydrogen peroxide: Competition

between the surface area and crystallinity of nanoparticles *J. Am. Chem. Soc.* **129**(35), 10929–10936.
Holthoff, E. L. and Bright, E. V. (2007) Molecularly templated materials in chemical sensing. *Anal. Ciem. Acta.* **594**(2), 147–161.
Hrbac, J., Halouzka, V., Zboril, R., Papadopoulos, K. and Triantis, T. (1997) Carbon electrodes modified by nanoscopic iron(III) oxides to assemble chemical sensors for the hydrogen peroxide amperometric detection. *Analys.* **122**(17), 985–989.
Huang, H. C., Lin, C. I., Joseph, A. K. and Lee, Y. D. (2004) Photo-lithographically impregnated and molecularly imprinted polymer thin film for biosensor applications. *J. Chromatogr. A* **1027**(1–2) 263–268.
Huang, H. C., Huang, S. Y., Lin, C. I. and Lee, Y. D. (2007) A multi-array sensor via the integration of acrylic molecularly imprinted photoresists and ultramicroelectrodes on a glass chip. *Anal. Chim. Acta* **582**(1), 137–146.
Huang, Y., Duan, X., Wei, Q. and Liber, C. M. (2001) Directed assembly of one-dimensional nanostructures into functional networks. *Science* **291**(5504), 630–633.
Huertas-Perez, J. F. and Garcia-Campana, A. M., (2008) Determination of N-methylcarbamate pesticides in water and vegetable samples by HPLC with post-column chemiluminescence detection using the luminol reaction. *Anal. Chim. Acta* **630**(2), 194–204.
Ivanov, A. N., Evtugyn, G. A., Gyurcsanyi, R. E., Toth, K. and Budnikov, H. C. (2000) Comparative investigation of electrochemical cholinesterase biosensors for pesticide determination. *Anal. Chim. Acta* **404**(1), 55–65.
Izaoumen, N. Bouchta, D. Zejli, H. Kaoutit, M. E., Stalcup, A. M. and Temsamani, K. R. (2005) Electrosynthesis and analytical performances of functionalized poly (pyrrole/beta-cyclodextrin) films. *Talanta* **66**(1), 111–117.
Ji, X., Zheng, J. Xu, J., Rastogi, V., Cheng, T. C., DeFrank J. J. and Leblanc, R. M. (2005) (CdSe)ZnS quantum dots and organophosphorus hydrolase bioconjugate as biosensors for detection of paraoxon. *J. Phys. Chem. B* **10**(9), 3793–3799.
Jimenez-Cadena, G., Riu, J. and Rius, F. X. (2007) Gas sensors based on nanostructured materials. *Analyst.* **132**(11), 1083–1099.
Joseph, R. L. (2006) *Principles of Fluorescence Spectroscopy*, Springer, 3rd Ed, New York, NY, USA.
Joshi, K. A., Tang, J., Haddon, R., Wang, J., Chen, W. and Mulchandani, A. (2005) A disposable biosensor for organophosphorus nerve agents based on carbon nanotubes modified thick film strip electrode. *Electroanal.* **17**(1), 54–58.
Jun, Y., Choi, J.-S. and Cheon, J. (2006) Shape control of semiconductor and metal oxide nanocrustals through nonhydrolytic colloidal routes. *Angew. Chem. Int. Ed.* **45**(21), 3411–3439.
Kamyabi, M. A. and Aghajanloo, F. (2008) Electrocatalytic oxidation and determination of nitrite on carbon paste electrode modified with oxovanadium(IV)-4-methyl salophen. *J. Electroanal. Chem.* **614**(1–2), 157–165.
Kan, J., Pan, X. and Chen, C. (2004) Polyaniline-uricase biosensor prepared with template process. *Biosens. Bioelectron.* **19**(12), 1635–1640.
Kerman, K., Saito, M, Tamiya, E. Yamamura, S. and Takamura, Y. (2008) Nanomaterial-based electrochemical biosensors for medical application, *Trends Anal. Chem.* **27**(7), 585–592.

Khan, R. and Dhayal, M. (2008) Nanocrystalline bioactive TiO$_2$–chitosan impedimetric immunosensor for ochratoxin-A. *Comm.* **10**(3), 492–495.
Kim, A., Li, C. Jin, C., Lee, K. W., Lee, S., Shon, K., Park, N., Kim, D., Kang, S., Shim, Y. and Park, J. (2007) A Sensitive and reliable quantification method for Bisphenol A based on modified competitive ELISA method. *Chemosphere* **68**(7), 1204–1209.
Kim, G. Y., Shim, J., Kang, M. S. and Moon, S. H. (2008) Optimized coverage of gold nanoparticles at tyrosinase electrode for measurement of a pesticide in various water sample. *J. Hazard. Mater.* **156**(1–3), 141–147.
Kurosawa, S., Aizawa, H. and Park, J.-W. (2005) Quartz crystal microbalance immunosensor for highly sensitive 2,3,7,8-tetrachlorodibenzo-p-dioxin detection in fly ash from municipal solid waste incinerators. *Analyst.* **130**(11), 1495–1501.
Kuwahara, T., Oshima, K., Shimomura, M. and Miyauchi, S. (2005) Immobilization of glucose oxidase and electron-mediating groups on the film of 3-methylthiophene/thiophene-3-acetic acid copolymer and its application to reagentless sensing of glucose. *Polymer* **46**(19), 8091–8097.
Lange, U., Roznyatovskaya, N. V. and Mirsky, V. M. (2008) Conducting polymers in chemical sensors and arrays. *Anal. Chim. Acta.* **614**(1), 1–26.
Law, K. A. and Higson, J. (2005) Sonochemically fabricated acetylcholinesterase microelectrode arrays within a flow injection analyser for the determination of organophosphate pesticides. *Biosens. Bioelectron.* **20**(10), 1914–1924.
Lei, Y., Mulchandani, P., Chen, W. and Mulchandani, A. (2005) Direct determination of p-nitrophenyl substituent organophosphorus nerve agents using a recombinant Pseudomonas putida JS444-modified Clark oxygen electrode. *J. Agric. Food Chem.* **53**(3), 524–527.
Lei, Y., Mulchandani, P., Chen, W., Mulchandani, A. (2007) Biosensor for direct determination of fenitrothion and EPN using recombinant Pseudomonas putida JS444 with surface-expressed organophosphorous hydrolase. 2. Modified carbon paste electrode. *Appl. Biochem. Biotechnol.* **136**(3), 243–50.
Leung, M. K. P., Chow, C. F. and Lam, M. H. W. (2001) A sol-gel derived molecular imprinted luminescent PET sensing material for 2,4-dichlorophenoxyacetic acid. *J. Mater. Chem.* **11**(12), 2985–2991.
Li, C., Wang, C., Guan, B., Zhang, Y. and Hu, S. (2005) Electrochemical sensor for the determination of parathion based on *p-tert*-Butylcalix[6]arene-1,4-crown-4 sol-gel film and its characterization by electrochemical methods. *Sens. Actuators B* **107**(1), 411–417.
Li, Y. F., Liu, Z. M., Liu, Y. Y., Yang, Y. H., Shen, G. L. and Yu, R. Q. (2006) A mediator-free phenol biosensor based on immobilizing tyrosinase to ZnO nanoparticles. *Anal. Biochem.* **349**(1), 33–40.
Liao, Y., Wang, W. and Wang, B. (1999) Building fluorescent sensors by template polymerization: The preparation of a fluorescent sensor for l-tryptophan. *Bioorg. Chem.* **27**(6), 463–476.
Lin, H. Y., Hsu, C. Y. Thomas, J. L. Wang, S. E., Chen, H. C. and Chou, T. C. (2006) The microcontact imprinting of proteins: The effect of cross-linking monomers for lysozyme, ribonuclease A and myoglobin. *Biosens. Bioelectron.* **22**(4), 534–543.
Lin, J. M. and Yamada, M. (2001) Chemiluminescent flow-through sensor for 1,10-phenanthroline based on the combination of molecular imprinting and chemiluminescence. *Analyst* **126**(6), 810–815.

Liu, G. and Lin, Y. (2007) Nanomaterial labels in electrochemical immunosensors and immunoassays. *Talanta* **74**(3), 308–317.

Liu, N., Cai, X., Lei, Y., Zhang, Q., Chan-Park, M. B., Li, C., Chen, W. and Mulchandani, A. (2007) Single-walled carbon nanotube based real-time organophosphate detector. *Electroanalysis* **19**(5), 616–619.

Liu, K., Wei, W., Zeng, J. X., Liu, X. Y. and Gao, Y. P. (2006) Application of a novel electrosynthesized polydopamine-imprinted film to the capacitive sensing of nicotine. *Anal. Bioanal. Chem.* **38**(4), 724–729.

Liu, M., Hashi, Y., Pan, F., Yao, J. , Song, G. and Lin, J. (2006a) Automated on-line liquid chromatography-photodiode array-mass spectrometr method with dilution line for the determination of bisphenol A and 4-octylphenol in serum . *J. Chromatogr. A* **1133**(1–2), 142–148.

Liu, Z., Liu, Y., Yang, H., Yang, Y., Shen, G. and Yu, R. (2005a) A phenol biosensor based on immobilizing tyrosinase to modified core-shell magnetic nanoparticles supported at a carbon paste electrode. *Anal. Chim. Acta* **533**(1), 3–9.

Liu, J. F., Wang, X., Peng, Q. and Li, Y. D. (2005b) Vanadium pentoxide nanobelts: Highly selective and stable ethanol sensor materials. *Adv. Mater.* **17**(6), 764–767.

López, M. S.-P., López-Cabarcos, E. and López-Ruiz, B. (2006) Organic phase enzyme electrodes. *Biomol. Eng.* **23**(4), 135–147.

Lu, P., Teranishi, T., Asakura, K., Miyake, M. and Toshima, N. (1999) Polymer-protected Ni/Pd bimetallic nano-clusters: Preparation, characterization and catalysis for hydrogenation of nitrobenzene. *J. Phys. Chem. B* **10**(44), 9673–9682.

Marchesini, G. R., Koopal, K., Meulenberg, E., Haasnoot, W. and Irth. H (2007) Spreeta-based biosensor assays for endocrine disruptors, *Biosen. Bioelectron.* **22**(9–10), 1908–1915.

Marchesini, G. R., Meulenberg, E., Haasnoot, W. and Irth, H. (2005) Biosensor immunoassays for the detection of bisphenol A. *Anal. Chim. Acta* **528**(1), 37–45.

Marchesini, G. R., Meulenberg, E., Haasnoot, W., Mizuguchi, M. and Irth, H. (2006) Biosensor recognition of thyroid-disrupting chemicals using transport proteins. *Anal. Chem.* **78**(4), 1107–1114.

Matsui, J., Akamatsu, K., Nishiguchi, S., Miyoshi, D., Nawafune, H., Tamaki, K. and Sugimoto, N. (2004) Composite of Au nanoparticles and molecularly imprinted polymer as a sensing material. *Anal. Chem.* **76**(5), 1310–1315.

Matsui, J., Higashi, M. and Takeuchi, T. (2000) Molecularly imprinted polymer as 9-ethyladenine receptor having a porphyrin-based recognition center. *J. Am. Chem. Soc.* **12**(21), 5218–5219.

Mauriz, E., Calle, A., Lechuga, L. M., Quintana, J., Montoya, A. and Manclus, J. J. (2006a) Real-time detection of chlorpyrifos at part per trillion levels in ground, surface and drinking water samples by a portable surface plasmon resonance immunosensor. *Anal. Chim. Acta* **561**(1–2), 40–47.

Mauriz, E. Calle, A. Manclus, J. J., Montoya, A., Escuela, A. M., Sendra, J. R. and Lechuga, L. M. (2006b) Single and multi-analyte surface plasmon resonance assays for simultaneous detection of cholinesterase inhibiting pesticides. *Sens. Actuators B* **118**(1–2), 399-407.

Melikhova, E. V., Kalmykova, E. N., Eremin, S. A. and Ermolaeva, T. N. (2006) Using a piezoelectric flow immunosensor for determining sulfamethoxazole in environmental samples. *J. Anal. Chem.* **61**(7), 687-693.

Migdalski, J., Blaz, T., Paczosa, B. and Lewenstam, A. (2003) Magnesium and calcium-dependent membrane potential of poly(pyrrole) films doped with adenosine triphosphate. *Microchim. Acta* **143**(2–3), 177-185.

Milson, E. V., Novak, J., Oyama, M. and Marken, F. (2007) Electrocatalytic oxidation of nitric oxide at TiO2-Au nanocomposite film electrodes. *Electrochem. Commun.* **9**(3), 436–442.

Mita, D. G., Attanasio, A., Arduini, F., Diano, N., Grano, V., Bencivenga, U., Rossi, S., Amine, A. and Moscone, D. (2007) Enzymatic determination of BPA by means of tyrosinase immobilized on different carbon carriers. *Biosens. Bioelectron.* **23**(1), 60–65.

Mottaleb, M. A., Usenko, S., O'Donnell, J. G., Ramirez, A. J., Brooks, B. W. and Chambliss, C. K. (2009) Gas chromatography–mass spectrometry screening methods for select UV filters, synthetic musks, alkylphenols, an antimicrobial agent, and an insect repellent in fish. *J. Chromatogr. A* **1216**(5), 815–823.

Mousavi, Z., Alaviuhkola, T., Bobacka, J., Latonen, R. M., Pursiainen, J. and Ivaska, A. (2008) Electrochemical characterization of poly (3,4-ethylenedioxythiophene) (PEDOT) doped with sulfonated thiophenes. *Electrochim. Acta* **53**(11), 3755-3762.

Mulchandani, A., Mulchandani, P., Chauhan, S., Kaneva, I. and Chen, W. (1998a) A potentiometric microbial biosensor for direct determination of organophosphate nerve agents. *Electroanalysis* **10**(11), 733-737.

Mulchandani, A., Mulchandani, P., Kaneva, I. and Chen, W. (1998b) Biosensor for direct determination of organophosphate nerve agents using recombinant Escherichia coli with surface-expressed organophosphorus hydrolase. 1. Potentiometric microbial electrode. *Anal. Chem.* **70**(19), 4140-4145.

Nakao, Y. and Kaeriyama, K. (1989) Adsorption of surfactant-stabilized colloidal noble metals by ion-exchange resins and their catalytic activity for hydrogenation. *J. Colloids Interf. Sci.* **131**(1), 186–191.

Nakata, H. Kannan, K., Jones, P. D. and Giesy, J. P. (2005) Determination of fluoroquinolone antibiotics in wastewater effluents by liquid chromatography–mass spectrometry and fluorescence detection, *Chemosphere* **58**(6), 759–766.

Navarro-Villoslada, F., Urraca, J. L., Moreno-Bondi, M. C. and Orellana, G. (2007) Zearalenone sensing with molecularly imprinted polymers and tailored fluorescent probes. *Sens. Actuators B: Chem.* **12**(1), 67-73.

Nistor, C., Rose, A., Farré, M., Stoica, L., Wollenberger, U., Ruzgas, T., Pfeiffer, D., Barcelo, D., Gorton, L. and Emneus, J. (2002) In-field monitoring of cleaning efficiency in waste water treatment plants using two phenol-sensitive biosensors. *Anal. Chim. Acta* **456**(1), 3–17.

Park, J., Kurosawa, S., Aizawa, H., Goda, Y., Takai, M. and Ishihara, K. (2006) Piezoelectric immunosensor for bisphenol A based on signal enhancing step with 2-methacrolyloxyethyl phosphorylcholine polymeric nanoparticles. *Analyst* **31**(1), 155–162.

Petrovic, M., Hernando, M. D., Diaz-Cruz, M. S. and Barcelo, D. (2005) Liquid chromatography–tandem mass spectrometry for the analysis of pharmaceutical residues in environmental samples: A review, *J. Chromatoga. A* **1067**(1–2), 1–14.

Pillay, J. and Ozoemena, K. I. (2007) Efficient electron transport across nickel powder modified basal plane pyrolytic graphite electrode: Sensitive detection of sulfohydryl degradation products of the V-type nerve agents. *Electrochemistry* **9**(7), 1816–1823.

Pita, M. T. P., Reviejo, A. J., Manuel de Villena, F. J., Pingarron, J. M. (1997) Amperometric selective biosensing of dimethyl- and diethyldithiocarbamates based on inhibition processes in a medium of reversed micelles. *Anal. Chim. Acta* **340**(1–3), 89–97.

Portnov, S., Aschennok, A., Gubskii, A., Gorin, D., Neveshkin, A., Klimov, B., Nefedov, A. and Lomova, M. (2006) An automated setup for production of nanodimensional coatings by the polyelectrolyte self-assembly method. *Instrum. Exp. Technol.* **4**(6), 849–854.

Potyrailo, R. A. and Mirsky, V. M. (2008) Combinatorial and high-throughput development of sensing materials: The first 10 years. *Chem. Rev.* **108**(2), 770–813.

Prakash, R. Srivastava, R. and Pandey, P. (2002) Copper(II) ion sensor based on electropolymerized undoped conducting polymers. *J. Solid State Electrochem.* **6**(3), 203–208.

Prasad, K., Prathish, K. P., Gladis, J. M., Naidu, G. R. K. and Prasada Rao, T. (2007) Molecularly imprinted polymer (biomimetic) potentiometric sensor for atrazine. *Sens. Actuators B: Chem.* **123**(1), 65–70.

Pringsheim, E. Terpetschnig, E. Piletsky, S. A. and Wolfbeis, O. S. (1999) A polyaniline with near-infrared optical response to saccharides. *Adv. Mater.* **11**(10), 865–868.

Pumera, M., Sanchez, S., Ichinose, I. and Tang, J. (2007) Electrochemical nanobiosensors. *Sens. Actuators, B* **123**(2), 1195–1205.

Quist, A. P., Pavlovic, E. and Oscarsson, S. (2005) Recent advances in microcontact printing. *Anal. Bioanal. Chem.* **381**(3), 591–600.

Rajasekar, S., Rajasekar, R. and Narasimham, K. C. (2000) Acetobacter peroxydans based electrochemical biosensor for hydrogen peroxide. *Bull. Electrochem.* **16**(1), 25–28.

Rahman, M. A., Kwon, M.-S., Won, Choe E. S. and Shim, Y.-B. (2005) Functionalized conducting polymer as an enzyme-immobilizing substrate: an amperometric glutamate microbiosensor for in vivo measurements. *Anal. Chem.* **77**(15), 4854–4860.

Rahman, M. A. R., Shiddiky, M. J. A., Park, J.-S. and Shim, Y.-B. (2007) An impedimetric immunosensor for the label-free detection of bisphenol A. *Biosen. Bioelectron.* **22**(11), 2464–2470.

Rico, M. A. G., Olivares-Marin, M. and Gil, E. P. (2009) Modification of carbon screen-printed electrodes by adsorption of chemically synthesized Bi nanoparticles for the voltammetric stripping detection of Zn(II), Cd(II) and Pb(II), *Talanta* **80**(9), 631–635.

Rivas, G. A., Rubianes, M. D., Rodriguez, M. C., Ferreyra, N. E., Luque, G. L., Pedano, M. L., Miscoria, S. A. and Parrado, C. (2007) Carbon nanotubes for electrochemical biosensing. *Talanta* **74**(3), 291–307

Rodrigues, A. M., Ferreira, V., Cardoso, V. V., Ferreira, E. and Benoliel, M. J. (2007) Determination of several pesticides in water by solid-phase extraction, liquid chromatography and electrospray tandem mass spectrometry. *J. Chromatogr. A* **1150**(1–2), 267–278.

Rodriguez-Mozaz, S., Lopéz de Alda, M. J. and Barcelo, D. (2006a) Biosensors as useful tools for environmental analysis and monitoring, *Anal. Bioanal. Chem.* **386**(4), 1025–1041.

Rodriguez-Mozaz, S., Lopéz de Alda, M. J. and Barcelo, D. (2006b) Fast and simultaneous monitoring of organic pollutants in a drinking water treatment plant by a multi-analyte biosensor followed by LC-MS validation. *Talanta* **69**(2), 377–384.

Rodriguez-Monaz, S. Lopéz de Alda, M. and Barcelo, D. (2005) Analysis of bisphenol A in natural waters by means of an optical immunosensor. *Water Res.* **39**(20), 5071–5079.

Sacks, V., Eshkenazi, I., Neufeld, T., Dosoretz, C. and Rishpon, J. (2000) Immobilized parathion hydrolase: An amperometric sensor for parathion. *Anal. Chem.* **72**(9), 2055–2058.

Salimi, A., Hallaj, R., Soltanian, S. and Mamkhezri, H. (2007) Nanomolar detection of hydrogen peroxide on glassy carbon electrode modified with electrodeposited cobalt oxide nanoparticles. *Anal. Chim. Acta* **594**(1), 24–31.

Sanchez-Acevedo Z. C., Riu, J. and Rius, F. X. (2009) Fast picomolar selective detection of bisphenol A in water using a carbon nanotube field effect transistor functionalized with estrogen receptor-α. *Biosen. Bioelectron.* **24**(9), 2842–2846.

Saraji, M. and Esteki, N. (2008) Analysis of carbamate pesticides in water samples using single-drop microextraction and gas chromatography-mass spectrometry. *Anal. Bioanal. Chem.* **391**(3), 1091–1100.

Sellergren, B. (2001) *Molecularly Imprinted Polymers: Man-made Mimics of Antibodies and their Application in Analytical Chemistry.* Elsevier, Amsterdam, Netherlands.

Seol, H., Shin, S. C. and Shim, Y.-B. (2004) Trace analysis of Al(III) ions based on the redox current of a conducting polymer. *Electroanalysis* **16**(24), 2051–2057.

Sha, Y., Qian, L., Ma, Y., Bai, H. and Yang, X. (2006) Multilayer films of carbon nanotubes and redox polymer on screen-printed carbon electrodes for electrocatalysis of ascorbic acid. *Talanta* **70**(3), 556–560.

Sharma, S. K., Singhal, R., Malhotra, B. D., Sehgal, N. and Kumar, A. (2004) Langmuir-Blodgett film based biosensor for estimation of galactose in milk. *Electrochim. Acta* **49**(15), 2479–2485.

Shelke, C. R., Kawtikwar, P. K., Sakarkar, D. M. and Kulkarni, N. P. (2008) Synthesis and characterization of MIPs – a viable commercial venture. *Pharmaceutical Reviews* **6**(5) (http://www.pharmainfo.net/reviews/synthesis-and-characterization-mips-viable-commercial-venture) (Accessed on Jan, 2010).

Shiddiky, M. J. A., Rahman, M., Cheol, C. S. and Shim, Y. B. (2008) Fabrication of disposable sensors for biomolecule detection using hydrazine electrocatalyst. *Anal. Biochem.* **379**(2), 170.

Shughart, E. L., Ahsan, K., Detty, M. R. and Bright, F. V. (2006) Site selectively templated and tagged xerogels for chemical sensors. *Anal. Chem.* **78**(9), 3165–3170.

Singh, S., Chaubey, A. and Malhotra, B. D. (2004) Amperometric cholesterol biosensor based on immobilized cholesterol esterase and cholesterol oxidase on conducting polypyrrole films. *Anal. Chim. Acta* **502**(2), 229–234.

Singh, S., Solanki, P. R., Pandey, M. K. and Malhotra, B. D. (2006) Covalent immobilization of cholesterol esterase and cholesterol oxidase on polyaniline films for application to cholesterol biosensor. *Anal. Chim. Acta* **568**(1–2), 126–132.

Skretas, G. and Wood, D. W. (2005) A bacterial biosensor of endocrine modulators. *J. Mol. Biol.* **349**(3), 464–474.

Somers, R. C., Bawendi, M. G. and Nocera, D, G. (2007) CdSe nanocrystal based chem-bio-sensors. *Chem. Soc. Rev.* **36**(4), 579–591.

Sreenivasan, K. (2007) Synthesis and evaluation of multiply templated molecularly imprinted polyaniline. *J. Mater. Sci.* **42**(17), 7575-7578.

Suri, C. R., Boro, R., Nangia, Y., Gandhi, S., Sharma, P., Wangoo, N., Rajesh, K. and Shekhawat, G. S. (2009) Immunoanalytical techniques for analyzing pesticides in the environment, *Trends Anal. Chem.* **28**(1), 29–39.

Székács, A., Trummer, N., Adányi, N., Váradi, M. and Szendro, I. (2003) Development of a non-labeled immunosensor for the herbicide trifluralin *via* optical waveguide light mode spectroscopic detection. *Anal. Chim. Acta* **487**(1), 31–42.

Taranova, L. A., Fesay, A. P., Ivashchenko, G. V., Reshetilov, A. N., Winther-Nielsen, M. and Emneus, J. (2004) Comamonas testosteroni strain TI as a potential base for a microbial sensor detecting surfactants. *Appl. Biochem. Microbiol.* **40**(4), 404–408.

Thompson, J. M. and Bezbaruah, A. N. (2008) *Selected Pesticide Remediation with Iron Nanoparticles: Modeling and Barrier Applications.* Technical Report No. ND08-04. North Dakota Water Resources Research Institute, Fargo, ND.

Trojanowicz, M. (2006) Analytical applications of carbon nanotubes: a review. *Trends Anal. Chem.* **25**(5), 480–489.

Tudorache, M. and Bala, C. (2007) Biosensors based on screen-printing technology, and their applications in environmental and food analysis. *Anal. Bioanal. Chem.* **388**(3), 565–578.

USEPA (2010a) What are endocrine disruptors? http://www.epa.gov/endo/pubs/edspoverview/whatare.htm Accessed January 2010.

USEPA (2010b) Drinking water contaminants http://www.epa.gov/safewater/contaminants/index.html Accessed January 2010.

Vandevelde, F., Leichle, T., Ayela, C., Bergaud, C., Nicu, L. and Haupt, K. (2007) Direct patterning of molecularly-imprinted microdot arrays for sensors and biochips. *Langmuir* **23**(12), 6490–6493.

Vaseashta, A., Vaclavikova, M., Vaseashta, S., Gallios, G., Roy, P. and Pummakarnchana, O. (2007) Nanostructures in environmental pollution detection, monitoring, and remediation. *Sci Tech. Adv. Mat.* **8**(1–2), 47–59.

Vazquez, M., Bobacka, J., and Ivaska, A. (2005) Potentiometric sensors for Ag+ based on poly(3-octylthiophene) (POT). *J. Solid State Electrochem.* **9**(12), 865–873.

Vedrine, C., Fabiano, S. and Tran-Minh, C. (2003) Amperometric tyrosinase based biosensor using an electrogenerated polythiophene film as an entrapment support. *Talanta* **59**(3), 535–544.

Vieno, N. M., Tuhkanen, T. and Kronberg, L. (2006) Analysis of neutral and basic pharmaceuticals in sewage treatment plants and in recipient rivers using solid phase extraction and liquid chromatography–tandem mass spectrometry detection, *J. Chromatogr. A* **1134**(1–2), 101–111.

Voicu, R., Faid, K., Farah, A. A., Bensebaa, F., Barjovanu, R., Py, C. and Tao Y. (2007) Nanotemplating for two-dimensional molecular imprinting. *Langmuir* **23**(10), 5452–5458.

Volf, R., Kral, V., Hrdlicka, J., Shishkanova, T. V., Broncova, G., Krondak, M., Grotschelova, S., St'astny, M., Kroulik, J., Valik, M., Matejka, P. and Volka, K., (2002) Preparation, characterization and analytical application of electropolymerized films. *Solid State Ionics* **154**(Part B Sp. Iss. SI), 57–63.

Wallace, G. G., Spinks, G. M., Kane-Maguire, L. A. P. and Teasdale, P. R. (2003) *Conductive Electroactive Polymers: Intelligent Materials Systems.* CRC Press, Boca Raton, FL, USA.

Wang, J. and Lin, Y. (2008) Functionalized carbon nanotubes and nanofibres for biosensing application, *Trends Anal. Chem.* **27**(7), 619–626.

Wang, J., Musameh, M. and Laocharoensuk, R. (2005) Magnetic catalytic nickel particles for on-demand control of electrocatalytic processes. *Electrochem. Commun.* **7**(7), 652–656.

Wang, J., Scampicchio, M., Laocharoensuk, R., Valentini, F., Gonzalez-Garcia, O. and Burdick, J. (2006) Magnetic tuning of the electrochemical reactivity through controlled surface orientation of catalytic nanowires. *J. Am. Chem. Soc.* **128**(4), 4562–4563.

Wang, S., Wu, Z., Zhang, F. Q. S., Shen, G.and Yu, R. (2008) A novel electrochemical immunosensor based on ordered Au nano-prickle clusters, *Biosens. Bioelectron.* **24**(4), 1026–1032.

Wen, Y., Zhou, B., Xu, Y., Jin, S. and Feng, Y. (2006) Analysis of estrogens in environmental waters using polymer monolith in-polyether ether ketone tube solid-phase microextraction combined with high-performance liquid chromatography. *J. Chromatogr. A* **1133**(1–2), 21–28.

Whitcombe, M. J. and Vulfson, E. N. (2001) Imprinting polymer. *Adv. Mater.* **13**(7), 467–478.

Whitcombe, M. J., Rodriguez, M. E., Villar, P. and Vulfson, E. N. (1995) A new method for the introduction of recognition site functionality into polymers prepared by molecular imprinting: Synthesis and characterization of polymeric receptors for cholesterol. *J. Am. Chem. Soc.* **117**(27), 7105–7111.

Widstrand, C., Yilmaz, E., Boyd, B., Billing, J. and Rees, A. (2006) Molecularly imprinted polymers: A new generation of affinity matrices. *American Lab.* **38**(19), 12–14.

Wilmer, M., Trau, D., Renneberg, R. and Spener, F. (1997) Amperometric immunosensor for the detection of 2,4-dichlorophenoxyacetic acid (2,4-D) in water. *Anal. Lett.* **30**(3), 515–525.

Wulff, G., Vesper, W., Grobe-Einsler, R. and Sarhan, A. (1977) Enzyme-analogue built polymers, 4. On the synthesis of polymers containing chiral cavities and their use for the resolution of racemates. *Makromol. Chem.* **178**(10), 2799–2816.

Yan, M. and Kapua, A. (2001) Fabrication of molecularly imprinted polymer microstructures. *Anal. Chim. Acta* **435**(1), 163–167.

Yan, M. and Ramstrom, O. (2005) *Molecularly Imprinted Materials: Science and Technology.* Marcel Dekker, New York, NY.

Yang, L., Wanzhi, W., Xia, J., Tao, H. and Yang, P. (2005) Capacitive biosensor for glutathione detection based on electropolymerized molecularly imprinted polymer and kinetic investigation of the recognition process. *Electroanalysis* **17**(11) 969–977.

Yao, S., Xu, J., Wang, Y., Chen, X., Xu, Y. and Hu, S. (2006) A highly sensitive hydrogen peroxide amperometric sensor based on MnO2 nanoparticles and dihexadecyl hydrogen phosphate composite film. *Anal. Chim. Acta* **557**(1–2), 78–84.

Yin, H. S., Zhou, Y. and Ai, S.-Y. (2009) Preparation and characteristic of cobalt phthalocyanine modified carbon paste electrode for bisphenol A detection. *J. Electroanalytical Chem.* **626**(1–2), 80–88.

Yu, J. C. C., Krushkova, S., Lai, E. P. C. and Dabek-Zlotorzynska, E. (2005) Molecularly-imprinted polypyrrole-modified stainless steel frits for selective solid phase preconcentration of ochratoxin A. *Anal. Bioanal. Chem.* **382**(7), 1534–1540.

Yuan, J., Guo, W. and Wang, W. (2008) Utilizing a CdTe quantum dots-enzyme hybrid system for the determination of both phenolic compounds and hydrogen peroxide. *Anal. Chem.* **80**(4), 1141–1145

Zacco, E., Pividori, M. I., Alegret, S., Galve, R. and Marco, M.-P. (2006) Electrochemical magnetoimmunosensing strategy for the detection of pesticides residues. *Anal. Chem.* **78**(6), 1780–1788.

Zanganeh, A. R. and Amini, M. K. (2007) A potentiometric and voltammetric sensor based on polypyrrole film with electrochemically induced recognition sites for detection of silver ion. *Electrochim. Acta* **52**(11), 3822–3830.

Zhang, W. X. (2003) Nanoscale iron particles for environmental remediation: An overview. *J. Nanopart. Res.* **5**(3–4), 323–332.

Zhang, J., Wang, H., Liu, W., Bai, L., Ma, N. and Lu, L. (2008) Synthesis of molecularly imprinted polymer for sensitive penicillin determination in milk. *Anal. Lett.* **41**(18), 3411–3419.

Zhang, L., Zhou, Q., Liu, Z., Hou, X., Li, Y. and Lv, Y. (2009) Novel Mn_3O_4 Micro-octahedra: Promising cataluminescence sensing material for acetone, *Chem. Mater.* **21**(21), 5066–5071.

Zhang, Z., Haiping, L., Li, H., Nie, L. and Yao, S. (2005) Stereoselective histidine sensor based on molecularly imprinted sol-gel films. *Anal. Biochem.* **336**(1), 108–116.

Chapter 4
Nanofiltration membranes and nanofilters

C. Y. Chang

4.1 INTRODUCTION OF NANOFILTRATION

Membrane separation is addressed as a pressure driven process. Pressure driven processes are commonly divided into four overlapping categories of increasing selectivity: microfiltration (MF), ultrafiltration (UF), nanofiltration (NF) and hyperfiltration or reverse osmosis (RO). MF can be used to remove bacteria and suspended solids with pore sizes of 0.1 to micron. UF will remove colloids, viruses and certain proteins with pore size of 0.0003 to 0.1 microns. NF relies on physical rejection based on molecular size and charge. Pore sizes are in the range of 0.001 to 0.003 microns. RO has a pore size of about 0.0005 microns and can be used for desalination. A membrane filtration spectrum is shown in Figure 4.1.

©2010 IWA Publishing. *Treatment of Micropollutants in Water and Wastewater.* Edited by Jurate Virkutyte, Veeriah Jegatheesan and Rajender S. Varma. ISBN: 9781843393160. Published by IWA Publishing, London, UK.

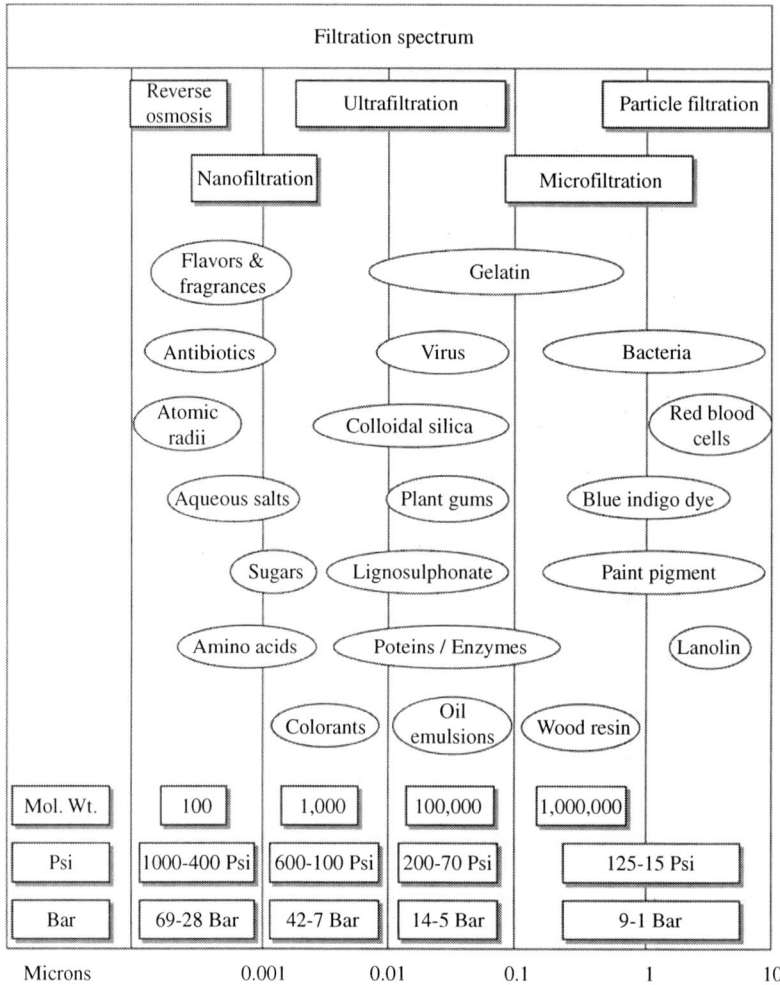

Figure 4.1 Reverse osmosis, nanofiltration, ultrafiltration and microfiltration are all related processes differing principally in the average pore diameter of the membrane and driving pressure

High pressures are required to cause water to pass across the membrane from a concentrated to dilute solution. In general, driving pressure increases as selectivity increases. Clearly it is desirable to achieve the required degree of separation (rejection) at the maximum specific flux (membrane flux/driving pressure). Separation is accomplished by MF membranes and UF membranes via mechanical sieving, while capillary flow or solution diffusion is responsible for separation in NF membranes and RO membranes.

Driven by more stringent water and wastewater treatment standards, applications employing membrane processes are increasing rapidly. Nanofiltration (NF), in particular, has been increasingly considered as an ecologically suited, reliable and affordable technique for the production of high quality water from unconventional sources such as brackish water, polluted surface water, and secondary treated effluent where micro-pollutants are to be removed (Dueom and Cabassud, 1999; Nghiem et al., 2004; Van der Bruggen et al., 2008).

The growth of nanofiltration study and application can be explained by a combination of (1) growing demand of water quantity and quality, (2) better manufacturing technology of the membranes, (3) lower prices of membranes due to wide variety of applications, and (4) more stringent standards, e.g., in the drinking water industry.

The history of nano-filtration (NF) dates back to the 1970s when reverse osmosis (RO) membranes with a reasonable water flux operating at relatively low pressures were developed. Hence, the high pressures traditionally employed in reverse osmosis resulted in a considerable energy cost. However, the permeate quality of RO was very good, and often even beyond expectations. Therefore, membranes with lower rejections of dissolved components, but with higher water permeability would be a great improvement for separation technology. Such "low-pressure reverse osmosis membranes" became known as nanofiltration membranes.

By the second half of the 1980s, nanofiltration had become established, and the first applications were reported (Conlon and McClellan, 1989; Eriksson, 1988). Nowadays NF membrane is often referred to as "softening" membrane because it is very good at rejecting hardness while letting smaller ions like sodium and chloride pass (Duran and Dunkelberger, 1995; Fu et al., 1994). Since then, the application range of nanofiltration has extended tremendously. New possibilities were discovered for drinking water as well as wastewater treatment and process water production, providing answers to new challenges such as arsenic removal (Waypa, 1997; Brandhuber and Amy, 1998; Urase et al., 1998; Košutić et al., 2005; Shih, 2005; Xia et al., 2007), removal of pesticides, endocrine disruptors and chemicals (Nghiem et al., 2004; Causserand et al., 2005; Jung et al., 2005; Košutić et al., 2005; Xu et al., 2005; Zhang et al., 2006; Yoon et al., 2007), and partial desalination (Al-Sofi et al., 1998; Hassan et al., 1998; Hassan et al., 2000; Semiat, 2000).

4.2 NANOFILTRATION MEMBRANE MATERIALS

NF membranes are generally classified into two major groups, organic and ceramic, according to their material properties. Today, organic NF membranes derived from polymeric materials are commercially available. They are applied in various fields such as drinking, process and wastewater treatment.

Two different techniques have been adopted for the development of polymer NF membranes: the phase-inversion method for asymmetric membranes (Jian *et al.*, 1999; Kim *et al.*, 2001), and interfacial polymerization (Rao *et al.*, 1997; Roh *et al.*, 1998; Jegal *et al.*, 2002; Kim *et al.*, 2002; Lu *et al.*, 2002; Song *et al.*, 2005; Verissimo *et al.*, 2005) and other coating techniques for thin film composite (TFC) membranes (Dai *et al.*, 2002; Moon *et al.*, 2004). The TFC membrane approach has some key advantages relative to asymmetric membrane approach because the selective layer and the porous support layer can be optimized separately (Petersen, 1993). Many commercial TFC membranes were prepared on polysulfone (PSf) substrate, because it has excellent chemical resistance and mechanical strength.

In general, hydrophobic polymers, such as polysulfones, polypropylene, and PVDF are widely used as membrane materials due to their good chemical resistance, superior thermal and mechanical properties. However, the affinity of organics for the membranes in a feed solution can readily cause fouling of hydrophobic materials (Ying *et al.*, 2003).

Various means have been employed in order to minimize fouling. Basically, there are either operational procedures applied during the membrane processes or fundamental modifications targeting the membrane material itself. Improvement of membrane productivity by higher flux membranes has been achieved through the development of thin film composite with very thin selective skin layer. Other methods which have been used to improve membrane performance include development of mixed-matrix membrane materials including hybrid organic-inorganic materials and surface modification of membranes. These improvements are not only limited to higher flux at lower operating pressures but also in terms of less fouling propensity, higher chlorine tolerance and increased solvent resistance (Nunes and Peinemann, 2001).

Several methods have been reported with the potential to reduce or eliminate adhesive fouling by changing the membrane surface chemistry. These methods include: (1) physically coating water soluble polymers or charged surfactants onto the membrane surface for temporary surface modification (Kim *et al.*, 1988; Jönsson and Jönsson, 1991), (2) forming ultrathin films on the membrane using Langmuir–Blodgett (LB) techniques (Kim *et al.*, 1989), (3) coating hydrophilic polymers on the membrane using heat curing (Stengaard, 1988;

Hvid *et al.*, 1990), (4) grafting monomers to the membranes by electron beam irradiation (Keszler *et al.*, 1991; Kim *et al.*, 1991), and (5) photografting monomers to the membrane using UV irradiation (Nystrom and Jarvinen, 1991; Yamagishi *et al.*, 1995; Ulbricht *et al.*, 1996).

However, for NF membranes in particular, the role of membrane charge in separations of ions has also been studied considerably. Further understanding of the exact nature of charge formation, as well as how they contribute towards rejection (Tay *et al.*, 2002) will definite leads towards significant improvement in NF membrane performances.

The development of ceramic membrane structure includes a multi-step synthesis procedure, generally based on sol–gel techniques (Burggraaf and Keizer, 1991). Firstly, a macroporous ceramic membrane support is coated with several mesoporous membrane layers in order to correct its surface roughness. Then, the last mesoporous membrane layer is modified with a very thin microporous top-layer, the actual NF layer with a cut-off value below 1000. Silicon, alumina and titanium oxide-based nanoporous membranes are chemically inert and mechanically stable. They have highly uniform and well-defined pore structures (Desai *et al.*, 1999; Martin *et al.*, 2005; Paulose *et al.*, 2008). Al_2O_3, ZrO_2 and TiO_2 are mostly considered as ceramic membrane materials (Luyten *et al.*, 1997).

Soria and Cominotti (1996) mentioned the commercialization of ceramic NF membranes, consisting of macroporous α-Al_2O_3 supports with TiO_2 top-layers, with an MWCO of ca. 1000. Larbot *et al.* (1994), Alami-Younssi *et al.* (1995) and Baticle *et al.* (1997) reported the preparation and characterization of microporous γ-Al_2O_3 membranes.

Van Gestel *et al.* (2002a) prepared and characterized a porous ceramic multilayer nanofiltration (NF) membrane from high-quality macroporous supports of α- Al_2O_3 (Figure 4.2). Three types of colloidal sol–gel derived mesoporous interlayers including Al_2O_3, TiO_2 and mixed Al_2O_3–TiO_2 were used. Corrosion measurements showed that application of a multilayer configuration including weakly crystallized α-Al_2O_3 layers is restricted to mild aqueous media (pH 3–11) or non-aqueous media (organic solvents). Optimized α-Al_2O_3/γ-Al_2O_3/anatase and α-Al_2O_3/anatase/anatase multilayer configurations show high retentions for relatively small organic molecules (molecular weight cut-off <200).

Zeolite membranes may offer an alternative choice for produced water treatment. Zeolites are crystalline aluminosilicate materials with uniform sub-nanometer- or nanometer-scale pores. For example, the MFI-type zeolite has a three-dimensional pore system with straight channels in the *b*-direction (5.4A × 5.6 A) and sinusoidal channels in the *a*-direction (5.1A × 5.5 A).

Due to the inert property of aluminosilicate crystal, zeolite membranes have superior thermal and chemical stabilities, hence holding great potential for application in difficult separations such as produced water purification and radioactive wastewater treatment. Research results demonstrated that zeolite membrane could separate several kinds of ions from water, methanolic, and ethanolic electrolyte solutions (Murad *et al.*, 1998; Lin and Murad, 2001a; Murad *et al.*, 2004; Murad and Nitche, 2004). Kumakiri *et al.* (2000) reported using an A-type zeolite membrane (pore size 0.42 nm) in RO separation of water–ethanol mixtures. The hydrophilic A-type zeolite membrane showed 44% rejection of ethanol and a water flux of 0.058 kgm^{-2} h^{-1} under an applied feed pressure of 1.5 MPa.

Polymeric toplayer (anatase)

Colloidal interlayer (anatase)

Main support (α-Al$_2$O$_3$)

Figure 4.2 FESEM cross-section (50,000×) of a multilayer membrane (Van Gestel *et al.*, 2002a)

Since their discovery by Iijima (1991), carbon nanotubes have received an increasing scientific interest because of their exceptional physical properties. In the past decade, carbon nanotubes (CNTs) have been proposed for use in numerous applications, including electronics, composite materials, fuel cells, sensors, optical devices, and biomedicine. Their use in NF membranes manufacturing, however, is still nascent, with few applications proposed or investigated so far.

Previous studies in this field have focused on the use of CNTs or CNTs functionalized with inorganic nanoparticles for adsorption of inorganic pollutantts and toxic metals from water (Long and Yang, 2001; Li *et al.*, 2002; Li *et al.*, 2003; Peng *et al.*, 2003; Agnihotri *et al.*, 2005; Lu *et al.*, 2005; Di *et al.*, 2006; Gauden *et al.*, 2006; Yang *et al.*, 2006).

A limited number of studies have explored the use of CNTs for filtration and separation applications. Srivastava *et al.* (2004) developed a cylindrical membrane filter composed of radially aligned multi-walled carbon nanotubes

(MWNTs) that formed a CNT layer several hundreds of micrometers thick. It was shown that the MWNT filter was effective in removing hydrocarbons from petroleum wastes as well as bacteria and viruses. Wang et al. (2005) prepared a composite polymeric ultrafiltration membrane, with oxidized MWNTs incorporated into the top layer. The composite ultrafiltration membrane demonstrated high retention of oil/water emulsions.

Carbon nanotube growth methods can be classified based on the number of walls in a given tube. First, both multiwalled nanotubes and single-walled nanotubes have been grown via arc-discharge carried out in an inert gas atmosphere between carbon or catalyst-containing carbon electrodes. Nowadays, carbon nanotubes and related materials are produced via a wide variety of processes. Three different processes are usually employed for the manufacture of carbon nanotube membranes including electric arc discharge, laser ablation and Chemical Vapour Deposition (CVD). Out of these, the CVD method is widely used because of its certain characteristics (Terrones, 2004).

Choi et al. (2006) prepared multi-walled carbon nanotubes (MWNTs)/ polysulfone (PSf) blend membranes by a phase inversion process using N-methyl-2-pyrrolidinone (NMP) as a solvent and water as a coagulant. Because of the hydrophiic MWNTs, the surface of the MWNTs/PSf blend membranes appeared to be more hydrophilic than a just PSf membrane. The PSf membrane with 4.0% of MWNTs showed higher flux and rejection than the PSf membrane without MWNTs. The order of the flux according to the contents of MWNTs of the blend membranes was 1.5% > 1.0% > 2.0% > 0.5% > 0.0% > 4.0%. The order of the pore size with the contents of MWNTs was: 4.0% < 0.0% < 0.5% < 2.0% < 1.0% < 1.5%.

Zhang et al. (2006) conducted a study for the removal of sodium dodecylbenzene sulfonate (SDBS) using silica/titania nanorods/nanotubes composite membrane with photocatalytic capability. XRD patterns confirmed that the embedding of amorphous silica into nanophase titania matrix helped to increase the thermal stability of titania and control the size of titania particles. SPM micrograph (Figure 4.3) shows silica/titania particles with rod-shaped homogenously distributed on the support of alumina and the silica/titania nanorod is about 5 nm high (the thickness of silica/titania layer). But most (95%) of the pore volume is located in mesopores of diameters ranging from 1.4 to 10 nm. It is these mesopore structures that allow rapid diffusion of various products during UV illumination and enhance the rate of photocatalytic reaction. The results showed that the removal of SDBS achieved 89% after 100 min by combining the photocatalysis with membrane filtration techniques.

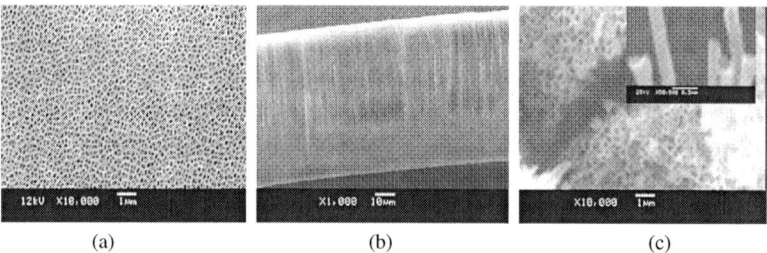

Figure 4.3 SEM photograph of: (a) the surface of 20%-silica/titania composite membrane; (b) the cross-section of 20%-silica/titania composite membrane; (c) 20%-silica/titania nanotubes (×10,000 and ×50,000)

Zhang et al. (2008) prepared titanium dioxide (TiO_2) nanotube membrane by grafting anatase TiO_2 nanotubes in the channels of alumina microfiltration (MF) membrane using TiF_4 solution through liquid-phase deposition. The experiment results of continuous filtration under UV irradiation showed that not only HA was rejected and photodegraded by the TiO_2 nanotube membrane, but also the membrane fouling was alleviated dramatically.

Tang et al. (2009) prepared chitosan/MWNTs porous membranes and found that the water flux of composite membrane with 10 wt% MWNTs (128.1 L/m^2 h) is 4.6 times that of neat one (27.6 L/m^2 h). In addition, a greatly improved tensile strength of chitosan porous membranes has been achieved by adding MWNTs.

4.3 SEPARATION AND FOULING OF NANOFILTRATION

The extent to which NF membranes are capable of retaining components depends on the NF membrane molecular weight cut-off (MWCO), surface morphology and the nature of the membrane material. Usually, the MWCO is defined as the MW of a solute that was rejected at 90 percent by a specific membrane (Van der Bruggen et al., 1999). However, the value of MWCO generally will be affected by test protocols including solute characteristics, solute concentration, solvent characteristics, as well as flow conditions such as dead-end versus cross-flow filtration. Normally, sieving effects due to steric hindrance will increase for larger molecule and the molecule is rejected by the membrane more often than a smaller molecule. However, the MWCO is only capable of providing a rough estimate of the sieving effect (Mohammad and Ali, 2002; Van der Bruggen and Vandecasteele, 2002). It may also be related to diffusion since a bigger molecule will diffuse more slowly than a smaller molecule.

The desalting degree of a membrane is also frequently used to describe the rejection characteristics of a membrane. Kiso *et al.* (1992; 2000) reported that the membranes with the highest desalting degree showed the highest pesticide rejection. Membrane surface morphology such as porosity and roughness has been regarded as another useful parameter in previous studies to estimate organic compound separation (Košutić *et al.*, 2000; Košutić and Kunst, 2002; Lee *et al.*, 2002). Košutić *et al.* (2000) reported that the membrane porous structure was the dominant parameter in determining the membrane performance, and that solute rejection could be explained by membrane pore size distribution (PSD) and effective number (N) of pores in the upper membrane layer. Scanning electron microscopy (SEM), atomic force microscopy (AFM) and field emission scanning electron microscopy (FESEM) have commonly been used for characterizing membrane surface morphology (Hirose *et al.*, 1996; Chung *et al.*, 2002).

The effects of solute characteristics on the membranes performance have been investigated by many researchers (Kiso *et al.*, 1992; Berg *et al.*, 1997; Van der Bruggen *et al.*, 1998; Kiso *et al.*, 2001a, b; Košutić and Kunst, 2002; Ozaki and Li, 2002; Schutte, 2003). The findings from those studies can be concluded that a quantification of the molecular size (and geometry) of non-charged and non-polar compounds coupled with the pore size of a membrane might be a better descriptor of the rejection than MWCO, MW, or desalting degree since steric hindrance may be an important driving factor in the rejection of molecules by NF membranes (Kiso *et al.*, 2001a; Košutić and Kunst, 2002; Ozaki and Li, 2002; Schutte, 2003). In addition, non-charged compound s with a higher number of methyl groups were reportedly rejected at higher levels than ones with lower numbers of methyl groups (Berg *et al.*, 1997).

Furthermore, several studies confirmed that molecular size parameters such as molecular width, Stokes radii, and molecular mean size have been shown to be a better predictor of steric hindrance effects upon the rejection of solutes by NF membranes than MW (Kiso *et al.*, 1992; Berg *et al.*, 1997; Van der Bruggen *et al.*, 1998; Van der Bruggen *et al.*, 1999; Kiso *et al.*, 2001a; Kiso *et al.*, 2001b; Ozaki and Li, 2002).

Electrostatic interactions between charged solutes and a porous membrane have been frequently reported to be an important rejection mechanism (Wang *et al.*, 1997; Bowen and Mohammad, 1998; Xu and Lebrun, 1999; Childress and Elimelech, 2000; Bowen *et al.*, 2002; Mohammad and Ali, 2002; Wang *et al.*, 2002).

For most thin-film composite (TFC) membranes, in order to minimize the adsorption of negatively charged foulants present in membrane feed waters and increase the rejection of dissolved salts, the membrane skin functionally carries a negative charge (Xu and Lebrun, 1999; Deshmukh and Childress, 2001; Shim *et al.*, 2002). Zeta potentials for most membranes have been observed in many studies to become increasingly more negative as pH is increased and functional groups (such as sulfonic and/or carboxylic acid groups which are deprotonated at neutral pH) deprotonate (Braghetta *et al.*, 1997; Hagmeyer and Gimbel, 1998; Deshmukh and Childress, 2001; Ariza *et al.*, 2002; Lee *et al.*, 2002; Tanninen *et al.*, 2002; Yoon *et al.*, 2002).

Charged organics and dissolved ion rejections by TFC NF membranes are heavily dependent upon the membrane surface charge and therefore feed water chemistry (Berg *et al.*, 1997; Wang *et al.*, 1997; Hagmeyer and Gimbel, 1998; Yoon *et al.*, 1998; Xu and Lebrun, 1999; Childress and Elimelech, 2000; Ozaki and Li, 2002; Wang *et al.*, 2002). Increasing the pH increased the negative surface charge of the membrane as confirmed by others (Braghetta *et al.*, 1997; Deshmukh and Childress, 2001; Lee *et al.*, 2002; Tanninen *et al.*, 2002; Yoon *et al.*, 2002), which results in increased electrostatic repulsion between a negatively charged solute and membrane and consequently rejects more solute from water. Conversely, it was determined that the presence of counter ions (Na^+, K^+, Ca^{2+}, and Mg^{2+}) in feed water can decrease the membrane rejection of negatively charged solute (Braghetta *et al.*, 1997; Ariza *et al.*, 2002; Yoon *et al.*, 2002).

However, the influence of pH and membrane surface charge on membrane pore structure and the rejection of uncharged organics as well as permeate flux is somewhat contradictory (Berg *et al.*, 1997; Braghetta *et al.*, 1997; Yoon *et al.*, 1998; Childress and Elimelech, 2000; Freger *et al.*, 2000; Boussahel *et al.*, 2002; Lee *et al.*, 2002; Ozaki and Li, 2002).

Since most high-pressure membranes are considered hydrophobic, the adsorption of hydrophobic compounds onto membranes may be an important factor in the rejection of micropollutants during membrane applications. In fact, many studies have conformed that hydrophobic–hydrophobic interactions between solute and membrane are an important factor for the rejection of hydrophobic compounds and that steric hindrance may also contribute to rejection (Kiso *et al.*, 2001a; Nghiem *et al.*, 2004; Van der Bruggen *et al.*, 2001a; Van der Bruggen *et al.*, 2001b; Nghiem *et al.*, 2002; Van der Bruggen *et al.*, 2002a; Agenson *et al.*, 2003; Kimura *et al.*, 2003a; Wintgens *et al.*, 2003). Among those studies, several parameters such as octanol-water distribution coefficient (K_{ow}), Taft and Hammett numbers (effect of the substituent group on polarity) and Dvs (measure of the stretching of the OH bond) were found to correlate with the rejection of these compounds.

Verliefde *et al.* (2007) offered a qualitative prediction of nanofiltration rejection for the selected priority micropollutants in Flemish and Dutch water sources. The qualitative prediction was based on the values of key solute and membrane parameters in nanofiltration. The qualitative predictions are roughly in agreement with literature values and may provide very quick and useful technique to assess the implementation of nanofiltration as a treatment step for organic micropollutants in drinking water plant design.

Feed water composition can certainly have a significant effect upon adsorption effects and rejection. Many studies have reported that the complexity of rejection mechanism and the effect feed water composition might have on solute rejection. (Tödtheide *et al.*, 1997; Kiso *et al.*, 2001a; Majewska-Nowak *et al.*, 2002; Schäfer *et al.*, 2002).

Fouling is one of the main problems in any membrane separation. Issues associated with membrane fouling remain quite problematic with respect to not only volume production but also permeate quality. Fouling may occur in pores of membranes by partial pore size reduction caused by foulants adsorbing on the inner pore walls, pore blockage and surface fouling such as cake and gel layer formation. The presence of the fouling layer can drastically alter the characteristics of the membrane surface including surface charge and hydrophobicity (Childress and Elimelech, 2000; Xu *et al.*, 2006). Consequently, in addition to a reduction of flux, membrane fouling can lead to considerable variation in the membrane separation efficiency. It has been reported that membrane fouling can either improve or deteriorate permeate quality but the negative consequences of fouling are obvious including the need for pretreatment, membrane cleaning, limited recoveries and feed water loss, and short lifetimes of membranes. A wide spectrum of constituents in process waters contribute to fouling. These include inorganic solutes, dissolved and macromolecular organic, suspended particles and biological solids.

Scaling usually refers to the formation of deposits of inverse-solubility salts. The greatest scaling potential species in NF membrane are $CaCO_3$, $CaSO_4 \cdot 2H_2O$ and silica, while the other potential scaling species are $BaSO_4$, $SrSO_4$, $Ca(PO4)_2$, ferric and aluminium hydroxides (Faller, 1999; Al-Amoudi and Lovitt, 2007). Inorganic scale formation can even lead to physical damage of the NF membrane, and it is difficult to restore NF membrane performance due to the difficulties of scale removal and irreversible membrane pore plugging (Jarusutthirak *et al.*, 2002; Al-Amoudi and Lovitt, 2007).

Organic fouling can be influenced by: membrane characteristics (Elimelech *et al.*, 1997; Schäfer *et al.*, 1998; Van der Bruggen *et al.*, 1999; Mänttäri, 2000;

Van der Bruggen *et al.*, 2002b), including surface structure as well as surface chemical properties, chemistry of feed solution including ionic strength (Ghosh and Schnitzer, 1980; Elimelech *et al.*, 1997), pH (Childress and Elimelech, 1996; Childress and Deshmukh, 1998; Schäfer *et al.*, 1998; Mänttäri, 2000; Schäfer *et al.*, 2004); the concentration of monovalent ions and divalent ions (Elimelech *et al.*, 1997; Schäfer *et al.*, 1998, 2004); the properties of NOM, including molecular weight and polarity (Van der Bruggen *et al.*, 1999, 2002b; Bellona *et al.*, 2004); the hydrodynamics and the operating conditions at the membrane surface including permeate flux (Van der Bruggen *et al.*, 2002b), pressure (Schäfer *et al.*, 1998; Le Roux *et al.*, 2005), concentration polarization (Schäfer *et al.*, 1998), and the mass transfer properties of the fluid boundary layer.

Organic fouling could cause either reversible or irreversible flux decline. The reversible flux decline, due to NOM fouling, can be restored partially or fully by chemical cleaning (Al-Amoudi and Farooque, 2005). However, the irreversible flux decline can not be restored at all even by rigorous chemical cleaning is applied to remove NOM (Roudman and DiGiano, 2000).

Different types of fouling may occur simultaneously and can influence each other (Flemming, 1993). Scaling by inorganic compounds is usually controlled using a scale inhibitor, such as a polymer or an acid. Particulate fouling can be controlled by pretreatment, such as ultrafiltration. Thus, all types of fouling except biofouling and organic fouling – likely related types of fouling – are controllable. Numerous authors describe biofouling problems in membrane installations (Flemming, 1993; Tasaka *et al.*, 1994; Ridgway and Flemming, 1996; Baker and Dudley, 1998; Schneider *et al.*, 2005; Karime *et al.*, 2008).

Biofouling is hard to quantify because no univocal quantification methods related to biofouling and operational problems are described. Pressure drop is generally used as a good parameter for evaluating fouling. An increase of pressure drop is, however, not conclusively linked to biofouling, since other factors may influence the pressure drop as well. Additionally, the pressure drop measurement may not be sensitive enough for early detection of biofouling. Biofouling can be controlled by (1) removal of degradable components from the feed water, (2) ensuring the relative purity of the chemicals dosed and (3) performing effective cleaning procedures (Ridgway and Flemming, 1996; Baker and Dudley, 1998; Jarusutthirak *et al.*, 2002).

A quick scan of membrane fouling is necessary to give the conclusive information about the types and extent of fouling as well as the fouling control in the membrane filtration process.

The Silt Density Index (SDI) and Fouling Index (FI) are presently used to measure the colloidal fouling potential of feedwater, but their limitations have been evidenced by several studies (Schippers and Verdouw, 1980; Boerlage *et al.*, 2003). Schippers and Verdouw (1980) developed the $MFI_{0.45}$ which was unable to take into account the influence of the colloidal particles. Boerlage *et al.* (1997) proposed a MFI–UF using a polyacrylonitrile membrane (PAN) with a MWCO of 13 kDa as a reference membrane and to measure the fouling potential of the feed water. The MFI-UF was found to be a promising tool for measuring the colloidal fouling potential for RO NF and UF systems. Roorda and Van der Graaf (2001) defined Normalized MFI–UF and proposed to give the results under standard conditions (1 m^2 membrane area and 1 bar trans-membrane pressure). Rabie *et al.* (2001) developed a method for the optimization of long-term operation of a membrane unit using the analysis of initial performance. However, the results obtained from Brauns *et al.* (2002) revealed that FI should be used as an intrinsic character of water but not as a parameter for design purposes.

As mentioned as above, MF and UF membranes are currently used for MFI. However, a fraction of colloids and solutes are not retained by these membranes and thus are not taken in account. For example, natural organic matter (NOM) and and even more of effluent organic matter (EfOM) contain a fraction of organics which molecular weight (MW) is around 1000Da (Abdessemed *et al.*, 2002; Jarusutthirak *et al.*, 2002).

Khirani *et al.* (2006) employed a loose, uncharged NF membrane with a MWCO of 500–1500 Daltons (hydrophylic polyether sulphone membranes with a thin-film oxidation resistant layer) to develop a NF-MFI. The limiting parameter of the MFI–UF is the duration of the test (more than 20 h). Khirani's method showed that the determination of a FI was possible in a short time (about 1 h) using a loose NF or a NF membrane. The hypothesis of Khirani's study is that a NF membrane is able to retain all the components responsible for fouling including small molecules (colloids and solutes) that are involved in membrane fouling. Obviously, the study showed that dissolved organics are responsible for fouling and have to be taken into account in determining the membrane fouling potential of feed water.

In addition to MFI, several methods developed for the diagnosis, prediction, prevention, and control of fouling have been proposed and applied in practice and have proven their value in controlling fouling. An overview of the coherent tools is shown in Table 4.1 (Vrouwenvelder *et al.*, 2003).

Table 4.1 Overview of tools available for determining the fouling potential of feed water and fouling diagnosis of NF and RO membranes used in water treatment (Vrouwenvelder et al., 2003)

Tools	Fouling diagnosis	Comment
Integrated diagnosis (autopsy)	Biofouling, inorganic, compounds and particles	Diagnosis of foulant in membrane elements
Biofilm monitor and AOC	Biofouling	Predictive and prevention of biofouling by determining the (growth) potential of water
SOCR	Biofouling	Non-destructive method for determining active biomass in membrane systems
MFI-UF	Particulate	Particulate fouling potential of water
ScaleGuard	Scaling	Optimizing recovery, acid dose and anti-scalant dose

4.4 NANOFILTRATION OF MICROPOLLUTANTS IN WATER

Nanofiltration membranes were initially developed for softening purposes. Nowadays, although softening is still a major application, nanofiltration is a rapidly developing technology with promising applications to remove pesticides and other organic contaminants from surface and ground waters to help insure the safety of public drinking water supplies.

Softening is a typical process for groundwater treatment. The traditional methods for water softening include lime-soda and ion exchange processes. In contact with an aqueous solution, most NF membranes become positively or negatively charged due to the presence of ionizable groups. Therefore, NF membranes can also be used to remove small ionic components or inorganic salts. The softening of ground water using NF has been studied by many investigators.

A comparison between lime softening and nanofiltration for groundwater treatment in Florida has been carried out by Bergman (1995). Cost evaluation was done under several operating modes in the study. The cost of membrane softening can be even lower than for lime softening if additional treatment processes are added to lime softening to match the better membrane softening permeate quality, or if some water can be bypassed around the membranes and blended to produce water comparable to the finished water in the lime softening plant. Definitely, NF membrane softening is becoming an attractive alternative presenting many advantages: superior water product quality, no sludge disposal, ease of operation as well as reduction of overall plant construction and O&M costs.

Sombekke *et al.* (1997) made a comparison between nanofiltration and pellet softening (combined with granular activated carbon (GAC) adsorption for organics removal) based on a life cycle analysis (LCA). The environmental impact of a product in its entire life cycle and all the extractions from and emissions into the environment were involved in the LCA study. Both treatment schemes were found to have a comparable impact. However, nanofiltration was advantageous for quality and health aspects.

The need for partial softening of raw waters as well as the removal of organic micropollutants, has led to the adaptation of nanofiltration for serving this dual purpose. An example is the treatment of water from a lake in Taiwan (Yeh *et al.*, 2000), where hardness, taste and odor problems had to be solved at the same time. Yeh *et al.* used different methods such as a conventional process followed by ozone, GAC and pellet softening, and an integrated process of membrane process (UF/NF) and conventional process. Softening was satisfactory for all processes, but water produced by the membrane process had the best quality as measured by turbidity, dissolved organics, biostability and organoleptic parameters.

The development of membrane with high rejection of organics but low hardness rejection was carried out in 1997 (De Witte, 1997). De Witte demonstrated that NF 200 (Filmtec) membrane performance remains good after repetitive cleaning and energy consumption is still low. NF 200 (Filmtec) membrane was successfully used at Debden Road water works, Saffron Walden, England (Wittmann *et al.*, 1998).

Fu *et al.* (1994, 1995) used NTR 7450 membrane made by Nitto-Denko to remove organics without removing much of the inorganics and the permeability of the membrane were superior compared to traditional NF membranes used for softening. The NTR 7450 could be operated at a recovery of 90% and a flux of 34 $l/m^2 h$ and organics removal was nearly complete.

Nanofiltration is also considered as an alternative that can be used to meet regulations for lowered arsenic concentrations in drinking water (Kartinen and Martin, 1995). Saitúa *et al.* (2005) reported that arsenic rejection by nanofiltration was independent of transmembrane pressure, crossflow velocity and temperature. The co-occurrence of dissolved inorganics does not significantly influence arsenic rejection. Waypa *et al.* (1997) studied the arsenic removal from synthetic freshwater and from surface water sources by NF. They presented that both As(V) and As(III) were effectively removed from the water by NF membrane over a range of operating conditions. The NF membrane can achieve rejection of 99%. The result also showed that the removal of As(V) and As(III) was comparable, with no preferential rejection of As(V) over As(III). The authors have concluded that size exclusion governed their separation behavior and not the charge interaction. Seidel et al. (2001) studied the difference in rejection between As(V) and As(III) using loose (porous) NF membranes. The removal of As(V) was varied between 60% and 90%, whereas As(III) was below 30%. Sato et al. (2002) also investigated the performance of nanofiltration for arsenic removal. In their studies, NF membranes could remove over 95% of pentavalent arsenic and more than 75% of trivalent arsenic could be removed without any chemical additives. Furthermore, both As(V) and As(III) removal by NF membranes was not affected by source water chemical compositions. An overview on the removal of arsenic from surface water and ground water by nanofiltration of is also reported by Van der Bruggen and Vandecasteele (2003).

Gestel et al. (2002b) employed a multilayer TiO_2 membrane for the retention of five types of salts. The membrane showed aminimal salt retention at pH 6 and a fairly high retention at alkaline pH (R(NaCl) = 85%, R(KCl) = 87%, R(LiCl) = 90%). For salts containing divalent ions, high retentions were again achieved (R(Na2SO4) >95%; R(CaCl2) = 78%).

El-Sheikh et al. (2007) used different kinds of multi-walled carbon nanotube (MWCNT) for enrichment of metal ions (Pb^{2+}, Cd^{2+}, Cu^{2+}, Zn^{2+} and MnO_4^-) from environmental waters prior to their analysis. It was found that long MWCNT of length 5–15 μm and external diameter 10–30 nm gave the highest enrichment efficiency towards MnO_4^-, Cu^{2+}, Zn^{2+} and Pb^{2+}; but not for Cd^{2+} due to its low recovery.

Lin and Murad (2001b) reported that 100% Na^+ rejection could be achieved on a perfect (single crystal), ZK-4 membrane through RO. The separation mechanism of the perfect ZK-4 zeolite membranes is the size exclusion of hydrated ions, which have kinetic sizes significantly larger than the aperture of the ZK-4 zeolite.

Li et al. (2004a, b) used MFI-type zeolite membranes in RO separation and showed 77% rejection of Na^+. In a complex feed solution containing 0.1M NaCl + 0.1M KCl + 0.1M NH_4Cl + 0.1M $CaCl_2$ + 0.1M $MgCl_2$, rejections of Na^+, K^+, NH_4^+, Ca^{2+}, and Mg^{2+} were 58.1%, 62.6%, 79.9%, 80.7%, and 88.4%, respectively.

Choi et al. (2008) investigated the effect of co-existing ions on the removal of several anions using negative surface charge nanomembrane. The results showed that sulfate would be rejected most among other ions in the groundwater. The chloride ions were rejected more than nitrate and fluoride ions. The experiment indicated that the electric repulsion between the nanomembrane and chloride ion was so high that it could even push some divalent sulfate ions through the membrane. Fluoride was less affected by the surface charge of the membrane than nitrate. The hydration effect of nitrate would be stronger at a membrane with lower surface potential. The experiment indicated that calcium ions shielded membrane charges more effectively than magnesium ions. However, despite the charge shielding effect, the rejection rates against the divalent anion were high, and more ions were rejected by the membrane that has a high negative surface potential.

Removal of natural organic matter (NOM) and disinfection byproduct (DBP) precursors from water sources by NF membranes have been studied by many investigators (Agbekodo *et al.*, 1996; Ericsson *et al.*, 1996; Alborzfar *et al.*, 1998; Visvanathan *et al.*, 1998; Cho *et al.*, 1999; Levine *et al.*, 1999; Escobar *et al.*, 2000; Everest and Malloy, 2000; Khalik and Praptowidodo, 2000). It is obvious that the best results were obtained with membranes with a MWCO around 200.

Visvanathan *et al.* (1998) evaluated the effects of interference parameters include operating pressure, feed THMPs concentration, pH, presence of other ions (Ca^{2+} and Mg^{2+}), and suspended solids on the performance of nanofiltration for removal of trihalomethane precursors (THMPs). Generally rejection was found to be greater than 90% for a precompacted membrane. Experimental results also showed that higher pressure, feed THMP concentration, and suspended solids increased rejection and divalent ions reduced the rejection capacity.

NOM essentially consists of molecules from a large range of molecular weights, this confirms the need for a membrane with a low MWCO for complete organics removal. Agbekodo *et al.* (1996a) reported that about 60% of the remaining DOC in NF permeate is caused by amino acids as well as lower fractions of fatty-aromatic acids and aldehydes.

The removal of anthropogenic micropollutants from water by NF membrane has been studied by many researchers (Agbekodo *et al.*, 1996b; Montovay *et al.*, 1996; Van der Bruggen *et al.*, 1998; Ducom and Cabassud, 1999; Kiso *et al.*, 2000; Kimura *et al.*, 2003; Causserand *et al.*, 2005; Plakas *et al.*, 2006; Lee *et al.*, 2008). Since the majority of the compounds categorized as pesticides have molecular weights (MW) of more than 200 Da, nanofiltration (NF) seems to be a promising option for their removal from contaminated water sources. However, the results came from those studies showed the removal efficiencies largely depend on the membranes used and on the micropollutants that have to be removed.

Agbekodo *et al.* (1996b) demonstrated the influence of natural organic matter (NOM) on the retention of atrazine and simazine. The removal efficiency of the NF70 membranes can vary from 50 to 100% as a function of the DOC level in the feed water. Montovay *et al.* (1996) found an 80% removal of atrazine and a 40% removal of metazachlor, which is insufficient. Van der Bruggen *et al.* (1998) showed that the NF70 membrane can reject pesticides such as atrazine, simazine, diuron and isoproturon over 90%. However, relatively low rejections were found for diuron and isoproturon for two other membranes (NF45 and UTC-20). Ducom and Cabassud (1999) studied the removal of trichloroethylene, tetrachloroethylene and chloroform by nanofiltration. The removal of trichloroethylene and tetrachloroethylene can be achieved using several different types of NF membranes, but chloroform rejection was significantly lower. However, good removal efficiencies of chloroform were obtained by Waniek et al. (2002). Kiso et al. (2000) studied the removal of 12 pesticides. Rejections obtained with three of these membranes were too low; the rejections with the fourth membrane were very high (over 95%), but this membrane appears to be a reverse osmosis membrane, given the high NaCl rejection. Kimura et al. (2003) studied the rejection of disinfection by-products (DBPs), endocrine disrupting compounds (EDCs), and pharmaceutically active compounds (PhACs) by nanofiltration (NF) and reverse osmosis (RO) membranes as a function of their physico-chemical properties and initial feed water concentration. Experimental results indicated that negatively charged compounds could be rejected very effectively (i.e., $>90\%$) regardless of other physico-chemical properties of the tested compounds and rejection of the compounds were not time-dependency. Contrarily, rejection of non-charged compounds was generally lower ($<90\%$ except for one case) and influenced mainly by the molecular size of the compounds. A clear time-dependency was observed for rejection of non-charged compounds, attributable to compound adsorption on the membrane. Causserand et al. (2005) reported the performance of polyamide membrane was much better that that of cellulose acetate membrane in removing 2,4-dichloroaniline. Plakas et al. (2006) investigated the role of organic matter and calcium concentration on the removals of atrazine, isoproturon and prometryn. The results showed that nanofiltration of water where herbicides are present together with humic substances results in increased herbicide retention. This trend is less evident in the presence of calcium ions due to their possible interference with the humic substances–herbicides interactions. Lee et al. (2008) demonstrated the effects of membrane properties and solution chemistries on removal efficiencies of TCEP and perchlorate.

Basically, the size of bacteria (0.5–10 μm) and protozoan cysts and oocysts (3–15 μm) are larger than the pore size of UF membranes. Both of them can be removed with at least 4 log units using UF membranes. The size of a virus varies between 20 and 80 nm, whereas UF membranes have pores of approximately 10 nm and more, so that complete removal of virus by UF membranes is theoretically possible. For NF membranes which have pore sizes below 1 nm, the smaller viruses may be rejected. In fact, NF membranes were found to be able to remove viruses and bacteria from surface water quite successfully [11,38–41].

Yahya et al. (1993) compared the performances of slow sand filtration and NF membranes in a 76 m^3/d surface water pilot plant for the removal of two bacteriophages (MS-2, 28 nm, and PRD-1, 65 nm). The slow sand filters removed

99 and 99.9% of the bacteriophages, respectively, and the NF membranes (NF70 – Filmtec; Desal 5 DK and Desal 5 SG–Osmonics) removed 4–6 log units of the test viruses. Otaki *et al.* (1998) reported a log 7 removal of poliomyelitis virus vaccine and a log 6 removal of coliphage Q beta from river water in the Tokyo area by the NTR-729HFS4 membrane (Nitto-Denko). Reiss *et al.* (1999) used an integrated membrane system of microfiltration with NF to remove *Bacillus subillus* spores from 5.4 to 10.7 log. Urase *et al.* (1996) used MF, UF and NF membranes for removal of the model viruses Q beta and T4. The removal of the test viruses ranged from 2 to 6 log units through different membranes. It can be concluded that 100% virus retention cannot be obtained with pressure driven membrane processes due to the leakage of viruses through 'abnormally large' pores.

Brady-Estevez *et al.* (2008) demonstrate the use of a single-walled carbon nanotube (SWNT) for the effective removal of bacterial and viral pathogens from water at low pressures. The filter was developed using a poly(vinylidene fluoride) (PVDF)-based microporous membrane (5 mm pore size) covered with a thin layer of SWNTs. Such a hybrid filter would be exceptionally robust, permitting reuse, as the high thermal resistance of carbon nanotubes and ceramics would allow for simple thermal regeneration of the filter. For bacterium removal study, Escherichia coli K12 was selected as a model bacterium. The study showed that viruses can be completely removed by a depth-filtration mechanism (Figure 4.4), that is, capture by nanotube bundles inside the SWNT layer. For virus rejection study, a model virus particle, MS2 bacteriophage, diameter 27 nm, was selected. The results showed that virus removal from the 10^7 virus particles per mL initial concentration was complete, without any viral particles detected by the PFU (plaque forming unit) method at the filter outlet.

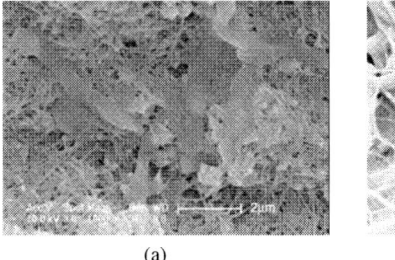
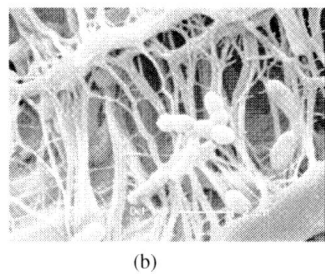

(a)   (b)

Figure 4.4 Retention of E. coli by SWNT filter. (a) SEM image of E. coli cells retained on SWNT filter. (b) SEM image of E. coli cells on the base membrane (5 mm pore PVDF membrane)

Normally, NF is not thought of as a pure disinfection process and post-treatment of the NF permeate will always be necessary. Moreover, chlorination might be required for the NF permeate, but would likely be low in DBPs formation, to prevent bacterial regrowth in the distribution network (Laurent et al., 1999). Another alternative is to integrate with RO for the further improvement of disinfection.

A case study at Lake Arrowhead, California using a combination of coagulation/flocculation/sedimentation, sand filtration, ozonation (in two stages), GAC filtration, UF/NF, and RO showed that the integrated system can remove 21–22 log units of bacteriophage and 8–10 log units of Giardia and Cryptosporidium (Madireddi et al., 1997).

Among MF, UF, NF and RO membranes, NF in particular offers a comprehensive approach to meeting multiple water quality objectives including removal of dissolved organics and inorganic contaminants. In addition, NF has obvious advantages compared to RO mainly: (1) lower operating pressure and (2) selective rejection between monovalent and multivalent ions.

Feed pretreatment is one of the major factors determining the success or failure of a desalination process. Pre-treatment of seawater feed to RO/thermal processes using nanofiltration is expected to lower the required pressure to operate RO plant by reducing seawater feed TDS as well as to reduce the energy the energy consumption (Redondo, 2001).

The main drawback of conventional pretreatment based on chemical and mechanical treatments (i.e., coagulation, flocculation, acid treatment, pH adjustment, addition of anti-sealant and mediafiltration) is known to be complex, labor intensive and space consuming (Sikora et al., 1989; Van Hoop et al., 2001). Another problem in using conventional pretreatment is corrosion and corrosion products since the acid dosing system is commonly used in the conventional process (Sikora et al., 1989; AI-Ahmad and Adbul Aleem, 1993).

NF was used for the first time by Hassan et al. (1998) as pretreatment for seawater reverse osmosis (SWRO), multistage flash (MSF), and seawater reverse osmosis rejected in multistage flash (SWRO$_{rejected}$-MSF) processes. The NF applicaton made it possible to operate a SWRO and MSF pilot plant at a high recovery of 70% and 80%, respectively.

Regarding the feed water quality improvement of SWRO desalination plants, NF pretreatment can be beneficial to (1) prevent SWRO membrane fouling by the removal of turbidity and bacteria, (2) prevent scaling (both in SWRO and MSF) by removal of scale forming hardness ions, and (3) lower required pressure to operate SWRO plants by reducing seawater feed TDS by 30–60%, depending on the type of NF membrane and operating conditions (Al-Sofi et al., 1998; Cfiscuoli and Drioli, 1999; Hassan et al., 2000; Al-Sofi, 2001; Drioli et al., 2002; Mohesn et al., 2003; Pontié et al., 2003).

A promising approach for pretreatment of seawater make-up feed to MSF and SWRO desalination processes using nanofiltration (NF) membranes has been introduced by the R&D Center (RDC) of SWCC. NF membranes are capable to reduce significantly scale forming ions from seawater, allow high temperature operation of thermal desalination processes, and subsequently increase water productivity (Hamed, 2005).

Hafiarle et al. (2000) used a TFC-S NF membrane to remove chromate from an aqueous solution. The results showed that the rejection depended on the ionic strength and pH. Better retention was obtained at basic pH (up to 80% at a pH of 8). Results also showed that NF is a very promising method of treatment for wastewater charged with hexavalent chromium.

Ku et al. (2005) studied the effect of solution composition on the removal of copper ions by nanofiltration. The results indicated that the rejection of copper ions increases with increasing the charge valence of co-anions present in aqueous solution. The surfactant presented in aqueous solution was adsorbed by the membrane to form a secondary filtration layer on the membrane surface, therefore, influenced the surface charge characteristic of the membrane.

Choi et al. (2006) evaluated the application potential of nanofiltration membranes for the rejection of organic acids in wastewaters. The rejection of succinic and citric acids, which have molecular weights (M_Ws) larger than or closer to the molecular weight cutoffs (MWCOs) of employed NF membranes, was over 90% irrespective of operating pressure. Contrarily, the rejection of organic acids with M_Ws much smaller than MWCOs of the NF membranes increased gradually with increasing the applied pressure. The increase of DOC rejection with filtration time could be explained by increase of electrostatic repulsion between membrane and dissociated organic acid and by membrane fouling. NF process showed considerable potential as an advanced wastewater treatment process for removing organic acids in wastewaters.

Kim et al. (2007) reported the nitrate rejection of nanomembrane (NTR 729HF) from stainless steel industry. The results showed that the rejection rate of NF was decreased as pH decreased and Ca_2^+ concentration increased indicating that charge repulsion is one of the major rejection mechanisms.

Ortega et al. (2008) used two commercial nanofiltration membranes to remove metal ions from an acidic leachate solution generated from a contaminated soil using H_2SO_4 as a soil washing agent. Two types of thin-film commercial NF membranes, Desal5 DK and NF-270, were studied for their permeation and ionic selectivity. Desal5 DK is a polymeric membrane in which a polyamide selective layer is supported on a polysulfone layer. NF-270 is a semi-aromatic piperazine-based polyamide layer on top of a polysulphone micro-porous support reinforced with a polyester non-vowen backing layer. The results demonstrated the effectiveness and feasibility of the application of nanofiltration treatments in the cleaning-up of contaminated water residues generated during soil washing processes.

The application of high temperature resistant membranes is increasingly gaining attention in industry because they have many advantages comparing to most commercial polymer NF membranes which can only be applied under 45–50°C. High temperature resistant membranes can be used in the treatment of various hot fluid streams without strict temperature control. What is more, the enhanced flux due to high temperature may allow certain reduction of operating pressure, which further saves operating cost.

Tang and Chen (2002) used NF to treat textile wastewater which was highly colored with a high loading of inorganic salts. The results showed that the rejection of dye was 98% under an operating pressure of 500 kPa, and the NaCl rejection was less than 14%. Results also showed that NF is a very promising method for water reuse of textile industries.

Wu et al. (2009) prepared a novel thermal stable composite NF membrane by interfacial polymerization of piperazine (PIP) and trimesoyl chloride (TMC) on the substrate of thermal stable PPEA UF membrane. The purification experiments were accomplished effectively with a rejection of 99.3% for dyes Congo red (CGR) and Acid chrome blue K (ACBK) at 1.0 MPa, 80°C.

Voigt et al. (2001) reported a new TiO_2-NF ceramic membrane to decolor textile wastewater using an integrated pilot plan. The results show that it is possible to treat textile wastewater with dye retention varying from 70% to 100%, COD reduction of 45–80%, and salt retention of 10–80%.

Reverse osmosis is an effective technology to remove organic compounds from water bodies, especially for those that contain low concentration and low molecular weight organic compounds. Traditional RO membrane is limited due to high operational cost and maintenance as RO involves requirement of high pressure to the system and need extensive pretreatment. In addition, more and more current water and wastewater treatment plants are requested to higher their water recovery rate, which should be close to 100%. To overcome the limitations of RO, many researchers investigated an integrated membrane system (IMS) and many evidences indicated that NF membranes could be an alternative for the integration system (Nederlof et al., 2000; Huiting et al., 2001; Kimura et al., 2003b; Zhao et al., 2005; Bellona et al., 2007; Jacob et al., 2009; Simon et al., 2009). Since NF membranes are generally supplied in the same configurations as RO membranes, utilities could replace RO with NF spiral-wound elements without the need for significant additional capital investment.

A general schematic plot of an integrated membrane system (IMS) for tap-water supply designed and installed by KINTECH Technology Co. Ltd. is depicted in Figure 4.5. The plant located in southern Taiwan was established by Taiwan Water Corporation in 1972 and initially operated on conventional process consisted of coagulation, flocculation, setting, air stripping and sand filtration. To meet new water quality requirement, the plant was upgraded in 2007 by integrating IMS (UF-NF/RO, Figure 4.6) into the conventional system. A flexible operating mode was adopted by mixing sand filtration treated water or UF permeate with RO permeate to save the operating cost and meet the water quality standard items of turbidity (turb. < 0.2 mg/L), total hardness (TH < 150 mg/L as $CaCO_3$), total dissolved solid (TDS < 250 mg/L), *Escherichia coli* (E. coli = 0.0 CFU/100ml) and total trihalomethanes (TTHMs < 30μg/L). The full capacity of this plant is 303,400 m^3/day includes 170,000 m^3/day of integrated membrane system. The total land area of IMS is 1380 m^2 and the recovery of IMS is 90%. Table 4.2 shows the raw and treated water quality of Caotan water purification plant.

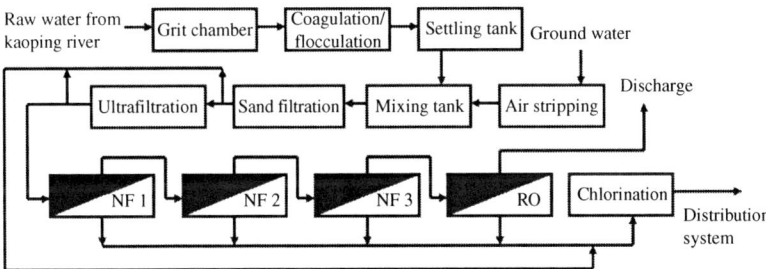

Figure 4.5 Schematic Flow Diagram of Caotan water purification plant showing the arrangement of UF, NF and RO membranes

Figure 4.6 Full-scale integrated membrane system (IMS) of Caotan water purification plant for tap water supply located at Kaohsiung County, Taiwan (a) Ultrafiltration system; (b) NF-LPRO system. The total land area of IMS is 1380 m^2. (Photos provided by KINTECH Technology Co., Ltd.)

Table 4.2 Water quality of Caotan water purification plant

River water	Ground water	RO permeate		Mixed treated water after chlorination
Turbidity (mg/L)	15–15000	2–5300	–	0.11
TH (mg/L as $CaCO_3$)	190–310	320–480	20	135
TDS (mg/L)	260–550	470–680	–	240
E. coli (CFU/100ml)	–	–	–	0
TTHMs (µg/L)	–	–	–	10

4.5 REFERENCES

Abdessemed, D., Nezzal G. and Ben Aim, R. (2002) Fractionation of a secondary effluent with membrane separation. *Desalination* **146**(1–3), 433–437.

Agbekodo, K. M., Legube B. and Cote, P. (1996a) Organics in NF permeate. *J. AWWA* **88**(5), 67–74.

Agbekodo, K. M., Legube, B. and Dard, S. (1996b) Atrazine and simazine removal mechanisms by nanofiltration: Influence of natural organic matter concentration. *Water Res.* **30**(11) 2535–2542.

Agenson, K. O., Oh, J.-H. and Urase, T. (2003) Retention of a wide variety of organic pollutants by different nanofiltration/reverse osmosis membranes: controlling parameters of process. *J. Membr. Sci.* **225**(1–2), 91–103.

Agnihotri, S., Rood, M. J. and Rostam-Abadi, M. (2005) Adsorption equilibrium of organic vapors on single-walled carbon nanotubes. *Carbon* **43**(11), 2379–2388.

AI-Ahmad, M. and Adbul Aleem, F. (1993) Scale formation and fouling problems effect on the performance of MSF and RO desalination, plants in Saudi Arabia. *Desalination* **93**(1–3), 287–310.

Alami-Younssi, S., Larbot, A., Persin, M., Sarrazin, J. and Cot, L. (1995) Rejection of mineral salts on a gamma alumina nanofiltration membrane: Application to environmental processes. *J. Membr. Sci.* **102**, 123–129.

Al-Amoudi, A. and Lovitt, R. W. (2007) Fouling strategies and the cleaning system of NF membranes and factors affecting cleaning efficiency. *J. Membr. Sci.* 303(1–2), 4–28.

Al-Amoudi, A. S. and Farooque, A. M. (2005) Performance, restoration and autopsy of NF membranes used in seawater pretreatment. *Desalination* **178**(1–3), 261–271.

Alborzfar, M., Escande, K. and Allen, S. J. (1998) Removal of natural organic matter from two types of humic ground waters by nanofiltration. *Water Res.* **32**(10), 2970–2983.

Al-Sofi, M. A. K., Hassan, A. M., Mustafa, G. M., Dalvi, A. G. I. and Kither, M. N. M. (1998) Nanofiltration as a means of achieving higher TBT of ⩾120 degrees C in MSF. *Desalination* **118**(1–3), 123–129.

Al-Sofi, M. A.-K. (2001) Seawater desalination – SWCC experience and vision. *Desalination* **135**(1–3), 121–139.

Ariza, M. J., Canas, A., Malfeito, J. and Benavente, J. (2002) Effect of pH on electrokinetic and electrochemical parameters of both sub-layers of composite polyamide/polysulfone membranes. *Desalination* **148**(1–3), 377–382.

Baker, J. S. and Dudley, L. Y. (1998) Biofouling in membrane systems – a review. *Desalination* **118**(1–3), 81–90.

Baticle, P., Kiefer, C., Lakhchaf, N., Larbot, A., Leclerc, O., Persin, M. and Sarrazin, J. (1997) Salt filtration on gamma alumina nanofiltration membranes fired at two different temperatures. *J. Membr. Sci.* **135**(1), 1–8.

Bellona, C. and Drewes, J. E. (2007) Viability of a low-pressure nanofilter in treating recycled water for water reuse applications: A pilot-scale study. *Water Res.* **41**(17), 3948–3958.

Bellona, C., Drewes, J. E., Xu, P. and Amy, G. (2004) Factors affecting the rejection of organic solutes during NF/RO treatment – a literature review. *Water Res.* **38**(12), 2795–2809.

Berg, P., Hagmeyer, G. and Gimbel, R. (1997) Removal of pesticides and other micro-pollutants by nanofiltration. *Desalination* **113**(2–3), 205–208.

Bergman, R. A. (1995) Membrane softening versus lime softening in Florida-a cost comparison update. *Desalination* **102**(1–3), 11–24.

Boeflage, S., Kennedy, M., Bonne, P. A. C., Galjaard G. and Schippers, J. (1997) Prediction of flux decline in membrane systems due to particulate fouling. *Desalination* **113**(2–3), 231–233.

Boerlage, S. F. E., Kennedy, M., Aniye, M. P. and Schippers, J. C. (2003) Applications of the MFI-UF to measure and predict particulate fouling in RO systems. *J. Membr. Sci.* **220**(1–2), 97–116.

Boussahel, R., Montiel, A. and Baudu, M. (2002) Effects of organic and inorganic matter on pesticide rejection by nanofiltration. *Desalination* **145**(1–3), 109–114.

Bowen, W. R. and Mohammad A. W. (1998) Diafiltration by nanofiltration: prediction and optimization. *AIChE J* **44**(8), 1799–1811.

Bowen, W. R., Welfoot, J. S. and Williams, M. (2002) Linearized transport model for nanofiltration: Development and assessment. *AIChE J* **48**(4), 760–771.

Brady-Estévez, A. S., Kang, S. and Elimelech, M. (2008) A single-walled-carbon-nanotube filter for removal of viral and bacterial pathogens. *Small* **4**(4), 481–484.

Braghetta, A., Digiano, F. A. and Ball, W. P. (1997) Nanofiltration of natural organic matter: pH and ionic strength effects. *J. Environ. Eng.* **123**(7), 628–640.

Brandhuber, P. and Amy, G. (1998) Alternative methods for membrane filtration of arsenic from drinking water. *Desalination* **117**(1–3), 1–10.

Brauns, E., Van Hoof, E., Molenberghs, B., Dotremont, C., Doyen, W. and Leysen, R. (2002) A new method of measuring and presenting the membrane fouling potential. *Desalination* **150**(1), 31–43.

Burggraaf, A. J. and Keizer, K. (1991) Synthesis of inorganic membranes. In Inorganic Membranes: Characterization and Applications (ed. Bhave, R. R.), Van Nostrand Rheinhold, New York, pp. 10–63.

Causserand, C., Aimar, P., Cravedi, J. P. and Singlande, E. (2005) Dichloroaniline retention by nanofiltration membranes. *Water Res.* **39**(8), 1594–1600.

Cfiscuoli A. and Drioli, E. (1999) Energetic and exergetic analysis of an integrated membrane desalination system. *Desalination* **124**(1–3), 243–249.

Childress, A. E. and Deshmukh, S. S. (1998) Effect of humic substances and anionic surfactants on the surface charge and performance of reverse osmosis membranes. *Desalination* **118**(1–3), 167–174.

Childress, A. E. and Elimelech, M. (1996) Effect of solution chemistry on the surface charge of polymeric reverse osmosis and nanofiltration membranes. *J. Membr. Sci.* **119**(2), 253–268.

Childress, A. E. and Elimelech, M. (2000) Relating nanofiltration membrane performance to membrane charge (electrokinetic) characteristics. *Environ. Sci. Technol.* **34**(17), 3710–3716.

Cho, J. W., Amy, G. and Pellegfino, J. (1999) Membrane filtration of natural organic matter: initial comparison of rejection and flux decline characteristics with ultrafiltration and nanofiltration membranes. *Water Res.* **33**(11), 2517–2526.

Choi, J.-H., Fukushi, K. and Yamamoto, K. (2008) A study on the removal of organic acids from wastewaters using nanofiltration membranes. *Sep. Purif. Technol.* **59**(1), 17–25.

Choi, J.-H., Jegal, J. and Kim, W.-N. (2006) Fabrication and characterization of multi-walled carbon nanotubes/polymer blend membranes. *J. Membr. Sci.* **284**(1–2), 406–415.

Choi, S., Yun, Z., Hong, S. and Ahn, K. (2001) The effect of co-existing ions and surface characteristics of nanomembranes on the removal of nitrate and fluoride. *Desalination* **133**(1), 53–64.

Chung, T.-S., Qin, J.-J., Huan, A. and Toh, K.-C. (2002) Visualization of the effect of shear rate on the outer surface morphology of ultrafiltration membranes by AFM. *J. Membr. Sci.* **196**(2), 251–266.

Conlon, W. J. and McClellan, S. A. (1989) Membrane softening: treatment process comes of age. *J. AWWA* **81**(11), 47–51.

Dai, Y., Jian, X., Zhang, S. and Guiver, M. D. (2002) Thin film composite (TFC) membranes with improved thermal stability from sulfonated poly(phthalazinone ether sulfone ketone) (SPPESK). *J. Membr. Sci.* **207**(2), 189–197.

De Witte, J. P. (1997) Surface water potabilisation by means of a novel nanofiltration element. *Desalination* **108**(1–3), 153–157.

Desai, T. A., Hansford, D. and Ferrari, M. (1999) Characterization of micromachnined silicon membranes for immunoisolation and bioseparation applications. *J. Membr. Sci.* **159**(1–2), 221–231.

Deshmukh, S. S. and Childress, A. E. (2001) Zeta potential of commercial RO membranes: Influence of source water type and chemistry. *Desalination* **140**(1), 87–95.

Di, Z.-C., Ding, J., Peng, X.-J., Li, Y.-H., Luan, Z.-K. and Liang, J. (2006) Chromium adsorption by aligned carbon nanotubes supported ceria nanoparticles. *Chemosphere* **62**(5), 861–865.

Drioli, E., Criscuoli A. and Curcioa, E. (2002) Integrated membrane operations for seawater desalination. *Desalination* **147**(1–3), 77–81.

Dueom, G. and Cabassud, C. (1999) Interests and limitations of nanofiltration for the removal of volatile organic compounds in drinking water production. *Desalination* **124**(1–3), 115–123.

Duran F. E. and Dunkelberger G. W. (1995) A comparison of membrane softening on three South Florida groundwaters. *Desalination* **102**(1–3), 27–34.

Elimelech, M., Zhu, X., Childress, A. E., and Hong, S. (1997) Role of membrane surface morphology in colloidal fouling of cellulose acetate and composite aromatic polyamide reverse osmosis membranes. *J. Membr. Sci.* **127**(1), 101–109.

El-Sheikh, A. H., Sweileh, J. A. and Al-Degs, Y. S. (2007) Effect of dimensions of multi-walled carbon nanotubes on its enrichment efficiency of metal ions from environmental waters. *Anal. Chim. Acta* **604**(2), 119–126.

Ericsson, B., Hallberg, M. and Wachenfeldt, J. (1996) Nanofiltration of highly colored raw water for drinking water production. *Desalination* **108**(1–3), 129–141.

Eriksson, P. (1988) Nanofiltration extends the range of membrane filtration. *Environ. Prog.* **7**(1), 58–62.

Escobar, I. C., Hong, S. and Randall, A. (2000) Removal of assimilable and biodegradable dissolved organic carbon by reverse osmosis and nanofiltration membranes. *J. Membr. Sci.* **175**(1), 1–17.

Everest, W. R. and Malloy, S. (2000) A design/build approach to deep aquifer membrane treatment in Southern California. *Desalination* **132**(1–3), 41–45.

Faller, K. A. (1999) Reverse Osmosis and Nanofiltration. *AWWA Manual of Water Supply Practice,* M46.

Flemming, H. C. (1993) Mechanistic aspects of reverse osmosis membrane biofouling and prevention. In Reverse Osmosis: Membrane Technology, Water Chemistry and Industrial Applications (ed. Amjad, Z.), Van Nostrand Reinhold, New York, pp. 163–209.

Freger, V., Arnot, A. C. and Howell, J. A. (2000) Separation of concentrated organic/in organic salt mixtures by nanofiltration. *J. Membr. Sci.* **178**(1–2), 185–193.

Fu, P., Ruiz, H., Lozier, J., Thompson, K. and Spangenberg, C. (1994) Selecting membranes for removing NOM and DBP precursors. *J. AWWA* **86**(12), 55–72.

Fu, P., Ruiz, H., Lozier, J., Thompson, K. and Spangenberg, C. (1995) A pilot study on groundwater natural organics removal by low-pressure membranes. *Desalination* **102**(1–3), 47–56.

Gauden, P. A., Terzyk, A. P., Rychlicki, G., Kowalczyk, P., Lota, K., Raymundo-Pinero, E., Frackowiak, E. and Beguin, F. (2006) Thermodynamic properties of benzene adsorbed in activated carbons and multi-walled carbon nanotubes. *Chem. Phys. Lett.* **421**(4–6), 409–414.

Ghosh, K. and Schnitzer, M. (1980) Macromolecular structures of humic substances. *Soil Sci.* **129**(5), 266–276.

Hafiarle, A., Lemordant, D. and Dhahbi, M. (2000) Removal of hexavalent chromium by nanofiltration. *Desalination* **130**(3), 305–312.

Hagmeyer, G. and Gimbel, R. (1998) Modelling the salt rejection of nanofiltration membranes for ternary ion mixtures and for single salts at different pH values. *Desalination* **117**(1–3), 247–256.

Hamed, O. A. (2005) Overview of hybrid desalination systems – current status and future prospects. *Desalination* **186**(1–3), 207–214.

Hassan, A. M., Al-Sofi, M. A. K., Al-Amoudi, A. S., Jamaluddin, A. T. M., Farooque, A. M., Rowaili, A., Dalvi, A. G. I., Kither, N. M., Mustafa, G. M. and Al-Tisan, I. A. R. (1998) A new approach to thermal seawater desalination processes using nanofiltration membranes (Part 1). *Desalination* **118**(1–3), 35–51.

Hassan, A. M., Farooque, A. M., Jamaluddin, A. T. M., Al-Amoudi, A. S., Al-Sofi, M. A., Al-Rubaian, A. F., Kither, N. M., Al-Tisan, I. A. R. and Rowaili, A. (2000) A demonstration plant based on the new NF-SWRO process. *Desalination* **131**(1–3), 157–171.

Hirose, M., Ito, H. and Kamiyama, Y. (1996) Effect of skin layer surface structures on the flux behavior of RO membrane. *J. Membr. Sci.* **121**(2), 209–215.

Huiting, H., Kappelhof, J. W. N. M. and Bosklopper, Th. G. J. (2001) Operation of NF/RO plants: from reactive to proactive. *Desalination* **139**(1–3), 183–189.

Hvid, K. B., Nielsen, P. S. and Stengaard, F. F. (1990) Preparation and characterization of a new ultrafiltration membrane. *J. Membr. Sci.* **53**(3), 189–202.

Iijima, S. (1991) Helical microtubules of graphitic carbon. *Nature* **354**, 56–58.

Jacob, M., Guigui, C., Cabassud, C., Darras, H., Lavison, G. and Moulin, L. (2009) Performances of RO and NF processes for wastewater reuse: Tertiary treatment after a conventional activated sludge or a membrane bioreactor. *Desalination* **250**(2), 833–839.

Jarusutthirak, C., Amy G. and Croué, J.-P. (2002) Fouling characteristics of wastewater effluent organic matter (EfOM) isolates on NF and UF membranes. *Desalination* **145**(1–3), 247–255.

Jegal, J., Min, S. G. and Lee, K.-H. (2002) Factors affecting the interfacial polymerization of polyamide active layers for the formation of polyamide composite membranes. *J. Appl. Polym. Sci.* **86**(11), 2781–2787.

Jian, X., Dai, Y., He, G. and Chen, G. (1999) Preparation of UF and NF poly(phthalazine ether sulfone ketone) membranes for high temperature application. *J. Membr. Sci.* **161**(1–2), 185–191.

Jönsson, A. and Jönsson, B. (1991) The influence of nonionic surfactants on hydrophobic and ultrafiltration membranes. *J. Membr. Sci.* **56**(1), 49–76.

Jung, Y. J., Kiso, Y., Othman, R. A. A. B., Ikeda, A., Nishimura, K., Min, K. S., Kumano, A. and Ariji, A. (2005) Rejection properties of aromatic pesticides with a hollow-fiber NF membrane. *Desalination* **180**(1–3), 63–71.

Karime, M., Bouguecha, S. and Hamrouni, B. (2008) RO membrane autopsy of Zarzis brackish water desalination plant. *Desalination* **220**(1–3), 258–266.

Kartinen, E. O. and Martin, C. J. (1995) An overview of arsenic removal processes. *Desalination* **103**(1–2), 79–88.

Keszler, B., Kovács, G., Tóth, A., Bertóti, I. and Hegyi, M. (1991) Modified polyethersulfone membranes. *J. Membr. Sci.* **62**(2), 201–210.

Khalik, A. and Praptowidodo, V. S. (2000) Nanofiltration for drinking water production from deep well water. *Desalination* **132**(1–3), 287–292.

Khirani, S., Ben Aim, R. and Manero, M.-H. (2006) Improving the measurement of the Modified Fouling Index using nanofiltration membranes (NF–MFI). *Desalination* **191**(1–3), 1–7.

Kim, I. C., Lee, K.-H. and Tak, T.-M. (2001) Preparation and characterization of integrally skinned uncharged polyetherimide asymmetric nanofiltration membrane. *J. Membr. Sci.* **183**(2), 235–247.

Kim, I.-C., Jegal, J. and Lee, K.-H. (2002) Effect of aqueous and organic solutions on the performance of polyamide thin-film-composite nanofiltration membranes. *J. Polym. Sci.: Part B* **40**(19), 2151–2163.

Kim, K. J., Fane, A. G. and Fell, C. J. D. (1988) The performance of ultrafiltration membranes pretreated by polymers. *Desalination* **70**(1–3), 229–249.

Kim, K. J., Fane, A. G. and Fell, C. J. D. (1989) The effect of Langmuir Blodgett layer pretreatment on the performance of ultrafiltration membranes. *J. Membr. Sci.* **43**(2–3), 187–204.

Kim, M., Saito, K. and Furusaki, S. (1991) Water flux and protein adsorption of a hollow fiber modified with hydroxyl groups. *J. Membr. Sci.* **56**(3), 289–302.
Kim, Y.-H., Hwang, E.-D., Shin, W. S., Choi, J.-H., Ha, T. W. and Choi, S. J. (2007) Treatments of stainless steel wastewater containing a high concentration of nitrate using reverse osmosis and nanomembranes. *Desalination* **202**(1–3), 286–292.
Kimura, K., Amy, G., Drewes, J. and Watanabe, Y. (2003b) Adsorption of hydrophobic compounds onto NF/RO membranes: An artifact leading to overestimation of rejection. *J. Membr. Sci.* **221**(1–2), 89–101.
Kimura, K., Amy, G., Drewes, J., Heberer, T., Kim, T.-U. and Watanabe, Y. (2003a) Rejection of organic micropollutants (disinfection by-products, endocrine disrupting compounds, and pharmaceutically active compounds) by NF/RO membranes. *J. Membr. Sci.* **227**(1–2), 113–121.
Kiso, Y., Kitao, T., Kiyokatsu, J. and Miyagi M. (1992) The effects of molecular width on permeation of organic solute through cellulose acetate reverse osmosis membrane. *J. Membr. Sci.* **74**(1–2), 95–103.
Kiso, Y., Kon, T., Kitao, T. and Nishimura, K. (2001a) Rejection properties of alkyl phthalates with nanofiltration membranes. *J. Membr. Sci.* **182**(1–2), 205–214.
Kiso, Y., Nishimura, Y., Kitao, T. and Nishimura, K. (2000) Rejection properties of non-phenylic pesticides with nanofiltration membranes. *J. Membr. Sci.* **171**(2), 229–237.
Kiso, Y., Sugiura, Y., Kitao, T. and Nishimura, K. (2001b) Effects of hydrophobicity and molecular size on rejection of aromatic pesticides with nanofiltration membranes. *J. Membr. Sci.* **192**(1–2), 1–10.
Košutić, K. and Kunst, B. (2002) Removal of organics from aqueous solutions by commercial RO and NF membranes of characterized porosities. *Desalination* **142**(1), 47–56.
Košutić, K., Furač, L., Sipos, L. and Kunst, B. (2005) Removal of arsenic and pesticides from drinking water by nanofiltration membranes. *Sep. Purif. Technol.* **42**(2), 137–144.
Košutić, K., Kaštelan-Kunst, L. and Kunst, B. (2000) Porosity of some commercial reverse osmosis and nanofiltration polyamide thin-film composite membranes. *J Membr Sci* **168**(1–2), 101–108.
Ku, Y., Chen, S.-W. and Wang, W.-Y. (2005) Effect of solution composition on the removal of copper ions by nanofiltration. *Sep. Purif. Technol.* **43**(2), 135–142.
Kumakiri, I., Yamaguchi, T. and Nakao, S. (2000) Application of a zeolite A membrane to reverse osmosis process. *J. Chem. Eng. Jpn.* **33**, 333.
Larbot, A., Alami-Younssi, S., Persin, M., Sarrazin, J. and Cot, L. (1994) Preparation of a γ-alumina nanofiltration membrane. *J. Membr. Sci.* **97**, 167–173.
Le Roux, I., Krieg, H. M., Yeates, C. A. and Breytenbach, J. C. (2005) Use of chitosan as an antifouling agent in a membrane bioreactor. *J. Membr. Sci.* **248**(1–2), 127–136.
Lee, S., Park, G., Amy, G., Hong, S.-K., Moon, S.-H., Lee, D.-H. and Cho, J. (2002) Determination of membrane pore size distribution using the fractional rejection of nonionic and charged macromolecules. *J. Membr. Sci.* **201**(1–2), 191–201.
Lee, S., Quyet, N., Lee, E., Kim, S., Lee, S., Jung, Y. D., Choi, S. H. and Cho, J. (2008) Efficient removals of tris(2-chloroethyl) phosphate (TCEP) and perchlorate using NF membrane filtrations. *Desalination* **221**(1–3), 234–237.

Levine, B. B., Madireddi, K., Lazarova, V., Stenstrom, M. K. and Suffet, M. (1999) Treatment of trace organic compounds by membrane processes: At the lake arrowhead water reuse pilot plant. *Water. Sci. Technol.* **40**(4–5), 293–301.
Li, L., Dong, J., Nenoff, T. M. and Lee, R. (2004a) Desalination by reverse osmosis using MFI zeolite membranes. *J. Membr. Sci.* **243**(1–2), 401–404.
Li, L., Dong, J., Neoff, T. M. and Lee, R. (2004b) Reverse osmosis of ionic aqueous solutions on a MFI zeolite membrane. *Desalination* **170**(3), 309–316.
Li, Y.-H., Wang, S., Wei, J., Zhang, X., Xu, C., Luan, Z., Wu, D. and Wei, B. (2002) Lead adsorption on carbon nanotubes. *Chem. Phys. Lett.* **357**(3–4), 263–266.
Li, Y.-H., Wang, S., Zhang, X., Wei, J., Xu, C., Luan, Z. and Wu, D. (2003) Adsorption of fluoride from water by aligned carbon nanotubes. *Mater. Res. Bull.* **38**(3), 469–476.
Lin, J. and Murad, S. (2001a) The role of external electric fields in membrane-based separation processes: A molecular dynamics study. *Mol. Phys.* **99**(5), 463–469.
Lin, J. and Murad, S. (2001b) A computer simulation study of the separation of aqueous solution using thin zeolite membranes. *Mol. Phys.* **99**(14), 1175–1181.
Long, R. Q. and Yang, R. T. (2001) Carbon nanotubes as superior sorbent for dioxin removal. *J. Am. Chem. Soc.* **123**(9), 2058–2059.
Lu, C., Chung, Y.-L. And Chang, K.-F. (2005) Adsorption of trihalomethanes from water with carbon nanotubes. *Water Res.* **39**(6), 1183–1189.
Lu, X., Bian, X. and Shi, L. (2002) Preparation and characterization of NF composite membrane. *J. Membr. Sci.* **210**(1), 3–11.
Luyten, J., Cooymans, J., Smolders, C., Vercauteren, S., Vansant, E. F. and R. Leysen (1997) Shaping of multilayer ceramic membranes by dip-coating. *J. Eur. Ceram. Soc.* **17**(2–3), 273–279.
Madireddi, K., Babcock, R. W., Levine, B., Huo, T. L., Khan, E., Ye, Q. F., Neethling, J. B., Suffet, I. H. and Stenstrom, M. K. (1997) Wastewater reclamation at Lake Arrowhead, California: An overview. *Water Environ. Res.* **69**(3), 350–362.
Majewska-Nowak, K., Kabsch-Korbutowicz, M., Dodź M. and Winnicki, T. (2002) The influence of organic carbon concentration on atrazine removal by UF membranes. *Desalination* **147**(1–3), 117–122.
Martin, F., Walczak, R., Boiarski, A., Cohen, M., West, T., Cosentino, C. and Ferrari, M. (2005) Tailoring width of microfabricated nanochannels to solute size can be used to control diffusion kinetics. *J. Control. Release* **102**(1), 123–133.
Mänttäri, M., Puro, L., Nuortila-Jokinen, J. and Nyström, M. (2000) Fouling effects of polysaccharides and humic acid in nanofiltration. *J. Membr. Sci.* **165**(1), 1–17.
Mohammad, A. W. and Ali, N. (2002) Understanding the steric and charge contributions in NF membranes using increasing MWCO polyamide membranes. *Desalination* **147**(1–3), 205–212.
Mohesn, M. S., Jaber J. O. and Afonso, M. D. (2003) Desalination of brackish water by nanofiltmtion and reverse osmosis. *Desalination* **157**(1–3), 167.
Montovay, T., Assenmacher, M. and Frimmel, F. H. (1996) Elimination of pesticides from aqueous solution by nanofiltration. *Magyar Kémiai Folyóirat* **102**(5), 241–247.
Moon, E. J., Seo, Y. S. and Kim, C. K. (2004) Novel composite membranes prepared from 2,2 bis [4-(2-hydroxy-3-methacryloyloxy propoxy) phenyl] propane, triethylene glycol dimethacrylate, and their mixtures for the reverse osmosis process. *J. Membr. Sci.* **243**(1–2), 311–316.

Murad, S. and Nitche, L. C. (2004) The effect of thickness, pore size and structure of a nanomembrane on the flux and selectivity in reverse osmosis separations: A molecular dynamics study. *Chem. Phys. Lett.* **397**(1–3), 211–215.

Murad, S., Jia, W. and Krishnamurthy, M. (2004) Ion-exchange of monovalent and bivalent cations with NaA zeolite membranes: a molecular dynamics study. *Mol. Phys.* **102**(19), 2103–2112.

Murad, S., Oder, K. and Lin, J. (1998) Molecular simulation of osmosis, reverse osmosis, and electro-osmosis in aqueous and methanolic electrolyte solutions. *Mol. Phys.* **95**(3), 401–408.

Nederlof, M. M., Kruithof, J. C., Taylor, J. S., Van Der Kooij, D. and Schippers, J. C. (2000) Comparison of NF/RO membrane performance in integrated membrane systems. *Desalination* **131**(1–3), 257–269.

Nghiem, L. D., Schäfer, A. I. and Elimilech M. (2004) Removal of natural hormones by nanofiltration membranes: Measurement, modeling, and mechanisms. *Environ. Sci. Technol.* **38**(6), 1888–1896.

Nghiem, L. D., Schäfer, A. I. and Waite, T. D. (2002) Adsorption of estrone on nanofiltration and reverse osmosis membranes in water and wastewater treatment. *Water Sci. Technol.* **46**(4–5), 265–272.

Nunes, S. R. and Peinemann, K. V. (2001) In *Membrane Technology in the Chemical Industry*, 1st edn (eds. Nunes, S. R. and Peinemann, K. V.), Wiley-VCH, Germany, pp. 1–53.

Nystrom, M. and Jarvinen, P. (1991) Modification of polysulfone ultrafiltration membranes with UV irradiation and hydrophilicity increasing agents. *J. Membr. Sci.* **60**(2–3), 275–296.

Ortega, L. M., Lebrun, R., Blais, J.-F. and Hausler, R. (2008) Removal of metal ions from an acidic leachate solution by nanofiltration membranes. *Desalination* **227**(1–3), 204–216.

Otaki, M., Yano, K. and Ohgaki, S. (1998) Virus removal in a membrane separation process. *Water Sci. Technol.* **37**(10), 107–116.

Ozaki, H. and Li, H. (2002) Rejection of organic compounds by ultralow pressure reverse osmosis membrane. *Water Res.* **36**(1), 123–130.

Paulose, M., Peng, L., Popat, K. C., Varghese, O. K., LaTempa, T. L., Bao, N., Desai, T. A. and Grimes, C. A. (2008) Fabrication of mechanically robust, large area, polycrystalline nanotubular/ porous TiO_2 membranes. *J. Membr. Sci.* **319**(1–2), 199–205.

Peng, X., Li, Y., Luan, Z., Di, Z., Wang, H., Tian, B. and Jia, Z. (2003) Adsorption of 1,2-dichlorobenzene from water to carbon nanotubes. *Chem. Phys. Lett.* **376**(1–2), 154–158.

Petersen, R. J. (1993) Composite reverse osmosis and nanofiltration membranes. *J. Membr. Sci.* **83**(1), 81–150.

Plakas, K. V., Karabelas, A. J., Wintgens, T. and Melin, T. (2006) A study of selected herbicides retention by nanofiltration membranes – The role of organic fouling. *J. Membr. Sci.* **284**(1–2), 291–300.

Pontié, M., Diawara, C., Rumeau, M., Aurean D. and Hemmerey, P. (2003) Seawater nanofiltration (NF): Fiction or reality? *Desalination* **158**(1–3), 277–280.

Rabie, H. R., Côté P. and Adams, N. (2001) A method for assessing membrane fouling in pilot- and full-scale systems. *Desalination* **141**(3), 237–243.

Rao, A. P., Desai, N. V. and Rangarajan, R. (1997) Interfacially synthesized thin film composite RO membranes for seawater desalination. *J. Membr. Sci.* **124**(2), 263–272.
Redondo, J. A. (2001) Lanzarote IV, a new concept for two-pass SWRO at low O&M cost using the new high-flow FILMTEC SW30-380. *Desalination* **138**(1–3), 231–236.
Reiss, C. R., Taylor, J. S. and Robert, C. (1999) Surface water treatment using nanofiltration – pilot testing results and design considerations. *Desalination* **125**(1–3), 97–112.
Ridgway, H. F. and Flemming, H. F. (1996) Membrane biofouling. In *Water Treatment Membrane Processes* (eds. Mallevialle, J. Odendaal, P. E. and Wiesner, M. R.), McGraw-Hill, New York, pp. 6.1–6.62.
Roh, I. J., Park, S. Y., Kim, J. J. and Kim, C. K. (1998) Effects of the polyamide molecular structure on the performance of reverse osmosis membranes. *J. Polym. Sci.: Part B* **36**(11), 1821–1830.
Roorda J. H. and Van der Graaf, J. H. J. M. (2001) New parameter for monitoring fouling during ultrafiltration of WWTP effluent. *Water Sci. Technol.* **43**(10), 241–248.
Roudman, A. R. and DiGiano, F. A. (2000) Surface energy of experimental and commercial nanofiltration membranes: Effects of wetting and natural organic matter fouling. *J. Membr. Sci.* **175**(1), 61–73.
Saitúa, H., Campderrós, M., Cerutti S. and Pérez Padilla, A. (2005) Effect of operating conditions in removal of arsenic from water by nanofiltration membrane. *Desalination* **172**(2), 173–180.
Sato, Y., Kang, M., Kamei T. and Magara, Y. (2002) Performance of nanofiltration for arsenic removal. *Water Res.* **36**(13), 3371–3377.
Schäfer, A. I., Fane, A. G. and Waite, T. (1998) Nanofiltration of natural organic matter: Removal, fouling and the influence of multivalent ions. *Desalination* **118**(1–3), 109–122.
Schäfer, A. I., Mastrup, M. and LundJensen, R. (2002) Particle interactions and removal of trace contaminants from water and wastewaters. *Desalination* **147**(1–3), 243–250.
Schäfer, A. I., Pihlajamäki, A., Fane, A. G., Waite, T. D. and Nyström M. (2004) Natural organic matter removal by nanofiltration: Effects of solution chemistry on retention of low molar mass acids versus bulk organic matter. *J. Membr. Sci.* **242**(1–2), 73–85.
Schippers, J. C. and Verdouw, J. (1980) The modified fouling index, a method of determining the fouling characteristics of water. *Desaleation* **32**, 137–148.
Schneider, R. P., Ferreira, L. M., Binder, P., Bejarano, E. M., Góes, K. P., Slongo, E., Machado, C. R. and Rosa, G. M. Z. (2005) Dynamics of organic carbon and of bacterial populations in a conventional pretreatment train of a reverse osmosis unit experiencing severe biofouling. *J. Membr. Sci.* **266**(1–2), 18–29.
Schutte, C. F. (2003) The rejection of specific organic compounds by reverse osmosis membranes. *Desalination* **158**(1–3), 285–294.
Seidel, A., Waypa, J. J. and Elimech, M. (2001) Role of charge (Donnan) exclusion in removal of arsenic from water by a negatively charged porous nanofiltration membrane. *Environ. Eng. Sci.* **18**(2), 105–113.
Semiat, R. (2000) Desalination: Present and future. *Water Internet.* **25**(1), 54–65.
Shih, M. C. (2005) An overview of arsenic removal by pressure-driven membrane processes. *Desalination* **172**(1), 85–97.

Shim, Y., Lee, H.-G., Lee, S., Moon, S.-H. and Cho, J. (2002) Effects of natural organic matter and ionic species on membrane surface charge. *Environ. Sci. Technol.* **36**(17), 3864–3871.

Sikora, J., Hansson C. H. and Ericsson, B. (1989) Pre-treatment and desalination of mine drainage water in a pilot plant. *Desalination* **75**, 363–373.

Simon, A., Nghiem, L. D., Le-Clech, P., Khan, S. J. and Drewes, J. E. (2009) Effects of membrane degradation on the removal of pharmaceutically active compounds (PhACs) by NF/RO filtration processes. *J. Membr. Sci.* **340**(1–2), 16–25.

Sombekke, H. D. M., Voorhoeve, D. K. and Hiemstra, P. (1997) Environmental impact assessment of groundwater treatment with nanofiltration. *Desalination* **113**(2–3), 293–296.

Song, Y. J., Liu, F. and Sun, B. H. (2005) Preparation, characterization, and application of thin film composite nanofiltration membranes. *J. Appl. Polym. Sci.* **95**(3), 1251–1261.

Soria, R. and Cominotti, S. (1996) Nanofiltration ceramic membrane. In *Proc. of the International Conference on the Membranes and Membrane Processes,* Yokohama, Japan, 18–23 August.

Srivastava, A., Srivastava, O. N., Talapatra, S., Vajtai, R. and Ajayan, P. M. (2004) Carbon nanotube filters. *Nat. Mater.* **3**, 610–614.

Stengaard, F. F. (1988) Characteristics and performance of new types of ultrafiltration membranes with chemically modified surfaces. *Desalination* **70**(1–3), 207–224.

Tang, A. and Chen, V. (2002) Nanofiltration of textile wastewater for water reuse. *Desalination* **143**(1), 11–20.

Tang, C., Zhang, Q., Wang, K., Fu, Q. and Zhang, C. (2009) Water transport behavior of chitosan porous membranes containing multi-walled carbon nanotubes (MWNTs). *J. Membr. Sci.* **337**(1–2), 240–247.

Tanninen, J. and Nystrom, M. (2002) Separation of ions in acidic conditions using NF. *Desalination* **147**(1–3), 295–299.

Tasaka, K., Katsura, T., Iwahori, H. and Kamiyama, Y. (1994) Analysis of RO elements operated at more than 80 plants in Japan. *Desalination* **96**(1–3), 259–272.

Tay, J.-H., Liu, J. and Sun, D. D. (2002) Effect of solution physico-chemistry on the charge property of nanofiltration membranes. *Water Res.* **36**(3), 585–598.

Terrones, M. (2004) Carbon nanotubes: Synthesis and properties, electronic devices and other emerging applications. *Int. Mater. Rev.* **49**(6), 325–377.

Tödtheide, V., Laufenberg, G. and Kunz, B. (1997) Waste water treatment using reverse osmosis: Real osmotic pressure and chemical functionality as influencing parameters on the retention of carboxylic acids in multi-component systems. *Desalination* **110**(3), 213–222.

Ulbricht, M., Matuschewski, H., Oechel, A. and Hicke, H. G. (1996) Photo-induced graft polymerization surface modification for the preparation of hydrophilic and low-protein-adsorbing ultrafiltration membranes. *J. Membr. Sci.* **115**(1), 31–47.

Urase, T., Oh, J. and Yamamoto, K. (1998) Effect of pH on rejection of different species of arsenic by nanofiltration. *Desalination* **117**(1–3), 11–18.

Urase, T., Yamamoto, K. and Ohgaki, S. (1996) Effect of pore structure of membranes and module configuration on virus retention. *J. Membr. Sci.* **115**(1), 21–29.

Van der Bruggen, B. and Vandecasteele, C. (2001b) Flux decline during nanofiltration of organic components in aqueous solution. *Environ. Sci. Technol.* **35**(17), 3535–3540.

Van der Bruggen, B. and Vandecasteele, C. (2002) Modeling of the retention of uncharged molecules with nanofiltration. *Water Res.* **36**(5), 1360–1368.
Van der Bruggen, B. and Vandecasteele, C. (2003) Removal of pollutants from surface water andgroundwater by nanofiltration: Overview of possible applications in the drinking water industry *Environ. Poll.* **122**(3), 435–445.
Van der Bruggen, B., Braeken, L. and Vandecasteele, C. (2002a) Evaluation of parameters describing flux decline in nanofiltration of aqueous solutions containing organic compounds. *Desalination* **147**(1–3), 281–288.
Van der Bruggen, B., Braeken, L. and Vandecasteele, C. (2002b) Flux decline in nanofiltration due to adsorption of organic compounds. *Sep. Purif. Technol.* **29**(1), 23–31.
Van der Bruggen, B., Everaert, K., Wilms, D. and Vandecasteele, C. (2001a) Application of nanofiltration for removal of pesticides, nitrate and hardness from groundwater: Rejection properties and economic evaluation. *J. Membr. Sci.* **193**(2), 239–248.
Van der Bruggen, B., Mänttäri, M. and Nyström, M. (2008) Drawbacks of applying nanofiltration and how to avoid them: A review. *Sep. Purif. Technol.* **63**(2), 251–263.
Van der Bruggen, B., Schaep, J., Maes, W., Wilms, D. and Vandecasteele, C. (1998) Nanofiltration as a treatment method for the removal of pesticides from ground waters. *Desalination* **117**(1–3), 139–147.
Van der Bruggen, B., Schaep J., Wilms D. and Vandecasteele C. (1999) Influence of molecular size, polarity and charge on the retention of organic molecules by nanofiltration. *J. Membr. Sci.* **156**(1), 29–41.
Van Gestel, T., Vandecasteele, C., Buekenhoudt, A., Dotremont, C., Luyten, J., Leysen, R., Van der Bruggen, B. and Maes, G. (2002a) Alumina and titania multilayer membranes for nanofiltration: Preparation, characterization and chemical stability. *J. Membr. Sci.* **207**(1), 73–89.
Van Gestel, T., Vandecasteele, C., Buekenhoudt, A., Dotremont, C., Luyten, J., Leysen, R., Van der Bruggen, B. and Maes, G. (2002b) Salt retention in nanofiltration with multilayer ceramic TiO_2 membranes. *J. Membr. Sci.* **209**(2), 379–389.
Van Hoop, S. C. J. M., Minnery, J. G. and Mack, B. (2001) Dead-end ultrafiltration as alternative pre-treatment to reverse osmosis in seawater desalination: A case study. *Desalination* **139**(1–3), 161–168.
Verissimo, S., Peinemann, K.-V. and Bordado, J. (2005) Thin-film composite hollow fiber membranes: An optimized manufacturing method. *J. Membr. Sci.* **264**(1–2), 48–55.
Verliefde, A., Cornelissen, E., Amy, G., Van der Bruggen, B. and Van Dijk, H. (2007) Priority organic micropollutants in water sources in Flanders and the Netherlands and assessment of removal possibilities with nanofiltration. *Environ. Pollut.* **146**(1), 281–289.
Visvanathan, C., Marsono B. and Basu, B. (1998) Removal of THMP by nanofiltration: Effects of interference parameters. *Water Res.* **32**(12), 3527–3538.
Voigt, I., Stahn, M., Wöhner, St., Junghans, A., Rost, J. and Voigt, W. (2001) Integrated cleaning of coloured wastewater by ceramic NF membranes. *Sep. Purifi. Technol.* **25**(1–3), 509–512.
Vrouwenvelder, J. S., Kappelhof, J. W. N. M., Heijman, S. G. J., Schippers J. C. and Van der Kooija, D. (2003) Tools for fouling diagnosis of NF and RO membranes and assessment of the fouling potential of feed water. *Desalination* **157**(1–3), 361–365.

Wang, X. F., Chen, X. M., Yoon, K., Fang, D. F., Hsiao, B. S. and Chu, B. (2005) High flux filtration medium based on nanofibrous substrate with hydrophilic nanocomposite coating. *Environ. Sci. Technol.* **39**(19), 7684–7691.

Wang, X.-L., Tsuru, T., Nakao, S.-I. and Kimura, S. (1997) The electrostatic and steric-hindrance model for the transport of charged solutes through nanofiltration membranes. *J. Membr. Sci.* **135**(1), 19–32.

Wang, X.-L., Wang, W.-N. and Wang, D.-X. (2002) Experimental investigation on separation performance of nanofiltration membranes for inorganic electrolyte solutions. *Desalination* **145**(1–3), 115–122.

Waniek, A., Bodzek, M. and Konieczny, K. (2002) Trihalomethanes removal from water using membrane processes. *Polish J. Environ. Studies* **11**(2), 171–178.

Waypa, J. J., Elimelech, M. and Hering, J. G. (1997) Arsenic removal by RO and NF membranes. *J. AWWA* **89**(10), 102–114.

Wintgens, T., Gallenkemper, M. and Melin, T. (2003) Occurrence and removal of endocrine disrupters in landfill leachate treatment plants. *Water Sci. Technol.* **48**(3), 127–134.

Wittmann, E., Cote, P., Medici, C., Leech, J. and Turner, A. G. (1998) Treatment of a hard borehole water containing low levels of pesticide by nanofiltration. *Desalination* **119**(1–3), 347–352.

Wu, C., Zhang, S., Yang, D. and Jian, X. (2009) Preparation, characterization and application of a novel thermal stable composite nanofiltration membrane. *J. Membr. Sci.* **326**(2), 429–434.

Xia, S. J., Dong, B. Z., Zhang, Q. L., Xu, B., Gao, N. Y. and Causserand, C. (2007) Study of arsenic removal by nanofiltration and its application in China. *Desalination* **204**(1–3), 374–379.

Xu, P., Drewes, J. E., Bellona, C., Amy, G., Kim, T.-U., Adam, M., and Heberer, T. (2005) Rejection of emerging organic micropollutants in nanofiltration-reverse osmosis membrane applications. *Water Environ. Res.* **77**(1), 40–48.

Xu, P., Drewes, J. E., Kim, T.-U., Bellona, C. and Amy, G. (2006) Effect of membrane fouling on transport of organic contaminants in NF/RO membrane applications. *J Membr. Sci.* **279**(1–2), 165–175.

Xu, Y. and Lebrun, R. E. (1999) Investigation of the solute separation by charged nanofiltration membrane: Effect of pH, ionic strength and solute type. *J Membr. Sci.* **158**(1–2), 93–104.

Yahya, M. T., Blu, C. B. and Gerha, C. P. (1993) Virus removal by slow sand filtration and nanofiltration. *Water Sci. Technol.* **27**(3–4), 445–448.

Yamagishi, H., Crivello, J. V. and Belfort, G. (1995) Development of a novel photochemical technique for modifying poly(arysulfone) ultrafiltration membranes. *J. Membr. Sci.* **105**(3), 237–247.

Yang, K., Zhu, L. and Xing, B. (2006) Adsorption of polycyclic aromatic hydrocarbons by carbon nanomaterials, *Environ. Sci. Technol.* **40**(6), 1855–1861.

Yeh, H.-H., Tseng, I-C., Kao, S.-J., Lai, W.-L., Chen, J.-J., Wang, G. T. and Lin, S.-H. (2000) Comparison of the finished water quality among an integrated membrane process, conventional and other advanced treatment processes. *Desalination* **131**(1–3), 237–244.

Ying, L., Zhai, G., Winata, A. Y., Kang, E. T. and Neoh, K. G. (2003) pH effect of coagulation bath on the characteristics of poly(acrylic acid)-grafted and poly(4-vinylpyridine)-grafted poly(vinylidene fluoride) microfiltration membranes. *J. Colloid Interface Sci.* **265**(2), 396–403.

Yoon, S.-H., Lee, C.-H., Kim, K.-J. and Fane, A. G. (1998) Effect of calcium ion on the fouling of nanofilter by humic acid in drinking water production. *Water Res.* **32**(7), 2180–2186.

Yoon, Y., Amy, G., Cho, J., Her, N. and Pellegrino, J. (2002) Transport of perchlorate (ClO_4^-) through NF and UF membranes. *Desalination* **147**(1–3), 11–17.

Yoon, Y., Westerhoff, P., Snyder, S. A., Wert, E. C. and Yoon, J. (2007) Removal of endocrine disrupting compounds and pharmaceuticals by nanofiltration and ultrafiltration membranes. *Desalination* **202**(1–3), 16–23.

Zhang, H., Quan, X., Chen, S., Zhao, H. and Zhao, Y. (2006) The removal of sodium dodecylbenzene sulfonate surfactant from water using silica/titania nanorods/nanotubes composite membrane with photocatalytic capability. *Appl. Surf. Sci.* **252**(24), 8598–8604.

Zhang, X., Du, A. J., Lee, P., Sun, D. D. and Leckie, J. O. (2008) Grafted multifunctional titanium dioxide nanotube membrane: Separation and photodegradation of aquatic pollutant. *Appl. Catal. B* **84**(1–2), 262–267.

Zhang, Y., Causserand, C., Aimar, P. and Cravedi, J. P. (2006) Removal of bisphenol A by a nanofiltration membrane in view of drinking water production. *Water Res.* **40**(20), 3793–3799.

Zhao, Y., Taylor, J. and Hong, S. (2005) Combined influence of membrane surface properties and feed water qualities on RO/NF mass transfer, a pilot study. *Water Res.* **39**(7), 1233–1244.

Chapter 5
Physico-chemical treatment of Micropollutants: Adsorption and ion exchange

N. Chubar

5.1 INTRODUCTION

Adsorptive phenomena are widely spread throughout nature. Mountain rocks and soils are just huge columns filled with adsorbents through which the water and gaseous solution flow. Lung tissue behaves very similarly to an adsorbent; it is a carrier for blood hemoglobin providing the transfer of oxygen to the organism. Many functions of the biological membranes of living cells are linked to their surface properties; for instance: the total square of biologically active membranes in organisms reaches a few thousand square metres. Even the senses of taste and smell of human beings depend on adsorption of the related molecules by the surface of the nose and tongue.

©2010 IWA Publishing. *Treatment of Micropollutants in Water and Wastewater*. Edited by Jurate Virkutyte, Veeriah Jegatheesan and Rajender S. Varma. ISBN: 9781843393160. Published by IWA Publishing, London, UK.

The complexity of the adsorptive phenomena and a variety of their applications in nature, society, sciences, and technology are reflected in slightly different definitions of adsorption which point at the process, its nature, or its application. *Sci-Tech Encyclopedia* writes that "adsorption is a process in which atoms or molecules move from a bulk phase (that is, solid, liquid, or gas) onto a solid or liquid surface". *Dental dictionary* explains that "adsorption is a natural process whereby molecules of a gas or liquid adhere to the surface of a solid". The *Britannica Concise Encyclopedia* defines adsorption as a capability of a solid substance (adsorbent) to attract to its surface molecules of a gas or solution (adsorbate) with which it is in contact. *Architecture dictionary* explains adsorption as "the action of a material in extracting a substance from the atmosphere (or a mixture of gases and liquids) and gathering it on the surface in a condensed layer". *Geography dictionary* gives a more practical: "... in soil science, Adsorption is the addition of ions or molecules to the electrically charged surface of a particle of clay or humus". *Veterinary dictionary* defines briefly that "adsorption is the action of a substance in attracting and holding other materials or particles on its surface". And the most practical definition of adsorption is given by business oriented website (TeachMeFinance.com) showing mostly the technological side of adsorption which is "[the] removal of a pollutant from air or water by collecting the pollutant on the surface of a solid material; e.g., an advanced method of treating waste in which activated carbon removes organic matter from waste-water".

The list of definitions is much longer but the few mentioned above are sufficient to conclude that adsorption is complex process involving different reactions (mechanisms); it is widespread in the environment, human society (medicine), and technological processes.

Going a little bit deeper on the scientific level we have to add that adsorption is a concentrating of gaseous or dissolved substances at the interface between two phases: on the surface of a solid or liquid. It is one of the surface phenomena taking place at the interface due to an excess of free energy. Adsorption is a spontaneous process which is accompanied by a decrease of Gibbs energy (at p = const.) or a decrease of Helmholtz energy (at V = const.). However, there is no equalization of concentrations in the whole volume. On the contrary, the difference between the concentrations in the gas/liquid and solid phases increases. At the same time, mobility of adsorbed material to the surface molecules reduces. Both of these factors lead to a decrease of entropy ($\Delta S < 0$).

At present, adsorption is the basis of many industrial operations and scientific investigations. One of the most important fields of research and application of adsorptive processes is the purification, concentrating, and separation of different substances, as well as, adsorptive gaseous and liquid chromatography.

This process is an important stage of heterogeneous catalysis and corrosion. Surface investigations are linked closely to the development of semi-conductor technology, medicine, construction, and military affairs. Also, Aadsorptive processes play key roles when strategic techniques of environmental protection must be chosen.

5.2 THE MAIN STAGES OF ADSORPTION & ION EXCHANGE SCIENCE DEVELOPMENT

The origin of Adsorption Science is dated to the second part of the 18th century which was also a break-though time in the development of chemistry and a transition period from Natural Philosophy and Alchemy (which were looking for universal catalysts (e.g., philosophy stone) as well as a universal solvent (able to transfer any metal into silver and gold) to modern Chemistry and Physical Chemistry. Adsorption science started from the studies of the three greatest chemists: F. Fontana, K.W. Sheele and J.T. Lovitz.

In 1777 looking for phlogiston (Conant, 1950) – like most chemists of that time – Italian chemist, Professor of Pisa University, Felice Fontana (who is also defined as the greatest Italian chemist of that century (Id. Science, 1991)) and Swedish chemist, Karl Wilhelm Scheele (Shectman, 2003) discovered independently the ability of wood coal to take up gases. Later, in 1785, academician Johann Tobias Lovitz (a naturalized German who had lived in St. Petersburg (Russia) since his childhood) discovered the phenomenon of adsorption by charcoal from water solutions (Zolotov, 1998). First, J.T. Lovitz noticed the ability of such coals to discolor the water solutions of organic acids, however, being a real scientist, he did not stop his study there and carried out many additional experiments investigating charcoal as an adsorbing agent. First of all, he tested carbon powders for purification of different contaminated water solutions. He found out that carbons purified a variety of duty (e.g., brown) solutions, made the color of different juices and honey much lighter and discolored the solution of painting substances. J.T. Lovitz studied the action of charcoal on the solution with some odors and discovered that charcoal purified spoilt, smelly water, making it suitable for drinking. He tested the carbon powder for garlic and even for bugs noticing that the charcoal reduced their unpleasant smell. J.T. Lovitz realized the practical applications quickly. In 1790 he published an article entitled: "Guide on the new method how to make a spoilt drinking water suitable for drinking during sea routes". His method was applied by the Russian navy.

The phenomenon of ion exchange was discovered around 150 years ago. In 1845 the English soil scientist H.M. Thompson was passing an ammonia-containing solution through some ordinary garden soil and discovered that the ammonia content of the liquid manure was greatly reduced (Thompson, 1850; Lucy, 2003). It was shown later that the reason for the ion exchange properties of soil was the presence of fine particles of a natural material called zeolite which had ion exchange properties. Since that time, the interest in ion exchange has not been weakened but has been growing permanently. The significance of these processes in agricultural chemistry and the chemistry of living (including human) organisms has been increasingly realized. The principles of ion exchange have been extensively applyied in agriculture, medicine, scientific investigations, and many industries including the treatment of drinking water. Such wide application of ion exchange process is due to two main factors: the heterogeneity of the system (e.g., a possibility of a simple phase separation like just passing a water solution through a column with the ion exchanger) and the ability of materials for ion exchange which provides a possibility to remove selectively ions from water or the separation of ions with a different charge, the value of charge or degree of hydration. The simplicity and efficiency of ion removal (either total or partial) from a solution, which could easily be realized in larger scales (from the laboratory columns to industrial filters), became a reason of its wide application in science and technology. The development of ion exchange and adsorption science has been conditioned by the requirements of different fields of practical application: the needs of agriculture, development of atomic energy production, hydrometallurgy, food and medical industries, environmental protection and water treatment. Such wide application of ion exchange processes was a result of the successful development of new ion exchange adsorbents.

We can summarize here that the development of adsorption and ion exchange sciences is based, first of all, on the success of material sciences, for example, the development of new adsorptive and/or ion exchange materials. Improved understating of the main mechanisms taking place at the interfaces of each system (the adsorbent and ion exchanger – the adsorbate) aids the total progress. Figure 5.1 shows the main types of adsorptive materials. Taking into account that the new adsorptive materials are a key moment (e.g., basis) of the progress in adsorption science, this chapter is built upon the characterization of each type of adsorptive materials (as mentioned in Figure 5.1) with some unfair attention distribution (due to experience of the author).

Figure 5.1 The main kinds of the adsorptive materials

5.3 CARBONS IN WATER TREATMENT AND MEDICINE

The discovery by J.T. Lovitz of the ability of carbon properties to purify water solutions from different admixtures and odors impressed the scientific community very much. Many great scientists tried to repeat the experiments with charcoal in order to find an explanation for the phenomenon of the material purifying properties. The discovery is still of great importance at the present time. Carbon adsorbents (activated, oxidized, pretreated etc.) are widely used in industries, medicine, and water treatment. Adsorption by carbons is an important part of any older (Voyutsky, 1964) or recent Colloid Chemistry book (Birdi, 2000).

Until the 20th century, carbon adsorbents (charcoals) were used mostly in the food and wine industries for the purification of water solutions. The next step in the development of carbon adsorbents was stimulated by the First World War when respirators were created due to the necessity to neutralize chemical agents. At the present time, the main area of the application of carbon adsorbents is within adsorptive purification of water solutions, separation and concentrating in gaseous and liquid phases. The role of carbon sorbents in drinking water and the treatment of waste streams has been growing and is reflected on the websites of the firms selling the materials, such as: sigma-aldrich.com/supelco. The field of carbon-based adsorbents application in medicine and pharmaceuticals has been widening too.

Originally, porous sorbents were synthesized mostly by the thermal treatment of woods, and later on, of coals. Today, carbon sorbents have been produced from a variety of carbon-containing raw materials: wood and cellulose (Malikov *et al.*, 2007), coals (Drozdnik, 1997), peat (Novoselova *et al.*, 2008), petroleum and coal pitches (Pokonova, 2001), synthetic polymeric materials (Mui *et al.*, 2004), liquid and gaseous hydrocarbons (Gunter and Werner, 1997; Likholobov 2007) and different organic wastes (Long *et al.*, 2007), (*http://home.att.net/ ~africantech/GhIE/ActCarbon.PDF*). Three-hundred types of activated carbons have been produced by the German company DonauCarbon for different types of water (drinking, swimming pools, aquariums) and waste water purification as well as carbon for medicine and industry (food industries, atomic energy production, carriers/support for catalysts).

Granular activated carbon (GAC) is a well known cost-effective, conventional adsorbent which has been widely used since the early 1990s for water and air purification. This porous material has a structure which contains both ordered and disordered carbon rings. In contrast graphite carbons consist of free porous space which is a three-dimensional labyrinth of pores (Voyutsky, 1964; Drozdnik, 1997; Birdi, 2000; Pokonova, 2001; Mui *et al.*, 2004; Malikov *et al.*, 2007; Novoselova, 2008). Many recent studies have looked at investigation to improve the understanding of the structural characteristics of carbon adsorbents, (Barata-Rodrigues *et al.*, 1998; Gun'ko *et al.*, 2003; Gun'ko and Mikhalovky, 2004). Generally the pore sizes of carbon adsorbents are different. Traditionally, they are grouped as follows:

- micropores ($\leq$ 2 nm);
- mesopores (2–50 nm);
- macropores ($>$ 50 nm).

Carbons are able to sorb (take up) different molecules and ions from liquid solutions and gases due to their highly developed porous structure and their surface in functional groups. The application of carbon adsorbents and their selectivity towards contaminants that are to be removed from water solutions depends, first of all, on the chemical composition and concentration of the surface functional groups. The main functional groups of carbons are oxygen-containing groups, such as, phenol (hydroxylic), carbonyl, carboxylic, ether, and lactones which are formed during oxidizing treatment of the surface (Voyutsky, 1964; Biniak *et al.*, 1997; Rodrigues-Reinoso and Molina-Sabio, 1998; Figuiero *et al.*, 1999; Yin *et al.*, 2007; Shen *et al.*, 2008). Different conditions around the carbon surface treatment (activation agents, temperature, time, precursors, and templates) can provide the carbon surfaces with functional groups containing nitrogen, sulfur, halogens, and phosphorus.

One of the first explanations of ion exchange ability of carbon adsorbents, (which were nonpolar and, at first sight, should not sorb polar and ionic substances), was given by N.A. Shilov and reflected in the Colloid Chemistry textbook (Voyutsky, 1964). If the raw materials for carbon synthesis contain inorganic admixtures, the source of ion exchange can be explained by the presence of the cations in the carbon structure which can be easily exchanged. If inorganic admixtures are absent in the composition of the raw materials chosen for carbon activation, cation and anion exchange groups can be formed correspondingly at the relatively lower and higher temperature as showed in Figure 5.2. Intermediate temperature creates the conditions when both type of ion exchange groups (cation and anion exchange) can be formed at the surface.

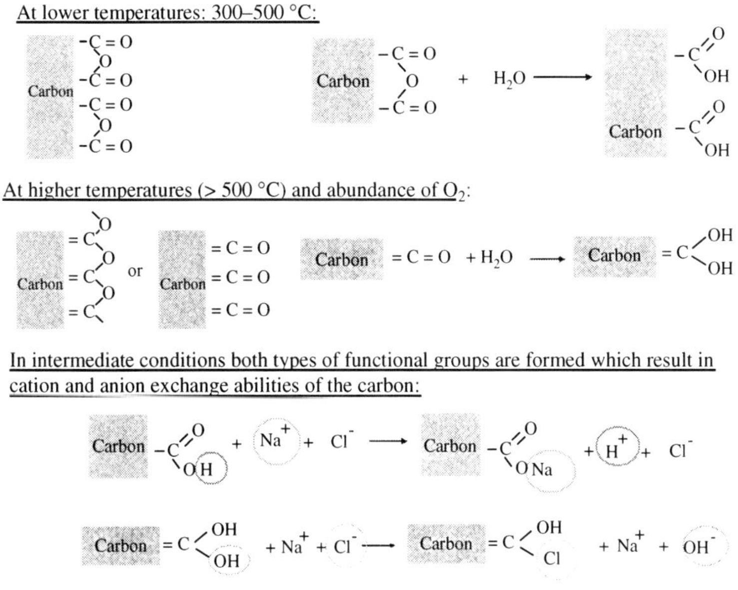

Figure 5.2 Mechanism of formation of ion exchange groups on the surface of carbon adsorbents as suggested by N.A. Shilov (Voyutsky, 1964)

Figure 5.3 shows (schematically) the main functional groups of carbon surface responsible for binding of molecules and particiles via complexation and ion exchange as shown in numerous publications (Biniak et al., 1997; Rodrigues-Reinoso and Molina-Sabio, 1998; Figuiero et al., 1999; Yin et al., 2007; Shen et al., 2008).

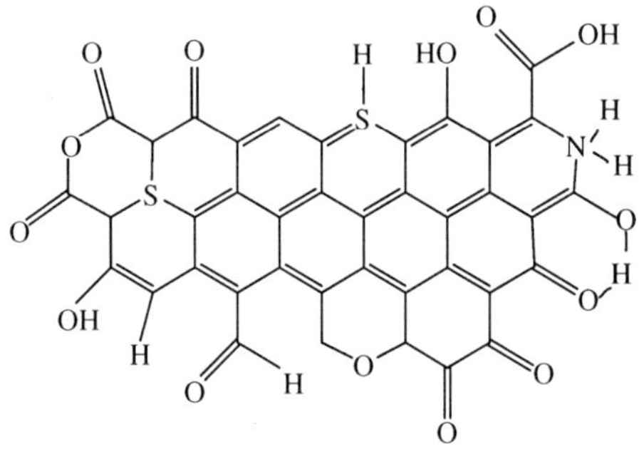

172 Treatment of Micropollutants in Water and Wastewater

Figure 5.3 Schematic presentation of carbon surface main functional groups

A variety of the functional groups, confirmed by spectroscopy and regular chemical analyses, makes the materials selective not only towards organic contaminants (Bhatnagar and Jain, 2006; Fletcher *et al.*, 2008; Anbia *et al.*, 2009; Mansoor *et al.*, 2009) but also towards cationic (Kononova *et al.*, 2001; Park *et al.*, 2007; Lach *et al.*, 2007; Namasivayam *et al.*, 2007; Valinurova *et al.*, 2008) and anionic (Mandich *et al.*, 1998; Lach *et al.*, 2007; Namasivayam *et al.*, 2007) substances. GAC is widely used to remove different natural chemical pollutants from the air or water streams, both in the field and in industrial processes such as spill cleanup, groundwater remediation, drinking water treatment, air purification, and the capture of volatile organic compounds (VOCs).

5.4 ZEOLITES (CLAYS)

The term "zeolite" was originally coined in the 18th century by the Swedish mineralogist A.F. Cronstedt, who observed that upon heating, the stone began to "dance about" (Colella and Gualtieri, 2007). This was due to the water adsorbed inside the zeolite's pores being driven off. The name came from Greek: "zein" means "to boil", "lithos" means "a stone". We still, of course, use zeolite today to remove unwanted ammonia from pond water (discovery of English soil scientist H.M. Thompson), however, the scientific community and the water industry have discovered (and developed) a variety of zeolite-based materials including some for water purification due to the unique structure of natural zeolites (which, in addition, are very beautify crystals) and a property to exchange substances from liquids and gases.

Zeolites are crystalline aluminosilicates with a microporous framework built from corner sharing SiO_4^{4-} and AlO_4^{5-} tetrahedral (with general formula: $M_n[(AlO_2)\times(SiO_2)y]*mH_2O$) which gives a large open framework structure.

The crystals are highly porous and are veined with submicroscopic channels. A large variety of different frameworks can be produced in this way. On March 10, 2009, the International Zeolite Society assigned 191 framework type codes of zeolite materials (*http://www.iza-structure.org/*). In 2007 the latest atlas of zeolite framework types (Baerlocher *et al.*, 2007) and a collection of XRD spectra (Treacy and Higgins, 2007) were published. Due to uniformity of the pore dimensions, these materials, often called Molecular Sieves, display many advantages. They are low-cost materials, thermally ($>800°C$) and chemically stable, of uniform size, and can develop specific surface area up to 690 m^2/g (Erdem-Şenatalar and Tatlıer, 2000), non-toxic and environmentally friendly. Due to the variety of these materials, adsorptive selectivity towards different molecules and particles is also wide. Zeolites have the ability to sort molecules selectively based primarily on a size exclusion process due to a very regular pore structure of molecular dimensions, for example, only molecules of a certain size can be absorbed by a given zeolite material, or pass through its pores, while bigger molecules cannot. The net negative charge in zeolites, within the symmetrical voids hold the cations for the cation exchange capacity. Ion exchangeable ions, such as potassium, calcium, magnesium and sodium, the major cations, are held electronically within the open structure (pore space) – up to 38% void space.

A.A.G. Tomlinson (1998) explained what zeolites are (history, classification, structure, nomenclature, application): "From being mere geological curiosities one hundred years ago, zeolites have progressed to their present status as indispensable absorbents and catalysts ..." (Henmi *et al.*, 1999).

From natural zeolites, water (and other) industries went to synthesis of the artificial zeolites. The first artificial zeolite was synthesized in the 1950s. Patents describing the new methods of zeolites synthesis and application have been issued on regular basis (Henmi *et al.*, 1999; Fiore *et al.*, 2009). Besides development of new zeolite-type materials, the theory of ion exchange has also been under focus (Petrus and Warchol, 2005; Melian-Cabrera *et al.*, 2005; Gorka *et al.*, 2008). The variety of natural zeolites and their origin is a source of curiosity for many researchers (Bartenev *et al.*, 2008).

5.5 ION EXCHANGE RESINS OR ION EXCHANGE POLYMERS

The term ion-exchange resins remains very commonly used to refer to ion exchange macromolecular materials based on different organic polymers despite the strong discouragement of IUPAC (IUPAC, 2004). The materials have a highly developed structure of pores on the surface and are abundant in functional groups capable to exchange cations and anions. The main advantages of ion-exchange resins are their high mechanical and chemical stabilities in combination with high adsorptive (e.g., ion exchange) capacity. If the ion-exchange material is an organic polymer (e.g., resin) the exchange of ions can take place in the whole volume of the resin due to the free penetration of the ions into the porous structure of the resin. Therefore, the most common usage of these materials has been water softening and purification. It is also employed in purification of liquid foods (Miers, 1995). Water softening was the first industrial application to use ion exchange. The process was firstly used by Robert Gans (working for German General Electric Company) in 1905 who suggested improvements in the type of materials and equipment, as well as water softening possibilities using permutite (Helfferich, 1962). Gans's process is still one of the simplest methods for softening water. The process of water softening means passing water that contains 'hardness' ions (mostly, calcium (Ca^{2+}) and magnesium (Mg^{2+})) through a column containing a strongly acidic cation exchange resin in sodium (Na^+) form. The calcium and magnesium ions are exchanged for the equivalent amount of sodium ions.

Ion-exchange resins have been obtained by condensation, polymerization, or from monomers, which already contain the active functional groups, or just introduce those groups in the previously synthesized resins. The ion-exchange capacity of a resin increases with an increasing number of functional groups. On the other hand, this increase in functional groups also leads to negative effects in resin properties: the resins are then liable to swell in water and become more soluble in water. The formation of cross-linking bonds can overcome the problem of swelling and solubility in water. Figure 5.4 shows(schematically) the structure of a cation-exchange resin with the typical functional groups of cation exchangers: $-SO_3H$, $-COOH$, $-OH$. Some exchangers contain only one type of functional group and others contain a combination of a few functional groups. Figure 5.4 shows cation-exchange resins in H^+ form where the exchangeable ions are H^+. For water softening, Na^+ form cation exchanger should be used.

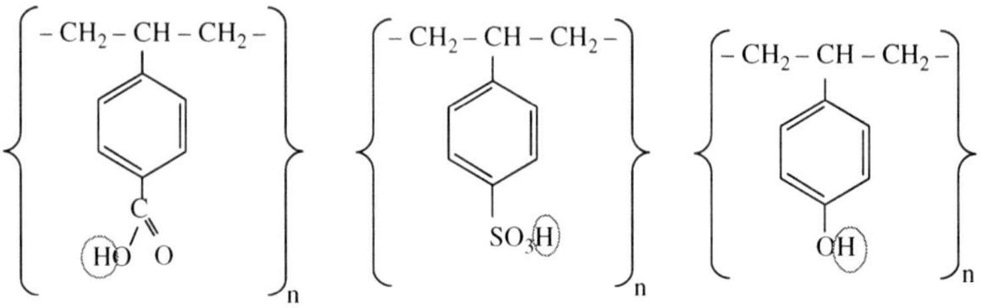

Figure 5.4 Repeating units of polymeric cation exchangers with the main functional groups (−SO3H, −COOH, −OH) and polystyrene matrix

Physico-chemical treatment of Micropollutants

The main active functional groups of anion exchangers are nitrogen-containing groups ($-NH_2$, $=NH$ and $\equiv N$) from aliphatic or aromatic amines. Figure 5.5 shows the typical structure of anion exchangers.

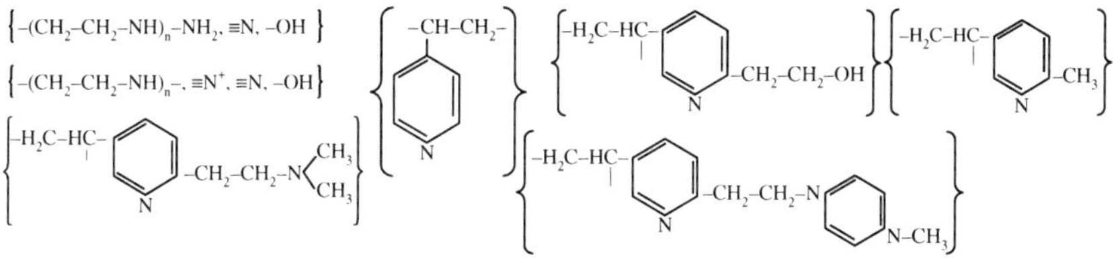

Figure 5.5 Typical repeating units of polymeric anion exchangers

Chemically active polymers have different properties: they are capable of ion exchange, complexation, reduction-oxidation and precipitation. Many colleagues divide ion-exchange resins into three groups (Sengupta and Sengupta, 1997). The first group includes nonfunctional strong acidic or strong basic groups having ion-exchange properties only. The second group collects cross-linked polymers of three-dimensional structure which show complexing or combined complexing and ion-exchange abilities due to the presence of electron-donor functional groups capable of forming complex substances with transition metals having vacant orbital. The third group includes cross-linked polymers which have the ability for redox transformation or combined ability for ion exchange and redox reactions.

The first ion-exchange resins were introduced in 40s of the last century as a more flexible alternative to natural or artificial zeolites (Irving et al., 1997; Whitehead, 2007). Irving M. Abrams and John R. Milk (1997) offer a very nice brief review of the history and development of macroporous ion-exchange resins, starting with the condensate polymers in the early 1940s, continuing with the addition of polymers in the late 1950s, and evolving into new adsorbents for a variety of interesting applications. They also pay some attention to the effect of different methods of resin preparation, and on the physical properties and internal structure of the (macroporous) resins synthesized. Ion-exchange capacity depends on the type of the active functional groups of the exchanger, the chemical nature of the exchanging ions, their concentration in the solutions, and the pH of the solutions.

Since the last almost 70 years, when ion-exchange resins were brought to water treatment (and purification technologies in general), the interest in this type of materials has not weakened but has been growing, this is reflected in a variety of scientific and technological applications, the creation of a journal *Reactive and Functional Polymers*, an increasing number of commercially available ion-exchange resins (*www.water.siemens.com*; Vollmer *et al.*, 2005) and the development of new ion exchangers with nontraditional morphology (Kunin, 1982) and from various raw materials (Wayne *et al.*, 2004).

Ion-exchange resins based on styrene and divinylbenzene are now dominating the international market due to their high quality, relatively low price and easy accessibility of the raw materials for synthesis. The development of porous copolymers by the introduction of solvents during polymerization has led to a considerable increase in the number of ion exchange resins with improved kinetics of ion exchange, high osmotic stability and sieve effect (Seniavin, 1981; Kunin, 1982) and the necessity to develop regulations for using those materials in the food, water, and beverages industries (Franzreb *et al.*, 1995).

German chemists working within the group of Wolfgang H. Hoell (Institute of Technical Chemistry from Karlsruhe) have been contributing considerably to the development of ion-exchange materials, and especially the theory of ion exchange in water treatment (Franzreb *et al.*, 1995; Hoell *et al.*, 2002, 2004). Ion-exchanger polymers have remained one of the most important groups of adsorptive materials which are widely employed for the removal of anions (Karcher *et al.*, 2002; Marshal *et al.*, 2004; Matulionytė *et al.*, 2007) and cations (Elshazly and Konsowa, 2003; Vollmer and Gross, 2005; Kim *et al.*, 2007) as well as polar molecules (Fettig, 1999; Cornelissen *et al.*, 2008) in drinking water treatment. Interesting results were obtained by Matulionytė *et al.* (2007) who studied the possibility of anion extraction from fixing process rinse waters contaminated by Br^-, $S_2O_3^{2-}$, SO_4^{2-}, SO_3^{2-}, CH_3COO^- and silver thiosulfate complex anions using Cl^- form of commercial anion-exchange resins: Amberlite IRA-93 RF, Purolite A-845, Purolite A-500 and AB-17-8.

5.6 INORGANIC ION-EXCHANGERS

Stricter regulations in water quality and the reduction of the maximum contaminant levels (for new toxicological investigations) caused the necessity of looking for new adsorptive materials with much higher selectivity towards the target contaminants. Also, if ion-exchange polymers do not meet the requirements of high chemical, thermal, and radioactive stability and an alternative is required, inorganic ion-exchange adsorbents have come under the focus of researchers and civil engineers. Thus, the need for inorganic ion-exchange adsorbents has been arising mostly when conventional ion-exchange resins (polymers) do not meet the requirements of the World Health Organization (WHO) (and their maximum permissible levels) and/or the requirements for physico-chemical characteristics of the materials in definite conditions of purifying solutions (high thermal, chemical and radioactive stability). Some of the inorganic ion-exchangers can keep radioactive irradiation up to 10^9 rad and higher without any damage to the adsorbent structure (Seniavin, 1981).

The possibility of choice and synthesis of inorganic adsorbents with the desired properties is almost unlimited due to (1) a variety of different classes of lower solubility inorganic compounds which are able to take up ions from water solutions by different sorptive actions; (2) the development of synthesis methods; and (3) the process of material modification/pretreatment during and after synthesis. Currently, the information on the adsorptive properties of the widely known classes of inorganic adsorbents, such as oxides, hydrous oxides, sulfides, phosphates, alumosilicates, and ferrocyanides have been studied extensively in the literature, and new methods of their synthesis have been regularly developing. New adsorbents based on the above mentioned substances (which often exceed the ion-exchange resins on ion-exchange capacity and selectivity, and in addition are characterized by higher chemical, thermal and radioactive stability) have been produced. The main classes of inorganic ion exchange adsorbents are discussed below.

5.6.1 Ferrocyanides adsorbents

These are based on simple (such as $Zn_2[Fe(CN)_6]$) or mixed (such as $K_2Zn_3[Fe(CN)_6]_2$) ferrocyanides. In the 1960s, this group of chemical compounds was recognized as ion exchangers with high selectivity towards cesium ions (Kawamura *et al.*, 1969; Seniavin, 1981). Ferrocyanides have cation and anion exchange capacities and can serve as scavengers for toxic cations and anions, as well as organic molecules (Ali *et al.*, 2004). Sorptive properties of ferrocyanides have been used not only in purification technologies but also to explain phenomena in geochemistry. Particularities of the interaction of two naturally occurring aromatic α-amino acids (namely tryptophan and phenylalanine) with zinc, nickel, cobalt, and copper ferrocyanides has been shown by Wang *et al.* (2006), and the hypothesis that metal ferrocyanides might have concentrated the biomonomers on their surface in primeval seas during the course of chemical evolution has been suggested. Recent studies have concentrated mostly on the immobilization of insoluble ferrocyanides on the surface of zeolites, clays, carbons, and hydrous oxides. Cesium removal from deionized water, seawater, and limewater was investigated using copper ferrocyanide incorporated into porous media including silica gel, bentonite, vermiculite, and zeolite as adsorbents (Huang and Wu, 1999). A granular inorganic cation exchanger showing high selectivity to cesium (Sharygin *et al.*, 2007) was a mixed nickel ferrocyanide distributed over zirconium hydroxide as an inorganic carrier having a trademark Termoxid-35. The structure of ferrocyanides is shown in Figure 5.6.

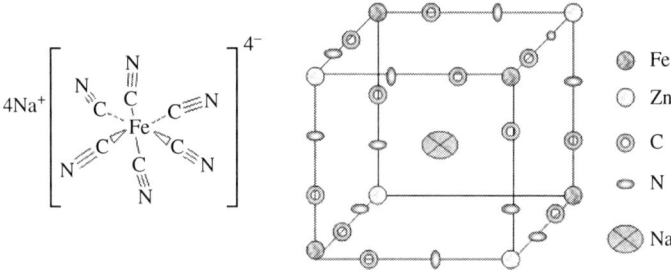

Figure 5.6 Schemes of the structure and a fragment of the crystal structure of ferrocyanide

The main sorption centers of ferrocyanides are as follows.

- Metal ions (Na^+) or H^+ existing in the structural emptiness of the crystals.
- The existence of charge due to the possible change of the oxidation state of the matrix metal or unsaturated coordination number (which can be temporarily saturated by aqua groups) due to vacancies in the matrix structure which can be filled due to sorption actions.
- Metals of the matrix can exchange cations from the solution or bind anions (including OH^-).
- The sorption of ions by ferrocyanides is linked to a few processes such as exchange for metal ions (e.g., Na^+) or H^+ existing in the structural emptiness (interstices) of the crystals, exchange with the metals ions of the matrix or non-ion exchange introduction into structural emptiness or micro-interstices with the following compensation of the excess charge or saturation of coordination number of the matrix atoms (e.g., exchange with aqua groups).

Of particular value with ferrocyanides is their exceptionally high selectivity towards ions of Cs^+, Rb^+, and Tl^+ and very little pH dependence (Kawamura *et al.*, 1969; Tananaev *et al.*, 1971; Volgin, 1979; Jain *et al.*, 1980; Clarke and Wai, 1998; Huang *et al.*, 1999; Ali *et al.*, 2004; Sharygin *et al.*, 2007). Ferrocyanides can selectively remove heavy alkaline metals and thallium from complex on the composition water solutions including strong acidic waste waters of color metallurgy and liquid radioactive wastes. Ferrocyanides have also been investigated for their uptake capacity towards aminopyridines (Kim *et al.*, 2000). Nickel ferrocyanide was found to be a better adsorbent as compared to cobalt ferrocyanide.

The other important type of inorganic ion-exchangers demonstrating both cation and anion exchange properties are based on *individual and mixed hydrous metal oxides*. The term hydrous oxides refers to all oxides (MO_x), hydroxides [$M(OH)_x$], and oxyhydroxides (MO_xOH_y) of a metal, M. Other terms that have been used include oxide hydrates, oxide-hydroxides, and sesquioxides among others. Hydrous oxides of Fe, Al, Mn and a combination of these metals with other metals are a predominant part of this class of inorganic ion-exchangers. The main known hydrous oxide compounds of these chemical elements are: *Goethite* (FeO(OH)), Hematite (Fe_2O_3), Magnetite (Fe_2O_3), *Maghemite* (Fe_2O_3, γ-Fe_2O_3), Goethite (FeO(OH)), *Lepidocrocite* (γ-Fe^{3+}O(OH)), Ferrihydrite (hydrous ferric oxyhydroxide minerals interpreted from XRD as α-Fe_2O_3 or δ-FeOOH), *Gibbsite* (γ-Al(OH)$_3$ but sometimes as α-Al(OH)$_3$), *Bayerite* (α-Al(OH)$_3$, but sometimes as β-Al(OH)$_3$), *Diaspore* (a native aluminium oxide hydroxide α-AlO(OH)), Hollandite (Ba($Mn^{4+}Mn^{2+}$)$_8O_{16}$, Todorokite (a rare complex hydrous manganese oxide mineral with formula (Mn,Mg,Ca, Ba,K,Na)$_2Mn_3O_{12} \cdot 3H_2O$), Corundum, *Pyrolusite,* Ramsdellite and Birnesite (minerals consisting essentially of manganese dioxide (MnO_2), Hausmannite (a complex oxide of manganese, $Mn^{2+}Mn_2^{3+}O_4$).

The main adsorptive active sites of these materials can be classified as follows.

- Aqua groups coordinated around metal atoms of the sorbent matrix can be exchanged with other polar molecules of the solutions (particularly, exchanged for neutral molecules and anions) due to dipole-dipole interaction.
- Surface hydroxyl groups (OH^-) can be exchanged with the solution anions which have higher coordination ability than OH^-.
- Hydroxyl groups in the bridge position: the bond forming bridges can be broken and the consequent steps are according to the possible mechanisms of adsorption.
- Oxygen atoms in the bridge or end-positions can be protonated, what lead to a break of the bridges and unsaturated coordination which will follow by binding of metal ion and anions, particularly, OH^-.
- Metal ions of the adsorbent matrix can bind aqua groups, anions including OH⁻ and can be exchanged directly for other metal ions of the contacting solution.
- If the structural emptiness of the matrix filled with cations or H^+, the latest can be exchanged with the ions of the contacting solution.
- Noncompensated charge or nonsaturated coordination number (due to a change of the matrix element(s) oxidation state or replacement of the matrix ions by ions of the other size can lead to binding the ions from the solution.

Cations of a solution can be bound to the surface of solid adsorbents due to: (1) exchange of hydrogen protons of protonated sorption sites; (2) exchange with metal ions (M^+) and H^+ existing in the emptiness of the crystal matrix; (3) non-ion exchange introduction into the structural emptiness and micro-emptiness with the following charge redistribution in the whole matrix using the closest environment of the emptiness; (4) direct ion exchange with the ions of the matrix; (5) binding of the ions or molecules to oxygen-containing sorption sites (contact adsorption) due to unsaturated coordination number of oxygen (such sorption sites can be formed due to preliminary uptake of oxy-anions (HSO_4^-, $H_2PO_4^-$ etc.); (6) direct biding to metal atoms of the sorbent matrix (contact adsorption).

Anions of a solution can be bound to the surface of an adsorbent due to: (1) direct exchange to the hydroxyl ions (OH^-) at the end position; (2) exchange with hydroxyl groups with the following binding to metal atoms of the matrix after a break of the bridge binds; (3) binding to metal ions of the matrix which have unsaturated coordination number; (4) exchange with the anions existing in the structural emptiness of the matrix; (5) exchange with the anionic groups of the sorbent matrix; (6) electrostatic interaction with the sorbent surface with high positive charge which can be generated due to the previous adsorption acts or due to change of the oxidation state of the matrix chemical elements.

Simultaneous uptake of cations and anions from a solution can take place due to few reasons: re-charge of the adsorbent surface due to the ion-exchange process with the following adsorption of the opposite charge ions; adsorption of the molecules such as MX_n, (where X is OH^-, Cl^- etc.) caused by the unsaturated coordination number of the atoms of the matrix or by the necessity to alternate cations and anions in the channels of the sorbent structure; uptake of cations and anions of the salts by metal hydrous oxides and formation of basic salts, and the other reasons.

The field of application of inorganic ion-exchangers based on individual and mixed hydrous oxides of metals is very broad due to the variety of possible mechanisms of sorption described above. They are explored as cation and anion exchangers.

For instance, the cation-exchange properties of manganese (hydrous) oxides were shown in the efficient adsorption of lithium, cadmium, potassium, radium etc. Manganese oxide adsorbents have been found to be very efficient in the removal of lithium from sea water (Ooi et al., 1986). Later studies by the same principle investigator led to the improvement of the syntheses methods and obtaining new manganese oxide adsorbent ($H_{1.6}Mn_{1.6}O_4$) from precursor $Li_{1.6}Mn_{1.6}O_4$ by heating $LiMnO_2$ at 400°C (Chitrakar et al., 2001). $LiMnO_2$ was prepared by two methods: hydrothermal and reflux. The new manganese oxides demonstrated higher adsorptive capacity (40 mg/g) as compared with other known adsorbents (Chirakar *et al.*, 2001). The cation exchange ability of manganese oxides was employed to remove cadmium (Hideki *et al.*, 2000), potassium (Tanaka and Tsuji, 1994) and cobalt (Manceau *et al.*, 1997). The addition of hydrous manganese oxide (HMO) has been identified by USEPA as an acceptable technology for radium removal in groundwater supplies (Valentine *et al.*, 1992; EPA, 2000; Radionuclides, 2004). This method of radium removal technology is particularly suited to small and medium-sized systems. This process relies on the natural affinity of radium to adsorb onto manganese oxides. Anion exchange properties of manganese oxides (natural, synthesis and biogenic) have been widely investigated with great attention to potential arsenic removal (Moore *et al.*, 1990; Manning *et al.*, 2002; Ouvrad *et al.*, 2002a, b; Katsoyiannis *et al.*, 2004), however, other anions have also been under focus (Ouvrad *et al.*, 2002a).

Arsenic contamination is widely known global problem (Wang and Wai, 2004). Exposure to high arsenic concentrations (>100 µg L^{-1}) can result in chronic arsenic poisoning called Arsenicosis or Black Foot disease. The most serious damage to health has been in Bangladesh and West Bengal, India (www.sos-arsenic.net; www.who.int/water_sanitation_health/dwq/arsenic/en/). However, long-term exposure to lower arsenic concentrations, such as the old drinking water standard for arsenic (50 µg L^{-1}), is also dangerous for humans as it can result in cardiovascular diseases, endocrine system disorders and can cause cancer. Therefore, the European Union (EU) and the United States established a new standard for levels of arsenic in drinking water of 10 µg L^{-1}. Since 1998, when this decision was made, new toxicological data suggest that this new standard is not safe either. As a result, some countries have adopted stricter arsenic guidelines for drinking water than the current WHO guidelines. In Denmark, the national guideline has already been lowered to 5 µg L^{-1} (Danish Ministry of the Environment, 2007), as well as in the American state of New Jersey (New Jersey Department of Environmental Protection, 2004). In addition, the American Natural Resources Defense 15 Council (Natural Resources Defense Council, 2000) advises that the drinking water standard should be set at 3 µg L^{-1}. Australia has a drinking water guideline for arsenic of 7 µg L^{-1} (National Health and Medical Research Center, 1996).

Activated alumina is one of the most popular adsorbents used on an industrial scale for toxic anion removal. It is obtained by thermal dehydration of aluminium hydroxide to retrieve the materials with high specific surface area and a distribution of macro- and micro-pores. Activated Al_2O_3 has been studied extensively for arsenic adsorption (Gupta and Chen, 1987; Singh and Pant, 2004; Manjare *et al.*, 2005) and has been effectively used for arsenic removal from drinking water at pH 5.5 at the Fallon Nevada, Naval Air Station (Hathaway and

Rubel, 1987). However, strong pH effect and often insufficient removal capacity of conventional activated alumina (Gupta and Chen, 1978; Manjare et al., 2005; Hathaway and Rubel, 1987; Singh and Pant, 2004) conditioned the development of better modification of this ion-exchange adsorbent. To prepare an ideal adsorbent with uniformly accessible pores, a three-dimensional pore system, a high surface area, fast adsorption kinetics and good physical and/or chemical stability mesoprous alumina with a large surface area (307 m^2/g) and uniform pore size (3.5 nm) was developed and tested for arsenic removal (Kim et al., 2004). The maximum As(V) uptake by mesoporous alumina was seven times higher (121 mg[As(V)]/g and 47 mg[As(III)]/g) than that of conventional activated alumina.

Hydrous oxides, oxides, oxyhydroxides of iron, amouphous hydrous ferric oxide (FeO-OH), goethite (α-FeO-OH), hematite (α-Fe$_2$O$_3$) (as well as modification by iron compounds) is another important class of substances investigated for their adsorptive affinity to arsenic (Ferguson and Gavis 1972; Wilkie and Hering 1996; Altundogan et al., 2000; , Roberts et al., 2004; Saha et al., 2005). Amorphous Fe(O)OH had the highest adsorption capability. At first glance, people try to explain the higher uptake capacity to arsenic by the highest surface area, however, for ion exchange, as the main or initiating mechanism of adsorption, this rule (surface area – adsorptive capacity) is not directly proportional. Most iron oxides are fine powders that are difficult to separate from solution afterwards. Therefore, some technological approaches are needed to make these materials more suitable for column conditions. For instance, the EPA proposed iron oxide-coated sand filtration as an emerging technology for arsenic removal at small water facilities (USEPA, 1999; Thirunavukkarasu et al., 2003). (Hydrous) oxides of iron have been also tested for their adsorptive capacity to the other anions such as phosphate, bromate (Kang et al., 2003; Bhatnagar et al., 2009).

The list of inorganic ion-exchangers based on metal oxides is much longer and also include zirconium (Kim, 2000) and titanium dioxides (Zhang et al., 2009), cerium dioxide (Watanabe et al., 2003; Bumajdad et al., 2009).

In the last decade mixed hydrous oxides attracted much more interest from researchers compared to individual hydrous oxides as their fitting and changing chemical composition gives more options to change the structure of the materials and to achieve the desired adsorptive properties and selectivity toward target ions and molecules. It was noticed that adsorption of anions onto mixed hydrous oxides of Fe and Al was less dependent on pH compared to individual ones (Chubar et al., 2005a, b, 2006). Venkatesan et al. (1996) were trying to establish the correlation between the surface properties of a silica-titania mixed hydrous oxide gel and adsorptive properties towards radioactive strontium.

Tomoyuki *et al.* (2007) synthesized ten Si-Fe-Mg mixed hydrous oxide samples with changing Si, Fe(III) and Mg molar ratio, detected the influence of the percentage of Mg on the structure of the materials and studied adsorption of arsenite, arsenate and phosphate on those materials.

Inorganic ion-exchangers based on *layered mixed hydrous oxides and hydrotalcite* type materials gained the reputation of being very competitive adsorptive materials capable of solving many tasks of purification technologies. Bruna *et al.* (2009) discovered that hydrotalcite-like compounds ($[Mg_3Al(OH)_8]Cl \cdot 4H_2O$; $[Mg_3Fe(OH)_8]Cl \cdot 4H_2O$; $Mg_3Al_{0.5}Fe_{0.5}(OH)_8]Cl \cdot 4H_2O$ (layered double hydrous oxides) and the calcined product of $[Mg_3Al(OH)_8]Cl \cdot 4H_2O$, $Mg_3AlO_{4.5}$ (hydrotalcites)) had a lot of potential as adsorbents of the herbicide MCPA [(4-chloro-2-methylphenoxy) acetic acid]. Sparks *et al.* (2008) developed new sulfur adsorbents (derived from layered double hydroxides) employed for COS (carbonyl sulfide) adsorption studies. It was found that selectivity of the layered mixed hydrous oxides towards arsenate was much higher as compared with balk double hydrous oxides (Chubar *et al.*, 2006). The reason for the layered materials higher selectivity to tetrahedral anions is the structure of the adsorbents (Figure 5.7) and the structural correspondence factor leading the adsorption process.

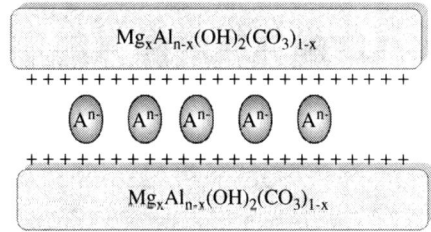

Figure 5.7 Schematic structure of hydrotalcite type substance (of general formula: Mg (Zn)2+6 Al3+2 (OH)16(CO3)) with adsorbed anion in the interlayer space

Natural (Hall and Stamatakis, 2000) and artificial (Frost *et al.*, 2005) hydrotalcites have been investigated for varying parameters for synthesis (Palmer *et al.*, 2009) and choosing the most advanced method of synthesis (in this case: sol-gel) (Lopes *et al.*, 1996) to develop novel hydrotalcite-type materials.

Inorganic ion-exchangers based on mixed and individual metal sulfides are a type of adsorbents with the chemical formula: $M_xE_yS_z$, where M and E are metals (for instance: ZnS, CuS or mixed: $Zn_x Cd_{1-x} S$, $Cd_x Mn_{1-x} S$, $Mn_x Zn_{1-x} S$ etc.). Sulfides differ from oxides as they are larger and have the stronger ability of S, Se, and Te for polarization as compared with oxygen, which leads to increased covalent bond formation (Breg and Klarinsgbull, 1967). Atoms of Se, Se, and Te are larger than metal atoms, therefore, the crystalline structure of these materials are characterized by strong packing when formation of the emptiness is almost impossible. Thus, the mechanism of particle adsorption by sulfides excludes participation of the structure emptiness. There is a similarity between hydrous oxides and sulfides. The role of aqua and hydroxyl groups can be executed by H_2S, HS^- and atoms of sulfur. Increased concentration of H_2S and HS^- ions on the surface of metal sulfides was defined experimentally (Seniavin, 1981). These groups participate in the process of interaction with the sulfide surface. There are many similarities in the adsorption mechanisms on metal oxides and sulfides but in the latter, adsorption is a pure surface process. Some recent studies focus on new synthesis methods for metal sulfides which would advance the application of the material. Yamamoto *et al.* (1990) developed the new organosols of nickel sulfides, palladium sulfides, manganese sulfide, and mixed metal sulfides, and studied their use in the preparation of semiconducting polymer-metal sulfide composites. Manolis *et al.* (2008) reported on the family of robust layered sulfides $K_{2x}Mn_xSn_{3-x}S_6$ ($x = 0.5$–0.95) (called KMS-1) which showed outstanding preference for strontium ions in highly alkaline solutions containing extremely large excesses of sodium cations as well as in an acidic environment where most alternative adsorbents with oxygen ligands are nearly inactive. They suggest the implication of these simple layered sulfides for the efficient remediation of certain nuclear wastes. The reason of the high selectivity to strontium was a feature hexagonal $[Mn_xSn_{3-x}S_6]^{2x-}$ slabs of the CdI_2 type that contain highly mobile K^+ ions in their interlayer space that are easily exchangeable with other cations and particularly strontium. A search for a cost-effective synthesis method brought Bagreev and Bandosz (2004) to the development of efficient hydrogen sulfide adsorbents obtained by the pyrolysis of sewage sludge derived fertilizer modified with spent mineral oil.

5.6.2 Synthesis of inorganic ion exchangers

Adsorptive and ion exchange properties of inorganic ion-exchangers depend on their composition and crystalline structure particularities. Technological characteristics of the materials are defined by their shape: spherical, grains of different shape, thin layers on the surface of the inert materials, membranes, fibers, cores, tubes, porous structures etc. Desired adsorptive and technological properties of inorganic ion-exchangers can be achieved by different synthesis methods. Therefore, the predominant attention of the researchers is given to the development of new synthesis methods and the improvement of existing ones. The main way of the obtaining inorganic adsorbents includes two stages which are often consequent stages: (1) synthesis of chemical compounds (or composition) or modification of the previously developed materials; and (2) preparation of the adsorbents in a shape suitable for technological application. Synthesis of inorganic adsorbents can be run as homogeneous precipitation from the solutions and hetero-phase reactions in the systems: solid – gas, solid – liquid and just solid phase reactions. Preparation of the materials in the desired shape can be carried out by:

(1) drying the synthesized material on the surface with definite shape;
(2) blending;
(3) spray drying;
(4) freezing;
(5) pressing as tablets;
(6) making grains by few methods;
(7) using binding materials;
(8) impregnation;
(9) precipitation inside of porous materials;
(10) making spherical particles by drop-wise methods including sol-gel method.

Amphlett (1964) and Clearfield (1982) gave detailed explanations on many basic methods of inorganic ion-exchanger syntheses. The last 50 years resulted in the discovery of a variety of ion-exchange adsorbents with high selectivity to cations and anions (Bengtsson *et al.*, 1996; Bortun *et al.*, 1997; Chubar *et al.*, 2005a, b, 2006; Liu *et al.*, 2006; Zhang *et al.*, 2007; Chen *et al.*, 2009; Zhuravlev *et al.*, 2006).

When searching for the most advanced synthesis approaches, scientists often look at the sol-gel method. The advantages of materials developed via this approach are (1) physically and chemically pure and uniform particles; (2) exceptional mechanical properties owing to their nanocrystalline structure; (3) homogeneous material, since mixing takes place on the atomic scale; and (4) low processing (often ambient) temperature. The sol-gel method of synthesis was first discovered in the late 18th century and has been extensively studied since the early 1930s. Interest in this technique was renewed in the early 1970s when monolithic inorganic gels were formed at low temperatures and converted to glass without a high-temperature melting process (www.psrc.usm.edu/mauritz/solgel.html). Through this process, homogeneous inorganic oxide materials with desirable properties of hardness, optical transparency, chemical durability, tailored porosity, and thermal resistance, can be produced at room temperatures, as opposed to the much higher melting temperatures required in the production of conventional inorganic glass. Traditional precursors for sol-gel syntheses are metal alkoxides. The most widely used metal alkoxides are the alkoxysilanes, such as tetramethoxysilane (TMOS) and tetraethoxysilane (TEOS). However, other substances such as aluminates, titanates, and borates are also commonly used in the sol-gel process, often mixed with TEOS. Traditional sol-gel synthesis can not be said to be environmentally friendly, cost-effective and technologically attractive due to the expensive and toxic precursors used to run the reactions. To overcome these disadvantages researchers have been trying to develop new methods of sol-gel synthesis which avoid using toxic and expensive alkoxides as raw materials. Thus, nontraditional sol-gel synthesis methods were developed that resulted in obtaining new cation (Bortun and Strelko, 1992; Zhuravlev *et al.*, 2002, 2004, 2005) and anion (Chubar *et al.*, 2005a, b, 2008a) exchange adsorbents. The methods employed only simple inorganic metal salts, mineral acids and bases for synthesis. There is no common synthesis approach suitable for all developed materials due to the unique properties of each chemical element. Fine inorganic synthesis reactions should be developed for each new adsorbent. The main approaches of these inorganic syntheses are: (1) preliminary synthesis of precursors; (2) partial neutralization of metal inorganic slats; (3) choosing the best fitting slat for each reaction (e.g., chlorides, sulfates, nitrates of Al, Fe(II), Fe(III), Zn, Mg, Mn, Zr); (4) choosing parameters for synthesis (e.g., metal salt concentration, temperature, pH, mixing regime); (5) using some additives (organic or inorganic); (6) choosing the best alkaline agent. Examples of the final materials (A) and the precursors (hydrogels) (B) are shown in Figure 5.8. Adsorptive properties of ZrO_2 were tested by Chubar *et al.* (2005b) and Meleshevych *et al.* (2007) (Figure 5.8a). The high affinity to arsenic of the recently developed layered mixed hydrous oxide of Mg-Al has been described in the EU provisional patent application (submitted by Utrecht University) dated September 14, 2009 (Figure 5.8b).

Figure 5.8 Sol-gel generated hydrous oxides: (a) ZrO2 (Chubar *et al.*, 2005b; Meleshevych *et al.*, 2007) and (b) intermediate product (hydrogel) of the mixed layered hydrous oxide of Mg and Al (Chubar not publ.)

The materials shown at Figure 5.8 were tested for arsenate adsorption. Isotherms of $H_2AsO_4^-$ by these ion exchangers (ZrO_2 and one of the ion exchangers prepared from hydrogel as shown in Figure 5.8b) are shown in Figure 5.9 (Chubar, not publ.). The isotherms were obtained at pH = 7 (the typical pH of treated drinking water) using background electrolyte 0.1 N NaCl. The concentration of solids (adsorbents) was 2 g_{dw}/L. Batch sorption experiments were running for 72 hours. New sol-gel generated hydrous oxide of Mg-Al demonstrated very competitive removal capacity (plateau of the isotherm) which reached 180 mg[As]/g_{dw} and exceptionally high affinity to arsenate (the plot is very close to axis Y). The latter means that the material should show very high selectivity at the lower (ppb) concentrations of this target ion and should be able to meet the very stringent forthcoming requirements of WHO (5 ppb).

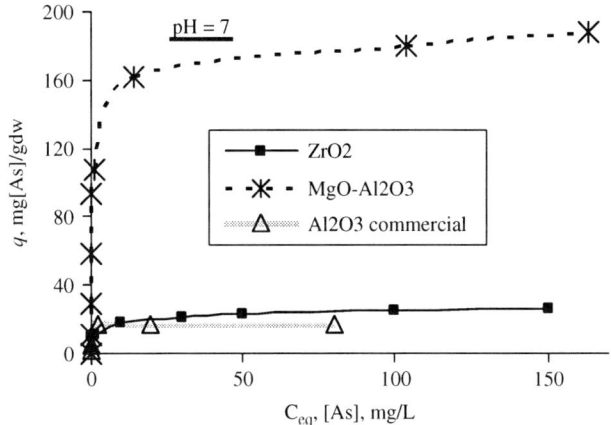

Figure 5.9 Isotherms of H2AsO4⁻ adsorption by ZrO2 and MgO-Al2O3 hydrous oxides (shown in Fig. 5.8). Conditions of the experiment: background electrolyte: 0.1 N NaCl, adsorbent concentration: 2 gdw/L, contact time: 72 hours, temperature: 22±2°C (Chubar not publ.)

5.7 BIOSORBENTS (BIOMASSES): AGRICULTURAL AND INDUSTRIAL BY-PRODUCTS, MICROORGANISMS

For the last more then two decades biosorption or concentrating of the substances by sorbents of the natural origin has attracted the attention of specialists working in the field of treatment technology. Biomasses, which in the most cases are waste (or side products) of industry or agriculture, have a variety of functional groups and often show high sorptive selectivity toward different substances to be removed from water solutions. The main advantage of these adsorptive materials is their lower costs. Additional attractive features of biomasses (such as: competitive removal performance, frequent selectivity of the materials to target xenobiotics, a possibility to improve adsorptive properties by different ways of pre-treatment – modification of biomasses surfaces, in principle the possibility to regenerate the biomaterials (if necessary)) makes this type of adsorptive material attractive for technological application. Thus, the main pilot tasks of biosorption are: (1) to search for new capable materials suitable for industrial application; and/or (2) the pre-treatment and modification of the already found biomaterials in order to make the adsorbents capable to solve the treatment technology problem.

The origin of biosorption science is dated to 1986 when a meeting was organized by the Solvent Engineering Extraction and Ion Exchange Group of the Society of Chemical Industry in the UK and biosorption was regarded as an emergent technology (Apel and Torma, 1993). Since that time a number of research groups have been appointed to define the best biosorbents. The attractive idea has been the utilization of industrial wastes and agricultural by-products in waste treatment. Numerous biomasses have been tested for removal capacity to, first of all, heavy metals. They include industrial large-scale fermentations (antibiotic enzymes, organic acid production processes etc.) (Volesky 1990a, b, c, 1995; Wang *et al.*, 2008), agricultural by-products (peat, cotton waste, rice husk, olive pomace, pectin-reach fruit wastes etc.) (Apel and Torma 1993; Ajmal *et al.*, 2000), algae and microorganisms (seaweeds, bacteria etc.) (Kogtev *et al.*, 1996; Patzak *et al.*, 2004; Cochrane *et al.*, 2006; Chubar *et al.*, 2008b), cork tree biomass (waste of bottle stoppers from the wine industry) (Annadurai *et al.*, 2003; Chubar *et al.*, 2003a, b, c, 2004), pectin-rich fruit waste (Schiewer *et al.*, 2008) and many others. Patents were also approved for many biosorbents developed using cheap biomass as raw materials (Kogtev *et al.*, 1996). Loredana Brinza (www.soc.soton.ac.uk/BIOTRACS/biotracs) characterized the list of biomasses for their uptake ability to heavy metal ions looking for the highest affinity to the metals: Macro-algae (Brown algae: *Ascophyllum nodosum*; *Fucus vesiculosus*; *Sargassum* spp. (numerous), *Laminaria digitata*; *Laminaria japonica*; *Ecklonia sp.*; *Padina pavonia*; *Petalonia fascia*; *Pilayella littoralis*, Red algae: *Corallina officinalis*; *Gracilaria fischeri*; *Porphyra columbina*, Green algae: *Cladophora crispata*; *Codium fragile*; *Ulva fascia*; *U. lactuca* and Micro-algae (Red algae: *Porphyridium purpureum*, Green algae: *Chlamydomonas reinhardtii*; *Chlorella salina*; *C. sorokiniana*; *C. vulgaris*; *Scenedesmus abundans*; *S. quadricauda*;

S. subspicatus; *Spirogyra spp.*, Diatoms: *Cyclotella cryptica*; *Phaeodactylum tricornutum*; *Chaetoceros minutissimus*, Cyanobacteria: *Lyngbya taylorii*; *Spirulina spp.*

Biosorbent contain a variety of functional groups capable of binding metal cations: carboxyl, imidazole, sulphydryl, amino, phosphate, sulfate, thioether, phenol, carbonyl, amide and hydroxyl moieties (Volesky 1990a, b, c, 1994, 2001, 2003, 2007; Chubar *et al.*, 2008b). They were very well characterized by Wang (2009). Most of them are similar to the main functional groups of carbons and ion exchange polymers shown in Figures 5.3–5.5.

Taking into account the similarity of the main surface functional groups of biomaterials, the methods which are traditionally used for the characterization of surface chemical properties of carbons, natural zeoliets and ion-exchange resins, can be successfully employed to study surface chemical properties of biomasses. Potentiometric titration (which shows relative cation and anion exchange capacities if the data are treated correctly and if the research has been conducted in batch conditions), Boehm method (which allows us to define the concentration of strong acidic carboxylic groups, weak acidic plus lactone groups and phenolic groups), spectroscopy (FTIR) and electrophoretic mobility measurements are very useful to characterize the surface chemical properties of cork tree biomass (Chubar *et al.*, 2003c, 2004; Psareva *et al.*, 2005) and even to detect the difference in the surface chemical properties of living (viable) and dead (autoclave inactivated) gram-negative microorganism *Shewanella putrefaciens* (Chubar *et al.*, 2008b). Living *Shewanella putrefaciens* showed a capacity to sorb Mn(II) ions over a 1 month period. Formation of Mn-containing precipitates and polymeric sugars accompanied the uptake process (Figure 5.10). Complementary spectroscopy techniques (FTIR, EXAFS, and XANES) and SEM allowed the characterization of Mn-containing mineral precipitates synthesized by viable cells as a function of temperature (range 5–30°C), the contact time (up to 20–30 days), the metal loading and the bacteria density (2–4 g_{dw}/L). These parameters predetermined whether only one or a few Mn-containing precipitates in different ratio were produced. $MnPO_4$, $MnCO_3$ and $MnOOH$ were the main precipitates formed by this bacterium (Chubar *et al.*, 2009).

Pretreatment of biomasses (surface modification) of the original biomass in order to improve or change their sorptive capacity and/or sorptive affinity towards the particles of interest is another important part of the current research of the biosorption scientists. Chemical, physical, biological, or mechanical treatment of the original biomass or a combination of the mentioned methods is usually used by the researchers. The most widely spread methods of biomass pretreatment are: (1) chemical (using acids under different concentration, different temperatures, pressure; using bases (alkali, ammonia, lime, carbonates etc), oxydative agents (oxygen, manganate, chromate, chlorates etc.), using different complexing agents; (2) physical (thermal, hydrothermal treatment, pressure, temperature etc or even production of carbons); (3) biological (enzyme and microbiological treatment). Different methods of biomass treatment are described by Walt (2003) and Kumar (2009). Patents have been also pending on the new methods of biomass treatment (Holtzapple and Davison, 1992; Hennessey et al., 2009).

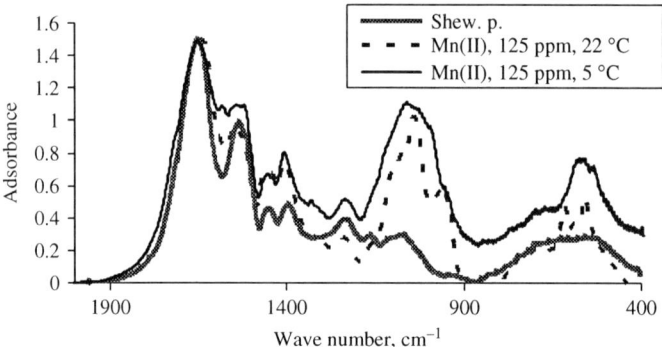

Figure 5.10 FTIR spectra of the (initially) viable Shewanella putrefaciens before and after sorption of Mn^{2+}. Batch conditions of Mn(II) sorption: contact time: 24 days, background electrolyte: 0.1N NaCl, temperatures: 5 and $22\pm2°C$, bacteria concentration: 2 g_{dw}/L (Chubar et al., not publ.)

Figure 5.11 shows an increase of Cu^{2+} sorption: (a) after treatment of cork biomass with NaClO at the initial concentration of Cu(II) of 200 and 300 mg/L and (b) of carbons produced from cork biomass at different conditions at the initial concentration of Cu(II) of 10 and 200 mg/L. Careful consideration is usually needed (taking into account biomass losses and the pretreatment costs) to conclude if a pretreatment is reasonable and cost-effective.

If an original and pretreated biomasses have been already characterized for their surface chemical and adsorptive properties, the next research steps should be the same as for any other type of adsorbents: sorption studies in column conditions, investigation of a possibility for regeneration, granulation of the biomasses showing the best performance and, finally, building a technological scheme for possible biosorption implementation using packed-bed columns for biosorption and desorption as was done by Davis et al. (2003) and Walsh, 2008.

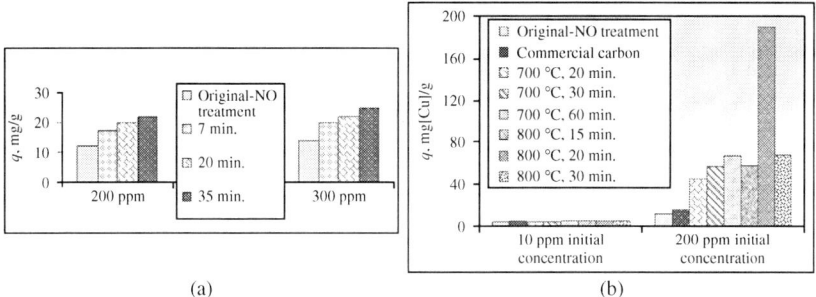

Figure 5.11 Cu^{2+} sorption on the original (not pretreated) cork tree biomass and (a) pretreated with NaClO biomass and (b) on carbons made from cork tree biomass at different temperatures (Chubar et al., 2004)

5.8 HYBRID AND COMPOSITE ADSORBENTS AND ION EXCHANGERS

Hybrid and composite adsorptive materials are the most recent and most fashionable field of researches in adsorption science and technology. There are two justifications for this situation: (1) the necessity to satisfy increasingly strong WHO maximum permissible levels and look for the materials which would combine surface chemical properties of a few types of adsorbents; and (2) scientific curiosity.

In spite of the presence of the theoretical works (mostly based on mathematical modeling), which are trying to explain how to design hybrid materials on the scientific basis, an exact definition of hybrid materials (Ashby and Bréchet, 2003) is still absent. The term (hybrid) came from biology (particularly genetics). *Encyclopedia Britannica* writes: "The term hybrid ... usually refers to animals or plants resulting from a cross between two races, breeds, strains, or varieties of the same species. There are many species hybrids in nature (in ducks, oaks, blackberries, etc.), and, although naturally occurring hybrids between two genera have been noted, most of these latter result from human intervention." (*www.britannica.com/*). It is also pointed that "... the process of hybridization is important biologically because it increases the genetic variety ... which is necessary for evolution to occur". *Cambridge Dictionary Online* gives a more brief definition: "a plant or animal that has been produced from two different types of plant or animal, especially to get better characteristics, or anything that is a mixture of two very different things" (*http://dictionary.cambridge.org/*). Other sources suggest similar definitions. We can conclude here that the key elements which can be used to give a definition of hybrid material, are: (1) new material with new properties; (2) made from two completely different origin materials; (3) appears due to "human intervention", that is, made by human beings; and (4) important for the society needs, that is, "... necessary for evolution to occur".

Hybrid adsorbents have been synthesized from two types of adsorbents in different possible combinations: carbons, ion-exchange polymers, zeolites and clays, inorganic ion exchangers and natural biomasses. The most advanced methods of synthesis typical for each main type of adsorptive material are usually chosen by the researchers. Recent developments in the field of hybrid and combined adsorbents were collected in a book edited by Loureiro *et al.* (2006). Numerous publications and patent storms regularly occur in journals and online. Blaney *et al.* (2007) reported the new anion exchanger for phosphate removal. Mrowiec-Białoń *et al.* (1997) applied the sol-gel method to develop an effective hybrid adsorbent for water vapor. A combination of the sol-gel processing and molecular imprinting approach were applied by Quirarte-Escalante *et al.* (2009) to develop the hybrid adsorbent with high removal capacity to lead. Many patents and patent applications are filed in the area of hybrid adsorbents (Misra and Genito 1993; Chang *et al.*, 2007).

5.9 COMMENTS ON THE THEORY AND FUTURE OF ADSORPTION AND ION-EXCHANGE SCIENCE

The term adsorption was introduced by Kayser in 1881 to describe the increase in concentration of gas molecules in neighboring solid surfaces, a phenomenon noted by Fontana and Scheele in 1777 (Gregg *et al.*, 1982). Kayser suggested the first empirical equation of adsorption isotherm ($V = a + bP$) and introduced the term adsorption. The term adsorption isotherm was introduced by Ostwald in 1885. McBain was the first one who separated the phenomena of adsorption and absorption in 1909 (Kiefer *et al.*, 2008). The next stages of the development of the theoretical basis of the adsorptive phenomena are widely known. At the end of the last century Gibbs developed the basis of thermodynamic theory of surface phenomena which is still officially called up as classical thermodynamic theory of heterogeneous systems. The first (famous) theoretical equation of adsorption isotherm was derived by Langmuir in 1916 which has been used actively by researchers up until today. Brunaer, Emmett and Teller used a derivation of the Langmuir isotherm to derive their BET adsorption isotherm (1938) also known as Brunaer-Emmett-Teller (BET) adsorption isotherm. Many famous names of scientists, who contributed considerably to the theoretical development of adsorption science, can be still mentioned. Their names and the main equations are summarized by Dabrowski (2001). However, the modern theory of adsorption is still developing.

It is impossible to predict a far future for adsorption science. The most serious "danger" on the way is new unexpected discoveries and (micro)revolutions which can cause multidimensional consequences. One examples of such a micro-revolution is the development of carbon-carbon composite materials composed from graphitized fibrous carbon (Buckley and Edie, 1993). The second micro-revolution is likened to the discovery of fullerene which is a special modification of carbon: a carbon allotrop composed entirely of carbon in the form of a hollow sphere (buckyball), an ellipsoid, a tube (nanotubes), or a plane (graphene) (Margadonna, 2008). It is difficult to make prognoses for the future of these materials even today.

The near future prognosis of adsorption science is pretty optimistic, first of all, due to the success of adsorption material science and the broadening variety of recently developed adsorptive materials and the considerably growing application of adsorptive materials in traditional fields and in new areas. New materials include relatively expensive adsorbents for catalysis, medicine, multifunctional materials as well as cheaper materials for domestic and industrial waste streams purification including biosorbents produced from industrial or agricultural waste for one time use. The promising recently developed materials are new classes of porous materials based on carbon, mineral, metallic, polymer matrix and their composite; new types of porous materials with strictly given pore distribution on radius, wide application of block constructions including on the composite carrier etc.

Fundamental scientific tasks of the near future for adsorption science are (1) the development of a modern theory of adsorption; and (2) a theory of controlled synthesis of porous materials, formulation of their structure and texture. Wide involvement of spectroscopy (including synchrotron based X-ray techniques), quantum-chemistry modeling, mathematic modeling of the adsorption processes and electron microscopy (Roddick-Lanzilotta *et al.*, 2002; Lefevre, 2004; Chubar *et al.*, 2009) is a reliable direction to improve understanding of the main processes taking place at the interface (adsorbent – adsorbat) and to develop a modern theory of adsorption and controlled syntheses of the adsorptive materials.

Wang and Chen (2009) think that the future of biosorption science and technology is facing great challenges. Many investigators think that the failure of the commercialization process is due to mainly nontechnical pitfalls involved in the commercialization of technological innovation. For the future of biosorption, it seems there are three trends to watch. One of them is using biosorbents in a hybrid technology as one of the adsorptive materials. The second (which would lead to using living microbial cells in the industrial scale treatment plants) is to improve understanding of the processes taking place at the interface: living microorganisms – metal ions (Chubar *et al.*, 2009) and to find the conditions where anionic target species (not only metal cations) can be also removed by living or nonliving microbs (Chubar *et al.*, 2008b). It was shown that living cells of *Shewanella putrefaciens* are capable to remove Mn^{2+} over month period and to form bioprecipiatates (or new biosorbents). The last trend is to develop a good commercial biosorbent just like a kind of ion-exchange resin, and to exploit the market with great endeavor (Volesky, 2007).

5.10 ACKNOWLEDGEMENT

King Abdullah University of Science and Technology (www.kaust.edu.sa) Center-in-Development Award to Utrecht University: www.sowacor.nl (award No. KUK-C1-017-12) is gratefully acknowledged.

5.11 REFERENCES

Abrams, I. M. and Milk, J. R. (1997) A history of the origin and development of macroporous ion-exchange resins. *Reactive & Functional Polymers* **35**, 7–22.

Ajmal, M., Rao, R. A. K., Ahmad, R. and Ahmad, J. (2000) Adsorption studies of *Citrus reticulata* (fruit peel of orange): Removal of Ni(II) from electroplating wastewater. *J. Hazard. Mater.* **79**(1–2), 117–131.

Ali, S. R. and Alam, T. Kamaluddin (2004) Interaction of tryptophan and phenylalanine with metal ferrocyanides and its relevance in chemical evolution. *Astrobiology* **4**, 420–426.

Altundogan, H. S., Altundogan, S., Tumen F. and Bildik, M. (2000) Arsenic removal from aqueous solutions by adsorption on red mud. *Waste Man.* **20**(8), 761–767.

Amphlett, C. B. (1964) *Inorganic Ion Exchangers*. Elsevier Pub. Co. Amsterdam, New York.

Anbia, M. and Ghaffari, A. (2009) Adsorption of phenolic compounds from aqueous solutions using carbon nanoporous adsorbent coated with polymer. *Applied Surface Science* **255**(23), 9487–9492

Annadurai, G., Juang, R. S. and Lee, D. J. (2003) Adsorption of heavy metals from water using banana and orange peel. *Water Sci. Technol.* **47**, 185–190.

Apel, M. L. and Torma, A. E. (1993) Immobilization of biomass for industrial application of biosorption. In *Biohydrometallurgical Technologies* (eds Torma, A. E., Apel, A. E. and Vrierley, C. L.) The Minerals, Metals and Materials Society, TMS Publication, Wyoming, USA, vol. II, pp. 25–33.

Ashby, M. F. and Bréchet, Y. J. M. (2003) Designing hybrid materials. *Acta Materialia*, **51**(19), 5801–5821.

Baerlocher, C., McCusker, L. B. and Olson, D. H. (2007) *Atlas of Zeolite Framework Types*, (revised edition). Elsevier, Amsterdam.

Bagreev, A. and Bandosz, T. (2004) Efficient hydrogen sulfide adsorbents obtained by pyrolysis of sewage sludge derived fertilizer modified with spent mineral oil. *Enviro. Sci. & Tech.* **38**(1), 345–351.
Barata-Rodrigues, P. M., Mays, T. J. and Moggridge, G. D. (2003) Structured carbon adsorbents from clays, zeolites and mesoporous aluminosilicate templates. *Carbon* **41**, 2231–2246.
Bartenev, B. K., Belchinskaya, L. I., Zhabin, A. B. and Khodosova, N. A. (2008) The approach to study the sorptive ability of mineral compounds as a function of their composition. *Vestnik of Voronesh University* **2**, 133–137 (in Russian).
Bengtsson, G. B., Bortun, A. I. and Strelko, V. V. (1996) Strontium binding properties of inorganic ion exchangers. *Journal of Radioanalytical and Nuclear Chemistry* **204**(1), 75–82.
Bhatnagar, A. and Jain, A. K. (2006) Column studies of phenols and Dyes removal from aqueous solutions utilizing fertilizer industry waste. *International J. Agr Res.* **1**(2), 161–168.
Bhatnagar, A., Choi, Y. H., Yoon, Y., Shin, Y., Jeon, B-H. and Kang, J-W. (2009) Bromate removal from water by granular ferric hydroxide (GFH). *J. Hazard. Mater.* **170**(1), 134–140.
Biniak, S., Szymanski, G., Siedlevski, J. and Swiatkovski, A. (1997) The characterization of activated carbons with oxygen and nitrogen surface groups. *Carbon* **35**, 1799–1810.
Birdi, K. S. (2000) *Surface and Colloid Chemistry: Principles and Applications.* CRC Press, Boca Raton, FL.
Blaney, L. M., Cinar, S. and SenGupta, A. K. (2007) Hybrid anion exchnager for trace phosphate removal from water and waste waters. *Water Res.* **41**, 1603–1613.
Bortun, A. I. and Strelko, V. V. (1992) Synthesis sorption properties and application of spherically granulated titanium and zirconium hydroxophosphates. In *Proc. of the 4th Intern. Conf. on Fundamentals of Adsorption*, Kyoto, 58–65.
Bortun, A. I., Bortun, L., Clearfield, A., Jaimez, E., Villa-García, M. A., García, J. R. and Rodríguez, R. (1997) Synthesis and characterization of the inorganic ion exchanger based on titanium 2-carboxyethylphosphonate, *Jour. Mater. Res.* **12**, 1122–1130.
Breg, U. and Klaringsbull, G. (1967) *Crytalline Structure of Mineral.* Moscow: Mir.
Bruna, F., Celis, R., Pavlovic, I., Barriga, C., Cornejo, J. and Ulibarri, M. A. (2009) Layered double hydroxides as adsorbents and carriers of the herbicide (4-chloro-2-methylphenoxy)acetic acid (MCPA): systems Mg-Al, Mg-Fe and Mg-Al-Fe. *J Hazard Mater.* **168**(2–3), 1476–81.
Buckley, J. D. and Edie, D. D. (1993) *Carbon-Carbon Materials and Composites.* William Andrew Publishing/Noyes.
Bumajdad, A., Eastie, J. and Mathew, A. (2009) Cerium oxide nanoparticles prepared in self-assemled systems. *Adv.Colloid Inter. Sci.* **147**(148), 56–66.
Chang, J.-S., Hwang, Y. K., Jhung, S. H., Hong, D-Y. and Seo, Y.-K. (2007) Porous organic-inorganic hybrid materials and adsorbent comprising the same, USPTO Application: 20090263621 dated Aug. 1, 2007.
Chen, X., Lam, K. F., Zhang, Q., Pan, B., Arruebo, M. and Yeung, K. L. (2009) Synthesis of highly selective magnetic mesoporous adsorbent. *J. Phys. Chem. C* **113**(22), 9804–9813.

Chitrakar, R., Kanoh, H., Miyai, Y. and Ooi, K. (2001) Recovery of lithium from seawater using manganese oxide adsorbent ($H_{1.6}Mn_{1.6}O_4$) derived from $Li_{1.6}Mn_{1.6}O_4$. *Ind. Eng. Chem. Res.* **40**(9), 2054–2058.

Chubar, N., Avramut, C., Behrends, T. and Van Cappellen, P. (2009) Long-term sorption of Mn^{2+} by viable and autoclaved *Shewanella putrefaciens:* FTIR, XAFS and SEM characterization of the precipitates synthesized by the (initially) live bacteria. ECASIA '09 European Conference on *Surface & Interface Analysis*, Antalya, Turkey, October 17–23. Abstract book, p. 12.

Chubar, N., Behrends, T. and Van Cappellen, P. (2008b) Biosorption of metals (Cu^{2+}, Zn^{2+}) and anions (F^-, H_2PO_4-) by viable and autoclaved cells of the gram-negative bacterium *Shewanella putrefaciens*. *Colloids and Surfaces B: Biointerfaces* **65**, 126–133.

Chubar, N. I., Kanibolotskiy, V. A., Strelko, V. V. and Shaposhnikova, T. O. (2008a) Adsorption of anions onto inorganic ion exchangers. In *Selective Sorption and Catalysis on Active Carbons and Inorganic Ion Exchangers*, Strelko, V. (ed). Press: Naukova Dumka, Kiev, Vol. 2.

Chubar, N. I., Kanibolotskiy, V. A., Strelko, V. V., Shaposhnikova, T. O., Milgrandt, V. G., Zhuravlev, I. Z., Gallios, G. G. and Samanidou, V. F. (2005b) Sorption of phosphate ions on the hydrous oxides. *Colloids and Surfaces: A* **255**(1–3), 55–63.

Chubar, N. I., Kouts, V. S., Kanibolotskiy, V. A. and Strelko, V. V. (2006) Adsorption of anions onto sol-gel generated hydrous oxides. In NATO ARW series book, *Viable Methods of Soil and Water Pollution Monitoring, Protection and Remediation: Development and Use*, (ed.) Twardowska, I. Kluwer Publisher, The Netherlands, **69**, 323–338.

Chubar, N. I., Kouts, V. S., Samanidou, V. F., Gallios, G. G., Kanibolotskiy, V. A. and Strelko, V. V. (2005a) Sorption of fluoride, bromide, bromate and chloride ions on the novel ion exchangers. *J. Colloid Interf. Sci.* **291**(1), 67–74.

Chubar, N. I., Machado, R., Neiva Correia, M. J. and Rodrigeus de Carvalho, J. M. (2003b) Biosorption of copper, zinc and nickel by grape-stalks and cork biomasses. In NATO ARW series book: *Role of Interfaces in Environmental Protection*, (ed. Barany, S.). Kluwer Publisher, The Netherlands, pp. 339–353.

Chubar, N. I., Neiva Correia, M. J. and Rodrigeus de Carvalho, J. M. (2003c) Cork biomass as biosorbent for copper, zinc and nickel. *Colloids and Surfaces: A* **23**(1–3), 57–66.

Chubar, N. I., Neiva Correia, M. J. and Rodrigeus de Carvalho, J. M. (2004) Heavy metals biosorption on cork biomass: effect of pretreatment. *Colloids and Surfaces: A* **238**(1–3), 51–58.

Chubar, N. I., Strelko, V. V., Rodrigeus de Carvalho, J. M. and Neiva Correia, M. J. (2003a) Cork biomass as sorbent for color metals. *Water Chemistry and Technology* **25**(1), 33–38.

Clarke, T. D. and Wai, C. M. (1998) Selective removal of cesium from acid solutions with immobilised copper ferrocyanide. *Analytic Chemistry* **70**(17), 3708–3711

Clearfield, A. (1982) *Inorganic Ion Exchange Materials*. CRC Press Inc., Boca Raton, Florida.

Cochrane, E. L., Lu, S., Gibb, S. W. and Villaescusa, I. (2006) A comparison of low-cost biosorbents and commercial sorbents for the removal of copper from aqueous media, *J. Hazard. Mater. B* **137**, 198–206.

Colella, C. and Gualtieri, A. F. (2007) Cronstedt's zeolite. *Microp. Mesop. Mater.* **105**, 213–221.

Conant, J. B. (1950) *The Overthrow of Phlogiston Theory: The Chemical Revolution of 1775–1789*. Harvard University Press, Cambridge.

Cornelissen, E. R., Moreau, N., Siegers, W. G., Abrahamse, A. J., Rietverld, L. C., Grefte, A., Dignum, M., Amy, G. and Wessels, L. P. (2008) Selection of anionic exchange resins for removal of natural organic matter (NOM) fractions. *Water Res.* **42**(1–2), 413–423.

Dabrowski, A. (2001) Adsorption – from theory to practice. *Advances in Colloid Interface Sci.* **93**(1–3), 135–224.

Danish Ministry of the Environment (2007) BEK 1449 from 11, App. 1b, 2007.

Davis, T. A., Volesky, B. and Mucci, A. (2003) A review of the biochemistry of heavy metal biosorption by brown algae (review). *Water Res.* **37**, 4311–4330.

Drozdnik, I .D. (1997) Properties of carbon sorbents from coals with various degrees of metamorphism. *Fuel and Energy Abstracts* **3**(1), 29.

Elshazly, A .H. and Konsowa, A. H. (2003) Removal of nickel ions from wastewater using a cation-exchange resin in a batch-stirred tank reactor. *Desalination* **158**(1–3), 189–193.

Environmental Protection Agency (EPA) (2000) *National Primary Drinking Water Regulations*, http://www.epa.gov/safewater/sdwa/current_regs.html

Erdem-Şenatalar, A. and Tatlıer, M. (2000) Effects of fractality on the accessible surface area values of zeolite adsorbents. *Chaos, Solutions & Fractals* **11**(6), 953–960.

Ferguson, J. F. and Gavis, J. (1972) A review of the arsenic cycle in natural waters. *Water Res.* **6**, 1259–1274.

Fettig, J. (1999) Removal of humic substances by adsorption/ion exchange. *Water Sci. Technol.* **40**(9), 173–182.

Figuiredo, J. L., Pereira, M. F. R., Treitas, M. M. A. and Orfao, J. J. M. (1999) Modification of the surface chemistry of activated carbons. *Carbon* **37**, 1379–1389.

Fiore, S., Cavalcante, F. and Belviso, C. (2009) Patent application title: *Synthesis of Zeolites from Fly Ash*. Patent application number: 20090257948.

Fletcher, A. J., Kennedy, M. J., Zhao, X. B., Bell, J. B. and Mark Thomas, K. (2008) Adsorption of organic vapour pollutants on activated carbon. In NATO Science for Peace and Security Series C, *Environmental Security: Recent Advances in Adsorption Processes for Environmental Protection and Security*. Springer Netherlands Press, 29–54.

Franzreb, M., Hoell, W. H. and Eberle, S. H. (1995) Liquid-phase mass transfer in multicomponent ion exchange. 2. Systems with irreversible chemical reactions in the film. *Ind. Eng. Chem. Res.* **34**(8), 2670–2675.

Frost, R., Musumeci, A. W., Bostrom, T., Adebajo, M. O., Weier, M. L. and Martens, W. (2005) Thermal decomposition of hydrotalcite with chromate, molybdate or sulfate in the interlayer. *Thermochim. Acta* **429**, 179–187.

Górka, A., Bochenek, B., Warchoł, I., Kaczmarski, K. and Antos, D. (2008) Ion exchange kinetics in removal of small ions. Effect of salt concentration on inter- and intraparticle diffusion. *Chem. Eng. Sci.* **63**(3), 637–650.

Gregg, S. J. and Sing, S. W. (1982) *Absorption, surface area and porosity*, 2nd Ed., Academic press, New York.

Gun'ko, V. M. and Mikhalovky, S. V. (2004) Evaluation of slitlike porocity of carbon adsorbents. *Carbon* **42**, 843–849.
Gun'ko, V. M., Turov, V. V., Skubiszewska-Zieba, J., Leboda, R., Tsarko, M. D. and Palijczuk, D. (2003) *Ap. Surf. Sci.* **214**, 178–189.
Gunter, D. and Werner, S. (1997) US patent 4132671. *Process for the preparation of carbon black pellets.* http://www.freepatentsonline.com/4132671.html.
Gupta, S. K. and Chen, K. Y. (1978) Arsenic removal by adsorption. *J. Water Pollut. Contr. Fed.* **50**(3), 493–506.
Hall, A. and Stamatakis, M. G. (2000) Hydrotalcite and an amorphous clay minerals in high-magnesium mudstones from the Kozani basin, Greece. *J. of Sedimentary Researcher* **70**(3), 549–558.
Hathaway, S. W. and Rubel, F. (1987) Removing arsenic from drinking water. *J. Am. Water Works Assoc.* **79**(8) 61–65.
Helfferich, F. (1962) *Ion Exchange.* McGraw Hill, New York.
Henmi, T. and Sakagami, E. *Method of Producing Artificial Zeolite. European Patent EP0963949.* Filling date: 06/11/1999. Publication Date: 04/14/2004.
Henmi, T., Nakamura, T., Ubukata, T., Matsuda, H. and Tada, S. (2009) *Method of manufacturing artificial zeolite.* IPC8 Class: AC01B3904FI, USPC Class: 423703.
Hennessey, S. M., Friend, J., Elander, R. T. and Tucker, M. P. (2009) *Biomass pretreatment,* US Government Patent application N 20090053770. http://www.freshpatents.com/-dt20090226ptan20090053770.php
Hideki, K., Shigeki, K., Liang, R. and Atsushi, U, (2000) Manganese Oxide(Mn_2O_3) as adsorbent for cadmium. *Journal of Japan Society on Water Environment* **23**(2) 116–121.
Hoell, W. H. and Kalinichev, A. (2004) The theory of formation of surface complexes and its application to the description of multicomponent dynamic sorption systems. *Russ. Chem. Rev.* **73**, 351–370.
Hoell, W. H., Zhao, X. and He, S. (2002) Elimination of trace heavy metals from drinking water by means of weakly basic anion exchangers. *J. Water SRT – Aqua* **51**, 165–172.
Holtzapple, M. T., Lundeen, J. E., Sturgiss, R., Lewis, J. E. and Dale, B. E. (1992) Pretreatment of lignocellulosic municipal solid waste by ammonia fiber explosion. *Appl. Biochem. Biotechnology* **34**, 5–21.
Huang, C.-T. and Wu, G. (1999) Improvement of Cs leaching resistance of solidified radwastes with copper ferrocyanide (CFC)-vermiculite. *Waste Man.* **19**(4), 263–268.
Id., *Science de l'air: Studi su Felice Fontana,* Brenner: Cosenza, 1991.
Irving M. Abrams and John R. Millar (1997) "A history of the origin and development of macroporous ion-exchange resins", Reactive & Functional Polymers, **35**, 7–22.
IUPAC Recommendations 2003 (2004) *Pure Appl. Chem.*, **76**(4), 889–906.
Jain, A. K., Agrawal, S. and Singh, R. P. (1980) Selective cation exchange separation of secium(I) on chromium ferricyanide gel. *Anal. Chem.* **52**, 1364–1366.
Kang, S.-K., Choo, K-H. and Lim, K-H. (2003) Use of iron oxide particles as adsorbents to enhance phosphorus removal from secondary wastewater effluent. *Sep. Sci. Technol.* **38**(15), 3853–3874.
Karcher, S., Kornmüller, A. and Jekel, M. (2002) Anion exchange resins for removal of reactive dyes from textile wastewaters. *Water Res.* **36**(19), 4717–4724.

Katsoyiannis, I. A., Zouboulis, A. I. and Jekel, M. (2004) Kinetics of bacterial As(III) oxidation and subsequent As(V) removal by sorption onto biogenic oxides during groundwater treatment. *Ind. Eng. Chem. Res.* **43**(2), 486–493.

Kawamura, S., Kurotaki, K. and Izawa, M. (1969) Preparation and ion-exchange behavior of potassium ferrocyanide. *Bulletin of the Chemical Society of Japan* **42**, 3003–3004.

Kiefer, S. and Robens, E. (2008) Some of intriguing items in the history of volumetic and gravimetric adsorption measurements. *Journal of Thermal Analysis and Calorimetry* **94**(3), 613–618.

Kim, B. K., Kim, S.-H. and Alam, T. Kamaluddin (2000) Interaction of 2-amino, 3-amino and 4-aminopyridines with nickel and cobalt ferrocyanides. *Engineering Aspects* **162**(1), 89–97.

Kim, Y., Kim, C., Choi, I., Rengaraj, S. and Yi, J (2004) Arsenic removal using mesoporous alumina prepared via a templating method. *Environ. Sci. Technol.* **38** (3), 924–931.

Kim, Yu-H. (2000) Adsorption characteristics of cobalt on ZrO_2 and Al_2O_3 adsorbents in high-temperature water. *Sep. Sci. Technol.* **35**(14), 2327–2341.

Kogtev, L., Park, J. K., Pyo, J. K. and Mo, Y. K. (1998) *Biosorbent for heavy metals prepared from biomass*, United States Patent 5789204.

Kononova, O. N., Kholmogorov, A. G., Lukianov, A. N., Kachin, S. V., Pashkov, G. L. and Kononov, Y. S. (2001) Sorption of Zn(II), Cu(II), Fe(III) on carbon adsorbents from manganese sulfate solutions. *Carbon* **39**, 383–387.

Kumar, P., Barrett, M. B., Delwiche, M. J. and Stroeve, P. (2009) Methods for pretreatment of Lignocellulosic Biomass for efficient hydrolysis and biofuel production. *Ind. Eng. Chem. Res.* **48**(8), 3713–3729.

Kunin, R. (1982) *Ion Exchange Resins*. Robert E. Krieger Publishing, Company, Melbourne, FL, pp. 3 and 130.

Lach, J., Okoniewska, E., Neczaj, E. and Kacprzak, M. (2007) Removal of Cr(III) cations and Cr(VI) anions on activated carbons oxidized by CO2. *Desalination* **206**, 259–269.

Lefevre, G. (2004) In situ Fourier-transform infrared spectroscopy studies of inorganic ions adsorption on metal oxides and hydroxides. *Adv. Colloid Interf. Sci.* **107**, 109–123.

Likholobov, V. A. (2007) Institute of Hydrocarbon Processing, Siberian Branch, Russian Academy of Sciences advances of science and practice in solving problems of chemical hydrocarbon processing. *Russian Journal of General Chemistry* **12**(17), 1070–3632.

Liu, Z., Wei, Y., Qi, Y., Liu, X., Zhao, Y. and Liu Z. (2006) Synthesis of ordered mesoporous Zr-P-Al materials with high thermal stability. *Microp. Mesop. Mater.* **91**, 225–232.

Long, C., Lu, J. D., Li, A., Hu, D., Liu, F. and Zhang, Q. (2008) Adsorption of naphthalene onto the carbon adsorbent from waste ion exchange resin: Equilibrium and kinetic characteristics. *J. Hazard. Mater.* **150**, 656–661.

Lopes, T., Ramos, B. E., Gomes, R., Novaro, O., Acosta, D. and Figueras, F. (1996) Synthesis and characterisation of sol-gel hydrotalcites, structure and texture. *Langmuir* **12**, 189–192.

Loureiro, J. M. and Kartel M. T. (2006) *Combined and Hybrid Adsorbents: Fundamentals and Application*. Springer-Verlag New York Inc.

Lucy, C. A. (2003) Evolution of ion-exchange: From Moses to the Manhattan project to modern times. *J. Chromatography A* **1000**(1–2), 711–724.

Malikov, I. N., Noskova, Yu, A., Karaseva, M. S. and Perederii, M. A. (2007) Granulated sorbents from wood wastes. *Solid Fuel Chemistry* **41**(2), 100–106.

Manceau, A., Drits, V. A., Silvester, E., Bartoli, C. and Lanson, B. (1997) Structural mechanism of Co^{2+} oxidation by the phyllomanganate buserite. *Am. Mineral.* **82**, 1150–1175.

Mandich, N. V., Lalvani, S. B., Wiltkowski, T. and Lalvani, L. S. (1998) Selective removal of chromate anion by a new carbon adsorbent. *Metal Finishing* **96**(5), 39–44.

Manjare, S. D., Sadique, M. H. and Ghoshal, A. K. (2005) Equilibrium and kinetics studies for As(III) adsorption on activated alumina and activated carbon. *Environ. Technol.* **26**(12), 1403–1410.

Manning, B. A., Fendorf, S. E., Bostick, B. and Suarez, D. L. (2002) Arsenic(III) oxidation and arsenic(V) adsorption reactions on synthetic birnessite. *Environ Sci. Technol.* **36**(5), 976–981.

Manos, Manolis J., Nan Ding and Kanatzidis, Mercouri, G. (2008) Layered metal sulfides: Exceptionally selective agents for radioactive strontium removal. *PNAS* **5**(10), 3696-3699.

Mansoor, A. and Moradi, S. I. (2009) Removal of naphthalene from petrochemical wastewater streams using carbon nanoporous adsorbent. *Ap. Surf. Sci.* **255**, 5041–5047.

Margadonna, S. (2008) *Fullerene-Related Materials: Recent Advances in Their Chemistry and Physics*, 1st edition, 700 pp, Springer.

Marshall, W. E. and Wartelle, L. H. (2004) An *anion exchange* resin from soy-bean hulls. *J. Technol. Biotechnol.* **79**, 1286–1292.

Matulionytė, J., Vengris, T., Ragauskas, R. and Padarauskas, A. (2007) Removal of various components from fixing rinse water by anion-exchange resins. *Desalination*, **208**, 81–88.

Meleshevych, I., Pakhovchyshyn, S., Kanibolotsky, V. and Strelko, V. (2007) Rheological properties of hydrated zirconium dioxide. *Colloids and Surfaces A* **298**, 274–279.

Melián-Cabrera, I., Kapteijn, F. and Moulijn, J. A. (2005) Innovations in the synthesis of Fe-(exchanged)-zeolites. *Catal. Today* **110**, 255–263.

Miers, J. A. (1995) Regulation of ion exchange resins for the food, water and beverage industries. *Reactive Polymers* **24**, 99–107.

Misra, C. and Genito, J. R. (1993) US Patent 5270278 – *Alumina coated with a layer of carbon as an absorbent*. Issued Dec. 14, 1993.

Moore, J. N., Walker, J. R. and Hayes, T. H. (1990) Reaction scheme for the oxidation of As(III) to arsenic(V) by birnessite. *Clays Clay Miner.* **38**, 549–555.

Mrowiec-Białoń, J., Jarzbski, A. J., Lachowski, A. I., Malinowski, J. J. and Aristov Y. I. (1997) Effective inorganic hybrid adsorbents of water vapor by the sol-gel method. *Chem. Mater.* **9**(11), 2486–2490.

Mui, E. L. K., Ko, D. C. K. and McKay, G. (2004) Production of active carbons from waste tyres – a review. *Carbon* **42**, 2789–2805.

Namasivayam, C., Sangeetha, D. and Gunasekaran, R. (2007) Removal of anions, heavy metals, organics and dyes from water by adsorption onto a new activated carbon from Jatropha Husk, in agro-industrial solid waste. *Process Safety and Environmental Protection* **85**(2), 181–184.

National Health and Medical Research Centre (1996) *Australian drinking water guidelines – Summary, Australian Water and Wastewater Association*, Artamon.

Natural Recourses Defense Council (2000) Arsenic and old laws: A scientific and public health analysis of arsenic occurrence in drinking water, its health effects, and EPA's outdated arsenic tap water standard, available at: ww.nrdc.org/water/drinking/arsenic/aolinx.asp.

New Jersey Department of Environmental Protection (2004) Safe drinking water act regulations N.J.A.C., **7**(10), 1–83.

Novoselova, L. Y. and Sirotkina, E. E. (2008) Peat-based carbons for purification of the contaminated environments. *Solid Fuel Chemistry* **42**(4), 251–262.

Ooi, K., Miyai, Y. and Katoh, S. (1986) Recovery of. lithium from seawater by manganese oxide adsorbent. *Sep. Sci. Technol.* **21**(8), 755–766.

Ouvrard, S., Simonnot, M. O. and Sardin, M. (2002a) Reactive behavior of natural manganese oxides toward the adsorption of phosphate and arsenate. *Ind. Eng. Chem. Res.* **41**, 2785–2791.

Ouvrard, S., Simonnot, M. O., Donato, P. and Sardin, M. (2002b) Diffusion-controlled adsorption of arsenate on a natural manganese oxide. *Ind. Eng. Chem. Res.* **41**, 6194–6199.

Palmer, S. and Frost, R. L. (2009) The effect of synthesis temperature on the formation of hydrotalcites in bayer liquor: A vibration spectroscopic analysis. *Applied Spectroscopy* **63**(7), 748–752.

Park, H. G., Kim, T. W., Chae, M. Y. and Yoo, I.-K. (2007) Activated carbon-containing alginate adsorbent for the simultaneous removal of heavy metals and toxic organics. *Process Biochem.* **14**(10), 1371–1377.

Patzak, M., Dostalek, P., Fogarty, R. V., Safarik, I. and Tobin, J. M. (2004) Development of magnetic biosorbents for metal uptake. *Biotechnol. Techniques* **11**(7), 483–487.

Petrus, R. and Warchol, I. (2005) Heavy metal removal by clinoptilolite. An equilibrium study in multi-component systems. *Water Res.* **39**(5), 819–830.

Pokonova, Y. V. (2001) Carbon adsorbents from coal pitch. *Chemistry and Technology of Fuels* **37**(3), 207–211.

Psareva, T. S., Zakutevskyy, O., Chubar, N.I., Strelko V. V. Shaposhnikova, T. O., Rodriges de Carvalho, J. M. and Neiva Correia, J. M. (2005) Uranium sorption on cork biomass. *Colloid and Surfaces: A* **252**(2–3), 231–236.

Quirarte-Escalante C. A., Soto, V., De La Cruz, W., Porras, G. R., Rangel, G., Manriques, R. and Gomez-Salazar, S. (2009) Synthesis of hybrid adsorbents combining sol-gel processing and molecular imprinting applied to lead removal from aqueous streams. *Chemistry of Materials* **21**(8), 1439–1450.

Radionuclides (2004) Final Rule, 40 CFR Parts 9, 141 and 142.

Roberts, L. C., Hug, S. J., Ruettimann, T., Khan, A. W. and Rahman, M. T. (2004) Arsenic removal with iron(II) and iron(III) in waters with high silicate and phosphate concentrations. *Environ. Sci. Technol.* **38**, 307–315.

Roddick-Lanzilotta, A. J., McQuillan, A. J. and Craw, D. (2002) Infrared spectroscopic characterization of arsenate(V) ion adsorption from mine waters, Macraes mine, New Zealand. *Appl. Geochem.* **17**, 445–454.

Rodrigues-Reinoso, F. and Molina-Sabio, M. (1998) Textural and chemical characterization of microporous carbons. *Adv. Colloid Interface Sci.* **76–77**, 271–294.

Saha, B., Bains, R. and Greenwood, F. (2005) Physicochemical characterization of granular ferric hydroxide (GFH) for arsenic(V) sorption from water. *Sep. Sci. Technol.* **40**(14), 2909–2932.

Schiewer, S. and Patil, P. B. (2008) Pectin-rich fruit wastes as biosorbents for heavy metal removal: Equilibrium and kinetics. *Biores. Technol.* **99**(6), 1896–1903.

Sengupta, S. and SenGupta, A. K. (1997) Heavy-metal separation from sludge using chelating ion exchangers with nontraditional morphology. *Reactive & Functional Polymers* **35**, 111–134.

Seniavin, M. M. (1981) *Ion Exchange*. Nauka Publishing, Moscow. (In Russian).

Sharygin, L., Muromskiy, A., Kalyagina, M. (2007) A granular inorganic cation-exchanger selective to cesium. *Journal of Nuclear Science and Technology* **44**(5), 767–773.

Shectman, J. (2003), *Groundbreaking Scientific Experiments, Inventions, and Discoveries of the 18th Century*. Connecticut: Greenwood Press.

Shen, W., Zhijie, L. and Liu, Y. (2008) Surface chemical functional groups modification of porous carbon. *Recent Patents on Chemical Engineering* **1**, 27–40.

Singh, T. S. and Pant, K. K. (2004) Equilibrium, kinetics and thermodynamic studies for adsorption of As(III) on activated alumina. *Sep. Purif. Technol.* **36**, 139–147.

Sparks, D. E., Morgan, T., Patterson, P. M., Adam, T., Morris, E. and Crocker, M. (2008) New sulfur adsorbents derived from layered double hydroxides I: Synthesis and COS adsorption. *Appl.Catal. B- Environ.* **82**(3–4), 190–198.

Tanaka, Y. and Tsuji, M. (1994) New synthetic method of producing α-manganese oxide for potassium selective adsorbent. *Materials Research Bulletin* **29**(11), 1183–1191.

Tananaev, I. B., Seifer, G. B., Kharitonov, Y. Y., Kuznetsov, B. G. and Korolkov A. P. (eds) (1971) *Chemistry of Ferrocyanides*Nauka press, Moscow.

Thirunavukkarasu, O. S., Viraghavan, T. and Suramanian, K.S. (2003) Arsenic removal from drinking water using iron-oxide coated sand. *Water Air Soil Pollut.* **142**, 95–111.

Thompson, H. S. (1850) Absorbent power of soils. *J. R. Agric. Soc. Engl.* **11**, 68.

Tomlinson, A. A. G. (1998) Modern Zeolites. Press: Trans Tech Publications Inc. Laubisrutistr, Switzerland.

Tomoyuki, K., Kosuke, A., Toshio S. and Yoshio O. (2007) Synthesis and characterization of Si-Fe-Mg mixed hydrous oxides as harmful ions removal materials. *J Soc Inorg Mater Jpn* **14**(327), 1345–3769.

Treacy, M. M. J. and Higgins, J. B. (2007) *Collection of Simulated XRD Powder Diffraction Patterns for Zeolites, 5th revised edition*. Elsevier, Amsterdam.

Kim, S. J., Park, Y. Q. and Moon, H. (2007) Removal of copper ions by a cation-exchange resin in a semifluidized bed. *Korean Journal of Chemical Engineering* **15**(4), 417–422.

United States Patent 5865898, http://www.freepatentsonline.com/5865898.html.

USEPA (1999) Technologies and Costs for Removal of Arsenic from Drinking Water, Draft Report, EPA-815-R-00-012, Washington, DC.

Valentine, R. L., Mulholland, T. S. and Splinter, R. C. (1992) *Radium removal using sorption to filter preformed hydrous manganese oxides*. Report for the American Water Works. Association Research Foundation.

Valinurova, E. R., Kadyrova, A. D., Sharafieva, L. R. and Kudasheva, F. Kh. (2008) Use of activated carbon materials for wastewater treatment to remove Ni(II), Co(II), and Cu(II) ions. *Russian Journal of Applied Chemistry* **81**(11), 1939–1941.

Venkatesan, K. A., SathiSasidharan, N. and Wattal, P. K. (1997) Sorption of radioactive strontium on a silica-titania mixed hydrous oxide gel. *Journ. of Analytical and Nuclear Chemistry* **20**(1), 55–58.

Volesky, B. (1990a) Biosorption by fungal biomass. In *Biosorption of Heavy Metals*. Volesky, B. (ed) Florida: CRC press, pp. 139–171.
Volesky, B. (1990b) Introduction. In *Biosorption of Heavy Metals*. Volesky, B. (ed) Florida: CRC press, pp. 3–5.
Volesky, B. (1990c) Removal and recovery of heavy metals by biosorption. In *Biosorption of Heavy Metals*. Volesky, B. (ed), Florida: CRC press, pp. 8–43.
Volesky, B. (1994) Advances in biosorption of metals — selection of biomass types. *FEMS, Microbiol Rev.* **14**, 291–302.
Volesky, B. (2001) Detoxification of metal-bearing effluents: biosorption for the next century. *Hydrometallurgy* **59**, 203–216.
Volesky, B. (2003) Biosorption process simulation tools. *Hydrometallurgy* **71**, 179–90.
Volesky, B. (2007) Biosorption and me. *Water Res.* **41**, 4017–29.
Volesky, B. and Holan, Z. R. (1995) Biosorption of heavy metals. *Biotechnol Prog.* **11**, 235–50.
Volesky, B. and Naja, G. (2005) Biosorption: application strategies. *16th Internat. Biotechnol. Symp. Compress Co., Cape Town, South Africa*.
Volesky, B, May, H. and Holan, Z. R. (1993) Cadmium biosorption by Saccharomyces cerevisiae. *Biotechnol Bioeng.* **41**, 826–829.
Volgin, V. V. (1979) News of the Academy of Sciences. *Inorganic Materials* **15**, 1084–1089.
Vollmer, D. L. and Gross, M. L. (2005) Cation-exchange resins for removal of alkali metal cations from oligonucleotide samples for fast atom bombardment mass spectrometry. *J. Mass Spectrometer* **30**(1), 113–118.
Voyutsky, S. S. (1964) *Colloid Chemistry*. Chemistry Press, Moscow.
Walsh, R. (2008) *Development of a biosorption column utilizing seaweed based biosorbents for the removal of metals from industrial waste streams*. PhD thesis, Waterford Institute of Technology: http://repository.wit.ie/1031/.
Walt, D. K. (2003) *Applied Biochemistry and Biotechnology*, Humana Press, **10**(1–3), Spring 2003.
Wang, J. and Chen, C. (2009) Biosorbents for heavy metals removal and their future. *Biotechnol. Adv.* **27**(2), 195–226.
Wang, J. S. and Wai, C. M. (2004) Arsenic in drinking water—a global environmental problem. *J. Chem. Educ.* **81**, 207–213.
Wang, L. K., Hung, Y-T., Lo, H. H. and Yapijakis, C. (eds) (2006) *Waste Treatment in the Process Industries*. Taylor & Francis Group, New York.
Watanabe, S. Velu, S. Ma, X. and Song, C. S. (2003) Preprint Paper – American Chemical Society, *Division Fuel Chemistry* **48**(2), 695–696.
Wayne, E. M. and Wartelle, L. (2004) An anion exchange resin from soybean hulls. *J. of Chemical Technol. Biotechnol.* **79**(11), 1286–1292.
Whitehead, P. (2007) Medicine from animal cell culture. In *Water Purity and Regulations*, 696 pp. Glyn Stacey and John Davis (eds), John Wiley & Sons.
Wilkie, J. A. and Hering, J. G. (1996) Adsorption of arsenic onto hydrous ferric oxide: Effects of adsorbate/adsorbent ratios and co-occurring solutes. *Colloid Surf. A: Physicochem. Eng. Aspects* **107**, 97–110.
Yamamoto T., Taniguchi A., Dev S., Kubota E., Osakada K. and Kubota K. (1990) New organosols of nickel sulfides, palladium sulfides, manganese sulfide, and mixed

metal sulfides and their use in preparation of semiconducting polymer-metal sulfide composites. *Colloid Polymer Science* **269**(10), 969–971.

Yin, C. Y., Arpna, M. K. and Daud, W. M. A. W. (2007) Review of modification of activated carbon for enhancing contaminant uptake from aqueous solutions. *Sep. Purif.Ttechnol.* **52**, 403–415.

Zhang, W., Zou, L.· and Wang, L. (2009) Photocatalytic TiO_2/adsorbent nanocomposites prepared via wet chemical impregnation for wastewater treatment: A review. *Appl. Catal A-General* **371**(1–2), 1–9.

Zhang, Y., Yang, H., Zhou, K. and Ping, Z. (2007) Synthesis of an affinity adsorbent based on silica gel and its application in endotoxin removal. *Reactive & Functional Polymers* **67**, 728–736.

Zhuravlev, I., Kanibolotsky, V., Strelko, V., Gallios, G. and Strelko, V. Jr. (2004) Novel high porous spherically granulated ferrophosphatesilicate gels. *Materials Research Bulletin* **39**(4–5), 737–744.

Zhuravlev, I. Z. (2005) *Sol-gel synthesis and properties of the ion exchangers based on composite phosphates of polyvalent metals and silica*, Ph.D. Thesis (Chemistry), Kiev, Ukraine.

Zhuravlev, I. Z. and Strelko V. V. (2006) Template effect of the M^{3+}-cations in the course of the synthesis of high dispersed titanium and zirconium phosphate. In *Combined and Hybrid Adsorbent, NATO Security through Science Series*, Springer, Netherlands. pp. 93–98.

Zhuravlev, I., Zakutevsky, O., Psareva, T., Kanibolotsky, V., Strelko, V., Taffet, M. and Gallios, G. (2002) Uranium sorption on amorphous titanium and zirconium phosphates modified by Al^{3+} or Fe^{3+} ions. *Journal of Radioanalytical and Nuclear Chemistry* **254**(1), 85–89.

Zolotov, Yu. A. (1998), Analytical Chemistry in Russia. *Fresenius J. Anal. Chem.* **361**, 223–226.

Chapter 6
Physico-chemical treatment of Micropollutants: coagulation and membrane processes

O. Lefebvre, Lai Yoke Lee and How Yong Ng

6.1 COAGULATION

Coagulation is one of the oldest processes commonly used for removal of suspended solids, while enhanced coagulation using inorganic or polymeric coagulants has been applied for micropollutant removal in water and wastewater treatment. Enhanced coagulation is a popular option for upgrading existing treatment facility to incorporate removal of selected micropollutants as it offers several advantages especially in terms of cost effectiveness, simplicity in design, and ease of operation. However, the major drawback in enhanced coagulation for micropollutant removal is the low to insignificant effect on some of the micropollutants in drinking water and wastewater treatment.

©2010 IWA Publishing. *Treatment of Micropollutants in Water and Wastewater.* Edited by Jurate Virkutyte, Veeriah Jegatheesan and Rajender S. Varma. ISBN: 9781843393160. Published by IWA Publishing, London, UK.

Improvement in the coagulation process for micropollutant removal has been reported using oxidation-coagulation/precipitation and coagulation-membrane separation (Bodzek and Dudziak, 2006; Lee *et al.,* 2009; Lim and Kim, 2009). Pharmaceutical compounds such as diclofenac using Fe (VI) oxidation-coagulation attained more than 95% removal at 5 mg Fe/L (Lee *et al.,* 2009) as compared with only more than 65% with enhanced coagulation using 50 mg/L $FeCl_3$ or $Al_2(SO_4)_3$ (Carballa *et al.,* 2005). Integrating coagulation-NF system was able to provide at least 18.5% increase in the removal of estrogen as compared with using NF system alone (Bodzek and Dudziak, 2006). Applications of enhanced coagulation and oxidation-coagulation for micropollutant removal, and mechanisms and controlling factors in these processes will be presented in the following sections.

6.1.1 Enhanced coagulation

In drinking water treatment using conventional filtration treatment, improved disinfection by-products (DBP) removal is achieved by applying enhanced coagulation. Enhanced coagulation in drinking water treatment is defined as the addition of excess coagulant and possibly accompanied by reduced coagulation pH (Edwards *et al.,* 2003). This has been introduced as a requirement in the Environmental Protection Agency (EPA) DBP Rule (Freese *et al.,* 2001).

A cost comparison between the advanced water treatment processes such as ozonation and granular activated carbon (GAC) with enhanced coagulation concluded that the latter is more cost-effective for smaller treatment works with capacity less than 175 million litres per day and for cleaner influent quality (i.e., with TOC concentration less than 5 mg/L) (Freese *et al.,* 2001). The cost of treatment of the clean influent water (TOC <5 mg/) using enhanced coagulation was at least 25% lower for a 175 million litres per day plant as compared to a plant with double of this capacity. Table 6.1 summarizes the removals of some of the micropollutants using enhanced coagulation method.

Sorption of micropollutants onto coagulant surface is the main removal mechanism in enhanced coagulation. The solid-liquid partitioning coefficient (K_d), which is the concentration ratio between solid and liquid phase at equilibrium conditions (Hemond and Fechner-Levy, 2000), can be used to determine the sorption behaviour of micropollutants. Two interactions between the micropollutants and the coagulants are involved; namely lipophilic interactions which are indicated by the octanol-water (K_{ow}) and organic carbon (K_{oc}) partition coefficients, and electrostatic interactions which are associated with the dissociation constant (pK_a) of the pollutants and the weak Van der Waals bonds with the coagulants (Stumm and Morgan, 1996). Therefore, the physical-chemical properties of the micropollutants and how they change with the aqueous environment are important factors that influence the micropollutants removal. Parameters that influence the physical-chemical properties of micropollutants in the aqueous environment include pH, alkalinity, choice of coagulants and their optimum concentrations (McGhee, 1991).

Table 6.1 Micropollutant removal using enhanced coagulation methods

Micropollutant	log K_{ow}	pK_a	Coagulant	Dosage	Removal Efficiency (%)
Anthropogenic					
Trihalomethane formation potential (THMFP)[1]	–	–	Iron (III) chloride	Up to 30 mg/L	Up to 40%
Fragrances					
Galaxolide[2]	5.9–6.3	–	Aluminum polychloride	17.5% w/w	63
Tonalide[2]	4.6–6.4	–	Aluminum polychloride	17.5% w/w	71
Pharmaceutical					
Diazepam[2]	2.5–3.0	3.3–3.4	Iron (III) chloride	50 mg/L	~25
Naproxen[2]	3.2	4.2	Iron (III) chloride	50 mg/L	~20
Diclofenac[2]	4.5–4.8	2.5–3.0	Iron (III) chloride	50 mg/L	>65
			Aluminum sulfate	50 mg/L	>65
Endocrine disruptors					
Bisphenol A (BPA)[3]		10.2[5]	Iron (III) chloride	Up to 200 mg/L	Up to 20
Diethylhexylphthalate (DEHP)[3]			Iron (III) chloride	Up to 200 mg/L	Up to 70
17β-Estradiol (E2)[4]	4.01	10.4[5]	Iron (III) sulfate	12.2 mg/L	15
			Polyaluminum chloride	5.4 mg/L	15

(*continued*)

Table 6.1 (continued)

Micropollutant	log K_{ow}	pK_a	Coagulant	Dosage	Removal Efficiency (%)
Estriol (E3)[4]	2.45	–	Iron (III) sulfate	12.2 mg/L	20
			Polyaluminum chloride	5.4 mg/L	30
Diethylstilbestrol (DES)[4]	5.07	–	Iron (III) sulfate	12.2 mg/L	25
			Polyaluminum chloride	5.4 mg/L	40

[1] Freese et al. (2001)
[2] Carballa et al. (2005)
[3] Asakura and Matsuto (2009)
[4] Bodzek and Dudziak (2006)
[5] Deborde et al. (2005)

a) Effects of physical-chemical properties of micropollutants

Micropollutants with high K_{ow} values are less soluble in water and would have a higher tendency to sorb onto particles (Hemond and Fechner-Levy, 2000). Hence, micropollutants with higher K_{ow} would have a higher tendency to be removed from the liquid phase. This could be observed from the results reported by Carballa et al. (2005) that with different coagulants, diclofenac (log K_{ow} of 4.5–4.8) experienced removal efficiencies of 50–70% as compared with naproxen (log K_{ow} of 3.2) which only achieved removal efficiencies in the range of 5–20% (Table 6.1).

Similarly, synthetic estrogens with higher K_{ow} values such as mestranol (log K_{ow} of 4.67) had about 40% higher sorption to sediment as compared with estriol (log K_{ow} of 2.81) (Table 6.2) (Lai et al., 2000). Competitive effects were also observed between the micropollutants for binding sites on the coagulants or sediments. Lai et al. (2000) reported that the sorption of other estrogens such as estriol and mestranol was suppressed by 89 and 31%, respectively, with the addition of a superhydrophobic synthetic estrogen, estradiol valerate (with higher log K_{ow} of 6.41). The suppression effect of this competitive binding was more significant for compounds with lower hydrophobicity as noted from the higher suppression experienced by estriol. Enhanced coagulation was also determined to be more effective for the removal of high molecular weight DBP compounds (> 30 kDa). At pH 7.5, adsorption was reported to be the main contributor for DBP compounds (>30 kDa) removal by the coagulants (Zhao et al., 2008). However, in the treatment of natural water and wastewater when a complex mixture of micropollutants is present, the effect of enhanced coagulation towards removal of micropollutants would differ and further study is required.

The sorption kinetics of the synthetic estrogen on sediments showed rapid sorption within the initial 0.5 hr followed by a slower sorption rate which subsequently achieved a steady decrease (Lai et al., 2000). The estrogen removal rate in a batch experiment followed a 3-stage sorption pattern. A general trend in the sorption experience in the batch test is shown in Figure 6.1. The initial stage 1 occurred at a rapid rate due to the availability of active binding sites. Upon gradual saturation of these sites, the rate was gradually reduced reaching close to a steady state. This could also be due to the lower availability of micropollutants in the aqueous phase. Subsequent desorption of micropollutants from the solids occurred, reducing the net amount of micropollutants sorbed by the solids. Hence, for an optimum removal through enhanced coagulation, a rapid mixing and gradual flocculation should fall within Stages 1 and 2 and removal of the flocs would be required before Stage 3 to avoid desorption of micropollutants back into the aqueous phase.

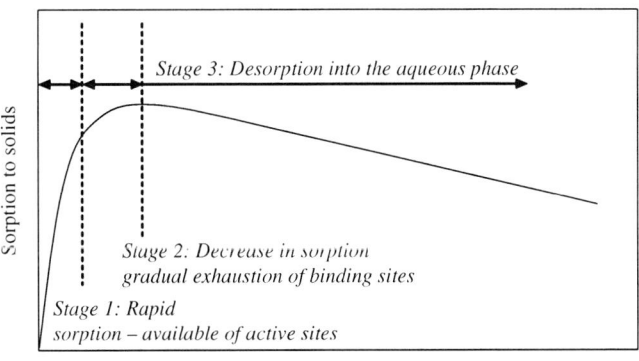

Figure 6.1 Typical sorption pattern of micropollutants onto solids

Similar to the effect of K_{ow} on the removal of micropollutants, synthetic estrogen with a higher K_{ow} was also reported to have a higher sorption rate onto the solids in Stage 1 and a lower desorption rate in Stage 3 (Lai et al., 2000). A summary on the physical-chemical properties and the sorption constant of selected synthetic estrogens determined by Lai et al. (2000) is given in Table 6.2.

Table 6.2 Physical-chemical properties and sorption characteristics of some selected synthetic estrogens (extracted from Lai et al., 2000)

Micropollutant	Molecular Weight	Water Solubility (mg/L)	log K_{ow}	Sorption Constant (1/n)	Sorption to Sediment (ng/g)
Estriol	288.39	13	2.81	0.57	3.2
Estradiol	272.39	13	3.94	0.67	4.1
Mestranol	310.42	0.3	4.67	0.78	5.5

b) Choice of coagulants and dosage

The coagulation-flocculation studies on removal of pharmaceutical and personal care products (PPCP) by Carballa *et al.* (2005) reported that the coagulant dose (using ferrous chloride ($FeCl_3$) of 250–350 mg/L, aluminum sulphate ($Al_2(SO_4)_3$) of 250–350 mg/L and aluminum polychloride (PAX) of 700–950 mg/L) and temperature (12 or 25°C) tested did not have a significant influence on the PPCPs removal. However, the removal efficiencies of specific PPCPs were influenced by the type of coagulants used. Overall, $FeCl_3$ (250 mg/L at 25°C) was able to provide more than 50% removal for galaxolide, tonalide and diclofenac, while about 20–25% removal for diazepam and naproxen. Table 6.3 summarises the highest removal efficiencies of different PPCPs achieved among the three coagulants tested at 25°C as reported by Carballa *et al.* (2005).

Table 6.3 Highest removal efficiencies of different PPCPs and the respective coagulants used (Carballa et al., 2005)

Type of PPCP	Type of Coagulant (Dosage in mg/L)	Removal Efficiency (%)
Galaxolide	PAX (850)	63
Tonalide	PAX (850)	71
Diclofenac	$FeCl_3$ (250)	70
Diazepam	$FeCl_3$ (250)	25
Naproxen	$FeCl_3$ (250)	20

Zorita *et al.* (2009) reported that by applying 0.07 mg/L $FeCl_3$ in a tertiary treatment comprised of coagulation and flocculation for sewage wastewater treatment system, a removal efficiency of more than 55% was achieved for the antibiotics tested, namely, ofloxacin, norfloxacin and ciprofloxacin. It was proposed that the main removal mechanism was due to adsorption of these micropollutants onto flocs. The addition of FeCl$_3$ did not significantly enhance the removal of acidic group of pharmaceutical compounds, such as ibuprofen, naproxen, diclofenac and clofibric acid in the wastewater treatment (less than 25% removal efficiency for this group of compounds). This could be due to the low FeCl$_3$ dose used which led to insignificant removals of naproxen and diclofenac (Carballa et al., 2005), whereas a higher dose of 250 mg/L of FeCl$_3$ used in the study by Zorita et al. (2009) induced removal efficiencies of 20 and 70% for naproxen and diclofenac, respectively. As these acidic compounds possess a physical-chemical property of pK_a 3–5, they would be partially ionized in aqueous phase. The effect of doubling the ferric chloride dose also provided an improvement in the removal of bisphenol A by enhanced coagulation from 5% to about 20% (Carbella et al., 2005). The addition of coagulants would enhance the binding of these compounds onto the suspended solids for subsequent removal from the aqueous phase (Carbella et al., 2005).

c) pH and alkalinity

Variation in pH could lead to the change in coagulants species and/or charge neutralization. Yan et al. (2008) and Zhao et al. (2008) demonstrated the changes in Al species with pH using ferron assay and electrospray ionization (ESI) mass spectrometry, respectively. Extensive researches have been performed by Yan et al. (2007, 2008) on the effect of pH/alkalinity on enhanced coagulation with PACls. The low pH in enhanced coagulation, however, is a major drawback in its applications which could lead to corrosion problems in plant infrastructures (Edwards et al., 2003). PACl is a type of pre-hydrolyzed coagulant which can be applied to reduce the pH drop after coagulation. The different hydrolyzed Al species are summarized in Table 6.4 (Yan et al., 2007).

Table 6.4 Hydrolyzed Al species (Yan et al., 2007)

Hydrolyzed Group	Species	Molecular Weight (Da)
Al$_a$	Monomer – Al^{3+}, Al(OH)$^{2+}$, Al(OH)$_2^+$ Dimer – Al$_2$(OH)$_2^{4+}$ Trimer – Al$_3$(OH)$_4^{5+}$ Small polymers	<500
Al$_b$	Tridecamer – Al$_{13}$O$_4$(OH)$_{24}^{7+}$ (more commonly known as Al$_{13}$)	500–3000
Al$_c$	Large polymer or colloidal species	>3000

The degree of hydrolysis is expressed using basicity (B) value which represents the ratio of hydroxide-to-aluminum ratio (Yan et al., 2008). Generally higher B value corresponds to lower Al_a and higher Al_b portion. The degree of hydrolysis is significantly influenced by the pH. Table 6.5 summarizes the dominant Al species at different pH region. Al_b has a tendency to form Al_c with aging. Together with pH, coagulant dose also plays an important role in determining the stability of the species present. Wang et al. (2007) reported that at higher PACl dosage above 2 mol Al/L, Al_b was not detected. However, using a different coagulant, such as the nano-Al_{13} coagulant, showed that Al_b species were relatively stable at coagulant concentrations of 0.11–2.11 mol Al/L even up to 30 days of aging (Wang et al., 2007). The presence of preformed Al_b in the pre-hydrolyzed PACl was noted to enhance natural organic matter (NOM) removal in the surface water (Yan et al., 2008).

Table 6.5 Dominant Al species at different pH region for 3 different types of coagulants (Yan et al., 2008)

Type of Coagulant	B	Dominant Al Species		
		pH < 5.0	5.5 < pH < 7.5	pH > 9.0
$AlCl_3$	0	Al_a	Al_b	Al_a
$PACl_l$	1.6	$Al_a \approx Al_b$	Al_b	$Al_a \approx Al_b$
$PACl_{20}$	2.0	Al_b	Al_b	Al_b

In addition to pH, the characteristics of the raw water also affect the performance of the coagulants. Yan et al. (2008) demonstrated that a higher coagulant dose was required to depress the pH of the Yellow River water, which had an alkalinity of 3.5 times higher than the water from the Pearl River, to a pH range (about pH 5.5–6.5 based on optimal NOM removal) favourable for coagulation.

Another controlling factor is the effect of pH on micropollutants. The change in reaction pH will affect dissociation of the micropollutants which is given by the dissociation constant, pK_a. The removal of DBP (such as Trihalomethane (THM) and Haloacetic acid (HAA)) precursors is affected by the pH of the reaction solution. Removal of THM and HAA precursors using PACl coagulation is mainly due to the nature of the THM and HAA precursors (Zhao et al., 2008). THM precursors contain more aliphatic structure while HAAs precursors are mainly aromatic. Charge neutralisation precipitation is responsible for the removal of the negatively-charged aliphatic THM precursors which could be achieved at pH 5.5. Two possible effects could occur: neutralization of the negatively charged compounds by the monomeric

Al species, or neutralization by the hydrogen ions and Al ions (Yan *et al.*, 2007). At a higher pH between 5.5–7.5, the presence of both Al_{13} and $Al(OH)_3$ would have resulted in simultaneous removal of THM precursors by charge neutralization precipitation and adsorption. In the case of HAA precursors, self-aggregations of the aromatic and hydrophobic functional groups occurred under acidic conditions, while under alkaline conditions, absorption onto flocs and subsequent removal through sweep flocculation occur.

6.1.2 Coagulation-oxidation

Removal of micropollutants can be further enhanced with the used of more advanced coagulation process such as oxidation-coagulation. Oxidizing agents have the potential of achieving both oxidation and coagulation, such as ferrate (Fe)(VI) for example through the formation of Fe(III) or ferric hydroxide in water (Lim and Kim, 2009) and wastewater treatment (Lee *et al.*, 2009). Fe(VI) could partially remove micropollutants by oxidation and subsequent coagulation process to enhance the micropollutants removal from the aqueous phase.

In water treatment, the high oxidative nature of Fe(VI) allows it to possess disinfection ability in addition to the removal of NOM through oxidation-coagulation process (Lim and Kim, 2009). In addition, the formation of non-toxic ferric ion as a by-product of the Fe(VI) reaction favors this chemical to be used in water treatment (Sharma, 2008). The NOM removal performance of Fe(VI) was reported comparable to the traditional coagulants such as alum and ferric sulphate, while a small dose of ferrate as a pretreatment could enhance the removal rate of humic acid by traditional coagulants (Lim and Kim, 2009). At a Fe(VI) dose of 2–46 mg/L, humic and fulvic acids (both with initial concentrations of 10 mg/L) removals were in the range of 21–74% and 48–78%, respectively.

In wastewater treatment, simultaneous removal of micropollutants and phosphate in secondary treated effluent could be achieved with Fe(VI) (Lee *et al.*, 2009). Higher reactivity between Fe(VI) with micropollutants containing phenolic-moiety was reported compared to other micropollutant compounds with amine and olefin moiety. More than 95% removal was achieved for 17β-estradiol, biosphenol A and 17α-ethinylestradiol (containing phenolic-moiety) using 2 mg/L Fe(VI) at pH 7–8 while more than twice this dose was required to achieve similar removal of sulfamethoxazole, diclofenac (containing aniline-moiety) and carbamazepine (containing olefine-moiety). In a study by Lee *et al.* (2009), the depletion of phosphate was only observed after elimination of micropollutants. Hence, a higher Fe(VI) dose would be required to achieve both micropollutants and phosphate removal. It was only at Fe(VI) of 7.5 mg/L that more than 80% phosphate (from 3.5 to less than 0.8 mg PO_4^{3-}-P/L) was achieved (Lee et al., 2009).

The Fe(VI) oxidation reaction with micropollutants followed a second-order rate constant (k) which was shown to increase with a decrease in the reaction pH (Lee et al., 2009). Similarly, under acidic condition, ferrate has three times higher redox potential (which is slightly above ozone). This shows the high potential of Fe(VI) as an oxidant (Lim and Kim, 2009). Under acidic condition, the fraction of Fe(VI) in the protonated form ($HFeO_4^-$) increases (Figure 6.2) (Sharma, 2008). $HFeO_4^-$ species is a strong oxidant as compared with the deprotonated Fe(VI) species ((FeO_4^{2-}), hence contributing to an increase in the oxidation rate at lower pH (Sharma, 2008).

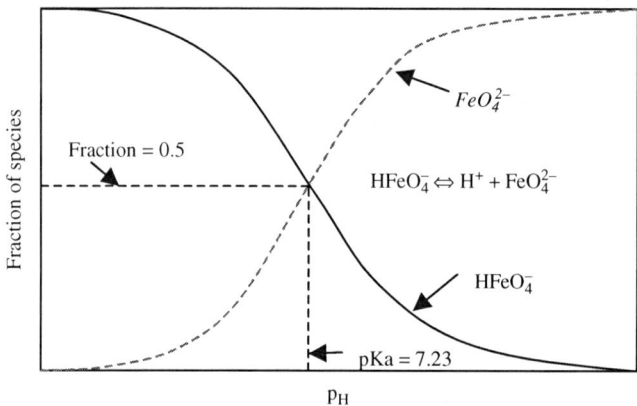

Figure 6.2 Changes in fraction of Fe(VI) species at different pH with pKa = 7.23 at 25°C (extracted from Sharma, 2008)

Fe(VI) also has a higher stability compared to ozone (O_3). Lee et al. (2009) showed that Fe(VI) persisted in the secondary treated effluent for more than 30 min compared with O_3 depletion in less than 5 min. The second-order rate constants of Fe(VI) for reaction with micropollutants were demonstrated to be at least 4–5 orders lower than O_3 at pH 7–8 (Lee et al., 2009). A summary on the second order rate constants for selected pollutants with Fe(VI) is given in Table 6.6. A longer reaction time would be required for Fe(VI) as compared with O_3 oxidation. Most pharmaceutical compounds have a half-life lower than 100 s when oxidized with O_3, while the half-life of sulfonamides is more than 2 times longer using Fe(VI) (Sharma, 2008). The application of Fe(VI) in environmental water samples may be dependent on Fe(VI) selectivity in oxidizing the micropollutants. Certain micropollutants with lower oxidation reactions with Fe(VI) such as ibuprofen (first order kinetic constant at 0.9×10^{-1} $M^{-1} \cdot s^{-1}$ at pH 8.0) could be removed by co-precipitation through the formation of ferric oxide (Sharma and Mishra, 2006).

Table 6.6 Second-order rate constant, k for selected micropollutant with Fe(VI) at pH 7

Compound	k ($M^{-1} \cdot s^{-1}$)	Reaction Temperature	Reference
Pharmaceuticals			
Diclofenac	1.3×10^2	23°C	Lee *et al.* (2009)
Carbamezepine	67	23°C	Lee *et al.* (2009)
Sulfamethazine	22.5	25°C	Sharma and Mishra (2006)
Endocrine disruptors			
Bisphenol A	6.4×10^2	23°C	Lee *et al.* (2009)
17β-estradiol	7.6×10^2	23°C	Lee *et al.* (2009)

6.2 MEMBRANE PROCESSES

Membrane processes make use of artificial semi-permeable or porous membranes to separate compounds based either on their size (pressure-driven operation) or on their electrical charge (electrodialysis). Pressure-driven processes can be divided into high-pressure membranes where compound separation occurs mainly by size exclusion and low-pressure membranes where adsorption and other phenomena are responsible for compound separation. Low-pressure membranes have pore sizes in the range of 0.001–0.1 (ultrafiltration) to 0.1–10 μm (microfiltration). These systems operate under driving pressure below 1,030 KN m^{-2} and are usually utilised to retain large molecules with molecular weight over 5,000 Da such as colloids and oil (Tchobanoglous *et al.*, 2003). Smaller molecules can usually be retained in the process of reverse osmosis (RO) at a driving pressure higher than the osmotic pressure of the solution to be filtered (up to 6,900 KN m^{-2}). This resulted in a higher operating cost for the RO process. On the other hand, electrodialysis works differently by using ion exchange membranes to separate ionic compounds. This is achieved by making an electrical current pass through the solution which causes cations to migrate towards a cation exchange membrane and anions towards an anion exchange membrane. Because cation- and anion-exchange membranes are spaced alternately, this results in alternated compartments of concentrated and diluted solutions.

This section will review the use of microfiltration (MF) and ultrafiltration (UF) as low pressure membrane processes for micropollutant removal, alone or in combination with other physico-chemical techniques (such as particulate activated carbon). It will also cover the incorporation of UF membranes into a biological reactor (known as membrane bioreactor). Rejection of micropollutants by the RO process – a high-pressure membrane – shall be presented but by nanofiltration will not be covered because it is the object of an earlier section of this book. Finally, an insight will be provided into electrodialysis applications for micropollutant removal.

6.2.1 Mechanisms of solute rejection during membrane treatment

Solute rejection in membrane treatment is directly determined by the interactions between the solute and the membrane. These interactions are of different natures, reflecting steric partition, adsorption and charge effects as well as biodegradation phenomena (Figure 6.3). Among these three categories of interactions, steric effects are the most straightforward to understand, directly related to the molecule size (or molecular weight) on the one hand and the membrane pore size on the other hand. If the molecule is bigger than the pores, it will be well rejected. If it is smaller, it will permeate through the membrane. Charge phenomena are related to the fact that most polymeric membranes are negatively charged. As a consequence, repulsive forces will keep away negatively-charged solutes, while positively-charged solutes will be attracted to the membranes. This results in better rejection of negatively-charged molecules by polymeric membranes. The charge of a molecule can be predicted by comparing its pKa value to the pH of the solution (Bellona and Drewes, 2005).

On top of the obvious sieving effect of membranes, hydrophobic solutes (i.e., having an octanol-water partitioning coefficient (K_{ow}) higher than 2) can be adsorbed onto the surface of the membrane. This results in increased rejection levels at the beginning of a membrane operation time. However, the adsorption sites on a membrane are not infinite and once the membrane adsorption capacity is reached, hydrophobic solutes will be able to pass through the membrane. This can lead to overestimation of hydrophobic solute rejection if tests are not done under steady state conditions, i.e., when all adsorption sites have been occupied (Kimura *et al.*, 2003a). Furthermore, adsorption of NOM onto the membrane can result in biodegradation phenomena, the importance of which in membrane systems depends on the frequency of backwashing and utilization of sanitisers (Laine *et al.*, 2002). Studies of biofouling have emphasised that the fouling layer acts like a second membrane and the overall impact on micropollutant removal depends on the specific affinity of each layer for the compound studied (Vanoers *et al.,* 1995). If the membrane rejects the solute better than the deposited layer, fouling might result in accumulation of solutes on the membrane surface. Ultimately, this might cause enhanced concentration polarization and decreased solute rejection. However, if the solute is rejected better by the deposited layer than by the membrane, solute rejection might improve.

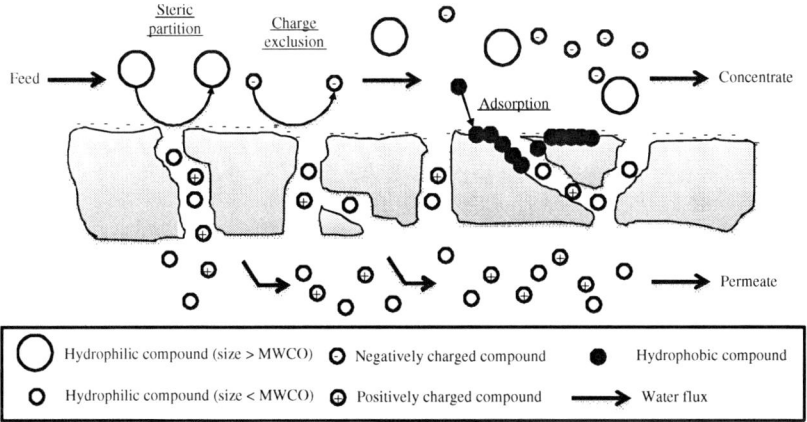

Figure 6.3 Mechanisms of solute partition during treatment by a negatively-charged membrane

6.2.2 Micropollutant removal by microfiltration

MF membranes are a type of low-pressure membranes that are characterized by a pore size in the range of 0.1–10 µm. Even though this pore size is adequate for removal of pathogens in drinking water, it is unlikely that micropollutants (< 1 kDa) can be retained by sieving effect. In one study, MF was used as a pre-treatment step before advanced treatment of a water reclamation plant tertiary effluent (Drewes *et al.,* 2003). The fate of trace organic compounds in the form of EDTA, nitrilotriacetic acid and alkylphenolpolyethoxycarboxylates was followed during MF (Osmonics/Desal EW4040F) operation at 3.8 L/min filtrate flow, 7.6 L/min concentrate flow, and a differential pressure of 21.6 kPa. Removals of nitrilotriacetic acid and alkylphenolpolyethoxycarboxylates were found to be ineffective. However, EDTA concentration attained 5.6 µg/L in the permeate as compared to 11.8 µg/L in the tertiary effluent (removal efficiency of 53%). In the absence of an understanding of the mechanism in EDTA removal by the MF membrane, there is a possibility that EDTA was merely adsorbed onto the MF membrane. MF inability to treat pharmaceutically active compounds (PhACs) and hormones was further established in two full-scale wastewater reclamation plants in Australia (Al-Rifai *et al.*, 2007).

6.2.3 Micropollutant removal by ultrafiltration
a) Ultrafiltration alone

Similarly as for MF, the molecular weight cut-off of UF membranes (10–100 kDa) does not allow sieving of most micropollutants. One study considered the role of fouling by NOM on the fate of estradiol and ibuprofen removal by UF membrane (Jermann *et al.*, 2009). Without NOM, retention by an hydrophilic UF membrane (regenerated cellulose, 100 kDa, Ultracel, Millipore, UK) was almost null for both micropollutants, whereas 80% of estradiol and 7% of ibuprofen were retained by a more hydrophobic membrane (polyethersulfone, 100 kDa, Biomax, Millipore, UK). The difference observed between the two types of membranes can be explained by adsorption phenomena on the hydrophobic membrane, as explained above. For ibuprofen, the retention rate was initially higher (25%) but decreased subsequently, which indicates that adsorption was mostly a temporary phenomenon until adsorption sites were saturated. The higher retention of estradiol could also be explained by the more hydrophobic nature of this compound, as compared to ibuprofen, as well as by its negative charge at the pH of the experiment. Most of estradiol could be recovered by sodium hydroxide (pH 12.3), which may have implications during membrane cleaning procedures. The effect of NOM varied depending on the nature of NOM, membrane and micropollutant. NOM could in some cases decrease the retention rate by occupying adsorption sites on the membrane (in the case of humic acids) or in other cases increase the retention of micropollutant by the formed cake layer which acts as a second membrane (in the case of alginate). Another study (Majewska-Nowak *et al.*, 2002) showed that atrazine separation by a polysulfone membrane (Ps, Sartorius) and a composite polysulfone/polyamide membrane (Ds-GS, Osmonics) was maximum at pH 7 (around 20% rejection). In the presence of humic acids at a moderate concentration (up to 20 mg/L), atrazine rejection was improved up to 80% due to atrazine adsorption on humic substances. However, higher doses of humic acids reduced the retention of atrazine, probably due to saturation of the membrane adsorption sites. These two experiments clearly show that the removal of micropollutants by UF membrane is mostly limited to adsorption phenomena, which is not a reliable treatment method on the long term. Surprisingly, one study showed higher rejection of chloroform by UF (93%) than by reverse osmosis or nanofiltration (81 and 75%, respectively) (Waniek *et al.*, 2002). However, the authors were unable to explain the reason of the better performance of UF and it is likely that adsorption is also the main mechanism involved in the retention of chloroform by UF in that experiment.

b) Combination of ultrafiltration and powdered activated carbon

Due to limited removal of micropollutants by UF alone, the combination of UF and powdered activated carbon (PAC) adsorption has attracted the attention of many researchers. The main advantages brought by the combined treatment are increased adsorption kinetics, increased mixing due to UF recirculation loop and flexibility (Laine *et al.,* 2000). However, specific information on the fate of micropollutants during the combined treatment is scarce. One study focused on the fate of three micropollutants (i.e., trichlorethene, tetrachlorethene, atrazine) present in karstic spring water during UF (Aquasource) combined with PAC (Envir-Link MV125) at the pilot scale (Pianta *et al.,* 1998). The rejection level was higher for atrazine (>99% at 5 mg/L of PAC) than for trichlorethene and tetrachlorethene. For the latter two, removal rates of 90% was only made possible by a PAC dose of 22 mg/L. The authors also emphasized on the importance of kinetics and demonstrated that the performance could be improved by reducing the membrane flux, which resulted in increasing contact time between the micropollutants and PAC.

Apart from the dosage of PAC, the way it is administered also has an impact on the efficacy of the UF-PAC treatment systems. In one study, 52% of atrazine was removed by the combined process when PAC was added continuously throughout the filtration cycle, but this percentage increased to 76% when the whole dose of PAC (8 mg/L) was added at the beginning of the cycle (Campos *et al.,* 1998). The authors explained that this is related to the kinetics of atrazine adsorption to PAC, whereby the contact time was being increased when all the PAC was dosed at one time in the beginning. However, another paper reaches the opposite conclusion (Ivancev-Tumbas *et al.,* 2008). In that study, an UF membrane operated in dead-end mode achieved a rejection of p-Nitrophenol by 39% when the PAC dose (10 mg/L) was added at the beginning of the cycle, which was inferior to the performance attained by continuous supply of the same PAC dose (rejection of 75%). The authors attributed the contradictory results to different experimental conditions in terms of membrane type and micropollutant nature and concentration. It is also possible that a single dosing of PAC resulted in the formation of a carbon monolayer inside the capillaries of the UF membrane that slowed the adsorption kinetics of the micropollutant onto the PAC and also reduced the adsorption capacity of the membrane.

c) Combination of ultrafiltration and biological module (membrane bioreactor)

An alternative to PAC in order to improve the performance of UF is the addition of a biological compartment to the UF membrane, resulting in a process known as membrane bioreactor (MBR). In an MBR, the biological component is responsible for the biological degradation of waste compounds while the membrane module accomplishes the solids-liquids separation (Choi and Ng, 2008; Ng and Hermanowicz, 2005). MBRs can be either submerged, i.e., membrane modules are directly immersed in the activated sludge tank, or side-stream, in which case membrane filtration is achieved outside of the activated sludge tank. MBR has attracted increasing interest in recent years in wastewater treatment because of the emergence of low-cost membranes with lower pressure requirements and higher permeate fluxes. One specific advantage that makes MBR attractive for micropollutant removal is the long sludge retention time (SRT) allowed by the technology (Tan et al., 2008).

An early study focused on the removal of polar sulphur aromatic micropollutants: naphthalene sulfonates and benzothiazoles in a large-scale side-stream MBR treating tannery wastewater (Reemtsma et al., 2002). Naphtalene monosulfonates were removed to a large extent (>99%) but disulfonate removal averaged only 44%. The same profile and variability depending on every specific compound were observed for the family of benzothiazoles with 2-mercaptobenzothiazole being almost entirely removed but 2-aminobenzothiazole being almost unaltered. It was found that a fraction of 2-mercaptobenzothiazole was oxidized into a byproduct benzothiazole-2-sulfonic acid that did not seem to be further altered. In that sense, MBR technology did not seem to produce results largely different from the conventional activated sludge process for some micropollutants. The same conclusion was reached in a comprehensive study comparing the fate of a variety of PPCPs and endocrine disrupting compounds (EDCs) in an MBR pilot plant and three conventional activated sludge wastewater treatment plants (Clara et al., 2005). Most of the compounds analysed were removed to the extent of 50 to 60% with the notable exception of bisphenol-A, ibuprofen and bezafibrate, for which the removal rate exceeded 90%. In all cases, no significant difference was observed between the MBR and conventional activated sludge process, showing that the UF membrane does not allow further removal of some micropollutants.

Excellent MBR removal of ibuprofen (up to 98%) was confirmed in another study along with naproxen (84%) and erythromycin (91%) and these performances were attributed by the authors to the exceptionally long SRT (44–72 d) achieved in their pilot-scale immersed MBR (Reif et al., 2007).

In addition, sulfamethoxazole and musk fragrances (galaxolide, tonalide, celestolide) were moderately removed (>50%) probably due to partial adsorption on the biomass. On the other hand, carbamazepine, diazepam, diclofenac and trimethoprim were poorly removed (<10%) due to poorer biodegradability. Another study confirmed the possibility of achieving good treatment of bisphenol-A (90%) due to both biodegradation and adsorption phenomena (Nghiem *et al.*, 2007). On the contrary, sulfamethoxazole removal was solely attributed to biodegradation, which can explain the poorer rejection level of this hydrophilic compound (50%). In the case of a decentralized MBR treating domestic wastewater spiked with PhACs, sulfonamide antibiotics, ibuprofen, bezafibrate, estrone and 17α-ethinylestradiol appeared to be easily biodegradable mostly under aerobic or anoxic conditions in the presence of nitrate (Abegglen *et al.*, 2009). However, macrolide antibiotics appeared poorly biodegradable and their elimination was mostly due to adsorption phenomenon. Another study confirmed the good biodegradability (>99%) of 17α-ethinylestradiol in an MBR enriched with nitrifiers (De Gusseme *et al.*, 2009).

The fate of metals in MBRs has also gained interest in the recent years. One paper studied the impact of SRT on the removal of various metals in an immersed MBR (Innocenti *et al.*, 2002). Again, high variability was observed with Ag, Cd and Sn being largely removed (>99%); Cu, Hg and Pb moderately removed (>50%); but B, Se and As poorly treated (<50%). Moreover, for As, Pb, Se and B, increasing SRT did not improve the removal of these species, which seemed to indicated that adsorption on the biomass was not involved in these metals removal. It is more likely that their charge effects with the UF membrane were directly linked to their rejection. Another study showed slightly different results with higher removal of As (65%), Hg and Cu (both over 90%) but on the other hand, slightly lower performance on Cd (>50% removal) (Fatone *et al.*, 2005). In the same study, the fate of a variety of polycyclic aromatic hydrocarbon (PAHs) was also studied, for which moderate removal in the range of 58 to 76% was observed.

Due to degradation in MBR being largely dominated by the biological component, bioaugmentation was shown to help achieving higher removal of compounds such as nonylphenol (Cirja *et al.*, 2009). In some cases, MBR can be further combined with PAC for improved micropollutant removal, such as pesticides (Laine *et al.*, 2000). A summary of these studies and additional data on micropollutant rejection by MBR technology are given in Table 6.7.

Table 6.7 Removal of micropollutants by the MBR technology

Classification	Micropollutant	Removal Efficiency (%)	Reference
Benzothiazole	2-Hydroxybenzothiazole	0	(Reemtsma et al., 2002)
	2-Mercaptobenzothiazole	>99	(Reemtsma et al., 2002)
	Benzothiazole	5	(Reemtsma et al., 2002)
	Benzothiazole-2-sulfonic acid	0	(Reemtsma et al., 2002)
	Methylthiobenzothiazole	50	(Reemtsma et al., 2002)
EDC	Nonylphenol	91	(Clara et al., 2005)
	Nonylphenol diethoxylate	94	(Clara et al., 2005)
	Nonylphenol monoethoxylate	99	(Clara et al., 2005)
	Nonylphenoxyacetic acid	0	(Clara et al., 2005)
	Nonylphenoxyethoxyacetic acid	0	(Clara et al., 2005)
	Octylphenol	>99	(Clara et al., 2005)
	Octylphenol diethoxylate	>99	(Clara et al., 2005)
	Octylphenol monoethoxylate	>99	(Clara et al., 2005)
	17α-Ethinylestradiol	99	(De Gusseme et al., 2009)
	Bisphenol-A	99	(Clara et al., 2005)
		90	(Nghiem et al., 2007)
Metal	Ag	>99	(Innocenti et al., 2002)
	Al	>99	(Innocenti et al., 2002)
		>96	(Fatone et al., 2005)
	As	65	(Fatone et al., 2005)
		35	(Innocenti et al., 2002)

(continued)

Table 6.7 (continued)

Classification	Micropollutant	Removal Efficiency (%)	Reference
Metal	B	30	(Innocenti et al., 2002)
	Ba	85	(Innocenti et al., 2002)
	Cd	>99	(Innocenti et al., 2002)
		>50	(Fatone et al., 2005)
	Co	80	(Innocenti et al., 2002)
	Cr	75	(Fatone et al., 2005)
	Cu	96	(Fatone et al., 2005)
		85	(Innocenti et al., 2002)
	Fe	>97	(Fatone et al., 2005)
		95	(Innocenti et al., 2002)
	Hg	99	(Innocenti et al., 2002)
		94	(Fatone et al., 2005)
	Mn	80	(Innocenti et al., 2002)
	Ni	79	(Fatone et al., 2005)
		60	(Innocenti et al., 2002)
	Pb	74	(Fatone et al., 2005)
		60	(Innocenti et al., 2002)
	Se	30	(Innocenti et al., 2002)
	V	90	(Innocenti et al., 2002)
	Zn	>90	(Fatone et al., 2005)
		80	(Innocenti et al., 2002)

(continued)

Table 6.7 (continued)

Classification	Micropollutant	Removal Efficiency (%)	Reference
Naphthalene sulfonate	1,7-Naphthalene disulfonate	40	(Reemtsma et al., 2002)
	2,6-Naphthalene disulfonate	90	(Reemtsma et al., 2002)
	2,7-Naphthalene disulfonate	0	(Reemtsma et al., 2002)
	1-Naphthalene monosulfonate	>99	(Reemtsma et al., 2002)
	2-Naphthalene monosulfonate	>99	(Reemtsma et al., 2002)
	1,5-Naphthalene disulfonate	0	(Reemtsma et al., 2002)
	1,6-Naphthalene disulfonate	80	(Reemtsma et al., 2002)
PAH	Acenafthene	76	(Fatone et al., 2005)
	Acenafthylene	>61	(Fatone et al., 2005)
	Fenanthrene	71	(Fatone et al., 2005)
	Fluoranthene	65	(Fatone et al., 2005)
	Fluorene	>54	(Fatone et al., 2005)
	Nafthalene	66	(Fatone et al., 2005)
	Pyrene	58	(Fatone et al., 2005)
PhAC	Diclofenac	50	(Clara et al., 2005)
		0	(Reif et al., 2007)
	Ibuprofen	99	(Reif et al., 2007)
		99	(Clara et al., 2005)
	Sulfamethoazole	61	(Clara et al., 2005)
		55	(Reif et al., 2007)
		50	(Nghiem et al., 2007)
	Carbamazepine	12	(Clara et al., 2005)
		10	(Reif et al., 2007)

(continued)

Table 6.7 (*continued*)

Classification	Micropollutant	Removal Efficiency (%)	Reference
PPCP	Bezafibrate	96	(Clara *et al.*, 2005)
	Diazepam	25	(Reif *et al.*, 2007)
	Erythromycin	90	(Reif *et al.*, 2007)
	Galaxolide	60	(Reif *et al.*, 2007)
		92	(Clara *et al.*, 2005)
	Naproxen	85	(Reif *et al.*, 2007)
	Roxithromycin	>99	(Clara *et al.*, 2005)
		80	(Reif *et al.*, 2007)
	Tonalide	85	(Clara *et al.*, 2005)
		40	(Reif *et al.*, 2007)
	Trimethoprim	30	(Reif *et al.*, 2007)
	Celestolide	50	(Reif *et al.*, 2007)

6.2.4 Micropollutant removal by reverse osmosis

RO membranes are characterized by having "pores" with size ranging between 0.22 and 0.44 nm (Kosutic and Kunst, 2002). As a consequence, RO – like nanofiltration – has the potential to result in partial or highly effective removal of most micropollutants (Jermann *et al.*, 2009). However, this is achieved at the cost of higher energy consumption (Jones *et al.*, 2007) and the development of ultra-low pressure reverse osmosis membranes, in which the water flux is facilitated by a hydrophilic support layer is of major concern (Ozaki and Li, 2002). In addition to steric effects, solute transport in RO membranes is influenced by diffusion and electrostatic phenomena throughout membrane pores and, as a consequence, the tightest pores do not always result in the best rejection of micropollutants (Kosutic and Kunst, 2002). In their study, this was notably the case for 2-butanone, better removed by looser membranes, and for the pesticide triadimefon that showed reduced rejection pattern (58–82%) as compared to atrazine (80–99%) in spite of its higher molecular width and weight. The main difference between RO and nanofiltration membranes is in terms of selectivity – RO being designed to remove all ions (including monovalent ions) wheareas nanofiltration are selective towards multivalent ions (Li, 2007).

Early studies of micropollutant removal by RO date back to the 1980s. It was found at that time that most metals were largely rejected by RO (>75%). However, organic micropollutants in the form of trihalomethanes, dichloromethane and alkylphenols were poorly removed (Hrubec *et al.*, 1983). Later on, a study investigated the performance of different types of RO membranes for micropollutant removal including several pesticides and chlorophenols (Hofman *et al.*, 1997). Ultra-low pressure RO membranes were found to compete well with polyamide membranes, almost all compounds tested being rejected below their detection level. However, cellulose-acetate displayed poorer performance.

In a very comprehensive study, the fate of DBPs, EDCs and PhACs by RO filtration was investigated (Kimura *et al.*, 2003b). In their study, adsorption was found to be an early mechanism in micropollutant removal but only temporarily after what steric and charge effects were mostly responsible for micropollutant removal. Notably, negatively charged molecules were found to be better rejected (>90%) than non-charged compounds (<90%). The exception was for the EDC Bisphenol-A that, in spite of being neutral (uncharged), could be removed up to 99% using an ultra-low pressure RO membrane (RO-XLE, Film-Tec, Vista, CA). In a later study, the overall better performance of this polyamide membrane as compared to a cellulose acetate one (SC-3100, Toray) was confirmed with the notable exception of the pharmaceutical compound sulfamethoxazole (Kimura *et al.*, 2004). The good performance of this polyamide membrane was also demonstrated on a variety of antibiotics that showed rejection rates of over 97% (Kosutic *et al.*, 2007).

The possibility of treating efficiently EDCs and PhACs over 90% by RO was further showcased in another study, in which most of the compounds studied were rejected to below detection levels (<25 ng/L) (Comerton *et al.*, 2008). Another study of two full-scale wastewater treatment plants further showed that RO was the most effective step for the treatment of PhACs and non-steroidal estrogenic compounds (Al-Rifai *et al.*, 2007). Total rejection of the antibiotic triclosan was also showcased during a 10-h filtration testing through a BW-30 (Dow FilmTec, Minneapolis, MN) RO membrane (Nghiem and Coleman, 2008). However, caffeine was found in the permeate of two full-scale RO facilities (Drewes *et al.*, 2005). This could be explained by its hydrophilic and non-ionic nature limiting its rejection to size exclusion effects. Similarly, the same study showed limited rejection of chloroform in the range of 50 to 85%.

As for other types of membranes, fouling by NOM was shown to increase the retention of hydrophobic non-ionic molecules by acting as a second membrane. However, in some cases, fouling by NOM resulted in membrane swelling and decreased rejection capacities (Xu et al., 2006). Again, the nature of the membrane was found to be of importance with cellulose triacetate and ultra-low pressure RO membranes being affected by fouling more than thin film composite membranes. The nature of the foulant is also important, with colloidal fouling being known to be particularly responsible for a decrease in water flux and permeate quality (Ng and Elimelech, 2004). In that study, colloidal fouling caused a decrease in RO (LFC-1, Hydranautics, Oceanside, CA) rejection of hormones (estradiol and progesterone) and, contrary as what was observed for salts and inert organic solutes, hormone rejection continued to decrease even after the water flux had stabilized. As a consequence, hormone removal constantly decreased from over 95% initially to values in the range of 75 to 85% in the presence of colloids (200 mg/L) after 110 h. This shows that hormone rejection by RO membrane was mostly due to adsorption phenomena, whereas larger organic molecules were removed by steric effects. In addition, in that crossflow experiment, hormone rejection was found to be about 10% higher when the height of the membrane filtration cell channel was reduced by a factor of 2, showing the importance of shear forces. Improved hormone rejection at higher shear rates could be explained by decreased concentration polarization. Comprehensive data on micropollutant removal by RO is provided in Table 6.8.

6.2.5 Electrodialysis

Recently, urine treatment by electrodialysis has brought to the light the potential of such membranes for micropollutant removal. Even though ethinylestradiol, a major compound excreted in urine, was found to be fully rejected in the long-term operation of laboratory electrodialysis membranes (Mega a.s., Prague, Czech Republic), good rejection was only temporary for other compounds such as diclofenac, carbamazepine, propranolol and ibuprofen, showing that most of the rejection mechanism was caused by adsorption for these molecules (Pronk et al., 2006). Overall, electrodialysis showed the best performance out of five treatment processes for urine, the other four being sequencing batch reactor, nanofiltration, struvite precipitation and ozone treatment, achieving up to 99.7% removal efficiency of estrogenicity (Escher et al., 2006).

Table 6.8 Removal of micropollutants by RO

Classification	Name of micropollutant	Removal (%)	Reference
Acid	Acetic acid	>99	(Ozaki and Li, 2002)
	Dichloroacetic acid	95	(Kimura et al., 2003b)
Alcohol	2-Propanol	86	(Kosutic et al., 2007)
	Benzyl alcohol	85	(Ozaki and Li, 2002)
	Erythritol	93	(Ng and Elimelech, 2004)
	Ethyl alcohol	40	(Ozaki and Li, 2002)
	Ethylene glycol	43	(Ng and Elimelech, 2004)
		50	(Ozaki and Li, 2002)
	Glycerol	92	(Kosutic et al., 2007)
		93	(Ng and Elimelech, 2004)
	Methyl alcohol	25	(Ozaki and Li, 2002)
	o-Nitrophenol	90	(Ozaki and Li, 2002)
	Phenol	75	(Ozaki and Li, 2002)
	p-Nitrophenol	95	(Ozaki and Li, 2002)
	Triethylene glycol	90	(Ozaki and Li, 2002)
Alkanes	Alkanes	90	(Hrubec et al., 1983)
Alkylbenzenes	C_1 Alkylbenzenes	8	(Hrubec et al., 1983)
	C_2 Alkylbenzenes	8	(Hrubec et al., 1983)
	C_3 Alkylbenzenes	80	(Hrubec et al., 1983)
	C_4 Alkylbenzenes	85	(Hrubec et al., 1983)
	C_5 Alkylbenzenes	90	(Hrubec et al., 1983)
Alkylindanes	Alkylindanes	90	(Hrubec et al., 1983)
Alkylnaphthalenes	Alkylnaphthalenes	15	(Hrubec et al., 1983)
Alkylphenols	Alkylphenols	70	(Hrubec et al., 1983)
Antibiotic	Triclosan	>99	(Nghiem and Coleman, 2008)
Antiepileptic	Primidone	>99	(Drewes et al., 2005)
Aromatic acid	2,4-Dihydroxybenzoic	90	(Xu et al., 2006)
Aromatic acid	2-Naphthalenesulfonic acid	90	(Xu et al., 2006)

(continued)

Table 6.8 (continued)

Classification	Name of micropollutant	Removal (%)	Reference
Aromatic amine	Aniline	75	(Ozaki and Li, 2002)
Aromatic hydrocarbon	Benzene	8	(Hrubec et al., 1983)
Carbohydrate	Glucose	95	(Ozaki and Li, 2002)
	Xylose	95	(Ng and Elimelech, 2004)
Chlorinated aliphatic	Dichloromethane	0	(Hrubec et al., 1983)
Chlorophenol	2,3,6-Trichlorophenol	>99	(Hofman et al., 1997)
	2,3-Dichlorophenol	95	(Ozaki and Li, 2002)
	2,4,5-Trichlorophenol	95	(Ozaki and Li, 2002)
	2,4,6-Trichlorophenol	>99	(Hofman et al., 1997)
	2,4-Dichlorophenol	90	(Ozaki and Li, 2002)
		>99	(Hofman et al., 1997)
	2,4-Dinitrophenol	95	(Ozaki and Li, 2002)
	2,6-Dichlorophenol	>99	(Hofman et al., 1997)
	4-Chlorophenol	65	(Ozaki and Li, 2002)
	Pentachlorophenol	99	(Ozaki and Li, 2002)
		>99	(Hofman et al., 1997)
Chlorophosphate	Chlorophosphates	60	(Hrubec et al., 1983)
Cyclic hydrocarbon	Cyclic hydrocarbons	80	(Hrubec et al., 1983)
	Cyclohexanone	98	(Kosutic et al., 2007)
Disinfection by-product	Bromoform	95	(Drewes et al., 2005)
	Trichloroacetic acid	96	(Kimura et al., 2003b)
EDC	Progesterone	85	(Ng and Elimelech, 2004)
	Testosterone	>91	(Drewes et al., 2005)
	Diethylstilbesterol	>99	(Comerton et al., 2008)
	Equilin	98	(Comerton et al., 2008)
	DEET	92	(Comerton et al., 2008)
EDC	Estrone	98	(Comerton et al., 2008)

(continued)

Table 6.8 (continued)

Classification	Name of micropollutant	Removal (%)	Reference
	17α-Estradiol	97	(Comerton et al., 2008)
	17β-Estradiol	83	(Kimura et al., 2004)
		96	(Comerton et al., 2008)
		>93	(Drewes et al., 2005)
	Estradiol	90	(Ng and Elimelech, 2004)
	Estriol	91	(Comerton et al., 2008)
		>80	(Drewes et al., 2005)
	17α-Ethynyl estradiol	98	(Comerton et al., 2008)
	Alachlor (Lasso)	95	(Comerton et al., 2008)
	Atraton	92	(Comerton et al., 2008)
	Carbaryl	79	(Kimura et al., 2004)
	Metolachlor	95	(Comerton et al., 2008)
	Bisphenol-A	66	(Drewes et al., 2005)
		83	(Kimura et al., 2004)
		95	(Comerton et al., 2008)
		99	(Kimura et al., 2003b)
	Oxybenzone	>99	(Comerton et al., 2008)
Flame retardant	Tris(1,3-dichloro-2-propyl)-phosphate	>99	(Drewes et al., 2005)
	Tris(2-chloroethyl)-phosphate	>99	(Drewes et al., 2005)
	Tris(2-chloroisopropyl)-phosphate	>99	(Drewes et al., 2005)
Metal	As	88	(Hrubec et al., 1983)
	Cd	75	(Hrubec et al., 1983)
	Cr	72	(Hrubec et al., 1983)
	Cu	72	(Hrubec et al., 1983)
	Hg	0	(Hrubec et al., 1983)
Metal	Mo	71	(Hrubec et al., 1983)
	Ni	85	(Hrubec et al., 1983)
	Pb	85	(Hrubec et al., 1983)

(continued)

Table 6.8 (continued)

Classification	Name of micropollutant	Removal (%)	Reference
	Zn	75	(Hrubec et al., 1983)
PAH	Hydronaphthalenes	75	(Hrubec et al., 1983)
	Naphthalene	15	(Hrubec et al., 1983)
Pesticide	Atrazine	99	(Kosutic and Kunst, 2002)
		>99	(Hofman et al., 1997)
	Bentazon	>99	(Hofman et al., 1997)
	Diuron	>99	(Hofman et al., 1997)
	DNOC	>99	(Hofman et al., 1997)
	MCPA	94	(Kosutic and Kunst, 2002)
		95	(Ozaki and Li, 2002)
		>99	(Hofman et al., 1997)
	Mecoprop	>99	(Hofman et al., 1997)
	Metalaxyl	>99	(Hofman et al., 1997)
	Metamitron	>99	(Hofman et al., 1997)
	Metribuzin	>99	(Hofman et al., 1997)
	Pirimicarb	>99	(Hofman et al., 1997)
	Propham	97	(Kosutic and Kunst, 2002)
	Simazin	>99	(Hofman et al., 1997)
	Triadimefon	83	(Kosutic and Kunst, 2002)
	Vinchlozolin	>99	(Hofman et al., 1997)
Petrochemical	1,2-ethanediol	62	(Kosutic and Kunst, 2002)
	2-Butanone	66	(Kosutic et al., 2007)
		78	(Kosutic and Kunst, 2002)
	Ethylacetate	75	(Kosutic and Kunst, 2002)
	Formaldehyde	31	(Kosutic and Kunst, 2002)

(continued)

Table 6.8 (continued)

Classification	Name of micropollutant	Removal (%)	Reference
	Indane	90	(Hrubec et al., 1983)
PhAC	Diclofenac	95	(Kimura et al., 2003b)
	Isopropylantipyrine	78	(Kimura et al., 2004)
	Phenacetine	71	(Kimura et al., 2003b)
		74	(Kimura et al., 2004)
	Primidone	84	(Kimura et al., 2003b)
		87	(Kimura et al., 2004)
		90	(Xu et al., 2006)
	Carbadox	90	(Comerton et al., 2008)
	Enrofloxacin	99	(Kosutic et al., 2007)
	Levamisole	99	(Kosutic et al., 2007)
	MBIK	97	(Kosutic et al., 2007)
	Oxytetracycline	99	(Kosutic et al., 2007)
	Praziquantel	99	(Kosutic et al., 2007)
	Sulfachloropyridazine	94	(Comerton et al., 2008)
	Sulfadiazine	99	(Kosutic et al., 2007)
	Sulfaguanidine	99	(Kosutic et al., 2007)
	Sulfamerazine	88	(Comerton et al., 2008)
	Sulfamethazine	99	(Kosutic et al., 2007)
	Sulfamethizole	93	(Comerton et al., 2008)
	Sulfamethoxazole	70	(Kimura et al., 2004)
		94	(Comerton et al., 2008)
	Trimethoprim	99	(Kosutic et al., 2007)
	Carbamazepine	91	(Kimura et al., 2004)
		91	(Comerton et al., 2008)
	Acetaminophen	82	(Comerton et al., 2008)
	Gemfibrozil	98	(Comerton et al., 2008)
	Caffeine	70	(Kimura et al., 2004)
		87	(Comerton et al., 2008)
Phenylphenol	4-Phenylphenol	61	(Kimura et al., 2004)
Phosphate	Organic phosphates	60	(Hrubec et al., 1983)

(continued)

Table 6.8 (*continued*)

Classification	Name of micropollutant	Removal (%)	Reference
Phthalate	Phthalates	55	(Hrubec et al., 1983)
Sulphonamide	Sulphonamides	10	(Hrubec et al., 1983)
Surrogate	2-Naphthol	43	(Kimura et al., 2003b)
	2-Naphthol	57	(Kimura et al., 2004)
	9-Anthracene carbonic acid	96	(Kimura et al., 2003b)
	Salicylic acid	92	(Kimura et al., 2003b)
Metabolite	Urea	30	(Ozaki and Li, 2002)
Trihalomethane	Bromoform	80	(Xu et al., 2006)
		13	(Hrubec et al., 1983)
	Bromodichloromethane	6	(Hrubec et al., 1983)
	Dibromochloroethane	>99	(Hrubec et al., 1983)
	Trichloroethylene	80	(Xu et al., 2006)
	Chloroform	5	(Hrubec et al., 1983)
		25	(Xu et al., 2006)
		85	(Drewes et al., 2005)

6.3 REFERENCES

Abegglen, C., Joss, A., McArdell, C. S., Fink, G., Schlusener, M. P., Ternes, T. A. and Siegrist, H. (2009) The fate of selected micropollutants in a single-house MBR. *Water Res.* **43**, 2036–2046.

Al-Rifai, J. H., Gabelish, C. L. and Schafer, A. I. (2007) Occurrence of pharmaceutically active and non-steroidal estrogenic compounds in three different wastewater recycling schemes in Australia. *Chemosphere* **69**, 803–815.

Asakura, H. and Matsuto, T. (2009) Experimental study of behaviour of endocrine-disrupting chemicals in leachate treatment process and evaluation of removal efficiency. *Waste Manage.* **29**, 1852–1859.

Bellona, C. and Drewes, J. E. (2005) The role of membrane surface charge and solute physico-chemical properties in the rejection of organic acids by NF membranes. *J. Membrane Sci.* **249**, 227–234.

Bodzek, M. and Dudziak, M. (2006) Elimination of steroidal sex hormones by conventional water treatment and membrane processes. *Desalination* **198**, 24–32.

Campos, C., Marinas, B. J., Snoeyink, V. L., Baudin, I. and Laine, J. M. (1998) Adsorption of trace organic compounds in CRISTAL (R) processes Conference on Membranes in Drinking and Industrial Water Production, Amsterdam, Netherlands, pp. 265–271.

Carballa, M., Omil, F. and Lema, J. M. (2005) Removal of cosmetic ingredients and pharmaceuticals in sewage primary treatment. *Water Res.* **39**, 4790–4796.

Choi, J. H. and Ng, H. Y. (2008) Effect of membrane type and material on performance of a submerged membrane bioreactor. *Chemosphere*, **71**, 853–859.

Cirja, M., Hommes, G., Ivashechkin, P., Prell, J., Schaffer, A., Corvini, P. F. X. and Lenz, M. (2009) Impact of bio-augmentation with Sphingomonas sp strain TTNP3 in membrane bioreactors degrading nonylphenol. *Appl. Microbiol. Biotechnol.* **84**, 183–189.

Clara, M., Strenn, B., Gans, O., Martinez, E., Kreuzinger, N. and Kroiss, H. (2005) Removal of selected pharmaceuticals, fragrances and endocrine disrupting compounds in a membrane bioreactor and conventional wastewater treatment plants. *Water Res.* **39**, 4797–4807.

Comerton, A. M., Andrews, R. C., Bagley, D. M. and Hao, C. Y. (2008) The rejection of endocrine disrupting and pharmaceutically active compounds by NF and RO membranes as a function of compound a water matrix properties. *J. Memb. Sci.* **313**, 323–335.

Deborde, M., Rabouan, S., Duguet, J. P. and Legube, B. (2005) Kinetics of aqueous ozone-induced oxidation of some endocrine disruptors. *Environ. Sci. Technol.* **39**, 6068–6092.

De Gusseme, B., Pycke, B., Hennebel, T., Marcoen, A., Vlaeminck, S. E., Noppe, H., Boon, N. and Verstraete, W. (2009) Biological removal of 17 alpha-ethinylestradiol by a nitrifier enrichment culture in a membrane bioreactor. *Water Res.* **43**, 2493–2503.

Drewes, J. E., Bellona, C., Oedekoven, M., Xu, P., Kim, T. U. and Amy, G. (2005) Rejection of wastewater-derived micropollutants in high-pressure membrane applications leading to indirect potable reuse. *Environ. Prog.* **24**, 400–409.

Drewes, J. E., Reinhard, M. and Fox, P. (2003) Comparing microfiltration-reverse osmosis and soil-aquifer treatment for indirect potable reuse of water. *Water Res.* **37**, 3612–3621.

Edwards, M., Scardina, P. and McNeil, L. S. (2003) Enhanced coagulation impacts on water treatment plant infrastructure. Awwa Research Foundation. IWA Publishing, London, UK.

Escher, B. I., Pronk, W., Suter, M. J .F. and Maurer, M. (2006) Monitoring the removal efficiency of pharmaceuticals and hormones in different treatment processes of source-separated urine with bioassays. *Environ. Sci. Technol.* **40**, 5095–5101.

Fatone, F., Bolzonella, D., Battistoni, P. and Cecchi, F. (2005) Removal of nutrients and micropollutants treating low loaded wastewaters in a membrane bioreactor operating the automatic alternate-cycles process European Conference on Desalination and the Environment, St Margherita, Italy, pp. 395–405.

Freese, S. D., Nozaic, D. J., Pryor, M. J., Rajogopaul, R., Trollip, D. L. and Smith, R. A. (2001) Enhanced coagulation: a viable option to advance treatment technologies in the South African context. *Water Sci. Technol: Water Supply* **1**(1), 33–41.

Hemond, H. F. and Fechner-Levy, E. J. (2000) Chemical Fate and Trasnport in the Environment. 2nd Edition. Academic Press, San Diego, USA.

Hofman, J., Beerendonk, E. F., Folmer, H. C. and Kruithof, J. C. (1997) Removal of pesticides and other micropollutants with cellulose-acetate, polyamide and ultra-low pressure reverse osmosis membranes Workshop on Membranes in Drinking Water Production – Technical Innovations and Health Aspects, Laquila, Italy, pp. 209–214.

Hrubec, J., Vankreijl, C. F., Morra, C. F. H. and Slooff, W. (1983) Treatment of municipal wastewater by reverse-osmosis and activated-carbon-removal of organic micropollutants and reduction of toxicity. *Sci. Tot. Environ.* **27**, 71–88.

Innocenti, L., Bolzonella, D., Pavan, P. and Cecchi, F. (2002) Effect of sludge age on the performance of a membrane bioreactor: influence on nutrient and metals removal International Congress on Membranes and Membrane Processes (ICOM), Taulouse, France, pp. 467–474.

Ivancev-Tumbas, I., Hobby, R., Kuchle, B., Panglisch, S. and Gimbel, R. (2008) p-Nitrophenol removal by combination of powdered activated carbon adsorption and ultrafiltration – comparison of different operational modes. *Water Res.* **42**, 4117–4124.

Jermann, D., Pronk, W., Boller, M. and Schafer, A. I. (2009) The role of NOM fouling for the retention of estradiol and ibuprofen during ultrafiltration. *J. Memb. Sci.* **329**, 75–84.

Jones, O. A. H., Green, P. G., Voulvoulis, N. and Lester, J. N. (2007) Questioning the excessive use of advanced treatment to remove organic micropollutants from wastewater. *Environ. Sci. Technol.* **41**, 5085–5089.

Kimura, K., Amy, G., Drewes, J. and Watanabe, Y. (2003a) Adsorption of hydrophobic compounds onto NF/RO membranes: An artifact leading to overestimation of rejection. *J. Memb. Sci.* **221**, 89–101.

Kimura, K., Amy, G., Drewes, J. E., Heberer, T., Kim, T. U. and Watanabe, Y. (2003b) Rejection of organic micropollutants (disinfection by-products, endocrine disrupting compounds, and pharmaceutically active compounds) by NF/RO membranes. *J. Memb. Sci.* **227**, 113–121.

Kimura, K., Toshima, S., Amy, G. and Watanabe, Y. (2004) Rejection of neutral endocrine disrupting compounds (EDCs) and pharmaceutical active compounds (PhACs) by RO membranes. *J. Memb. Sci.* **245**, 71–78.

Kosutic, K., Dolar, D., Asperger, D. and Kunst, B. (2007) Removal of antibiotics from a model wastewater by RO/NF membranes. *Sep. Purif. Technol.* **53**, 244–249.

Kosutic, K. and Kunst, B. (2002) Removal of organics from aqueous solutions by commercial RO and NF membranes of characterized porosities. *Desalination* **142**, 47–56.

Lai, K. M., Johnson, K. L., Scrimshaw, M. D. and Lester, J. N. (2000) Binding of waterborne steroid estrogens to solid phases in river and estuarine systems. *Environ. Sci. Technol.* **34**, 3890–3894.

Laine, J. M., Campos, C., Baudin, I. and Janex, M. L. (2002) Understanding membrane fouling: A review of over a decade of research. in: G. Hagmeyer, J.C. Schipper, R. Gimbel (Eds.), 3rd International Conference on Membranes in Drinking and Industrial Water Production, Mulheim, Germany, pp. 155–164.

Laine, J. M., Vial, D. and Moulart, P. (2000) Status after 10 years of operation – overview of UF technology today Conference on Membranes in Drinking and Industrial Water Production, Paris, France, pp. 17–25.

Lee, Y., Zimmermann, S. G., Kieu, A. T. and Von Gunten, U. (2009) Ferrate (Fe(VI)) application for municipal wastewater treatment: A novel process for simultaneous micropollutant oxidation and phosphate removal. *Environ. Sci. Technol.* **43**, 3831–3838.

Li, K. (2007) Ceramic membranes for separation and reaction. John Wiley and Sons.

Lim, M. and Kim, M. J. (2009) Removal of natural organic matter from river water using potassium ferrate (VI). *Water, Air, and Soil Poll.* **200**, 181–189.

Majewska-Nowak, K., Kabsch-Korbutowicz, M. and Dodz, M. (2002) Effects of natural organic matter on atrazine rejection by pressure driven membrane processes International Congress on Membranes and Membrane Processes (ICOM), Taulouse, France, pp. 281–286.

McGhee, T. J. (1991) Water Supply and Sewerage Engineering. 6th Edition. McGraw-Hill International Editions, Singapore.

Ng, H. Y., Elimelech, M. (2004) Influence of colloidal fouling on rejection of trace organic contaminants by reverse osmosis. *J. Memb. Sci.* **244**, 215–226.

Ng, H. Y. and Hermanowicz, S. W. (2005) Membrane bioreactor operation at short solids retention times: performance and biomass characteristics. *Water Res.* **39**, 981–992.

Nghiem, L. D. and Coleman, P. J. (2008) NF/RO filtration of the hydrophobic ionogenic compound triclosan: Transport mechanisms and the influence of membrane fouling. *Sep. Purif. Technol.* **62**, 709–716.

Nghiem, L. D., Tadkaew, N. and Sivakumar, M. (2007) Removal of trace organic contaminants by submerged membrane bioreactors 6th International Membrane Science and Technology Conference, Sydney, Australia, pp. 127–134.

Ozaki, H. and Li, H. F. (2002) Rejection of organic compounds by ultra-low pressure reverse osmosis membrane. *Water Res.* **36**, 123–130.

Pianta, R., Boller, M., Janex, M. L., Chappaz, A., Birou, B., Ponce, R. and Walther, J. L. (1998) Micro- and ultrafiltration of karstic spring water Conference on Membranes in Drinking and Industrial Water Production, Amsterdam, Netherlands, pp. 61–71.

Pronk, W., Biebow, M. and Boller, M. (2006) Electrodialysis for recovering salts from a urine solution containing micropollutants. *Environ. Sci. Technol.* **40**, 2414–2420.

Reemtsma, T., Zywicki, B., Stueber, M., Kloepfer, A. and Jekel, M. (2002) Removal of sulfur-organic polar micropollutants in a membrane bioreactor treating industrial wastewater. *Environ. Sci. Technol.* **36**, 1102–1106.

Reif, R., Suarez, S., Omil, F. and Lema, J. M. (2007) Fate of pharmaceuticals and cosmetic ingredients during the operation of a MBR treating sewage Conference of the European-Desalination-Society and Center-for-Research-and-Technology-Hellas, Halkidiki, Greece, pp. 511–517.

Sharma, V. K. and Mishra, S. K. (2006) Ferrate (VI) oxidation of ibuprofen: A kinetic study. *Environ. Chem. Lett.* **3**(4), 182–185.

Sharma, V. K. (2008) Oxidative transformations of environmental pharmaceuticals by Cl_2, ClO_2, O_3 and Fe (VI): Kinetic assessment. *Chemosphere* **73**, 1379–1386.

Stumm, W. and Morgan, J. J. (1996) Aquatic Chemistry – Chemical Equilibria and Rates in Natural Waters, 3rd Edition. Wiley Interscience, N.Y., USA.

Tan, T. W., Ng, H. Y. and Ong, S. L. (2008) Effect of mean cell residence time on the performance and microbial diversity of pre-denitrification submerged membrane bioreactors. *Chemosphere* **70**, 387–396.

Tchobanoglous, G., Burton, F. L. and Stensel, H. D. (2003) Wastewater Engineering: Treatment and Reuse. 4th Edition. McGraw Hill, N.Y., USA.

Vanoers, C. W., Vorstman, M. A. G. and Kerkhof, P. (1995) Solute rejection in the presence of a deposited layer during ultrafiltration. *Journal of Membrane Science*, **107**, 173–192.

Wang, D.S., Wu, X.H., Huang, L., Tang, H. X. and Qu, J. H. (2007) Nano-inorganic polymer flocculant: From theory to practice. In Chemical Water ad Wastewater Treatment IX. Edited by Hahn H.H., Hoffman E., Ódegaard H. IWA Publishing, London. pp. 181–188.

Waniek, A., Bodzek, M. and Konieczny, K. (2002) Trihalomethane removal from water using membrane processes. *Polish Journal of Environmental Studies* **11**, 171–178.

Xu, P., Drewes, J. E., Kim, T. U., Bellona, C. and Amy, G. (2006) Effect of membrane fouling on transport of organic contaminants in NF/RO membrane applications. *J. Memb. Sci.* **279**, 165–175.

Yan, M., Wang, D., Qu, J., He, W. and Chow, C. W. K. (2007) Relatively importance of hydrolyzed Al (III) species (Al_a, Al_b and Al_c) during coagulation with polyaluminum chloride: A case study with the typical micro-polluted source waters. *J. Coll. Interf. Sci.* **316**, 482–489.

Yan, M., Wang, D., Yu, J. Ni, J., Edwards, M. and Qu J. (2008) Enhanced coagulation with polyaluminum chlorides: Role of pH/alkalinity and speciation. *Chemosphere* **71**, 1665–1673.

Zhao, H., Hu, C., Liu, H., Zhao, X. and Qu, J. (2008) Role of aluminium speciation in the removal of disinfection byproduct precursors by a coagulation process. *Environ. Sci. Technol.* **42**, 5752–5758.

Zorita, S., Martensson, L. and Mathiasson L. (2009) Occurrence and removal of pharmaceuticals in a municipal sewage treatment system in the south of Sweden. *Sci. Tot. Environ.* **407**, 2760–2770.

Chapter 7
Biological treatment of Micropollutants

T. H. Ergüder and G. N. Demirer

7.1 INTRODUCTION

Municipal wastewater is the main source of micropollutants which are present in the household products (detergents, cosmetics and paints) and natural excretion of humans (drugs and metabolites, synthetic hormones) (Bruchet *et al.*, 2002; Bicchi *et al.*, 2009). After their use, pharmaceuticals are excreted intact and/or as metabolites with feces and urine; thus are introduced directly into wastewater (Löffler *et al.*, 2005). New sanitation concepts where wastewater streams are separated and treated according to their characteristics is thus an important source control strategy to avoid input of pharmaceuticals to the wastewaters and in turn to the environment.

©2010 IWA Publishing. *Treatment of Micropollutants in Water and Wastewater.* Edited by Jurate Virkutyte, Veeriah Jegatheesan and Rajender S. Varma. ISBN: 9781843393160. Published by IWA Publishing, London, UK.

7.2 MUNICIPAL SEWAGE AS THE SOURCE OF MICROPOLLUTANTS

Many studies reveal that varied micropollutants of different groups reach wastewater treatment plants (WWTPs). They are only partly eliminated during conventional wastewater treatment and have been measured in the effluent of various WWTPs (Desbrow *et al.*, 1998; Pickering and Sumpter, 2003). The typical measured concentrations for a variety of pharmaceuticals in the influent of WWTPs as well as the removal efficiencies obtained are shown in Table 7.1. The risks associated with pharmaceuticals, an important sub-group of micropollutants, ares provided in Table 7.2. Hammer *et al.* (2005) investigated pharmaceutical residues in terms of their occurrence in the environment and conducted a database assessment for this purpose. They reported significant concentrations of pharmaceuticals in raw as well as treated wastewaters (Table 7.3).

Table 7.1 Concentrations and removal efficiencies of some pharmaceuticals in WWTP (Wang, 2009)

Pharmaceutical	Influent (µg/L)	Removal (%)	Reference
Aspirin	0.34–3.1	81–88	Heberer (2002)
Bezafibrate	1.2–5.3	27–83	Ternes (1998)
Caffeine	230	99.9	Heberer (2002)
Carbamazepine	1.78–2.1	7–8	Heberer (2002)
Clofibric Acid	0.46–1.2	0–15	Heberer (2002)
Cyclophosphamide	0.007–0.143	0–94	StegerHartmann *et al.* (1997)
Diclofenac	0.035–3.02	69–98	Heberer (2002); Ternes (1998)
Fenofibric acid	0.5–1.03	6–64	Stumpf *et al.* (1999)
Gemfibrozil	0.35–0.9	16–69	Ternes (1998); Stumpf *et al.* (1999)
Ibuprofen	0.3–4.1	90	Ternes (1998); Stumpf *et al.* (1999)
Ketoprofen	0.6	48–69	Stumpf *et al.* (1999)
Naproxen	0.6–1.3	15–78	Ternes (1998)
Phenazone	0.3	33	Ternes (1998)

Bruchet et al. (2002) analyzed the raw sewage of three WWTPs in France and detected more than 200 compounds consisting of various endocrine disrupters, and pharmaceutical and personal care products (PPCPs). The significant pollutant groups identified were glycol ethers, phthalates, nonylphenols, drugs, hormones, personal care products (PCPs), detergents, additives and additional undesirable compounds such as 1,1,1-trichloroethane, phenol and p-cresol, 2,4-dichlorophenol, dimethyltrisulfide, pyridine, cyclamidomycin and luminol etc. Highly powerful endocrine disrupters estrone (E1) and 17β-estradiol (E2) and the active ingredient of most birth control drugs, 17α-ethinylestradiol (EE2) were detected in mechanical and biological units as well as in sewage-sludge treatment unit at a municipal German WWTP (Andersen et al., 2003). Over 80 compounds have been found in sewage effluents, surface waters, and even in ground waters (Heberer, 2002).

Table 7.2 Pharmaceuticals and their risks (Wang, 2009)

Drug	Examples	Risk (Exposure and Toxicity)
Antibiotics	penicillins, sulfamethoxazole	High volumes (about 23000 tons of antibiotics/year in the U.S.); concerns over toxicity and antibacterial resistance
Painkiller	ibuprofen, naproxen, ketoprofen	Very high prescription and over-the-counter volumes (70 million prescriptions and 30 billion over-the-counter doses sold annually in the U.S.); toxicity has not been claimed.
Antiepileptics	carbamazepine, phenobarbital	Detected in the environment; persistent
Lipid regulators	clofibric acid, gemfibrozil	Long-term prescriptions; commonly detected
β-blockers	propranolol, metoprolol	High volumes; detected in the environment
Antidepressants	fluoxetins, risperidone	High volumes; subject of toxicity testing
Antihistamines	loratadine, cetrizine	Commonly held nonprescription medicine
Others	contraceptive pills, estradiol	Associated with endocrine disruptions

Despite the increased number of investigations on the topic, due to the difficulty and the cost of the chemical analyses of the micropollutants, the information on the existence of micropollutants in sewage and their concentrations are not established for a number of countries. Carballa et al. (2008) examined the consumption and excretion rates of 17 pharmaceuticals, two musk fragrances and two hormones by the Spanish population in 2003 by three different models; (1) Extrapolation of the per capita use in Europe to the number of inhabitants of Spain for musk fragrances; (2) Annual prescription items multiplied by the average daily dose for pharmaceuticals and; (3) Excretion rates of different groups of population for hormones. The prediction of the expected concentrations entering the treatment plant was then compared to the concentrations measured in the raw sewage. It is striking to observe that predicted concentrations are consistent with the measured concentrations for half of the 21 substances selected (such as Carbamazepine, Diazepam, Ibuprofen, Naproxen, Diclofenac, Sulfamethoxazole, Roxithromycin, Erythromycin and 17α-ethinylestradiol).

Table 7.3 Pharmaceutical substances analyzed in wastewater (Hammer et al., 2005)

Name of Pharmaceutical Substance	Type of wastewater	Minimum conc*	Maximum conc*	Average conc*
Bezafibrate	effluent of WWTP	485	2610	1765
	raw wastewater	405	5600	2935
Carbamazepine	effluent of WWTP	920	2100	1605
	raw wastewater	1200	2200	1580
Diclofenac	effluent of WWTP	750	1500	1080
	raw wastewater	1625	1900	1808
Metoprolol	effluent of WWTP	615	1700	1015
	raw wastewater	425	7200	3041
Sotalol	effluent of WWTP	630	1320	975
	raw wastewater	800	1600	1200

* $ng\ L^{-1}$

Varied types of micropollutants which are either removed or remained in the effluent in WWTPs are well discussed in in Section 6.2. The micropollutants which are not biodegraded or removed are discharged into the receiving bodies via treated water or by disposal of the sludge on land. Thus, they remain to be the source of micropollutants and threat for the environment and living organisms. The information on occurrence of pharmaceuticals in the environment is usually limited to the parent compounds. Yet, the metabolites of some micropollutants might be discharged in higher amounts than the parent compounds. For example, the concentration of the metabolite of carbamazepine (10,11-dihydro-10,11-dihydroxycarbamazepine) was almost 3–4 times higher than the parent compound, carbamazepine, in the influent and effluent of a WWTP in Ontario and in the surface water samples (Miao and Metcalf, 2003).

The current concern of micropollutants in the receiving waters may also call for new approaches in wastewater treatment. Wastewater treatment plants are designed to deal with bulk substances that arrive regularly and in large quantities (primarily organic matter and the nutrients or nitrogen and phosphorus). Pharmaceuticals are entirely different with respect to their fate. They are single compounds with a unique behaviour in the treatment plant, and they represent only a minor part of the wastewater organic load (Larsen et al., 2004). The approach of decreasing the influent micropollutants load to the WWTPs through urine source separation is, thus, one of the new topics investigated for the removal of micropollutants and in turn their decreased amount of discharge to the environment (Henze, 1997; Larsen and Gujer, 2001; Larsen et al., 2004).

7.2.1 Urine source separation and possible advantages

A source separated sanitation concept, based on collection, transport and treatment of black water (toilet) separate from grey (shower, bath, kitchen and laundry) water, enables the recovery of energy and nutrients, and keeps micropollutants in a relatively small volume (Larsen et al., 2004; De Mes, 2007). The pharmaceutical residues investigated and the concentrations determined, if any, in urine from a public urinal at Hansaplatz, Hamburg (HH) and from the separation system Stahnsdorf, Berlin are depicted in Table 7.4.

There are different types of toilets and NoMix technology is a promising innovation for the urine source separation concept (in http://www.novaquatis.eawag.ch/index_EN). Rossi et al. (2009) conducted measurements regarding the toilet usage, flushing behavior and recovered urine. They found that toilet usage values are 5.2/person/day for weekdays and 6.3/person/day for weekends in households, 30–85% comprising the small flushes. They have calculated the amount of urine effectively recovered per voiding in NoMix toilets as 138 mL/flush in households and 309 mL/flush in women's toilets. The estimated urine recovery in the households was a maximum of 70–75% of the expected quantity.

Table 7.4 Concentrations of pharmaceutical residues (Tettenborn et al., 2008)

Pharmaceutical type	Pharmaceutical Concentration (µg/l)				
	Average	Standard Deviation	95% Confidence	Minimum	Maximum
Bezafibrate	362	241	211	192	846
ß-Sitosterol	32	11	10	18	52
Diclofenac	21	12	10	9	45
Carbamazepine	17	8	7	4	29
Phenacetin	7	9	8	1	23
Pentoxifylline	7	2	2	3	9
Phenazone	3	1	1	2	4

Urine source separation concept is based on the expectation that many anthropogenic pharmaceuticals are mainly excreted via urine. Accordingly, the release of micropollutants to the wastewater stream and, in turn, to the environment can be prevented at the source by urine separation systems. The excretion of pharmaceuticals from the human body can occur by three ways (Ternes and Joss, 2006) which are;

- Renal excretion via urine
- Biliary and intestinal via feces
- Pulmonary via exhaled air

It is generally assumed that renal excretion is the major pathway for most pharmaceuticals and this excretion can occur in metabolized or un-metabolized form. For the load assessment of treatment plants, un-metabolized compounds should be taken into consideration since conjugated compounds will be de-conjugated during biological treatment processes (Ternes and Joss, 2006). Lienert et al. (2007) conducted a quantitative screening of official pharmaceutical data to investigate the effect of urine source separation. They analyzed the excretion pathways of 212 pharmaceuticals' active ingredients (AI) and found that, on average, 64% (±27%) of each AI was excreted via urine, and 35% (±26%) via feces. In urine, 42% (±28%) of each AI was excreted as metabolites. Urine source separation was found to effectively remove the highly sold and non-degradable x-ray contrast media, 94% (±4%) of which are excreted via urine. However, for some pharmaceuticals, excretion via urine was 6–98% (such as cytostatics). Due to the serious variability in the numbers changing with respect to the compound type, it is advised to be careful about using urine separation technology solely because not all the compouds would be separated via urine (Lienert et al., 2007).

Despite the uncertainty about the effectiveness of the urine source separation for all types of micropollutants, it might have significant advantages. It prevents potentially hazardous micropollutants from entering the wastewater stream. The wastewater treatment plants operated currently are removing only some part of the micropollutants. The high cost requirement for the advanced treatment processes for the removal of pharmaceuticals has generated more attention to the separation at source (Environment Canada, 2009). Thus, urine source separation might reduce the treatment plant size and operation costs.

The other advantage of urine source separation is the relatively low volume produced for treatment. Since pharmaceuticals are present mainly in a small wastewater stream (black water), a target treatment can be accomplished by managing this small stream separately. The typical domestic wastewater production value is 200 L/person/day. However, the urine production values are 1.5 and 7.5 L/person/day, when collected undiluted (black water) and when vacuum toilet is applied, respectively. One of the advantages of this approach is that pharmaceuticals are present in the highest possible concentrations in this relatively small stream (Kujawa-Roeleveld et al., 2006 and 2008). Separation of this waste stream allows collection of a significant amount of the total pharmaceutical consumption at a concentration about 100 times higher than that in municipal wastewater (Ternes and Joss, 2006). Kujawa-Roeleveld et al. (2006) compared the concentrations of excreted pharmaceuticals in source separated urine, black water (calculated as maximum possible) and in an influent of a WWTP from a combined sewer (measured) as an argument to treat concentrated streams separately in order to remove pharmaceutical compounds (Table 7.5). They considered the worst case scenario assuming that all people administer a given pharmaceutical compound at a maximum defined daily dose (WHO, 2006) and a percentage excreted is a parent compound. Indeed, the data indicate the significant effect of source separation in concentrating the streams or in preventing the dilution of micropollutants and in turn the possible removal problems associated with their trace-level concentrations in wastewaters arriving WWTPs.

Table 7.5 Concentrations of the excreted pharmaceuticals (examples) in different sources (Kujawa-Roeleveld et al., 2006)

Example compound	Concentration (µg/L)		
	Source Separated*		Influent of WWTP**
	Urine	Black water	
Ibuprofen (analgesic)	80000	16000	27
Carbamazepine (anti-epileptic)	13000	2700	0.25–2.2

* Calculated assuming that all people administer a given pharmaceutical compound at a maximum defined daily dose and a percentage excreted is a parent compound.
** From a combined sewer.

Main reasons for using urine source separation are the possibility of improved water pollution control regarding nutrients and micropollutants and the possibility of closing the nutrient cycles. According to The Swiss Federal Institute of Aquatic Science and Technology (Eawag), urine source separation could serve as a main assistant for water pollution control (Bryner, 2007). Even though urine constitutes less than 1% of the wastewater volume, it contains most of the nutrients content up in wastewater and many micropollutants from the human metabolism. In most cases, efficient urine source separation would render nutrient removal at treatment plants obsolete; to obtain more stringent threshold values for phosphorus, merely a small technical effort would be necessary (Larsen and Gujer, 1996). Additional to water pollution control, urine source separation offers a promising solution to nutrient recycling, a sustainability issue especially for phosphorus (Driver *et al.*, 1999; Lienert *et al.*, 2003; Maurer *et al.*, 2003). Different techniques have been studied to remove and recover nutrients from urine, focusing mainly on nitrogen and phosphate (Udert *et al.*, 2003; Pronk *et al.*, 2006; Wilsenach and van Loosdrecht, 2006; Ronteltap *et al.*, 2007). Perhaps the most vital aspect of urine separation is the potential of flexible adaptation of the present wastewater system without losing capital bound in existing infrastructure (sewers and treatment plants). Larsen and Gujer (1996) suggested transition scenarios with storage of urine in households and subsequent release when nitrogen is required at the treatment plant. Furthermore, storage capacity could be chosen such that urine in combined sewer overflow (CSO, release of untreated wastewater to receiving waters during rain) could be avoided. In a typical Swiss wastewater treatment plant, Rauch *et al.* (2003) reported that with a very moderate storage capacity (10 L per toilet), which could possibly be integrated into the toilet itself, and a simple control strategy, a 30% increase of nitrification capacity and a 50% reduction of urine in CSOs can be achieved (Larsen *et al.*, 2004).

On the other hand, there are some limitations for the urine source separation. The treatment of the collected urine is, for example, feasible only if the dilution is very low (Ternes and Joss, 2006; Joss *et al.*, 2006). Besides, the pharmaceuticals, which are found in the feces, are generally more lipophilic and might adsorb well to fecal material and end up in sludge (Lienert *et al.*, 2007). The appropriate treatment technology should also be considered for sludge in terms of the presence of these micropollutants.

7.2.2 Biological degradation in source separated urine

Urine is a concentrated mixture containing a number of water-soluble waste products from the human metabolism. Due to rapid hydrolysis of urea, once urine has left the urinary tract, the concentration of ammonia/ammonium and pH rise rapidly in source separated urine. Research on biological treatment of urine has mainly concentrated on partial nitrification, either to stabilize urine for further treatment (decreasing the pH below 7 and thereby preventing stripping of ammonia) or as a first step in nitrogen removal by autotrophic denitrification (Udert *et al.*, 2003). From this experimental work, it is known that about 85% of the organic fraction in urine is biologically degradable (Udert, 2002).

The studies on biological degradation of micropollutants in source separated urine are very limited. First results pointed out that the half-life of natural estrogens in a biological reactor treating urine is less than 15 min. Compared to wastewater treatment plants, a higher degree of transformation/degradation of pharmaceuticals is expected due to significantly higher concentrations (100–500 times larger than in wastewater) and the possibility of obtaining a substantially higher solids retention time at very low costs (due to the low organic loading from urine). Substrate inhibition, which may possibly occur in wastewater treatment plants due to peak organic loadings, can be avoided more easily in urine-treating systems, but this will require some storage capacity (Larsen *et al.*, 2004).

Kujawa-Roeleveld *et al.* (2008) investigated the biodegradability of eight pharmaceutically active compounds under various environmental conditions (redox, temperature and type of seed). Selected compounds were characterized by different physical-chemical-biological properties in order to be able to extend the results of this research to the broader group of environmentally relevant micropollutants. The selected compounds were acetylsalicylic acid, bezafibrate, carbamazepine, clofibric acid, diclofenac, fenofibrate, iburpofen and metoprolol. Fate of some of these compounds (e.g., carbamazepine, diclofenac, ibuprofen) in wastewater treatment systems have been already described in literature to some extent. However, very little is known about the fate of micropollutants in source separation based wastewater treatment systems. The applied initial concentrations were significantly higher than those at the conventional treatment plant (low mg/L range against low µg/L range). The reason was the applicability of the results to treatment concepts for concentrated black water or urine. As seen in Table 7.6, the results of this study indicated that many of these compounds could be biodegraded under aerobic conditions. The extent of biodegradation depends, in many cases, on the exposure time of a biomass to a given compound. Aerobic biodegradation was reported to be faster than anoxic degradation and elevating operational temperatures speed up the biodegradation processes, as expected. Under anaerobic conditions and relatively extended hydraulic retention times (HRT = 30 d) some of these compounds could be degraded (acetylsalicylic acid, ibuprofen, fenofibrate) but at much lower rate than under aerobic or anoxic conditions. It was further reported that anaerobic pretreatment followed by aerobic main treatment and a tertiary physical or chemical polishing step should be used to eliminate persistent compounds for source separated wastewater.

Table 7.6 Comparison of biotransformation rate of the selected pharmaceuticals at different environmental conditions (Kujawa-Roeleveld et al., 2008)

Pharmaceuticals	Aerobic (20°C)	Aerobic (10°C)	Anoxic (20°C)	Anoxic (10°C)	Anaerobic (30°C)
Acetylsalicylic acid	+++	+++	++	++	+
Fenofibrate	+++	++	++	++	+
Ibuprofen	++	++	+	+	+
Metoprolol	++	+	+	−	−
Bezafibrate	±	±	+	−	−
Diclofenac	±	±	−	−	−
Carbamazepine	−	−	−	−	−
Clofibric acid	−	−	−	−	−

(Biotransformability: +++ high, ++ good, + moderate, ± only at HRTs >2 days, − not biotransformed)

Two natural hormones estrone (E1) and 17β-estradiol (E2) and the synthetic hormone 17α-ethinylestradiol (EE2) which are endocrine disrupting compounds (EDCs) that display the strongest estrogenic effects, are excreted by humans mainly in urine and a small amount in faeces. Therefore, nearly all of these estrogens are present in black water when source-separation is applied. Although they do not exhibit any estrogenic potency in this form, microbial enzymes can cleave these conjugates back to their original active form. Sulphate conjugates are more stable than glucuronide conjugates and no decrease in sulphate conjugates was observed in a septic tank for domestic wastewater treatment (D'Ascenzo et al., 2003).

Anaerobic biodegradation of estrogens is very slow or does not occur, and up to 60% is adsorbed to sludge (De Mes, 2007). Under anaerobic conditions, E1 and E2 are inter-convertible. Lee and Liu (2002) reported a 60% conversion of E2 into E1 after 20 days under anaerobic conditions at 21°C, spiked with E2 at an initial concentration of 2 mg/L. Similar findings were reported by Czajka and Londry (2006) for spiked lake sediment. Czajka and Londry (2006) investigated anaerobic transformation of E2 under methanogenic, nitrate-, sulphate- and iron-reducing conditions. In all cases the sum of both E1 and E2 decreased with only 10% over 383 days. However, all three compounds can potentially be removed under aerobic conditions (Ternes et al., 1999; Layton et al., 2000; De Mes, 2007).

The fate of E1, E2 and EE2 was investigated in a concentrated black water treatment system consisting of an upflow anaerobic sludge blanket (UASB) septic tank, with micro-aerobic post-treatment (De Mes, 2007). In UASB septic tank effluent, concentrations of natural total concentrations of 4.02 µg/L E1 and 18.69 µg/L E2, comprising the sum of conjugated ($>70\%$ for E1 and $>80\%$ for E2) and unconjugated forms, were measured. However, no EE2 was detected. In the effluent of the post-treatment, E1 and E2 were present in concentrations of 1.37 ± 1.45 µg/L and 0.65 ± 0.78 µg/L, respectively. A percentage of 77% of the measured unconjugated E1 and 82% of E2 was associated with particles (>1.2 µm) in the final effluent implying high sorption affinity of both compounds. When spiking the UASB septic tank effluent with E1, E2, EE2 and the sulphate conjugate of E2, removal in the micro-aerobic post-treatment was $>99\%$ for both E2 and EE2 and 83% for E1.

7.3 BIOLOGICAL TREATMENT OF MICROPOLLUTANTS

Since the last decade, the environmental fate and behaviour of micropollutants such as synthetic organic chemicals (e.g., some pesticides, polychlorinated biphenyls- PCBs), several industrial and household chemicals, pharmaceuticals, personal care products and natural hormones are of particular concern due to their potential endocrine disrupting activities. Among endocrine disrupting compounds (EDCs), pharmaceuticals and personal care products (such as antibiotics, fragrances, estrogens, antioxidants, veterinary growth hormones, disinfectants, antimicrobial compounds, fire retardants, psychiatric drugs, insect repellents, soaps, surfactants, x-ray contrast media), industrial and household chemicals (e.g., phthalates, solvents) and side products of industrial and household processes (e.g., polycyclic aromatic hydrocarbons- PAHs, dioxins) can be listed (Caliman and Gavrilescu, 2009). There is increasing evidence for the exposure to EDCs to lead to the reproductive and health effect on humans and other living organisms by disturbing the endocrine system via mimicking, blocking or also disrupting function of hormone (Joffe, 2001; WHO, 2002; Vajda et al., 2006; Bolong et al., 2009). Some of the micropollutants or their metabolites reaching to the wastewater treatment plants are not completely removed, and therefore are introduced to the surface and ground waters. Human exposure to those pollutants might occur through consumption of water and aquatic organisms that accumulated persistent micropollutants. Therefore, the exposure of humans and living organisms to these micropollutants via WWTP effluents should be prevented by applying appropriate treatment techniques. This requires the understanding the fate of different compounds and factors affecting their removal in biological units and correspondingly the improvement of the existing treatment systems or innovation of new systems.

7.3.1 Analysis of Micropollutants

Treatability studies performed to evaluate the degradation rate of a compound and/or the performance of a treatment unit require proper analysis and well-detection of the compound of concern. However, the measurement of micropollutants, especially those which are found in complex matrices such as wastewater and sludge is challenging:

- Micropollutants are found in trace levels, usually in units of µg/L or ng/L. Thus, extraction is required to concentrate the chemicals. However, the low level pollutants in complex matrices such as sludge and wastewater are difficult to analyze and could seriously affect their extraction and analysis. A highly sensitive measurement is essential (Bolong et al., 2009).
- There is a broad range of micropollutants with different physico-chemical characteristics. There is no standard or common method for monitoring and analyses. Each compound requires specific analysis by different techniques (Caliman and Gavrilescu, 2009).

Biological treatment of Micropollutants

- For many of the micropollutants, the required analytical techniques are expensive, difficult to employ; thus might require expertise. Few laboratories have the possibility and resources to perform them (Caliman and Gavrilescu, 2009).
- Improved and advanced analytical and bioanalytical technologies that enable the detection of more xenobiotics at an even lower range of concentrations are required (Bolong *et al.*, 2009).

7.3.1.1 Analytical techniques used for wastewater and sludge samples

The analytical techniques such as Gas Chromatography (Tandem) Mass spectrometry (GC-MS-MS), Liquid Chromatography combined with mass spectrometry (LC-MS-MS) and High Performance Liquid Chromatography (HPLC) are usually employed for the analyses of micropollutants. Several methods of screening PPCPs and EDCs are well-summarized by Caliman and Gavrilescu (2009).

GC-MS-MS is an often used technique for the detection and quantification of estrogens in surface water, wastewater or sludge, after the volatility of the compounds has been increased by a necessary derivatisation step. Also LC-MS-MS nowadays is an often used method for determination of the steroid estrogens (Zuehlke *et al.*, 2005; Richardson, 2006). The main advantage of this method is that derivatisation is not required, whilst still very low limits of detection can be achieved (around 1 ng/l) (Cui *et al.*, 2006). Moreover, it is possible to detect conjugated estrogens without applying a deconjugation step. New developments can be expected with HPLC-time of flight mass spectrometry. A TOF analyser can provide a more accurate mass estimation, which allows better distinction between estrogens and matrix (Reddy and Brownawell, 2005). Another way of quantification is HPLC followed by a UV-(ultra violet visible) detector (UV-VIS), Diode Array Detector (DAD) or Fluorescence detector. It was reported that the use of only HPLC with UV-detection has high detection limit and limited sensitivity and accuracy for estrogens; thus requires high spiking concentrations in unit of mg/L (De Mes, 2007). The use of fluorescence detector in series with UV detector improves the sensitivity and consequently lowers the limits of detection. The replacement of UV with DAD improves the selectivity. Yet, for measurement of estrogens without the need of spiked samples, highly selective and sensitive method (GC-MS) is essential. De Mes (2007) stated that GC-MS-MS does not improve the sensitivity of the analysis performed for the anaerobically treated concentrated black water samples.

The analytical methods, as previously mentioned, require specific procedures and sample purification which might vary with respect to the physico-chemical property of the micropollutant. To concentrate the chemicals, usually in assessing the PPCPs/EDCs, solid phase extraction (SPE) or solid phase micro extraction (SPME) are performed. For low molecular weight polar compounds such as glycol ethers, SPME coupled with GC-MS is the most suitable method to extract it from wastewaters (Bensoam *et al.*, 1999). SPME seems to be advantageous due to its comparably low cost, simple application (solvent free and fast) and successful use for a variety of compounds when combined with GC and GC-MS (Yang *et al.*, 2006). Other extraction methods can be as follows (Peck, 2006):

- Liquid liquid extraction (LLE) and steam distillation solvent extraction (SDE) used mainly for extraction of environmental contaminants from water;
- Pressurized fluid extraction (PFE), supercritical fluid extraction (SFE), sonication and Soxhlet extraction, generally used for contaminants present in sludge;

Alkylphenol polyethoxylates are polar compounds and were quantified only in the aqueous media but not in the sludge. Degradation of alkylphenol polyethoxylates resulted in formation of shorter and toxic alkylphenols which concentrate in the sludge. Due to the restrictions on analytical methods for quantification of those compounds, their determination is not possible when the matrix is sludge. (Janex-Habibi *et al.*, 2009)

7.3.1.2 Endocrine disrupting effect

Varied chemical analyses and biological assays have been performed to identify the compounds having endocrine disrupting effect/activity such as (Cespedes *et al.*, 2004; Chen *et al.*, 2007; Hashimoto *et al.*, 2007; Muller *et al.*, 2008; Kanda and Churchley, 2008; Caliman and Gavrilescu, 2009; Liu *et al.*, 2009a);

- *In vitro* and *in vivo* bioassays
- *In vitro* androgen receptor mediated transcriptional activity assays,
- *In vivo* Vitellogenin (VTG) assays
- Recombinant yeast assay (RYA),
- Yeast estrogen screen (YES) test
- E-screen test
- Estrogen-responsive reporter cell lines (MELN)
- An estrogen receptor (ER)/androgen receptor (AR) ligand competitive binding assay (ER/AR-binding assay)

Samples can also be exposed to a bioassay, e.g., a Yeast Estrogen Response, in which the total estrogenity of a sample is expressed as E2-equivalents (Witters et al., 2001; Murk et al., 2002). YES test was demonstrated to be very suitable for the identification of real-life effects in environmental samples, viz. surface waters, soils, sediments, wastewater (Murk et al., 2002; Onda et al., 2002; Saito et al., 2002; Tilton et al., 2002). However, it was found to be inappropriate to elucidate fate mechanisms during wastewater treatment, like adsorption and biodegradation (Cordoba, 2004). This can be attributed to a high background noise in the samples as well as toxicity of wastewater and sludge to the yeast.

The information on these bioassays is beyond the scope of this chapter. Yet, it should be noted that researches are important from the perspective of indicating the complexity of the mechanisms underlying the biological response, which is not only due to one chemical. The fractionation of samples and the existence of a chemical matrix as well as their concentration values in the samples should be taken into consideration (Cargouet et al., 2004). Bicchi et al. (2009) investigated the presence of estrogenic substances and performed E-screen assay on effluent of a WWTP serving four towns of a metropolitan area in Italy, on water samples of the receiving river and upstream and downstream of the plant by measuring the 17β-estradiol (E2) equivalency quantity (EEQ). They targeted the four groups; phenols (2,4-dichlorophenol, 4-t-butylphenol, 4-n-octylphenol, 4-n-nonylphenol, bisphenol A (BPA), E1, E2, ethinylestradiol and paracetamol), acids and amines (3,4-dichloroaniline, ketoprofen 2,4-D, 2,4,5-T, naproxen, ibuprofen, diclofenac, salicylic acid and acetylsalicylic acid), organotin compounds (tributyltin chloride, triphenyltin chloride) and apolar chemicals (atrazine, alachlor, 7,12-dinethylbenz[a]anthracene, bis (2-ethylhexyl) phthalate, dibutylphthalate, diethylphthalate, benzo[a]pyrene, cholesterol, coprostanol, sitosterol and stigmasterol). Mean EEQs were 4.7 ng/L ($\pm$2.7 ng/L) upstream and 4.4 ng/L ($\pm$3.7 ng/L) downstream of the plant, and 11.1 ng/L ($\pm$11.7 ng/L) in the effluent, indicating the little impact of the WWTP effluent on estrogenicity and the concentration of EDCs in the river water. The acute toxicity test (*Vibrio fischeri*) revealed no correlation between the toxicity and estrogenic disrupting activity (Bicchi et al., 2009). Yet, in another study, the estrogenic potential of 4 WWTPs' influents and effluents and the receiving waters in France were examined by *in vitro* estrogenicity bioassay for natural and synthetic estrogens, E1, E2, Estriol (E3) and EE2 (Cargouet et al., 2004). The concentrations ranged within 2.7–17.6 ng/L in WWTPs and 1.0–3.2 ng/L in rivers. EE2 was found to be resistant to biodegradation in all WWTPs and thus accounted for 35–50% of the estimated estrogenic activity in rivers. The results reveal the significant effect of chemical type and matrix structure on bioassay results as well as the effect of treatment unit type on the removal of chemicals from matrices.

7.3.2 Removal mechanisms of Micropollutants

The physico-chemical properties of a chemical define its removal mechanisms or fate in natural systems and engineered systems such as WWTPs. Some of the pollutants can be removed by sorption or destruction in wastewaters such as tonalide, galaxolide and celestolide, ibuprofen, diclofenac and mefenamic acid, E1 and E2. On the other hand, some of the micropollutants are very persistent that cannot be sorbed or transformed such as carbamazipine, lincomycin, tylosin, sulfamethoxazole and trimethoprim (Caliman and Gavrilescu, 2009). The removal mechanisms of PPCPs during wastewater treatment are given as follows (Caliman and Gavrilescu, 2009):

- Biological degradation
- Sorption onto sludge
- Stripping
- Chemical oxidation
- Photocatalysis

7.3.2.1 Sorption

Sorption behaviour can be estimated with the help of the sorption coefficient (K_d, solid-water distribution coefficient) and it is dependent on the characteristics of the compound as well as on the sludge type (Caliman and Gavrilescu, 2009). Hydrophobicity is the main property, which determines the bioavailability of a compound in aquatic environment; and it is involved in the removal of pollutants during the wastewater treatment through sorption and biodegradation processes (Garcia *et al.*, 2002; Ilani *et al.*, 2005; Yu and Huang, 2005). The K_{OW} values corresponded to the equilibrium of partitioning of the organic solute between the organic phase, i.e., octanol, and the water phase. High K_{OW} (high octanol-water partition coefficient) is characteristic for hydrophobic compounds, and indicates poor hydrosolubility and high tendency to sorb on organic material of the sludge matrix (Stangroom *et al.*, 2000; Yoon *et al.*, 2004). Compounds with log K_{OW} <2.5 are characterized by high bioavailability and their sorption to activated sludge is not expected to contribute significantly to the removal of the pollutants via excess sludge withdrawal. For chemicals having log K_{OW} between 2.5 and 4, moderate sorption is expected. Organic compounds with log K_{OW} values higher than 4.0 display high sorption potential (Rogers 1996; Ter Laak *et al.*, 2005). A wide-range of practical examples depicts the interdependency between the log K_{OW} of a compound and its properties of sorption to organic matter in wastewater. Yamamoto *et al.* (2003) showed that the fate of compounds with log K ranging from 2.5 to 4.5 (e.g., E2, EE2, octylphenol, bisphenol A (BPA), nonylphenol) is highly correlated to their respective log K. Log K_{OW} values of the compounds govern their sorption/ desorption behaviour and diffusion to remote sites. The latter are sometimes inaccessible to microorganisms and are often also nonextractable using classical chemical extraction procedure (Cirja, 2007).

Among PPCPs, polycyclic musk fragrances have the lowest solubility values in water (<2 mg/L), strong lipophilic character (log K_{OW} = 4.6–6.6) and increased log K_d values (3.3–4.2). Hormones are a little less hydrophobic than fragrances (log K_{OW} = 2.8–4.2) with lower sorption coefficients (log K_d = 2.3–2.6). The sorption potential of PPCPs is associated to the function of their lipophilic character (K_{OW}) and acidity (pKa) (Suarez *et al.*, 2008).

Pharmaceuticals. The main removal mechanisms for pharmaceuticals in WWTPs are sorption and degradation processes (Kummerer *et al.*, 2004). Their fate depends on their physico-chemical properties (e.g., chemical structure, aqueous solubility, K_{OW}, and Henry's law constants) as well as the operation of WWTPs. Sorption processes lead to the redistribution of pharmaceuticals after they enter WWTPs. For instance, in the case of ciprofloxacin (a common antibiotic), 20% sorbs onto primary sludge and 40% of the remainder partitions onto the secondary sludge (Gobel *et al.*, 2005). Carbamazepine, an antiepileptic drug, has been shown to be effectively removed by granular activated carbon (Ternes *et al.*, 2002). Moreover, acidic pharmaceuticals, such as the non-steroidal anti-inflammatories (e.g., ibuprofen), do not sorb onto sludge significantly. Pharmaceuticals are reported to be removed by sedimentation in surface waters (Tixier *et al.*, 2003). Pharmaceuticals which do not have functional groups (e.g., –OH, –COOH, or –NH$_2$ functional groups) tend to not be charged at neutral pH; hence, the sorption for this class of pharmaceuticals is probably caused by nonspecific sorption interactions. Besides pKa, K_{OW}, solid-water distribution coefficient (K_d), some characteristics such as liposome/water distribution coefficient (D_{lipw}) are also good indicators of a compound's ability to sorb. The higher these coefficients are, the more likely the pharmaceuticals is to partition to the sludge/biosolids along with the nonpolar fats and lipids, mineral oils, and greases generally present in wastewater (Krogmann *et al.*, 1999; Maurer *et al.*, 2007).

Estrogens. Estrogens present in discharged domestic WWTP effluents represent the most important input to the aquatic environment and constituent important point sources, especially in densely populated areas (Ternes et al., 1999). De Mes (2007) indicated that processes playing a role in the removal of estrogens from the aquatic phase are: adsorption, biodegradation, and photolytic degradation. Volatilization is not expected to play a significant role in the removal of E1, E2 and EE2. De Mes (2007) reported that at most 5% of E1, E2 and EE2 will ultimately be discharged with the sludge.

Several tests were conducted to assess the adsorption behaviour of estrogens on activated sludge, anaerobic sludge, sediments, soils or other organic materials. From the research regarding the sorption on sediments, three sorption phases were distinguished; a rapid sorption during 0 and 0.5 h, followed by a period of slower sorption up to 1 h and then with a period with desorption, explained by an increase in dissolved organic matter in the water phase (Lai et al., 2000). With activated carbon, an equilibrium for the sorption of E2 was attained after 50–180 minutes (Füerhacker et al., 2001). Jürgens et al. (1999) reported that an 'equilibrium' with the sorption on river sediments is nearly reached after two days, even though after 5 days the adsorbed amount still was increasing. According to results of Bowman et al. (2003) a final equilibrium only established after 50 days on river water sediments. The time required for establishment of an equilibrium is clearly related to the type of sorption material and the test conditions applied. Roughly, after a period of several hours more than 90% of the equilibrium concentration is already reached (De Mes, 2007).

Holthaus et al. (2002) conducted adsorption experiments in sediments under anaerobic conditions. Around 80–90% of the equilibrium was achieved within one day, but a complete equilibrium was only achieved after two days. EE2 indicated a high affinity to the bed sediments, with sorption K_d values two to three times higher than those determined for E2. The K_d values were ranging from 4–72 l/kg for E2.

Yamamoto et al. (2003) found sorption of E2 and EE2 to be highest on tannic acid, with a log K_{OC} values of 5.28 and 5.22, respectively, and the lowest for the polysaccharide algic acid, 2.62 and 2.53, respectively. It was concluded from this research that in natural waters containing total organic carbon (TOC) concentration of 5 mg/L, approximately 15 to 50% of the estrogens are bound. For sorption of E1 and E2 to activated sludge the highest observed percentage adsorbed was 23% at pH 8 and 55% at pH 2 (Jensen and Schäfer, 2001). At a concentration range of 5–500 ng/L radio labelled E1 and E2, adsorption to activated sludge was linear, indicating adsorption sites are in excess (Schäfer et al., 2002). The adsorption percentage is depending on the sludge concentration, as approximately 15% of E1 was adsorbed at approximately 2 g sludge/l and 30% was adsorbed at 8 g sludge/l (De Mes, 2007). Tetracyclines were highly sorbed to sludge and the sorption correlated well with the solid retention times during adsorption tests in a batch system (Sithole and Guy, 1987).

In wastewater, containing suspended solid (SS) concentration of 128 mg/l, spiked with radio labelled E2 to a concentration of 50 ng/L, 86% of the radioactivity remained in the liquid phase after 24 h (Fürhacker *et al.*, 1999). This research approached the fate in sewer systems, as raw municipal wastewater was spiked, with no addition of activated sludge, and incubation was without aeration. In a test with activated sludge at a concentration of 2–5 g SS/l, only 20% of labelled EE2 remained in the aqueous phase after one hour, when 20% mineralisation was observed, concluding that 60% was bound to the sludge (Layton *et al.*, 2000). During a biochemical oxygen demand (BOD) test, 28% of E2 and 68% of EE2 was calculated to be sorbed to sludge after 3 h incubation, which is greater than 20% and therefore considered to be significant (Kozak *et al.*, 2001).

7.3.2.2 Abiotic degradation and volatilization

Abiotic degradation comprises the degradation of organic chemicals via chemical (e.g., hydrolysis) or physical (e.g., photolysis) reactions (Doll and Frimmel, 2003; Iesce *et al.*, 2006) and can be man-induced and natural processes. Abiotic processes are not mediated by bacteria and have been found to be of fairly limited importance in wastewater relatively to the biodegradation of micropollutants (Stangroom *et al.*, 2000; Lalah *et al.*, 2003; Soares *et al.*, 2006; Katsoyiannis and Samara, 2007).

Photolysis. Photolysis has been proposed as a possible removal pathway for pharmaceuticals. Direct phototransformation has been shown as an elimination process in surface waters for diclofenac (Buser *et al.*, 1998). For ketoprofen and naproxen, direct phototransformation was considered as possible elimination processes in surface waters (Tixier *et al.*, 2003). Conventional activated sludge processes are located outside and the wastewater in the tanks is directly exposed to the sunlight. Therefore, photo-transformation can be responsible for the removal of some pharmaceuticals for the first layer of the tank (Zhang *et al.*, 2008). Hydrolysis is another possible removal pathway for pharmaceuticals although there are limited reports on this topic. It has been demonstrated that in some cases pharmaceuticals undergoing hydrolysis generate more reactive and toxic products than the parent compounds (Halling-Sorensen *et al.*, 1998).

Photolytic degradation of E2 and EE2 occurs; approximately 40% of the initial concentration was left after 144 h with a spectral distribution similar to natural sunlight, whilst no degradation in the dark controls was observed (Layton et al., 2000). The half-life is 124 h for E2 and 126 h for EE2, so it would take at least ten days to degrade the components to half the initial concentration and is therefore slow compared to the biodegradation of E2. For EE2 it might be more significant, since the half-life for biodegradation is 17 d in rivers (Layton et al., 2000). Experiments by Segmuller et al. (2000) to identify auto-oxidation and photodegradation products of EE2 indicate a series of isomeric dimeric oxidation products, a molecule that exists of two EE2 molecules. This molecule might have lost estrogenic properties, but no information was provided on its stability in the environment.

Volatilization. The elimination of trace pollutants by volatilization during the activated sludge process depends on the (dimensionless) Henry's constant (H) and K_{OW} of the analyzed trace pollutant, and becomes significant when the Henry's law constant ranges from 10^{-2} to 10^{-3} (Stenstrom et al., 1989). At very low H/K_{OW} ratio, the compound tends to be retained by particles (Rogers, 1996; Galassi et al., 1997). The rate of volatilization is affected by the gas flow rate and therefore, submerged aeration systems such as fine bubble diffusers should be used to minimize volatilization rates in WWTPs (Stenstrom et al., 1989).

Volatilization is not expected to play a significant role in the removal of E1, E2 and EE2, since compounds with Henry's law constant lower than 10^{-4} and a ratio of Henry's law constant to octanol-water partition coefficient (K_{OW}) (i.e., H_C/K_{OW}) ratio below 10^{-9} have low volatilization potential (Rogers, 1996). These processes were linked with each other for the three estrogens in model developed by Joss et al. (2004), describing the behaviour in WWTPs.

Most pharmaceuticals are large molecules with low Henry's law constants values (Maurin and Taylor 2000; Poiger et al., 2003). Thus, volatilization is usually considered negligible as an elimination mechanism and is rarely considered as a removal mechanism. For having a good a good solubility and low Henry's law constants, stripping is not suitable for these types of pollutants except WWTPs equipped with mechanical surface or coarse bubble aeration (Caliman and Gavrilescu, 2009).

7.3.2.3 Biodegradation

Biodegradation of pollutants varies with respect to the characteristics of compounds; thus, should be experimentally calculated. The biological degradation of 25 pharmaceuticals, hormones and fragrances was studied in batch experiments at typical concentration levels using activated sewage sludge originating from nutrient-eliminating municipal WWTPs (Joss et al., 2006).

According to this study, three groups were identified related to their degradation constants k_{biol}:

- Compounds with $k_{biol} < 0.1$ L/g SS.d are not removed to a significant extent ($<20\%$),
- Compounds with $k_{biol} > 10$ L/g SS.d are transformed by $>90\%$,
- Compounds in between (0.1 L/g SS.d $< k_{biol} <$ 10 L/g SS.d) have moderate removal (Joss *et al.*, 2006).

Joss *et al.* (2006), considering k_{biol} values found in their study, stated that biological degradation in municipal wastewater treatment contributes only to a limited extent to the overall load reduction of pharmaceuticals. It has been suggested that dilution of wastewater (e.g., rain or infiltration) reduces the degradation of the micropollutants; therefore, treatment at the source might be a favorable option and dividing of the available reactor volume into reactor cascades can appreciably improve performance.

In addition to the physico-chemical properties of the micropollutants, the type of the biological reactor as well as the operational parameters applied are of significance for biodegradation (and level of biodegradation) of the compounds. The factors affecting the biodegradation of the compounds and the removal performances of different bioreactor types are detailed mentioned in the following sections.

7.3.3 Factors affecting the biological removal efficiency

Treatment systems have different operational parameters designated specific to the type and characteristics of the wastewater treated, the location of the plant, the population size etc. The age of the sludge, the season, temperature and sunlight intensity, the hydraulic retention time (HRT), sludge retention time (SRT), nitrification environment and the location of the WWTP (downstream or upstream) are among the factors affecting the removal efficiency of micropollutants (Suarez *et al.*, 2008; Caliman and Gavrilescu, 2009; Liu *et al.*, 2009b; Wick *et al.*, 2009). In addition to the operational parameters and characteristics/units of a WWTP, the properties of a chemical also play a role in its biodegradation.

7.3.3.1 Compound structure

The structure of a compound can influence its fate in wastewater. A compound with simple chemical structure will be subject to removal via degradation during the wastewater treatment, while a compound with complex structure is likely to persist as parent compound or partially degraded compound in sewage water. For instance, pharmaceuticals are complex molecules and are notably characterized by their ionic nature. Compounds such as ketoprofen and naproxen were not eliminated during conventional wastewater treatment processes, while membrane bioreactor (MBR) treatment led to their elimination (Kimura et al., 2007). It was assumed that the poor removal in conventional wastewater treatment processes is due to their complex molecules including two aromatic rings making the compound more resistant to degradation processes. Compounds like clofibric acid and diclofenac are small molecules harbouring chlorine groups and were not efficiently removed by both conventional wastewater treatment processes and MBR. Therefore, the recalcitrance of these PPCPs was attributed to the presence of halogen groups, based on many observations with other chemicals substances. On the basis of the removal extent and the chemical structure the same authors proposed classification of PPCP as (Cirja, 2007);

- Compounds which are easily removed by both conventional wastewater treatment processes and MBR (i.e., ibuprofen),
- Compounds which are not efficiently removed in both systems (i.e., clofibric acid, diclofenac),
- Compounds which are not satisfactorily removed by conventional wastewater treatment processes but well removed by MBR (i.e., ketoprofen, mefenamic acid, and naproxen).

The extent of removal of polar compounds such as naphthalene sulphonates (anionic surfactants) during MBR treatment depends strongly on their respective molecular structure (Reemtsma et al., 2002). The removal of the naphthalene monosulphonates was almost complete, while it was limited to about 40% in the case of naphthalene disulfonates. Degradation and partitioning behaviour have also been reported to be a function of the polar to non-polar group ratio of the molecule, the presence of aromatic moieties (Chiou et al., 1998), and the organic carbon content (Yamamoto et al., 2003) which characterize the molecule. Linear alkylbenzene sulphonates (LAS) with long alkyl chains were preferentially adsorbed to the sludge matrix, while the short homologues of this anionic surfactant were found in the effluent in a comparative study in conventional treatment plant and MBR (Terzic et al., 2005). Long and branched chained compounds are more resistant whereas unsaturated aliphatic compounds are more biodegradable than saturated or complicated aromatic compounds (Jones et al., 2005).

7.3.3.2 Bioavailability

Bioavailability of xenobiotics to degrading microorganisms is one of the most important prerequisite for biodegradation of a trace pollutant in wastewater with activated sludge treatment (Vinken *et al.*, 2004; Burgess *et al.*, 2005). In general, bioavailability consists of the combination of physico-chemical aspects related to phase distribution and mass transfer, and of physiological aspects related to microorganisms such as the permeability of their membranes, the presence of active uptake systems, their enzymatic equipment and ability to excrete enzymes and biosurfactants (Wallberg *et al.*, 2001; Cavret and Feidt, 2005; Ehlers and Loibner, 2006). High bioavailability and thus potential for biological degradation of pollutants depends mainly on the solubility of these compounds in aqueous medium. The bioavailability of organic pollutants in aqueous environment is influenced by the presence of different forms of organic carbon like cellulose or humic acids (Burgess *et al.*, 2005). For instance, the extensive formation of associates consisting of one branched isomer of nonylphenol and humic acids was observed, whereas no interactions occurred when the compound was incubated with fulvic acids. It was assumed that the association between nonylphenol and humic acids occurs through rapid and reversible hydrophobic interactions (Vinken *et al.*, 2004). Another study indicated that the apparent solubility of nonlyphenol was enhanced through its association with humic acids and the compound was less prone to volatility (Li *et al.*, 2007). Consequently, the extent of mineralization of this xenobiotic was increased by 15% in presence of organic matter. Furthermore, bioavailability can be stimulated by the use of artificial surfactants. One practical application is the use of surfactants to enhance oil recovery by increasing the apparent solubility of petroleum components and efficient reduction of the interfacial tensions of oil and water *insitu* (Singh *et al.*, 2007). The desorption of polycyclic aromatic hydrocarbons (PAHs) was increased by addition of non-ionic surfactants in soil-water systems and the formation of dissolved organic matter (DOM)-surfactant complexes in the soil-water system is a possible reason to explain the enhanced desorption of PAHs (Cheng and Wong, 2006).

7.3.3.3 Dissolved oxygen and pH

Dissolved oxygen (DO) is found to be an important factor in removal of the micropollutants, especially of EDCs which is easily biodegraded under aerobic conditions compared to anaerobic counterpart (Furuichi *et al.*, 2006; Ermawati *et al.*, 2007).

Cirja (2007) reported that the acidity or alkalinity of an aqueous environment can influence the removal of organic micropollutants from wastewater by influencing the physiology of microorganisms, the activity of extracellular enzymes, and the solubility of micropollutants present in wastewater. Depending on their pKa values, pharmaceuticals can exist in various protonation states as a consequence of pH variation in the aquatic environment. For instance, at pH 6–7 tetracyclines are not charged and therefore, adsorption to sludge becomes an important removal mechanism (Kim *et al.*, 2005). Another study identified the pH value as critical parameter affecting the removal of micropollutants during the MBR treatment, pH values varied from neutral to acidic as nitrification became significant in the MBR (Urase *et al.*, 2005). At pH values below 6, high removal rates of up to 90% were reported for ibuprofen. Ketoprofen was removed from MBR up to 70% when the pH dropped down below 5, but obviously these are unique conditions, not applicable to the municipal wastewater treatment.

The sorption of E1 and E2 to the organic matrix was reported to be strongly dependent on the pH value (Jensen and Schäefer, 2001). In these studies, 23% of the steroid estrogens were sorbed to the activated sludge at a pH value of 8, while it was increased up to 55% when the pH value was maintained at 2 and it was shown that increasing pH value up to the compounds' *pKa* (pH > 9) lead to an increased desorption of steroids. A similar observation was noted in the study of Clara *et al.* (2004), where the compound's solubility increased at pH of 7–12. During the sludge treatment like sludge dewatering and conditioning with lime, the pH increases above 9 and the micropollutants can be desorbed from sludge solids. For instance, the recovery of Bisphenol A in the aqueous phase took place at pH > 12 and desorption was attributed to the increased hydrosolubility of the deprotonated form of Bisphenol A (Clara *et al.*, 2004; Ivashechkin *et al.*, 2004). The consequence of such high release was a high back-loading of the treatment plant via the recycling of the process water. In another study, the sludge-water partition coefficients of estrogens in activated sludge process were increased with decreasing pH for almost all the investigated compounds (Bisphenol A, E2, EE2) (Kikuta, 2004).

Cirja (2007) suggested that the control of the pH value might be a solution for the removal of micropollutants in WWTPs. The pH of industrial wastewater is often subject to variations and may also negatively influence the removal of the micropollutants. One solution would be the adjustment of the pH of the influent before biological treatment (Tchobanoglous *et al.*, 2004). Enhanced removal rates of deprotonable micropollutants from wastewater can be conducted at low pH values, as the protonation state influences both sorption and degradation processes. Acidic conditions are not usual in treatment plant or MBR, but it could be adapted for systems treating wastewater from highly contaminated sites or industrial wastewater in order to increase degradation rates.

7.3.3.4 Hydraulic and sludge retention time

The relationship between the physico-chemical properties of compounds among PPCPs, hydraulic retention time (HRT) and sludge retention time (SRT) of a plant on the biological removal is given below (Suarez et al., 2008):

- Compounds with high pseudo first-order biodegradation constants (k_{biol}) and low solid-water distribution coefficients (K_d) such as ibuprofen are efficiently transformed independently of SRT or HRT,
- Compounds with low k_{biol} and high K_d values, such as musks, are retained in the aeration tank by sorption and significantly biotransformed at sufficient SRT,
- Compounds with low k_{biol} and medium K_d values, such as E1 and E2, are moderately transformed independently of HRT and slightly dependent on SRT,
- Compounds with low k_{biol} and K_d values, such as carbamazepine, are neither removed nor biotransformed, no matter which SRT or HRT is used.

Hydraulic retention time. A long HRT allows more time for adsorption and biodegradation. Superior removal efficiencies of E1, E2 and EE2 were achieved at UK WWTPs with longer HRT of around 13 h compared to 2–5 h (Kirk et al., 2002). Concentrations of estrogens below the detection limit were obtained for a plant with an HRT of 20 h and for a plant which included a wetland with an HRT of 7 days (Svenson et al., 2003). Plants operating at HRTs of between 2–8 h achieved 58–94% removal compared to 99% removal which was achieved in a plant with an HRT of 12 h. Cargouet et al. (2004) found better removal for E1 (58%) and E2 (60%) in plants with an HRT of 10–14 h compared to a plant with an HRT of 2–3 h in which a removal of 44% for E1 and 49% of E2 was established (Cargouet et al., 2004). The behavior of E2, E1 and EE2 in 17 different WWTP across Europe was investigated. When E1 effluent values (as % of estimated influent concentration) were plotted against the different WWTP parameters on a double logarithmic scale, better E1 removal rates (i.e., lower percentage E1 remaining) appeared to be associated with longer total HRT and SRT ($r^2 = 0.39$, 0.28 respectively, $p < 0.5\%$) and longer HRT in the biological part ($r^2 = 0.16$, $p < 5\%$).

Despite the evidences that were presented, no strong statistical correlations could be established between HRT/SRT and hormone removal. Johnson et al. (2005) observed a weak significant ($\alpha = 5\%$) correlation between E1 removal and HRT or SRT (Johnson et al., 2005). In a Canadian study (Servos et al., 2005), there was little or no statistical correlation ($r^2 < 0.53$) between HRT or SRT and hormone or estrogenicity removal for 9 conventional secondary plants and 3 tertiary plants. This conclusion was reported despite the observations that plants with high SRT (>35 days) or HRT (>27 h) had relative high removal of E1 and E2 and reduction in estrogenicity while low SRT plants (2.7 and 4.7 days) had more variability and lower removal. These studies highlighted some evidence that increased HRT and SRT increases the amount of E1 and E2 removal and other endocrine disrupters alike within the WWTPs. It has also underlined the importance of biological activity associated with longer HRT and SRT (Koh, 2008).

A study investigating the removal efficiencies of different groups of micropollutants in 6 WWTPs with activated sludge systems of different HRTs was performed in Spain (Gros et al., 2007). It was found out that the plants with the highest HRTs (25–33 h) have the highest removal efficiencies, while the plant with an HRT of 8 h displayed no or poor removal of the majority of the compounds studied. Yet, due to the different characteristics of the compounds, a general removal trend was hard to establish. It was stated still that anti-inflammatories were removed by 50–90%; sulfonamide and fluoroquinolone antibiotics by 30–60% and lipid regulators, antihistamines and β-blockers on average by 50–60%. On the other hand, in some of the plants, lipid regulators, antihistamines and β-blockers, and macrolides and trimethoprim were not removed at all. It should be noted that in this study, the calculation of removal efficiencies were based on the influent and effluent chemical concentrations; thus removal was not limited to biological degradation.

Sludge retention time. Sufficiently high SRT is essential for the removal and degradation of micropollutants in wastewater (Joss et al., 2005). In activated sludge processes, biodegradation or biotransformation is possible only by the microorganisms that could reproduce themselves during the designed SRT (Zhang et al., 2008). The biodegradation potential of some micropollutants is increased at long SRTs. This was attributed to the increase in the diversity of the microorganisms with the increased SRT due to the increased diversification of the microbial community or a diversification of the available enzymes within the microbial community (Ternes et al., 2004; Clara et al., 2005a). High SRTs allow the enrichment of slowly growing bacteria and also the establishment of a more diverse biocoenosis able to degrade a large number of micropollutants. It was proposed that at short SRTs (<8 days) slowly-growing bacteria are removed from the system and in this case, the biodegradation is less significant and adsorption to sludge will be more important (Jacobsen et al., 1993).

SRT of at least 10 to 12.5 days has been suggested as the period required for the growth of organisms that decompose E2 and E1 (Saino et al., 2004). An increase in SRT may enhance the biodegradative and sorptive capacity of the activated sludge. A high SRT in a biological process allows for more diverse and specialized microbial communities to enrich, including slow growing microorganisms adapted to remove EDCs. The influence of increased SRT is illustrated by a German WWTP which has been upgraded from a conventional to a nutrient removal plant, with substantial higher sludge retention time, increasing from <4 days to 11–13 days. Batch experiments with sludge from the old plant did not show any reduction of EE2 (Ternes *et al.*, 1999), while at the increased SRT a reduction of around 90% was established in the full scale plant, which indicates the growth of microorganisms capable of degrading EE2 (Andersen *et al.*, 2003).

SRT is one of the parameters easy to handle to enhance the process efficiency to remove pharmaceuticals from wastewater. Two MBRs which were operated at high SRTs of 26 days demonstrated removal efficiency of 43% for benzothiazoles (Kloepfer *et al.*, 2004). By varying the SRT in MBRs, Lesjean *et al.* (2005) reported that the removal of pharmaceutical residues increased with a high sludge age of 26 days and inversely decreased at lower SRT of 8 days. SRT values between 5 and 15 days are required for biological transformation of some pharmaceuticals, i.e., benzafibrate, sulfamethoxazole, ibuprofen, and acetylsalicylic acid (Ternes *et al.*, 2004). Similarly, a critical SRT of at least 10 days was set for bisphenol-A, ibuprofen, bezafibrate and the natural estrogens (Clara *et al.*, 2005a). For SRTs of 2 days, Clara *et al.* (2005b) found out that none of the investigated pharmaceuticals, i.e., ibuprofen, benzothiazole, dichlorofenac and carbamazepine was eliminated, while applying SRT of 82 days in MBR removal rates above 80% were obtained. Nevertheless, removal rates of carbamazepine remained below 20% for all applied SRTs. Similar results were reported by Ternes *et al.* (2004) where carbamazepine and diazepam were not degraded even at SRT longer than 20 days. Even at SRTs up to 275 days or 500 days, carbamazepine could not be removed in a full-scale WWTP (Clara *et al.*, 2004; Clara *et al.*, 2005a). Diclofenac removal was increased up to 60% being independent of the increase in the SRT (Clara *et al.*, 2004; Clara *et al.*, 2005a). In MBR containing adapted sludge, the removal of diclofenac ranged from 44 to 85% at SRTs of 190–212 days (Gonzalez *et al.*, 2006).

The adsorption kinetics for tetracyclines was evaluated at various biomass concentrations in sequencing batch reactors at varying SRTs and HRTs (Kim *et al.*, 2005). Between 75 and 95% of the applied tetracyclines was adsorbed onto the sludge after 1 h. At long SRTs (10 days) the removal of tetracyclines was 85–86%, while the decrease of the SRT to 3 days led to in a lower removal rate (78%). The lower degradation rates were assigned to the reduced biomass concentrations in system with shortened SRT.

7.3.3.5 Organic load rate

Koh (2008) stated that sludge loading is a key parameter influencing the removal of estrogens from a WWTP. This was confirmed by a low degradation rate observed in the first compartments of the reactors monitored. This suggests that microorganisms prefer to degrade other organic compounds over estrogens. When sludge load is low, in terms of BOD, the microorganisms are forced to mineralise poorly degradable organic compounds. Therefore, researchers proposed that a reactor cascade provides better E1 and E2 removal rates than a completely stirred tank (Joss *et al.*, 2004). However, no clear correlation can be found within one WWTP with different organic loadings in relation to the removal of estrogens. Onda *et al.* (2003) could not establish a clear correlation between E1 removal and loadings, despite the tendency for E1 removal to be lower under higher loading in most cases (Onda *et al.*, 2003). Johnson *et al.* (2000) tried to find a correlation between the flow per head and the E2 removal. Data from Svenson *et al.* (2003) depicts a correlation of a decrease in total estrogen removal with increasing percentage of flow, therefore indicating higher loading (Svenson *et al.*, 2003). In the review on MBRs for wastewater treatment, the enhanced elimination efficiencies of MBR with respect to EDCs have been attributed to the low sludge loading among other possible factors (Melin *et al.*, 2006).

During periods of storm-induced elevated flow conditions, step feed operation conserves the biomass concentration in the activated sludge process and prevents hydraulic washout of the biomass. This operating mode conserves the microorganisms that are responsible for the degradation of alkylphenol polyethoxylates and its metabolites, and prevents the discharge of high concentrations of suspended solids that may contain adsorbed alkylphenol polyethoxylate or its metabolites (Melcer *et al.*, 2006).

7.3.3.6 Temperature

Biodegradation and sorption of organic micropollutants are dependent on temperature. The temperature affects the microbial activity in wastewater treatment systems as microbial growth rates strongly vary based on the applied temperature (Price and Sowers, 2004). For most compounds, sorption equilibrium decreases with increase of temperature, whereas the biodegradation is less efficient at lower temperatures. The temperature influences the solubility and further physicochemical properties of micropollutants as well as the structure of the bacterial community. Conventional treatment processes depict better stability than MBR during seasonal temperature variations with respect to removal of micropollutants. The larger surface of conventional treatment processes in comparison to MBR would attenuate the variations of temperature and thus, protect bacterial activity against temperature shock produced in the system. The temperature increase in microbial activity also favors higher biodegradation rates of micropollutants. Wastewater treatment systems located in areas with prolonged high temperature may efficiently eliminate micropollutants (Cirja, 2007).

The removal of pharmaceuticals, i.e., ibuprofen, benzafibrate, diclofenac, naproxen, and ketoprofen, was reported to enhance during the summer time when the temperature reached 17°C in comparison to the winter season when the water temperature was 7°C (Vieno et al., 2005). A temperature of 20°C was beneficial for the removal of pharmaceuticals in conventional treatment processes and MBR, and for instance more than 90% of bezafibrate was eliminated under the conditions applied (Clara et al., 2004). At low temperature during winter season, the degradation rates decreased. In the case of diclofenac, naproxen, and ibuprofen better performances of removal are reached when the systems are operated at 25°C than at 12°C (Carballa et al., 2005).

A comprehensive mass balance of E1 and E2 study was carried out on a Japanese WWTP (Nakada et al., 2006) in summer and winter. Similar characteristics were found in both seasons, there was little difference in the E1 flux in the influent between winter and summer survey periods and the sum of the E1 fluxes in the effluent of the final sedimentation tank and in the return sludge line was significantly higher than that in the aeration tank. However, sulphate-conjugated estrogens were not degraded in the treatment processes and persisted in the return sludge to a greater extent in the summer (15 ng /L and 16.5 ng /L, respectively) than in the winter (2.1 ng/L and 4.4 ng/L, respectively). Estrone was effectively removed in the aeration tank in the summer as compared to the winter. Short SRT (8.2 d) was attributed to the ineffective removal of E1 from the WWTP. However, high efficiency in summer with an SRT of 6 d was effective in reducing E1. This can be attributed to the higher water temperature (27°C) allowing rapid growth of E1-oxidising microorganisms. Removal efficiency was high for E2 in winter (70%) and summer (87%). In contrast, concentrations of E1 increased by 74% during the cold winter period and 10% in summer. It was also observed that the removal of total nitrogen was smaller in winter (26%) than in summer (60%), probably due to the effect of influent temperature on nitrification which could affect the diversity of microorganisms responsible for the degradation of these steroid estrogens (Nakada et al., 2006).

Temperature was reported to influence also the mineralization of E2 (Layton et al., 2000). An increase of 10°C leads to a duplication of microbial activity and mineralization rates. Changes of approximately 15°C had statistically significant effect on the mineralization rate of E2 present in the aqueous phase.

The adsorption of the antibiotic fluoroquinolone to the particles in the raw water is influenced by the temperature. Lindberg *et al.* (2006) observed 80% adsorption at 12°C, while Golet *et al.* (2003) reported an adsorption of 33% when the temperature was higher. Studies on the influence of high temperature on the removal of COD from wastewater of pharmaceutical industry led to the conclusion that temperature serves as pressure of selection for the bacterial community development during aerobic biological wastewater treatment (LaPara *et al.*, 2001). In the same time, it stimulates higher degradation rates of pharmaceuticals.

The temperature of wastewater treatment seems to play an important role concerning the removal of xenobiotics; WWTP in countries with average temperature of 15–20°C may better eliminate micropollutants than those in cold countries with a mean temperature mostly under 10°C (Cirja, 2007).

7.3.4 Biological treatment of Micropollutants in different processes

The rapid increase in the population, industrial development and the corresponding increase in the diversity and consumption of the chemicals lead to the increase in the type and amount of micropollutants load to the WWTPs. Advanced oxidation options, ozonation, UV-radiation, membrane filtration and activated carbon adsorption are potential treatment methods that will improve the removal of the micropollutants in WWTPs (Andersen *et al.*, 2003; Suarez *et al.*, 2008). However, these methods might increase the investment and operational cost of the treatment plant. Besides, most of the WWTPs involve biological treatment processes; thus the biodegradation possibility of varied micropollutants has been hitherto investigated in bench-, pilot- and full-scale studies under different operational conditions. In this section, the biological removal of the micropollutants in different types of bioreactors is discussed.

7.3.4.1 Activated sludge systems

Biological processes in conventional activated sludge reactors demonstrate low performances in removing EDCs such as β-blockers or psycho-active drugs (e.g., carbamazepine) in the wastewater. A study was performed in a WWTP in Germany with two activated sludge units followed by a post-denitrification unit (Wick *et al.*, 2009). Almost no removal was observed in the first activated sludge unit with an HRT and SRT of 1 d and 0.5 d, respectively. The removal efficiencies for all compounds detected in the plant were less than 60%, except for the natural opium alkaloids codeine and morphine being greater than 80%. The batch tests indicated that sorption mechanism is not effective in the removal of the compounds. Thus, the difference in the removal between the first and the second activated sludge units was attributed to the nitrification in the second unit together with the long SRT of 18 d and temperature effect. Yet, still some psycho-active drugs such as the antiepileptic carbamazepine could not be removed.

In aerobic batch experiments it was shown that after 1–3 hrs, more than 95% of E2 was oxidised to E1 (Ternes et al., 1999). In the same experimental set up, EE2 appeared to be stable. Norpoth et al. (1973) also found no degradation of EE2 in activated sludge after an incubation time of five days. The findings for the conversion of E2 to E1 were confirmed in experiments with river water samples, in which E2 was converted into E1 and mineralised according to first order kinetics (Jürgens et al., 2002).

The removal performance for E1 in an activated sludge process varied but removals were greater than 85% for E2 and EE2 (Johnson and Sumpter, 2001). Baronti et al. (2000) reported average E1, E2 and EE2 removals of 61%, 86% and 85% respectively in six WWTPs near Rome (Baronti et al., 2000). Ternes et al. (1999) found low elimination efficiencies for E1 and EE2 ($<10\%$), but approximately two thirds of E2 was eliminated in the WWTPs. This was in agreement with Komori et al. (2004) who observed a 45% reduction in E1 which was considerably less than that of E2 (Komori et al., 2004). The persistence of EE2 under aerobic conditions and rapid degradation of E1 and E2 were also found in the laboratory experiments using WWTP sludge (Ternes et al., 1999). In two pilot-scale municipal WWTPs, removal efficiencies for E1 and EE2 were 60% and 65% respectively, with elimination of more than 94% of the E2 entering the aeration tank (Esperanza et al., 2004). Laboratory experiments and field studies have indicated both biodegradation and sorption to biosolids as the main removal mechanisms (De Mes et al., 2005).

Among the slow growing bacteria, ammonia oxidizing bacteria (AOB) can co-metabolize organic pollutants. It was reported that the activity of AOB dominated the degradation of estrone, estradiol, estriol and ethinylestradio (Ren et al., 2007). Several studies also demonstrated some pharmaceuticals and EDCs exhibited higher removal in nitrifying activated sludge (Drewes et al., 2002; Kreuzinger et al., 2004; Vader et al., 2000). On the other hand, Carucci et al. (2006) reported that some pharmaceuticals inhibited nitrification in a laboratory-scale sequencing batch reactor, but the effect of pharmaceutically active compounds on the non-nitrifying microbial fraction was not investigated.

Natural estrogens are thought to be biodegraded via a pathway where bacteria can use the conversion for growth, as EE2 is thought to be biodegraded by co-metabolism, in which an organic compound is modified but not utilised for growth (Vader et al., 2000). Nitrifying sludge is held responsible for the conversion of EE2 by the use of the enzyme ammonium monooxygenase, which insert oxygen into C-H bonds. The nitrifying activated sludge transformed EE2 to more hydrophilic metabolites almost completely in about six days, while sludge with a very low nitrifying capacity did not transform EE2 (Vader et al., 2000). Using N-Allylthiourea (ATU), a chemical that inhibits the nitrification by blocking the ammonium monooxygenase enzyme, resulted in slower conversion of EE2, while the conversion rates of E1 and E2 remained the same. If ATU is applied on a pure culture of nitrifying bacteria the conversion is completely blocked, whereas in activated sludge it was only slowed down, suggesting that in activated sludge also other bacteria are able to convert EE2.

Indeed, the modified activated sludge systems, where biological nutrient removal is aimed, display significant removal efficiencies for some micropollutants. Muller et al. (2008) studied the removal of E1, E2, E3, EE2 and their conjugated forms in an activated sludge process including aerobic/anoxic/anaerobic stages with high HRT (3–5 d) and SRT (20 d) values to achieve biological nitrogen and phosphorus removal. Biological degradation was the main mechanism with 93–97% removal efficiency of the total estrogen load, while sorption remained to be 2–2.5%. Similarly, a municipal WWTP in Germany with an activated sludge system for nitrification and denitrification including sludge recirculation was found to eliminate natural and synthetic estrogens (E1, E2 and EE2) (Andersen et al., 2003). The natural estrogens (E1 and E2) were biodegraded by more than 98% in the denitrification and aerated nitrifying tanks, whereas EE2 was degraded by 90% only in the nitrifying tank. The portion of the estrogens that were sorbed to the digested sludge was only about 5%. It is suggested that conjugates (glucuronides and sulfates) of the estrogens were cleaved into the parent compounds mainly in the first denitrification tank.

The removal efficiencies of E1, E2 and E3 in 20 WWTPs involving oxidation ditches or conventional activated sludge processes in Japan were investigated (Hashimoto et al., 2007). In the conventional activated sludge plants with HRTs of 6 to 26 h and SRTs of 3 to 10 d, increments of E1 were observed frequently during the biological treatment while E2 and E3 were removed successfully in the process (86 and 99.5%, respectively). The increment of E1, which was not experienced in oxidation ditches, was attributed to the influent concentrations of conjugated estrogens and the rates of de-conjugation. High and stable removal efficiencies were obtained in the plants with high SRT and HRT values. The oxidation ditches (HRT: 21–67 h, SRT: 8–118 d) displayed better removal of natural estrogens and estrogenic activity than the conventional activated sludge plants with lower SRT and HRT. Yet, it was suggested that the biological nitrogen removal process with high SRT remove natural estrogens as effectively as the oxidation ditch plants. Moreover, SRT was proposed to be a more determining factor than HRT in the removal of natural estrogens.

The increase in the E1 was also observed by Ternes et al. (1999). They found that E2 is converted rapidly to E1, the removal of which is slower than the oxidation of E2 to E1. The conjugates of the estrogens in contact with activated sludge were cleaved by using commercial standard estrogenic substances in the batch experiment. In the study of Kanda and Churchley (2008), a nitrifying activated sludge plant in England was detected for E1, E2 and EE2 removal. Despite the high E1 and E2 removal (97 and 99%, respectively) and successful nitrification, EE2 was removed by only 3%. Natural and synthetic estrogens, which are excreted as inactive conjugated forms of sulphuric and glucuronic acids, may be re-transformed to active forms (free estrogens) in the raw sewage or WWTP by microorganisms that present glucuroidase and sulfatase activity (Isidori et al., 2007). The conjugated forms are likely to break down to release free EE2 during aerobic treatment. The disagreement between the finding of the Kanda and Churchley (2008) and of those that have observed high EE2 removal in aerated nitrifying reactor studies (e.g., Andersen et al., 2003; Hashimoto et al., 2007) was speculated to be due to the presence of EE2 in the influent in the conjugated form and prevention of its de-conjugation resultant of sample preparation method.

The use of nitrifying activity has been proposed for the removal of various EDCs in WWTPs, yet, there is limited information on the role of autotrophs in EDCs biodegradation. Nitrifier-enriched activated sludge was used in batch degradation of bisphenol A (BPA) and nonylphenol (NP) to investigate the effect of ammonia-oxidizers (Kim et al., 2007). It was observed that BPA and NP concentrations were simultaneously decreased with the oxidation of ammonium (NH_4^+) to nitrate (NO_3^-) by nitrifying sludge. Replacing the ammonium by nitrite (NO_2^-) required an acclimation period. In the presence of inhibitors such as allylthiourea or Hg_2SO_4, BPA and/or NP reduction decreased significantly, implying that removal of BPA and NP was mostly mediated by biological activity rather than by physic-chemical adsorption onto sludge flocs. In addition, ammonium-oxidizing activity rather than nitrite-oxidizing activity within the nitrifying sludge seems to be more closely related to the removal of BPA and NP. Kanda and Churchley (2008) similarly observed high removal of NP and nonyl phenolethoxylates (94 and 98%, respectively) in a nitrifying activated sludge plant in UKs.

In the study of Press-Kristensen K. (2007), it is found that concentration of BPA in the effluent of activated sludge system were independent from the influent concentration at steady-state. This activated sludge system includes an anaerobic BioP reactor and two alternating aerobic/anoxic BioN reactors. Moreover, a larger biomass/BPA ratio causes lower effluent BPA concentration. Biodegradation of BPA may be increased by the increased aerobic phase time.

In order to develop a biological process for the degradation of the fire retardant tetrabromobisphenol A (TBBPA) and to investigate the possible formation of endocrine disruptor BPA through reductive debromination, Brenner et al. (2006) conducted lab-scale reactors. The reactors were operated under solely aerobic, solely anaerobic or aerobic/anaerobic conditions, with different carbon sources. The reactors were fed with contaminated sediments that might have contained indigenous bacteria exposed to these compounds and TBBPA waste mixture containing also tribromophenol (TBP). Results indicated that TBBPA could not be biodegraded and BPA did not accumulate in any of the redox conditions. On the other hand, TBP was found to be easily biodegradable for aerobic microorganisms. It can be concluded that TBBPA can be considered as a refractory compound and will remain stable in natural systems due to its very low volatility, solubility and bioavailability.

The biodegradability of trimethoprim was investigated during different sewage treatment steps using batch systems (Perez et al., 2005). The main outcome of this study was that the activated sludge treatment comprising the nitrification process was the only treatment capable to eliminate trimethoprim. The capacity of nitrifying bacteria to break down trimethoprim was quite unexpected, because a subsequent study reported that this xenobiotic cannot be degraded by microorganisms (Junker et al., 2006).

High removals mainly resulting from biodegradation have been reported for ibuprofen (82%) in pilot nitrifying-denitrifying activated sludge system (Suarez et al., 2005). Moreover, it has been reported that 68% of naproxen was removed in a pilot nitrifying-denitrifying activated sludge system (Suarez et al., 2005) and 50 to 65% of ketoprofen in activated sludge municipal wastewater treatment (Quintana et al., 2005). Because the removal is unlikely from sorption (low sorption potential) and phototransformation (turbid environment in WWTPs), naproxen and ketoprofen removal is mostly attributed to biodegradation (Carballa et al., 2007).

7.3.4.2 Wetlands

Wetland treatment could be one of the treatment options for micropollutants. The mechanisms involved in the wetland treatment is not only the microbial degradation but also includes the photolysis, plant uptake and sorption to the soil (White et al., 2006).

In USA, many municipalities have chosen to treat wastewater using natural systems such as lagoons and wetlands, rather than conventional wastewater treatment technologies (Conkle et al., 2008). For example, in Manderville LA, a wastewater flow of 7600 m^3 per day containing 15 pharmaceutically active compounds are treated in a series of lagoons (basins), followed by a constructed wetland and UV disinfection before its discharge into a natural forest wetland. The concentrations of most compounds were reduced by more than 90% and further polishing in the natural forested wetland resulted in removal efficiencies of 96% on average for the whole system. Exceptions were encountered in the case of carbamazepine and sotalol that appeared to be more persistent, being reduced by only 51 and 82%, respectively. Yet, it was claimed that these two antiepileptics were removed in a higher degree in the wetland plant, compared to the activated sludge systems in the literature. The high removal efficiencies obtained for the micropollutants were related to the long retention times of almost 30 days. Despite the seasonal variability observed with greater concentrations entering the facility during the colder months, the removal efficiencies were similar for spring and fall period.

Horizontal subsurface flow constructed wetlands (SSFCWs) might be another option for the removal of micropollutants in small communities. In a study, carbamazepine, ibuprofen and clofibric acid behaviour was studied in two SSFCW systems with synthetic wastewater containing 2 different organic matters (dissolved glucose and particulate starch) (Matamoros *et al.*, 2008a). It was found out that removal efficiencies were independent of organic matter type. But it should be noted that the system fed with glucose reached the net removal before the system fed with starch due to higher biofilm development. Removal of carbamazepine by bed sorption was only 5% and that of ibuprofen was 51%. Ibuprofen biodegradation intermediates, carboxy and hydroxy derivatives, supported the elimination of ibuprofen under aerobic conditions. It is expected that ibuprofen was removed by biodegradation due to its low log K_{ow} value rather than by sorption. Aerobic conditions are necessary for the high removal of ibuprofen in subsurface flow constructed wetland systems, which can be achieved in subsurface flow constructed wetlands (Matamoros *et al.*, 2008a).

In another study, different types of reactors, filtralite-P filter units, biological sand filters, horizontal subsurface flow and vertical flow constructed wetlands were operated to study the removal of 13 PPCPs. More than 80% PPCP removal was achieved in all the systems, except for carbamazepine, diclofenac and ketoprofen which are more recalcitrant. The vegetated vertical flow constructed wetland has higher removal efficiency compared to other systems although statistically the results are not different. Vegetation and unsaturated water flow can be the appropriate methods to tolerate the fluctuations in loading rate and removal rate (Matamoros *et al.*, 2009). The vegetated vertical flow constructed wetlands, which have unsaturated flow and better oxygenation appeared consistently to perform best removal for PPCPs in domestic waters (Matamoros *et al.*, 2009). Similarly, 90% of the influent PPCPs, herbicide and a veterinary drug (flunixin) were removed in a surface flow constructed wetland with the exception of the recalcitrant carbamazepine and clofibric acid (30–47%) (Matamoros et al., 2008b). The seasonal variation (solar radiation and water temperature) affected the compounds such as naproxen and diclofenac that require longer period for biodegradation and photodegradation.

Wetlands are likely to be significant treatment units for the removal of micropollutant especially for the small communities when economical concerns are taken into consideration. Seasonal variation is effective on the biodegradable micropollutants.

7.3.4.3 Membrane bioreactors

Membrane bioreactors (MBRs) are considered as significant alternatives to the conventional treatment plants due to the fact that (Spring et al., 2007):

- They act as a complete barrier to solids onto which many EDCs are adsorbed;
- The membrane surface also retains the EDCs;
- The longer SRT in MBRs may facilitate additional biological transformation of these compounds since bacteria have more time to break down their structure.
- For combining the adsorption and biodegradation processes, MBRs might be a good compromise for simultaneous carbon, nitrogen and phosphorus and EDCs removal (Auriol et al., 2006).

Chen et al. (2008) has studied BPA removal by using a submerged MBR comparing with a conventional activated sludge system. In spite of the changes in sludge loadings ranging from 0.046 to 10.2 g kg^{-1} d^{-1}, MBR could remove BPA a little more effectively than conventional activated sludge system. It is also found that sludge adsorption assistance is very low for BPA removal in both systems. One metabolite of BPA biodegradation, 4-hydroxy-acetophenone, was detected. It was concluded that biodegradation dominated the BPA removal process (Chen et al., 2008). In another study carried out by Clara et al. (2005b) no differences in treatment efficiencies were reported between two treatment techniques. The SRT is an important design parameter and an SRT value that is suitable also for nitrogen removal (SRT > 10 days at 10°C) increases the removal potential of the selected micropollutants. In this study, different removals were monitored for different compounds. Some compounds such as carbamazepine were not removed in any of the treatment systems whereas compounds such as BPA, the analgesic ibuprofen or the lipid regulator bezafibrate were nearly completely removed (Clara et al., 2005b). On the other hand, in a lab scale MBR with high sludge concentration ranging between 20 and 30 g/L and an SRT of 37 days, bezafibrate was transformed (60%) but not mineralized and the metabolites were tentatively identified (Quintana et al., 2005). It was noted that naproxen was degraded over a period of 28 days and ibuprofen degradation started after 5 days and was complete after 22 days. Another study revealed that a membrane bioreactor (MBR) with 65 day SRT has better biodegradation of ketoprofen and diclofenac than the one with 15 day SRT (Kimura et al., 2007). This phenomenon may due to the growth of slower growing bacteria with longer SRT.

Membrane bioreactors are means for the removal of micropollutants such as pesticides, PPCPs, etc. Sipma et al. (2009) stated that MBR system is a good choice for slowly degradable pharmaceuticals due to the long sludge age, which allows the growth of different groups of microorganisms. Operation at high biomass concentrations is one the most important advantages of MBR systems as well (Witzig et al., 2002). In addition, in an MBR some pharmaceuticals can be removed by air stripping mechanism due to their relative high Henry's law constants and high rate of aeration in the MBR (Sipma et al., 2009).

7.3.4.4 Anaerobic treatment

Anaerobic biodegradation has many advantages over aerobic counterpart, including the low sludge production, low nutrient need, the dehalogenation efficiency of the highly-halogenated compounds, etc. In order to investigate the potential use of anaerobic treatment, or anaerobic bioreactors; anaerobic degradation studies were performed for micropollutants in wastewaters.

Little results have been reported on the fate of estrogens under anaerobic conditions. Bed sediment was used to examine anaerobic degradation of E2; which was fairly rapidly converted to E1 at 20°C, almost completely after an incubation of 2 days (Jürgens et al., 2002). In batch experiments with activated sludge supernatant under anaerobic conditions (purged with nitrogen gas), 50% of the spiked amount of E2 was converted into E1 after 7 days (Lee and Liu, 2002). No further degradation of E1 was observed, so E1 may accumulate as a by-product. Autoclaved samples were used as sterile controls. EE2 tested under anaerobic conditions in river water samples showed no degradation over 46 days (Jürgens et al., 1999). Under strict anaerobic conditions E1 is expected be reduced into E2, rather than E2 converted to E1. This pathway was demonstrated by Joss et al. (2004), who found a half-life of approximately 20 minutes and also a conversion of E2 with a half-life of 6 minutes in activated sludge and 2 minutes in MBR sludge. Thus, somehow under anaerobic conditions there are still electron acceptors available, like Fe^{3+} and various organic oxidative compounds, responsible for the conversion. Joss et al. (2004) even found conversion of EE2 in MBR sludge under anaerobic conditions with a rate of about 1.5 L/g.d, but this value is nearly the same as the degradation value derived from the blank experiment, where no sludge was present. Under anoxic conditions the conversion rates lay in between those under anaerobic and aerobic conditions. For example the calculated half-life of the degradation of EE2 was 11 h under anaerobic conditions, 2.8 h under aerobic and 5.6 h under anoxic conditions (Joss et al., 2004).

Packed bed anaerobic mesophilic reactor can be used for the biodegradation of chemical – pharmaceutical wastewater with high chemical oxygen demand (COD) values such as 23–31 g/L. Removal efficiencies of 80–98% was obtained with sand, anthracite and black tezontle in the reactor at low organic loads less than 3.6 kg/m^3 d. It has better results with granular activated carbon in the reactor (98% efficiency and 17 kg/m^3.d organic load) (Nacheva et al., 2006).

A study demonstrated that, in an upflow anaerobic sludge blanket (UASB) reactor, an average of 95% removal of the antibiotic tylosin can be achieved (Chelliapan et al., 2006) while only 63% tylosin was removed in a activated sludge treatment process (Watkinson et al., 2007).

Ejlersson et al. (1999) characterized the products of nonylphenol monoethoxylate degradation by mixed microbial communities in anaerobic batch tests. Within seven days of anoxic incubation, nonylphenol diethoxylate appeared as the major degradation product. After 21 days, nonylphenol was the main species detected, and was not degraded further even after 35 days. The anaerobic degradation of nonylphenol ethoxylates seems to be dependent on the nitrate source. The observed generation of nonylphenol coupled to nitrate reduction suggests that the microbial consortium possessed an alternative pathway for the degradation of nonylphenol monoethoxylate (Luppi et al., 2007). Minamiyama et al. (2006) spiked an anaerobic digestion testing apparatus that was operated at a retention time of approximately 28 days and a temperature of 35°C with thickened sludge with nonylphenol monoethoxylate. Approximately 40% of nonylphenol monoethoxylate was converted to nonylphenol.

Chang et al. (2005) investigated the effects of various parameters on the anaerobic degradation of nonylphenol in sludge. The optimal pH for nonylphenol degradation in sludge was 7 and the degradation rate was enhanced when the temperature was increased. The addition of aluminum sulfate (used for chemical treatment of wastewater) inhibited the nonylphenol degradation rate within 84 days of incubation. The same study revealed the theory that sulfate-reducing bacteria, methanogen, and eubacteria are involved in the degradation of nonylphenol as the sulfate-reducing bacteria are a major component of anaerobic treated sludge. It seems that more research is required to define the mechanisms involved in the anaerobic degradation of nonylphenol and derivates.

7.3.4.5 Other bioreactors

Trickling filter. Generally trickling filters are less effective than activated sludge systems in eliminating estrogenic activity (Svenson *et al.*, 2003). A study on the distribution of natural estrogens (E1 and E2) in 18 Canadian WWTPs reported that the trickling filter was ineffective in the removal of estrogens (Servos *et al.*, 2005). Ternes *et al.* (1999) reported inefficient removal of estrogens by a trickling filter system in Brazil (Ternes *et al.*, 1999). Spengler *et al.* (2001) reported that a trickling filter plant had elevated levels of estrogens in its effluent (Spengler *et al.*, 2001). This supports the work of Turan (1995) who found that estrogens, particularly synthetic compounds, are stable enough to withstand the sewage treatment process (Koh, 2008).

There are not many studies that report the fate and behavior of alkylphenol polyethoxylates in trickling filters. However, two studies indicated relatively high removal of alkylphenol polyethoxylates (68–77%) (Gerike, 1987) and circa (75%) removal (Brown *et al.*, 1987) from a domestic WWTP with trickling filter. This high removal capacity was attributed to its high COD removal as comparable to that of an activated sludge plant. In general, a mixture of long chain and short chain alkylphenol polyethoxylates, and alkylphenols are expected in a trickling filter plant, since the stratified layers and pockets of anaerobic zones within the fixed film reactor would create less oxidative metabolites i.e., short chain alkylphenol polyethoxylates and alkylphenols. High removal rates for these EDCs therefore can be achieved at plants with comprehensive treatment technologies such as combined biological and chemical removal of organic matter, nitrogen and phosphorus rather than WWTP with trickling filters (Koh, 2008).

Combination of bioreactors. Anaerobic-aerobic treatment combination could result in effective removal of organic matter in high-strength pharmaceutical wastewater.

Zhou *et al.* (2006) conducted a pilot-scale system composed of an anaerobic baffled reactor followed by a biofilm airlift suspension reactor. The influent concentrations of antibiotics ampicillin and aureomycin 3.2 and 1.0 mg/L respectively were partially degraded in the anaerobic baffled reactor. Their removal efficiencies were 16.4 and 25.9% at an HRT of 1.25 day, and 42.1 and 31.3% at HRT of 2.5 day, respectively (Zhou *et al.*, 2006).

7.3.5 Biological treatment of Micropollutants in sludge

Due to the specific physico-chemical characteristics, some micropollutants are able to significantly sorb onto sewage sludge. To estimate the effect of potential harms of digested or raw sludge as a micropollutant source to the environment, it is of significance to investigate the pollutants in the sludge. Some of these potential micropollutants that are reported to sorb onto sludge are brominated biphenyl ethers (flame retardants), nitro musks (synthetic fragrances), linear alkylbenzene sulfonates (detergents), pharmaceutical compounds (antibiotics and drugs), odorants (for sewage sludge odor control) and polyelectrolytes (for sewage sludge de-watering) (Caliman and Gavrilescu, 2009). Among these, some pollutants neither sorbed nor degraded, such as carbamazepine. Carbamazepine has a low K_d value (1.2 L/g SS) far from the value of value 500 L/g SS required for a significant sorption onto the sludge (Ternes et al., 2004; Zhang et al., 2008); therefore, bulk of carbamazepine remains associated with the aqueous phase.

Mesophilic anaerobic digestion is one of the widespread used processes for treatment of primary and secondary sludges of WWTPs. This process may be supplemented with thermophilic (55°C) treatment that contributes to acceleration of the biochemical reactions. Carballa et al. (2007) performed a pilot-scale reactor study operated under mesophilic (37°C) and thermophilic conditions. Very high removal efficiencies were achieved for naxopren, sulfamethoxazole, roxithromicin and estrogens (>85%) and for musks (galaxolide and tonalide) and diclofenac; while medium removal for diazepam and diclofenac (40–60%), low removal for iopromide and no removal for carbamazepine were obtained. Sludge stabilization increased when operating at low organic loading rate (OLR) and at high temperatures (Carballa et al., 2007).

7.3.6 Specific microorganisms/cultures used for biodegradation of Micropollutants

Nonylphenol can be degraded aerobically by specialized microorganisms like bacteria, yeast and fungi under aerobic conditions (Yuan et al., 2004; Chang et al., 2005). Most of the nonylphenol-degraders belong to sphingomonads. They were isolated from municipal wastewater treatment plants and identified as *Sphingomonas* sp. Strain TTNP3, *Sphingomonas cloacae, Sphingomonas xenophaga* strain Bayram (Tanghe et al., 1999; Fujii et al., 2001; Gabriel et al., 2005). These bacterial strains are known for their capability of utilizing nonylphenol as a carbon source. The rates of nonylphenol degradation are ranging from 29 mg nonylphenol/L · d (Tanghe et al., 1999) up to 140 mg nonylphenol/L·day (Gabriel *et al.*, 2005). The metabolites of nonylphenol degradation varied depending on the bacterial strain and nonylphenol isomer used in the different studies (e.g., *Candida aquaetextoris* degrades linear nonylphenol to 4-acetylphenol, *Sphingomonas xenophaga* Bayram degrades branched nonylphenol to the corresponding branched alcohol and benzoquinone) (Vallini *et al.*, 2001; Gabriel *et al.*, 2005).

The degradation pathways of nonylphenol by different Sphingomonads share similarities as the corresponding alcohol of the alkyl side-chain of nonylphenol are degradation intermediates produced by the isolated strains (Tanghe *et al.*, 2000; Fujii *et al.*, 2001; Gabriel *et al.*, 2005). Some scientists assumed that the degradation of para- nonylphenol starts with the fission of the aromatic ring (Tanghe *et al.*, 1999; Fujii *et al.*, 2001). Corvini *et al.* (2004) provided evidence that hydroquinone is the central metabolite in the degradation pathway of nonylphenol isomers. The degradation mechanism is a type II *ipso* substitution requiring a free hydroxyl group at the para position. Hydroquinone is formed via hydroxylation at C4 position of the alkyl chain, where alkylated quinol leaves the molecule as a carbocation or radical species (Corvini *et al.*, 2006b). Hydroquinone is further degraded to organic acids, such as succinate and 3,4-dihydroxy butanedioic acid (Cirja, 2007).

Several estrogen-degrading bacteria isolated from activated sludge of WWTP were reported (Fujii *et al.*, 2003; Yoshimoto *et al.*, 2004). However, nothing is known about the behavior of estrogen-degrading bacteria in activated sludge. Further study is required to clarify the relationship between the removal efficiency of natural estrogens and the behavior of estrogen-degrading bacteria (Hashimoto *et al.*, 2007).

In a German WWTP, an EE2 removal efficiency of 90% was reported and satisfying biodegradation occurred in the nitrifying tank (Andersen *et al.*, 2003). A significant degradation of EE2 was observed in nitrifying activated sludge, where the ammonia-oxidizing bacterium *Nitrosomonas europaea* was present (Shi *et al.*, 2004). These results were in agreement with those of previous studies indicating the fast degradation of EE2 with the concomitant formation of hydrophilic compounds in nitrifying activated sludge (Vader *et al.*, 2000). Additionally to reports on the positive influence of nitrifying bacteria, good degradation rates were obtained in axenic cultures of EE2-degrading fungi and bacteria isolated from activated sludge. In some cases, the microorganisms could degrade up to 87% of the added substrate (30 mg/L EE2) (Yoshimoto *et al.*, 2004; Haiyan *et al.*, 2007). The fungus *Fusarium proliferatum* has been isolated from a cowshed sample and is capable of converting 97% of EE2 at an initial concentration of 25 mg/L in 15 days at 30°C at an optimum pH of 7.2 (Shi *et al.*, 2002).

Shi et al. (2002) and Yoshimoto et al. (2004) reported the existence of EE2-degradation products but did not identify them. Horinouchi et al. (2004) proposed a pathway for the degradation of testosterone by *Comamonas Testosteroni* TA441, consisting of aromatic-compound degrading genes for seco-steroids and 3-ketosteroid dehydrogenase. Studies on the metabolisms of EE2 by the isolated bacterium *Sphingobacterium* sp. JCR5 led to the identification of three metabolites. Further metabolites, i.e., two acids (e.g., 2-hydroxy-2,4-dienevaleric acid and 2-hydroxy-2,4-diene-1,6-dionic acids) were the main catabolic intermediates for the formation of E1 (Shi et al., 2004; Haiyan et al., 2007). Furthermore, Shi et al. (2004) reported that E2 was most easily degraded via E1 by nitrifying bacteria and it is assumed as being also the degradation product of EE2. The catabolic pathway of EE2 by *Sphingobacterium Sp. JCR5* is proposed to start with the oxidation of C-17 to a ketone group and the hydroxylation and ketonization of C-9α followed by the cleavage of B ring. A ring is hydroxylated by the addition of molecular oxygen (Haiyan et al., 2007).

7.3.7 Formation of by-products during biodegradation

During wastewater treatment an increase in the concentrations of some pollutants may be observed due to biological degradation such as formation of alkylphenols from biodegradation of polyethoxylates (Isidori et al., 2007).

One metabolite of BPA biodegradation, 4-hydroxy-acetophenone, was detected in MBR and conventional activated sludge reactor (Chen et al., 2008). The behaviour of the two main ibuprofen biodegradation intermediates (carboxy and hydroxy derivatives), on the other hand, supported that the main ibuprofen elimination pathway which occurs under aerobic conditions (Matamoros et al., 2008a).

Nonylphenols were not detected in the raw wastewater of one of the plants but were generated by biodegradation of ethoxylated nonylphenols during later anaerobic stages of treatment. They were, however, not detected in the treated water and were likely eliminated with the sludge (Bruchet et al., 2002). Natural and synthetic estrogens, which are excreted as inactive conjugated forms of sulphuric and glucuronic acids, may be re-transformed to active forms (free estrogens) in the raw sewage or WWTP by microorganisms that present glucuroidase and sulfatase activity (Isidori et al., 2007). Besides, certain drinking water disinfection by-products (DBPs) may act as EDCs, while the oxidation of PPCPs could lead to the formation of by-products more toxic than their parent compounds. Therefore, it is important to determine the by-products of micropollutants and to develop the proper treatment technologies for their treatment, because, they might be even more toxic and persistant, or cannot be either sorbed or biotransformed like antiepileptics (Caliman and Gavrilescu, 2009).

7.4 REFERENCES

Andersen, H., Siegrist, H., Halling-Sørensen, B. and Ternes, T. A. (2003) Fate of estrogens in a municipal sewage treatment plant. *Environ. Sci. Technol.* **37**(18), 4021–4026.

Auriol, M., Filali-Meknassi, Y., Tyagi, R. D., Adams, C. D. and Surampalli, R. Y. (2006) Endocrine disrupting compounds removal from wastewater, a new challenge. *Process Biochem.* **41**(3), 525–539.

Baronti, C., Curini, R., D'-Ascenzo, G., Di-Corcia, A., Gentili, A. and Samperi, R. (2000) Monitoring natural and synthetic estrogens at activated sludge sewage treatment plants and in a receiving river water. *Environ. Sci. Technol.* **34**(24), 5059–5066.

Bensoam, J., Cicolella, A. and Dujardin, R. (1999) Improved extraction of glycol ethers from water by solid-phase micro extraction by carboxen polydimethylsiloxane-coated fiber. *Chromatographia* **50**(3–4), 155–159.

Bicchi, C., Schiliro, T., Pignata, C., Fea, E., Cordero, C., Canale, F. and Gilli, G. (2009) Analysis of environmental endocrine disrupting chemicals using the E-screen method and stir bar sorptive extraction in wastewater treatment plant effluents. *Sci. Tot. Environ.* **407**(6), 1842–1851.

Bolong, N., Ismail, A. F., Salim, M. R. and Matsuurad, T. (2009) A review of the effects of emerging contaminants in wastewater and options for their removal. *Desalination* **239** (1–3), 229–246.

Bowman, J. C., Readman, J. W. and Zhou, J. L. (2003) Sorption of the natural endocrine disruptors, oestrone and 17beta-oestradiol in the aquatic environment. *Environ. Geochem. Health.* **25**(1), 63–67.

Brenner, A., Mukmenev, I., AAbeliovich, A. and Kushmaro, A. (2006) Biodegradability of tetrabromobisphenol A and tribromophenol by activated sludge. *Ecotoxicology* **15**(4), 399–402.

Brown, D., de Henau, H., Garrigan J. T., Gerike, P., Holt, M., Kunkel, E., Matthijs, E., Waters J. and Watkinson, R. J. (1987) Removal of non-ionics in sewage treatment plants. II: Removal of domestic detergent non-ionic surfactants in a trickling filter sewage treatment plant. *Tenside Surfactants Detergents* **24**(1), 14–19.

Bryner, A. (2007) Urine source separation: A promising wastewater management option. http://www.innovations-report.com/html/reports/environment_sciences/report-80295.html (accessed October 2009).

Burgess, R. M., Pelletier, M. C., Gundersen, J. L., Perron, M. M. and Ryba, S. A. (2005) Effects of different forms of organic carbon on the partitioning and bioavailability of nonylphenol. *Environ. Toxicol. Chem.* **24**(7), 1609–1617.

Buser, H. R., Poiger, T. and Muller, M. D. (1998) Occurrence and fate of the pharmaceutical drug diclofenac in surface waters: Rapid photodegradation in a lake. *Environ. Sci. Technol.* **32**(22), 3449–3456.

Caliman, F. A. and Gavrilescu, M. (2009) Pharmaceuticals, Personal Care Products and Endocrine Disrupting Agents in the Environment – A Review. *Clean* **37**(4–5), 277–303.

Carballa, M., Omil, F. and Lema, J. M. (2007) Calculation methods to perform mass balances of micropollutants in sewage treatment plants. Application to pharmaceutical and personal care products (PPCPs). *Environ. Sci. Technol.* **41**(3), 884–890.

Carballa, M., Omil, F. and Lema, J. M. (2008a) Comparison of predicted and measured concentrations of selected pharmaceuticals, fragrances and hormones in Spanish sewage. *Chemosphere* **72**(8), 1118–1123.

Carballa, M., Omil, F., Lema, J. M., Llompart, M., Garcia, C., Rodriguez, I., Gomez, M. and Ternes, T. (2005) Behaviour of pharmaceuticals and personal care products in a sewage treatment plant of northwest Spain. *Wat. Sci. Technol.*, **52**(8), 29–35.

Cargouet, M., Perdiz, D., Mouatassim-Souali, A., Tamisier-Karolak, S. and Levi, Y. (2004) Assessment of river contamination by estrogenic compounds in Paris area (France). *Sci. Tot. Environ.* **324**(1–3), 55–66.

Carucci, A., Cappai, G. and Piredda, M. (2006) Biodegradability and toxicity of pharmaceuticals in biological wastewater treatment plants. *J. Environ. Sci. Health Part A-Toxic/Hazard. Subst. Environ. Eng.* **41**(9), 1831–1842.

Cavret, S. and Feidt, C. (2005) Intestinal metabolism of PAH: in vitro demonstration and study of its impact on PAH transfer through the intestinal epithelium. *Environ. Res.* **98**(1), 22–32.

Cespedes, R., Petrovic, M. and Raldua, D. (2004) Integrated procedure for determination of endocrine-disrupting activity in surface waters and sediments by use of the biological technique recombinant yeast assay and chemical analysis by LC-ESI-MS. *Anal. Bioanal. Chem.* **378**(3), 697–708.

Chang, B. V., Chiang, F. and Yuan, S. Y. (2005) Biodegradation of nonylphenol in sewage sludge. *Chemosphere* **60**(11), 1652–1659.

Chelliapan, S., Wilby, T. and Sallis, P. J. (2006) Performance of an up-flow anaerobic stage reactor (UASR) in the treatment of pharmaceutical wastewater containing macrolide antibiotics. *Wat. Res.* **40**(3), 507–516.

Chen, J., Huang, X. and Lee, D. (2008) Bisphenol A removal by a membrane bioreactor. *Process Biochemistry* **43**(4), 451–456.

Chen, P. J., Rosenfeldt, E. J., Kullman, S. W., Hinton, D. E. and Linden, K. G. (2007) Biological assessments of a mixture of endocrine disruptors at environmentally relevant concentrations in water following UV/H2O2 oxidation. *Sci. Tot. Environ.* **376**(1–3), 18–26.

Cheng, K. Y. and Wong, J. W. (2006) Effect of synthetic surfactants on the solubilization and distribution of PAHs in water/soil-water systems. *Environ. Technol.* **27**(8), 835–844.

Chiou, C. T., McGroddy, S. E. and Kile, E. (1998) Partition characteristics of polycyclic aromatic hydrocarbons on soils and sediments. *Environ. Sci. Technol.* **32**(2), 264–269.

Cirja, M. (2007) *Studies on the behaviour of endocrine disrupting compounds in a membrane bioreactor*. PhD thesis, RWTH Aachen University, Aachen, Germany.

Clara, M., Kreuzinger, N., Strenn, B., Gans, O. and Kroiss, H. (2005a) The solids retention time – a suitable design parameter to evaluate the capacity of wastewater treatment plants to remove micropollutants. *Wat. Res.* **39**(1), 97–106.

Clara, M., Strenn, B., Ausserleitner, M. and Kreuzinger, N. (2004) Comparison of the behaviour of selected micropollutants in a membrane bioreactor and a conventional wastewater treatment plant. *Wat. Sci. Technol.* **50**(5), 29–36.

Clara, M., Strenn, B., Gans, O., Martinez, E., Kreuzinger, N. and Kroiss, H. (2005b) Removal of selected pharmaceuticals, fragrances and endocrine disrupting compounds in a membrane bioreactor and conventional wastewater treatment plants. *Wat. Res.* **39**(19), 4797–4807.

Conkle, J., White, J. R. and Metcalfe, C. D. (2008) Reduction of pharmaceutically active compounds by a lagoon wetland wastewater treatment system in Southeast Louisiana. *Chemosphere* **73**(11), 1741–1748.

Cordoba, E. C. (2004) The fate of 17α-ethynylestradiol in aerobic and anaerobic sludge. Developing a methodology for adsorption. MSc thesis, Sub-Dept of Environmental Technology, Wageningen Univ, Wageningen, Netherlands.

Corvini P. F., Meesters R. J., Schaffer A., Schroder H. F., Vinken R. and Hollender J. (2004) Degradation of a nonylphenol single isomer by Sphingomonas sp. strain TTNP3 leads to a hydroxylation-induced migration product. *Appl. Environ. Microbiol.* **70**(11), 6897–6900.

Corvini P. F., Schaeffer A. and Schlosser D. (2006a) Microbial degradation of nonylphenol and other alkylphenols – our evolving view. *Appl. Microbiol. Biotechnol.* **72**(2), 223–243.

Corvini, P. F., Hollender, J., Ji R., Schumacher, S., Prell, J., Hommes, G., Priefer, U., Vinken, R. and Schaffer, A. (2006b) The degradation of alpha-quaternary nonylphenol isomers by Sphingomonas sp. strain TTNP3 involves a type II ipso-substitution mechanism. *Appl. Microbiol. Biotechnol.* **70**(1), 114–122.

Cui, C. W., Ji, S. L. and Ren, H. Y. (2006) Determination of steroid estrogens in wastewater treatment plant of a controceptives producing factory. *Environ. Monit. Assess.* **121**(1–3), 407–417.

Czajka, C. P. and Londry, K. L. (2006) Anaerobic biotransformation of estrogens. *Environ. Sci. Total.* **E367**(2–3), 932–941.

D'Ascenzo, G., Di Corcia, A., Gentili, A., Mancini, R., Mastropasqua, R., Nazzari, M. and Samperi, R. (2003) Fate of natural estrogen conjugates in municipal sewage transport and treatment facilities. *Sci. Total. Environ.* **302**(1–3), 199–209.

De Mes, T. (2007) *Fate of estrogens in biological treatment of concentrated black water.* PhD thesis, Wageningen University, Wageningen, Netherlands.

De Mes, T., Zeeman, G. and Lettinga, G. (2005) Occurrence and fate of estrone, 17-beta-estradiol and 17-alpha-ethynylestradiol in STPs for domestic wastewater. *Reviews in Environ. Sci. Biotechnol.* **4**(4), 275–311.

Desbrow, C., Routledge, E.J., Brighty, G.C., Sumpter J.P. and Waldock M., (1998) Identification of estrogenic chemicals in STW effluent. 1. Chemical fractionation and in vitro biological screening. *Environ. Sci. Technol.* **32**(11), 1549–1558.

Doll, T.E. and Frimmel, F.H. (2003) Fate of pharmaceuticals – photodegradation by simulated solar UV-light. *Chemosphere* **52**(10), 1757–1769.

Drewes, J. E., Heberer, T. and Reddersen, K. (2002) Fate of pharmaceuticals during indirect potable reuse. *Wat. Sci. Technol.* **46**(3), 73–80.

Driver, J., Lijmbach, D. and Steen, I. (1999) Why recover phosphorus for recycling, and how? *Environ. Technol.* **20**(7), 651–662.

Ehlers, G. A. and Loibner, A. P. (2006) Linking organic pollutant (bio)availability with geosorbent properties and biomimetic methodology: A review of geosorbent characterisation and (bio)availability prediction. *Environ. Pollut.* **141**(3), 494–512.

Ejlersson, J., Nilsson, M., Kylin H., Bergman, A., Karlson L., Oequist M. and Svensson, B. (1999) Anaerobic degradation of of nonylphenol mono- and diethoxylates in digestor sludge, landfill municipal solid waste, and landfilled sludge. *Environ. Sci. Technol.* **33**(2), 301–306.

Environment Canada (2009) Pharmaceuticals and Personal Care Products in the Canadian Environment: Research and Policy Directions, Workshop Proceedings NWRI Scientific Assessment Report Series No. 8, http://www.ec.gc.ca/inre-nwri/default.asp?lang = En&n = C00A589F-1&offset = 23&toc = show (accessed October 2009).

Ermawati, R., Morimura, S., Tang, Y. Q., Liu, K. and Kida, K. (2007) Degradation and behavior of natural steroid hormones in cow manure waste during biological treatments and ozone oxidation. *J. Biosci. Bioeng.* **103**(1), 27–31.

Esperanza, M., Suidan, M. T., Nishimura, F., Wang, Z. M., Sorial, G. A., Zaffiro, A., McCauley, P., Brenner, R. and Sayles, G. (2004) Determination of sex hormones and nonylphenol ethoxylates in the aqueous matrixes of two pilotscale municipal wastewater treatment plants, *Environ. Sci. Technol.* **38**(11), 3028–3035.

Füerhacker, M., Dürauer, A. and Jungbauer, A. (2001) Adsorption isotherms of 17β-estradiol on granular activated carbon (GAC). *Chemosphere* **44**(7), 1573–1579.

Fujii, K., Satomi, M., Morita, N., Motomura, T., Tanaka, T. and Kikuchi, S. (2003) Novosphigobium tardaugens sp. vov., an oestradiol-degrading bacterium isolated from the activated sludge of a sewage treatment plant in Tokyo. *Int. J. Syst. Evol. Microbiol.* **53**(Pt 1), 47–52.

Fujii, K., Urano, N., Ushio, H., Satomi, M. and Kimura, S. (2001) Sphingomonas cloacae sp. nov., a nonylphenol-degrading bacterium isolated from wastewater of a sewage-treatment plant in Tokyo. *Int. J. Syst. Evol. Microbiol.* **51**(Pt 2), 603–610.

Fürhacker, M., Breithofer, A. and Jungbauer, A. (1999) 17β-estradiol: Behavior during waste water analyses. *Chemosphere* **39**(11), 1903–1909.

Furuichi, T., Kannan, K., Suzuki, K., Giesy, J. P. and Masunaga, S. (2006). Occurrence of estrogenic compounds in and removal by a swine farm waste treatment plant. *Environ. Sci. Technol.* **40**(24), 7896–902.

Gabriel, F. L., Giger, W., Guenther, K. and Kohler, H. P. (2005) Differential degradation of nonylphenol isomers by Sphingomonas xenophaga Bayram. *Appl. Environ. Microbiol.* **71**(3), 1123–1129.

Galassi, S., Valescchi, S. and Tartari, G. A. (1997) The distribution of PCB's and chlorinated pesticides in two connected Himalayan lakes. *Water Air Soil Pollut.* **99**(1–4), 717–725.

Garcia, M. T., Campos, E., Dalmau, M., Ribosa, I. and Sanchez-Leal, J. (2002) Structure-activity relationships for association of linear alkylbenzene sulfonates with activated sludge. *Chemosphere* **49**(3), 279–286.

Gerike, P. (1987) Removal of nonionics in sewage treatment plants; II, *Tenside Surfactants Detergents* **24**, 14–19.

Gobel, A., Thomsen, A., Mcardell, C. S., Joss, A. and Giger, W. (2005) Occurrence and sorption behavior of sulfonamides, macrolides, and trimethoprim in activated sludge treatment. *Environ. Sci. Technol.* **39**(11), 3981–3989.

Golet, E. M., Xifra, I., Siegrist, H., Alder, A. C. and Giger, W. (2003) Environmental exposure assessment of fluoroquinolone antibacterial agents from sewage to soil. *Environ. Sci. Technol.* **37**(15), 3243–3249.

Gonzalez, S., Muller, J., Petrovic, M., Barcelo, D. and Knepper, T. P. (2006) Biodegradation studies of selected priority acidic pesticides and diclofenac in different bioreactors. *Environ. Pollut.* **144**(3), 926–932.

Gros, M., Petrovic, M. and Barcelo, D. (2007) Wastewater treatment plants as a pathway for aquatic contamination by pharmaceuticals in the Ebro River Basin (Northeast Spain). *Environ. Toxicol. Chem.* **26**(8), 1553–1562.

Haiyan, R., Shulan, J., Ud din Ahmed, N., Dao, W., Chengwu, C. (2007) Degradation characteristics and metabolic pathway of 17alpha-ethynylestradiol by *Sphingobacterium sp. JCR5*. *Chemosphere.* **66**(2), 340–346.

Halling-Sorensen, B., Nielsen, S. N., Lanzky, P. F., Ingerslev, F., Lutzhoft, H.C.H. and Jorgensen S. E. (1998) Occurrence, fate and effects of pharmaceutical substances in the environment – A review. *Chemosphere* **36**(2), 357–394.

Hammer, M., Tettenborn, F., Behrendt, J., Gulyas, H. and Otterpohl, R. (2005) Pharmaceutical residues: Database assessment of occurrence in the environment and exemplary treatment processes for urine. *IWA 1st National Young Researchers Conference – Emerging Pollutants and Emerging Technologies*, October 27–28, 2005, RWTH Aachen University, Germany.

Hashimoto, T., Onda, K., Nakamura, Y., Tada, K., Miya, A. and Murakami, T. (2007) Comparison of natural estrogen removal efficiency in the conventional activated sludge process and the oxidation ditch process. *Wat Res.* **41**(10), 2117–2126.

Heberer, T. (2002) Occurrence, fate, and removal of pharmaceutical residues in the aquatic environment: A review of recent research data. *Toxicology Letters* **131**(1–2), 5–17.

Henze, M. (1997) Waste design for households with respect to water, organics and nutrients. *Wat. Sci. Technol.* **35**(9), 113–120.

Holthaus, K. I., Johnson, A. C., Jurgens, M. D., Williams, R. J., Smith, J. J. and Carter, J. E. (2002) The potential for estradiol and ethinylestradiol to sorb to suspended and bed sediments in some English rivers. *Environ. Toxicol. Chem.* **21**(12), 2526–35.

Iesce, M.R., della Greca, M., Cermolal, F., Rubino, M., Isidori, M and Pascarella, L. (2006) Transformation and ecotoxicity of carbamic pesticides in water. *Environ. Sci. Pollut. Res. Int.* **13**(2), 105–109.

Ilani, T., Schulz, E. and Chefetz, B. (2005). Interactions of organic compounds with wastewater dissolved organic matter: Role of hydrophobic fractions. *Environ. Qual.* **34**(2), 552–562.

Isidori, M., Lavorgna, M., Palumbo, M., Piccioli, V. and Parrella, A. (2007) Influence of alkylphenols and trace elements in toxic, genotoxic, and endocrine disruption activity of wastewater treatment plants. *Environ. Toxicol. Chem.* **26**(8),1686–94.

Ivashechkin, P., Corvini, P., Fahrbach, M., Hollender, J., Konietzko, M., Meesters, R., Schröder, H.F. and Dohmann, M. (2004) *Comparison of the elimination of endocrine disrupters in conventional wastewater treatment plants and membrane bioreactors.* In Proceedings of the 2nd IWA Leading-Edge Conference on Water and Wastewater Treatment Technologies – Prague, Part Two: Wastewater Treatment, Water Environment Management Series (WEMS), IWA Publishing.

Jacobsen, B. N., Nyholm, N., Pedersen, B. M., Poulsen, O. and Ostfeldt, P. (1993) Removal of organic micropollutants in laboratory activated sludge reactors under various operating conditions: sorption. *Wat. Res.* **27**(10), 1505–1510.

Janex-Habibi, M. L., Huyard, A., Esperanza, M. and Bruchet, A. (2009) Reduction of endocrine disruptor emissions in the environment: The benefit of wastewater treatment. *Wat. Res.* **43**(6), 1565–1576.

Jensen, R. L. and Schäfer, A. I. (2001) Adsorption of estrone and 17β-estradiol by particulates-activated sludge bentonite, hemalite and cellulose. Recent Advances in Water Recycling Technologies, Brisbane, Australia.

Joffe, M. (2001) Are problems with male reproductive health caused by endocrine disruption? *Occup. Environ. Med.* **58**(4), 281–288.

Johnson, A. C. and Sumpter, J. P. (2001) Removal of endocrine-disrupting chemicals in activated sludge treatment works. *Environ. Sci. Technol.* **35**(24), 4697–4703.

Johnson, A. C., Aerni, H.-R., Gerritsen, A., Gibert, M., Giger, W., Hylland, K., Jurgens, M., Nakari, T., Pickering, A. and Suter, M. J. F. (2005) Comparing steroid estrogen, and nonylphenol content across a range of European sewage plants with different treatment and management practices. *Wat. Res.* **39**(1), 47–58.

Johnson, A. C., Belfroid, A. and Di Corcia, A. (2000) Estimating steroid oestrogen inputs into activated sludge treatment works and observations on their removal from the effluent. *Sci. Tot. Environ.* **256**(2–3), 163–173.

Jones, O. A. H., Voulvoulis, N. and Lester, J. N. (2005) Human pharmaceuticals in wastewater treatment processes. *Crit. Rev. Environ. Sci. Technol.*, **35**(4) 401–427.

Joss, A., Andersen, H., Ternes, T., Richle, P. R. and Siegrist, H. (2004) Removal of estrogens in municipal wastewater treatment under aerobic and anaerobic conditions: Consequences for plant optimization. *Environ. Sci. Technol.* **38**(11), 3047–3055.

Joss, A., Keller, E., Alder, A. C., Goebel, A., McArdell, C. S., Ternes, T. and Siegrist, H. (2005) Removal of pharmaceuticals and fragrances in biological wastewater treatment. *Wat. Res.* **39**(14), 3139–3152.

Joss, A., Zabczynski, S., Göbel, A., Hoffmann, B., Löffler, D., McArdell, C. S., Ternes, T. A., Thomsen, A. and Siegrist, H. (2006) Biological degradation of pharmaceuticals in municipal wastewater treatment: Proposing a classification scheme. *Wat Res.* **40**(8), 1686–1696.

Junker, T., Alexy, R., Knacker, T. and Kummerer, K. (2006) Biodegradability of ^{14}C-labeled antibiotics in a modified laboratory scale sewage treatment plant at environmentally relevant concentrations. *Environ. Sci. Technol.* **40**(1), 318–324.

Jürgens, M. D., Holthaus, K. I. E., Johnson, A. C., Smith, J. J. L., Hetheridge, M. and Williams, R. J. (2002) The potential for estadiol and ethinylestradiol degradation in English rivers. *Environ. Toxicol. Chem.* **21**(3), 480–488.

Jürgens, M. D., Williams, R. J. and Johnson, A. C. (1999) *Fate and Behaviour of Steriod Oestrogens in Rivers: A Scoping Study*. Oxon, Institute of Hydrology, p. 80.

Kanda, R. and Churchley, J. (2008) Removal of endocrine disrupting compounds during conventional wastewater treatment. *Environ. Technol.* **29**(3), 315–323

Katsoyiannis, A. and Samara, C. (2007) Comparison of active and passive sampling for the determination of persistent organic pollutants (POPs) in sewage treatment plants. *Chemosphere* **7**(7), 1375–1382.

Kikuta, T. (2004) *Modelling of Degradation of Organic Micropollutants in Activated Sludge Process Focusing on Partitioning between Water and Sludge Phases.* Msc thesis, Thesis Institute of Technology, Dept of Civil and Environmental Engineering. Tokyo, Japan.

Kim, J. Y., Ryu, K., Kim, E. J., Choe, W. S., Cha, G. C. and Yoo, I. K. (2007) Degradation of bisphenol A and nonylphenol by nitrifying activated sludge. *Process Biochem.* **42**(10), 1470–1474.

Kim, S., Eichhorn, P., Jensen, J. N., Weber, A. S. and Aga, D. S. (2005) Removal of antibiotics in wastewater: Effect of hydraulic and solid retention times on the fate of tetracycline in the activated sludge process. *Environ. Sci. Technol.* **39**(15), 5816–5823.

Kimura, K., Hara, H. and Watanabe, Y. (2007) Elimination of selected acidic pharmaceuticals from municipal wastewater by an activated sludge system and membrane bioreactors. *Environ. Sci. Technol.* **41**(10), 3708–3714.

Kirk, L. A., Tyler, C. R., Lye, C. M. and Sumpter, J. P. (2002) Changes in estrogenic and androgenic activities at different stages of treatment in wastewater treatment works. *Environ. Toxicol. Chem.* **21**(5), 972–979.

Kloepfer, A., Gnirss, R., Jekel, M. and Reemtsma, T. (2004) Occurrence of benzothiazoles in municipal wastewater and their fate in biological treatment. *Wat. Sci. Technol.* **50**(5), 203–208.

Koh, Y. K. K., (2008) *An Evaluation of the Factors Controlling Biodegradation of Endocrine Disrupting Chemicals During Wastewater Treatment*, PhD thesis, Imperial College, London, UK.

Komori, K., Tanaka, H., Okayasu, Y., Yasojima, M. and Sato, C. (2004) Analysis and occurrence of estrogen in wastewater in Japan. *Wat. Sci. Technol.* **50**(5), 93–100.

Kozak, R. G., D'Haese, I. and Verstraete, W. (2001) *Pharmaceuticals in the Environment: Focus on 17α-ethinyloestradiol. Pharmaceuticals in the Environment, Source, Fate, Effects and Risks.* (ed) Kümmerer K., Springer-Verlag, Berlin, Germany. pp. 49–65.

Kreuzinger, N., Clara, M., Strenn, B. and Kroiss, H. (2004) Relevance of the sludge retention time (SRT) as design criteria for wastewater treatment plants for the removal of endocrine disruptors and pharmaceuticals from wastewater. *Wat. Sci. Technol.* **50**(5), 149–156.

Krogmann, U., Boyles, L. S., Bamka, W. J., Chaiprapat, S. and Martel, C. J. (1999) Biosolids and sludge management. *Wat. Environ. Res.* **71**(5), 692–714.

Kujawa-Roeleveld, K., Schuman, E., Grotenhuis, T., Kragić, D., Mels, A. and Zeeman, G. (2008) *Biodegradability of Human Pharmaceutically Active Compounds (PhAC) in Biological Systems Treating Source Separated Wastewater Streams.* Third SWITCH Scientific Meeting Belo Horizonte, Brazil.

Kujawa-Roeleveld, K., Zeeman, G. and Mels, A. (2006) *Elimination of pharmaceuticals from concentrated wastewater.* First SWITCH Scientific Meeting, University of Birmingham, UK.

Kummerer, K., Alexy, R., Huttig, J. and Scholl, A. (2004) Standardized tests fail to assess the effects of antibiotics on environmental bacteria. *Wat. Res.* **38**(8), 2111–2116.

Lai, K. M., Johnson, K. L., Scrimshaw, M. D. and Lester, J. N. (2000) Binding of waterborne steroid estrogens to solid phases in river and estuarine systems. *Environ. Sci. Technol.* **34**(18), 3890–3894.

Lalah, J. O., Schramm, K. W., Henkelmann, B., Lenoir, D., Behechti, A., Guenther, K., Kettrup, A. (2003) The dissipation, distribution and fate of a branched 14C-nonylphenol isomer in lake water/sediment system. *Environ. Pollut.* **122**(2), 195–203.

Lapara, T. M., Nakatsu, C. H., Pantea, L. M. and Alleman, J. E. (2001) Aerobic biological treatment of a pharmaceutical wastewater: Effect of temperature on cod removal and bacterial community development. *Wat. Res.* **35**(18), 4417–4425.

Larsen, T. A. and Gujer, W. (1996) Separate management of anthropogenic nutrient solutions (human urine). *Wat. Sci. Technol.* **34** (3–4), 87–94.

Larsen, T. A. and Gujer, W. (2001) Waste design and source control lead to flexibility in wastewater management. *Wat. Sci. Technol.* **43**(5), 309–318.

Larsen, T. A., Lienert, J., Joss, A. and Siegrist, H. (2004) How to avoid pharmaceuticals in the aquatic environment. *J Biotechnol.* **113**(1–3), 295–304.

Layton, A. C., Gregory, B. W., Seward, J. R., Schultz, T. W. and Sayler, G. S. (2000) Mineralization of steroidal hormones by biosolids in wastewater treatment systems in Tennessee U.S.A. *Environ. Sci. Technol.* **34**(18), 3925–3931.

Lee, H. B. and Liu, D. (2002) Degradation of 17ß-estradiol and its metabolites by sewage bacteria. *Water Air Soil Pollut.* **134**(1–4), 353–368.

Lesjean, B., Gnirss, R., Buisson, H., Keller, S., Tazi-Pain, A. and Luck, F. (2005) Outcomes of a 2-year investigation on enhanced biological nutrients removal and trace organics elimination in membrane bioreactor (MBR). *Wat. Sci. Technol.* **52**(10–11), 453–460.

Li, C., Ji, R., Vinken, R., Hommes, G., Bertmer, M., Schäffer, A. and Corvini, P. F. (2007) Role of dissolved humic acids in the biodegradation of a single isomer of nonylphenol by Sphingomonas sp. *Chemosphere.* **68**(11), 2172–2180.

Lienert, J., Bürki, T. and Escher, B. I. (2007) Reducing micropollutants with source control: substance flow analysis of 212 pharmaceuticals in faeces and urine. *Wat. Sci. Technol.* **56**(5), 87–96.

Lienert, J., Haller, M., Berner, A., Stauffacher, M. and Larse, T. A. (2003) How farmers in Switzerland perceive fertilizers from recycled anthropogenic nutrients (urine). *Wat. Sci. Technol.* **48**(1), 47–56.

Lindberg, R. H., Wennberg, P., Johansson, M. I., Tysklind, M. and Andersson, B. A. (2005) Screening of human antibiotic substances and determination of weekly mass flows in five sewage treatment plants in Sweden. *Environ. Sci. Technol.* **39**(10), 3421–3429.

Liu, Z., Ito, M., Kanjo, Y. and Yamamoto, A. (2009a) Profile and removal of endocrine disrupting chemicals by using an ER/AR competitive ligand binding assay and chemical analyses. *J. Environ. Sci.* **21**(7), 900–906.

Liu, Z., Kanjo, Y. and Mizutani, S. (2009b) Removal mechanisms for endocrine disrupting compounds (EDCs) in wastewater treatment – physical means, biodegradation, and chemical advanced oxidation: A review. *Sci. Tot. Environ.* **407**(2), 731–748.

Luppi, L. I., Hardmeier, I., Babay, P. A., Itria, R. F. and Erijman, L. (2007) Anaerobic nonylphenol ethoxylate degradation coupled to nitrate reduction in a modified biodegradability batch test. *Chemosphere* **68**(11), 2136–2143.

Matamoros, V., Carlos, A. and Brix, H. (2009) Preliminary screening of small-scale domestic wastewater treatment systems for removal of pharmaceutical and personal care products. *Wat Res.* **43**(1), 55–62.

Matamoros, V., Garcia, J. and Bayona, J. M. (2008b) Organic micropollutant removal in a full-scale surface flow constructed wetland fed with secondary effluent. *Wat Res.* **42**(3), 653–660.

Matamoros, V., Osorio, A. C. and Bayone, J. M. (2008a) Behaviour of pharmaceutical products and biodegradation intermediates in horizontal subsurface flow constructed wetland. A microcosm experiment. *Sci. Tot. Environ.* **394**(1), 171– 176.

Maurer, M., Escher, B. I., Richle, P., Schaffner, C. and Alder, A. C. (2007) Elimination of beta-blockers in sewage treatment plants. *Wat. Res.* **41**(7), 1614–1622.

Maurer, M., Schwegler, P. and Larsen, T. A. (2003) Nutrients in urine: Energetical aspects of removal and recovery. *Wat. Sci. Technol.* **48**(1), 37–46.

Maurin, M. B. and Taylor, A. (2000) Variable heating rate thermogravimetric analysis as a mechanism to improve efficiency and resolution of the weight loss profiles of three model pharmaceuticals. *J. Pharm. Biomed. Anal.* **23**(6), 1065–1071.

Melcer, H., Monteith, H., Staples, C. and Klecka, G. (2006) The removal of alkylphenol ethoxylate surfactants in activated sludge systems. In Proceedings *Water Environment Federation, WEFTEC 2006, Session 21–30*, pp. 1695–1708.

Melin, T., Jefferson, B., Bixio, D., Thoeye, C., De Wilde, W., De Koning, J., van der Graaf, J. and Wintgens, T. (2006) Membrane bioreactor technology for wastewater treatment and reuse. *Desalination* **187**(1–3), 271–282.

Miao, X. S. and Metcalfe, C. D. (2003) Determination of carbamazepine and its metabolites in aqueous samples using liquid chromatography-electrospray tandem mass spectrometry. *Anal. Chem.* **75**(15), 3731–3738.

Minamiyama, M., Ochi, S. and Suzuki, Y. (2006) Fate of nonylphenol polyethoxylates and nonylphenoxy acetic acids in an anaerobic digestion process for sewage sludge treatment. *Wat. Sci. Technol.* **53**(11), 221–226.

Muller, M., Rabenoelina, F., Balaguer, P., Patureau, D., Lemenach, K., Budzinski, H., Barcelo, D., Lopez de Alda, M., Kuster, M., Delgenés, J. P. and Hernandez-Raquet, G. (2008) Chemical and biological analysis of endocrine-disrupting hormones and estrogenic activity in an advanced sewage treatment plant. *Environ. Toxicol Chem.* **27**(8), 1649–1658.

Murk, A. J., Legler, J., van Lipzig, M. M. H., Meerman, J. H. N., Belfroid, A. C., Spenkelink, A., van der Burg, B., Rijs, G. B. J. and Vethaak, D. (2002) Detection of estrogenic potency in wastewater and surface water with three in vitro bioassays. *Environ. Toxicol. Chem.* **21**(1), 16–23.

Nacheva, P. M., Peña-Loera, B. and Moralez-Guzmán, F. (2006) Treatment of chemical-pharmaceutical wastewater in packed bed anaerobic reactors. *Wat. Sci. Technol.* **54**(2), 157–163.

Nakada, N., Yasojima, M., Okayasu, Y., Komori, K., Tanaka, H. and Suzuki,Y. (2006) Fate of oestrogenic compounds and identification of oestrogenicity in a wastewater treatment process. *Wat. Sci. Technol.* **53**(11), 51–63.

Norpoth, K., Nehrkorn, A., Kirchner, M., Holsen, H. and Teipel, H. (1973) Investigations on the problem of solubility and stability of steroid ovoulation inhibitors in water, waste water and activated sludge. *Zbl. Bakt. Hyg., I. Abt Orig.* **156**(6), 500–511.

Novaquatis (2008) A Cross-cutting Eawag Project. http://www.novaquatis.eawag.ch/ index_EN (accesses October 2009).

Onda, K., Nakamura, Y., Takatoh, C., Miya, A. and Katsu, Y. (2003) The behavior of estrogenic substances in the biological treatment process of sewage. *Wat. Sci. Technol.* **47**(9), 109–116.

Onda, K., Yang, S. Y., Miya, A. and Tanaka, T. (2002) Evaluation of estrogen-like activity on sewage treatment processes using recombinant yeast. *Wat. Sci. Technol.* **46**(11–12), 367–373.

Peck, A. M. (2006) Analytical methods for the determination of persistent ingredients of personal care products in environmental matrices. *Anal. Bioanal. Chem.* **386**(4), 907–939.

Pickering, A. D. and Sumpter, J. P. (2003) Comprehending endocrine disrupters in aquatic environments. *Environ. Sci. Technol.* **37**(17), 331A–336A.

Poiger, T., Buser, H. R., Muller, M. D., Balmer, M. E. and Buerge, I. J. (2003) Occurrence and fate of organic micropollutants in the environment: Regional mass balances and source apportioning in surface waters based on laboratory incubation studies in soil and water, monitoring, and computer modeling. *Chimia* **57**(9), 492–498.

Press-Kristensen, K., Lindblom, E. and Henze, M. (2007) Modeling as a tool when interpreting biodegradation of micro pollutants in activated sludge systems. *Wat. Sci. Technol.* **56**(11), 11–16.

Price, P. B. and Sowers, T. (2004) Temperature dependence of metabolic rates for microbial growth, maintenance, and survival. *Microbiology.* **101**(13), 4631–4636.

Pronk, W., Biebow, M. and Boller, M. (2006). Treatment of source separated urine by a combination of bipolar electrodialysis and a gas transfer membrane. *Wat. Sci. Technol.* **53**(3), 139–146.

Quintana, J. B., Weiss, S. and Reemtsma, T. (2005) Pathway's and metabolites of microbial degradation of selected acidic pharmaceutical and their occurrence in municipal wastewater treated by a membrane bioreactor. *Wat. Res.* **39**(12), 2654–2664.

Rauch, W., Brockmann, D., Peters, I., Larsen, T. A. and Gujer, W. (2003) Combining urine separation with waste design: an analysis using a stochastic model for urine production. *Wat. Res.* **37**(3), 681–689.

Reddy, S. and Brownawell, B. J. (2005) Analysis of estrogens in sediments from a sewage impacted urban estuary using high-performance liquid chromatography/time-of-flight mass spectrometry. *Environ. Toxicol. Chem.* **24**(5), 1041–1047.

Reemtsma, T., Zywicki, B., Stueber, M., Kloepfer, A. and Jekel, M. (2002) Removal of sulfur-organic polar micropollutants in a membrane bioreactor treating industrial wastewater. *Environ. Sci. Technol.* **36**(5), 1102–1106.

Ren, Y. X., Nakano, K., Nomura, M., Chiba, N. and Nishimura, O. (2007) Effects of bacterial activity on estrogen removal in nitrifying activated sludge. *Wat. Res.* **41**(14), 3089–3096.

Richardson, S. D. (2006) Environmental mass spectrometry: Emerging contaminants and current issues. *Anal. Chem.* **78**(12), 4021–4046.

Rogers, H. R. (1996) Sources, behaviour and fate of organic contaminants during sewage treatment and sewage sludges. *Sci. Tot. Environ.* **185**(1–3), 3–26.

Ronteltap, M., Maurer, M. and Gujer, W. (2007) The behaviour of pharmaceuticals and heavy metals during struvite precipitation in urine. *Water Res.* **41**(9), 1859–1868.

Rossi, L, Lienert, J. and Larsen, T. A. (2009) Real-life efficiency of urine source separation. *J. Environ. Management* **90**(5), 1909–1917.
Saino, H., Jamagata, H., Nakajima, H., Shigemura, H. and Suzuki, Y. (2004) Removal of endocrine disrupting chemicals in wastewater by SRT control. *J. Japan Society Water Environ.* **27**, 61–68.
Saito, M., Tanaka, H., Takahashi, A. and Yakou, Y. (2002) Comparison of yeast-based estrogen receptor assays. *Wat. Sci. Technol.* **46**(11–12), 349–354.
Segmuller, B. E., Armstrong, B. L., Dunphy, R. and Oyler, A. R. (2000) Identification of autoxidation and photodegradation products of ethynylestradiol by on-line HPLC-NMR and HPLC-S. *J. Pharm. Biomed. Anal.* **23**(5), 927–37.
Servos, M. R., Bennie, D. T., Burnison, B. K., Jurkovic, A., McInnis, R., Neheli, T., Schnell, A., Seto, P., Smyth, S. A. and Ternes, T. A. (2005) Distribution of estrogens, 17b-estradiol and estrone, in Canadian municipal wastewater treatment plants. *Sci. Tot. Environ.* **336**(1–3), 155–170.
Shi, J., Fujisawa, S., Nakai, S. and Hosomi, M. (2004) Biodegradation of natural and synthetic estrogens by nitrifying activated sludge and ammonia-oxidizing bacterium Nitrosomonas europea. *Water Res.* **34**(9), 2323–2330.
Shi, J. H., Suzuki, Y., Lee, B. D., Nakai, S. and Hosomi, M. (2002) Isolation and characterization of the ethynylestradiol-biodegrading microorganism Fusarium proliferatum strain HNS-1. *Water Sci. Technol.* **45**(12), 175–179.
Singh, A., Van Hamme, J. D. and Ward, O. P. (2007) Surfactants in microbiology and biotechnology: Part 2. Application aspects. *Biotechnol. Adv.* **25**(1), 99–121.
Sipma, J., Osuna, B., Collado, N., Monclús, H., Ferrero, G., Comas, J. and Rodriguez-Roda, I. (2009) Comparison of removal of pharmaceuticals in MBR and activated sludge systems. *Desalination* **250**(2), 653–659.
Sithole, B. B. and Guy, R. D. (1987) Models for tetracycline in aquatic environments: II. Interaction with humic substances. *Water Air Soil Pollut.* **32**(3–4), 315–321.
Soares, A., Murto, M., Guieysse, B. and Mattiasson, B. (2006) Biodegradation of Nonylphenol in a continuous bioreactor at low temperatures and effects on the microbial population. *Appl. Microbial. Biotechnol.* **69**(5), 597–606.
Spengler, P., Korner, W. and Metzger, J. W. (2001) Substances with estrogenic activity in effluents of sewage treatment plants in southwestern Germany. 1. Chemical analysis. *Environ. Toxicol. Chem.* **20**(10), 2133–2141.
Spring, A. J., Bagley, D. M., Andrews, R. C., Lemanik, S. and Yang, P. (2007) Removal of endocrine disrupting compounds using a membrane bioreactor and disinfection. *J. Environ. Eng. Sci.* **6**(2), 131–137.
Stangroom, S. J., Collins, C. D. and Lester, J. N. (2000) Abiotic behaviour of organic Micropollutants in soils and the aquatic environment. A review: 2 Transformations. *Environ. Technol.* **21**(8), 865–882.
StegerHartmann, T., Kummerer, K. and Hartmann, A. (1997) Biological degradation of cyclophosphamide and its occurrence in sewage water. *Ecotoxicology Environ. Safety* **36**(2), 174–179.
Stenstrom, M. K., Cardinal, L. and Libra, L. (1989) Treatment of hazardous substances in wastewater treatment plants. *Environ. Progr.* **8**(2), 107–112.
Stumpf, M., Ternes, T. A., Wilken, R. D., Rodrigues, S. V. and Bauman, W. (1999) Polar drug residues in sewage and natural waters in the state of Rio de Janeiro, Brazil. *Sci. Tot. Environ.* **225**(1–2), 135–141.

Suarez, S., Carballa, M., Omil, F. and Lema, J. M. (2008) How are pharmaceutical and personal care products (PPCPs) removed from urban wastewaters? *Rev. Environ. Sci. Biotechnol.* **7**(2), 125–138.

Suarez, S., Ramill, M., Omil, F. and Lema, J. M. (2005) Removal of pharmaceutically active compounds in nitrifying-denitrifying plants. *Wat. Sci. Technol.* **52**(8), 9–14.

Svenson, A., Allard, A. S. and Ek, M. (2003) Removal of estrogenicity in Swedish municipal sewage treatment plants. *Wat. Res.* **37**(18), 4433–4443.

Tanghe, T., Dhooge, W. and Verstraete, W. (1999) Isolation of a bacterial strain able to degrade branched nonylphenol. *Appl. Environ. Microbiol.* **65**(2), 746–751.

Tchobanoglous, G., Burton, F. L. and Stensel, H. D. (2004) *Wastewater Engineering, Treatment and Reuse*. 4th Ed, McGrawHill Company, Inc., New York.

Ter Laak, T. L., Durjava, M., Struijs, J. and Hermens, J. L. (2005) Solid phase dosing and sampling technique to determine partition coefficients of hydrophobic chemicals in complex matrixes. *Environ. Sci. Technol.* **39**(10), 3736–3742.

Ternes, T. A. (1998) Occurrence of drugs in German sewage treatment plants and rivers. *Wat. Res.* **32**(11), 3245–3260.

Ternes, T. A. and Joss, A. (2006) *Human Pharmaceuticals, Hormones and Fragrances: The Challenge of micropollutants in urban water management*. International Water Association (IWA), IWA Publishing.

Ternes, T. A., Joss A. and Siegrist, H. (2004) Scrutinizing pharmaceuticals and personal care products in wastewater treatment. *Environ. Sci. Technol.* **38**(20), 392–399.

Ternes, T. A., Kreckel, P. and Mueller, J. (1999). Behaviour and occurrence of estrogens in municipal sewage treatment plants. II. Aerobic batch experiments with activated sludge. *Sci. Total. Environ.* **225**(1–2), 91–99.

Ternes, T. A., Meisenheimer, M., McDowell, D., Sacher, F., Brauch, H. J., Gulde, B. H., Preuss, G., Wilme, U. and Seibert, N. Z. (2002) Removal of pharmaceuticals during drinking water treatment. *Environ. Sci. Technol.* **36**(17), 3855–3863.

Terzic, S., Matosic, M., Ahel, M. and Mijatovic, I. (2005) Elimination of aromatic surfactants from municipal wastewaters: Comparison of conventional activated sludge treatment and membrane biological reactor. *Wat. Sci. Technol.* **51**(8), 447–453.

Tettenborn, F., Behrendt, J. and Otterpohl, R. (2008) Pharmaceutical residues in source separated urine and their fate during nutrient-recovery processes, In *Conference Proceeding of IWA World Water Congress and Exhibition*, IWA, Vienna, Austria.

Tilton, F., Benson, W. H. and Schlenk, D. (2002) Evaluation of estrogenic activity from a municipal wastewater treatment plant with predominantly domestic input. *Aquat Toxicol.* **61**(3–4), 211–224.

Tixier, C., Singer, H. P., Oellers, S. and Muller, S. R. (2003) Occurrence and fate of carbamazepine, clofibric acid, diclofenac, ibuprofen, ketoprofen, and naproxen in surface waters. *Environ. Sci. Technol.* **37**(6), 1061–1068.

Turan, A. (1995) Excretion of natural and synthetic estrogens and their metabolites: Occurrence and behaviour in water. In Endocrinally *Active Chemicals in the Environment*. German Federal Environment Agency, Berlin, Germany, pp. 15–50.

Udert, K. M. (2002) The fate of nitrogen and phosphorus in source separated urine. Dissertation ETH, No. 14847. ETH-Zurich, CH.

Udert, K. M., Fux, C., Munster, M., Larsen, T. A., Siegrist, H. and Gujer, W. (2003) Nitrification and autotrophic denitrification of source separated urine. *Wat. Sci. Technol.* **48**(1), 119–130.

Urase, T. and Kikuta, T. (2005) Separate estimation of adsorption and degradation of pharmaceutical substances and estrogens in the activated sludge process. *Wat. Res.* **39**(7), 1289–1300.

Vader, J. S., van Ginkel, C. G., Sperling, F. M. G. M., de Jong, J., de Boer, W., de Graaf, J. S., van der Most, M. and Stokman, P. G. W. (2000) Degradation of ethinyl estradiol by nitrifying activated sludge. *Chemosphere* **41**(8), 1239–1243.

Vajda, A., Barber, L. B., Gray, J. L., Lopez, E. M., Woodling, J. D. and Norris, D. O. (2008) Reproductive disruption in fish downstream from an estrogenic wastewater effluent. *Environ. Sci. Technol.* **42**(9), 3407–3414

Vallini, G., Frassinetti, S., D'Andrea, F., Catelani, G. and Agnolucci, M. (2001) Biodegradation of 4-(1-nonyl)phenol by axenic cultures of the yeast Candida Aquaetextoris: Identification of microbial breakdown products and proposal of a possible metabolic pathway. *Int. Biodeterior. Biodegrad.* **47**(3), 133–140.

Vieno, N. M., Tuhkanen, T. and Kronberg, L. (2005) Seasonal variation in the occurrence of pharmaceuticals in effluents from a sewage treatment plant and in the recipient water. *Environ. Sci. Technol.* **39**(21), 8220–8226.

Vinken, R., Höllrigl-Rosta, A., Schmidt, B., Schaffer, A. and Corvini, P. F. X. (2004) Bioavailability of a nonylphenol isomer in dependence on the association to dissolved humic substances. *Wat. Sci. Technol.* **50**(5), 277–283.

Wallberg, P., Jonsson, P. R. and Andersson, A. (2001) Trophic transfer and passive uptake of a polychlorinated biphenyl in experimental marine microbial communities. *Environ. Toxicol. Chem.* **20**(10), 2158–2164.

Wang, S. (2009) *Microbial Impacts of Selected Pharmaceutically Active Compounds Found in Domestic Wastewater Treatment Plants*. Ph.D. thesis, Dept of Civil and Environmental Engineering, Duke Univ, Durham, NC, USA.

Wang, S. (2009) *Microbial Impacts of Selected Pharmaceutically Active Compounds Found in Domestic Wastewater Treatment Plants*. PhD thesis, Dept of Civil and Environmental Engineering, Duke Univ, Durham, NC, USA.

Watkinson, A. J., Murby, E. J. and Costanzo, S. D. (2007) Removal of antibiotics in conventional and advanced wastewater treatment: Implications for environmental discharge and wastewater recycling. *Wat. Res.* **41**(18), 4164–4176.

White, J. R., Belmont, M. A. and Metcalfe, C. D. (2006) Pharmaceutical compounds in wastewater: Wetland treatment as a potential solution. *The Scientific World J* **6**, 1731–1736.

WHO (2002) *International Programme on Chemical Safety – Global Assessment of the State-of-the-science of Endocrine Disruptors*, (eds. T. Damstra, S. Barlow, A. Bergman, R. Kavlock , G. Van Der Kraak). World Health Organization.

WHO (2006) *ATC classification index with DDDs*. Collaborating Centre for Drug Statistics Methodology, WHO, Oslo.

Wick, A., Fink, G., Joss, A., Hansruedi, S. and Ternes, T. A. (2009) Fate of beta blockers and psycho-active drugs in conventional wastewater treatment. *Wat Res.* **43**(4), 1060–1074.

Wilsenach, J. and van Loosdrecht, M. (2006) Integration of processes to treat wastewater and source-separated urine. *J. Environ. Eng.* **132**(3), 331–341.

Witters, H. E., Vangenechten, C. and Berckmans, P. (2001) Detection of estrogenic activity in Flemish surface waters using an in vitro recombinant assay with yeast cells. *Wat. Sci. Technol.* **43**(2), 117–123

Witzig, R., Manz, W., Rosenberger, S., Kruger, U., Kraume, M. and Szewzyk, U. (2002) Microbiological aspects of a bioreactor with submerged membranes for aerobic treatment of municipal wastewater. *Wat. Res.* **36**(2), 394–402.

Yamamoto, H., Liljestrand, H. M., Shimizu, Y. and Morita, M. (2003) Effects of physical-chemical characteristics on the sorption of selected endocrine disruptors by dissolved organic matter surrogates. *Environ. Sci. Technol.* **37**(12), 2646–2657.

Yang, L., Lan, C., Liu, H., Dong, J. and Luan, T. (2006) Full automation of solid-phase microextraction/on-fiber derivatization for simultaneous determination of endocrine-disrupting chemicals and steroid hormones by gas chromatography-mass spectrometry. *Anal. Bioanal.Chem.* **386**(2), 391–397

Yoon, Y., Westerhoff, P., Yoon, J. and Snyder, S. A. (2004) Removal of 17β-estradiol and fluoranthene by nanofiltration and ultrafiltration. *J. Environ. Eng.* **130**(12), 1460–1467.

Yoshimoto, T., Nagai, F., Fujimoto, F., Watanabe, K., Mizukoshi, H., Makino, T., Kimura, K., Saino, H., Sawada, H. and Omura, H. (2004) Degradation of estrogens by Rhodococcus zopfii and Rhodococcus equi isolates from activated sludge in wastewater treatment plants. *Appl. Environ. Microbiol.* **70**(9), 5283–5289.

Yu, Z. and Huang, W. (2005) Competitive sorption between 17alpha-ethinyl estradiol and naphthalene/phenanthrene by sediments. *Environ. Sci. Technol.* **39**(13), 4878–4885.

Yuan, S. Y., Yu, C. H. and Chang, B. V. (2004) Biodegradation of nonylphenol in river sediment. *Environ Pollut.* **127**(3), 425–430.

Zhang, Y., Geißen, S. U. and Gal, C. (2008) Carbamazepine and diclofenac: Removal in wastewater treatment. plants and occurrence in water bodies. *Chemosphere* **73**(8), 1151–1161

Zhou, P., Su C., Li, B. and Qian, Y. (2006) Treatment of high-strength pharmaceutical wastewater and removal of antibiotics in anaerobic and aerobic biological treatment processes. *J. Envir. Engrg.* **132**(1), 129–136.

Zuehlke, S., Duennbier, U. and Heberer, T. (2005) Determination of estrogenic steroids in surface water and wastewater by liquid chromatography-electrospray tandem mass spectrometry. *J. Sep. Sci.* **28**(1), 52–58.

Chapter 8
UV irradiation for Micropollutant removal from aqueous solutions in the presence of H_2O_2

Tuula A. Tuhkanen and Paula Cajal Mariñosa

8.1 INTRODUCTION

One of the driving forces for the development of Advanced Oxidation Techniques to treat water and waste water is the EC Water Framework Directive (WFD), Article 4 (EC, 2000). (Belgiarno *et al.*, 2007) WFD is meant to ensure achievement and maintenance of "good status" for all the community waters including inland surface and ground waters, transitional and coastal waters by 2015. In this framework, a number of micropollutants which are continuously released into the environment, such as polycyclic aromatic, hydrocarbons (PAHs), alkylphenols (APs), organotins(OTs), volatile organic compounds (VOCs), pesticides and heavy metals have been listed as priority substances.

©2010 IWA Publishing. *Treatment of Micropollutants in Water and Wastewater.* Edited by Jurate Virkutyte, Veeriah Jegatheesan and Rajender S. Varma. ISBN: 9781843393160. Published by IWA Publishing, London, UK.

A recent proposal issued by the EU Parliament and Council regulates their concentrations in surface waters. The priority list of WFD includes 33 substances characterizedby high toxicity, high environmental persistence and high lipophilicity leading to bioaccumulation in food webs and by increased risk for the environment and human health. Several of them (e.g., PAHs, APs, OTs, brominated flame retardants) have been already proven to be or are potential endocrine disruptors (EDCs). Another emerging group of micropollutants that has caused concern of their possible ecological effects are Pharmaceutical and Personal Care Products (PPCPs) (Haberer, 2002, Ternes et al., 2003).

UV/H_2O_2 process have been widely studied and applied particularly in water and wastewater purification and treatment of contaminated groundwater. Continued advances in the technology have made it practicable method micropollutant removal. (Watts et al., 1990, 1991; AOP Handbooks, 1997). During the last year's full-scale applications of UV/H_2O_2 have become commercially available. Depending on the constituents in the water, chemical oxidation may be used in conjugation with other technologies. These are stripping of volatile organics, granular activated carbon treatment (Wang et al., 1990b) or biological oxidation through biologically activated carbon columns (DeWaters et al., 1990) or other processes depending on the identity and concentration of organics in the water.

Usually biological treatment is the most cost-effective technology for organic matter removal from the wastewaters. Some micropollutants, however, are either non-biodegradable or are toxic either to the biological process or aquatic life and must therefore be pre- or post-treated. In many cases, pre-treatment by chemical oxidation will remove the toxicity as well as enhanced biodegradation (Adams et al., 1993; Wang, 1991). On the other hand, sometimes the effluent is still toxic after biological oxidation due to the presence of non-degradable constituents or biological by-products such as residuals of pharmaceutical and personal care products in treated wastewater.

Recently a growing interest has been observed around UV-based techniques due to the introduction of more cost effective and low-energy lamps, development of non-contact reactors that avoid the fouling of UV-lamps and combination of UV-radiation to the use of hydrogen peroxide, ozonation or some other techniques that enable the formation of highly reactive radicals, mainly hydroxyl radicals.

8.2 THEORY OF UV/H_2O_2

8.2.1 General

UV/oxidation processes generally involve generation of OH• radical through UV photolysis of hydrogen peroxide or/and ozone. Most of the information related to these techniques is produced by laboratory scale experiments. The most direct method for generation of hydroxyl radicals is through the cleavage of hydrogen peroxide. Photolysis of hydrogen peroxide yields hydroxyl radicals by a direct process with a yield of two radicals formed per photon absorbed by 254 nm (Baxendale et al., 1957):

$$H_2O_2 + h\nu \rightarrow OH\bullet + OH\bullet \tag{7.1}$$

The quantum yield for the photolysis of hydrogen peroxide by 253.7 nm light in 0.1 N perchloric acid is found to be 1.00 at 25°C and it is independent of concentration and light intensity over the range 2×10^{-5} M to 0.1 M peroxide and 4.5×10^{-7} Einstein.l^{-1}min^{-1} to 5×10^{-4} Einstein.l^{-1}min^{-1}. The hydroxyl radicals generated in water have a very high oxidizing capacity and they attack the organic compounds relatively non-selectively with rate constants ranging from 10^6 to 10^{10} M^{-1} s^{-1} (Buxton et al., 1988)

The molar extinction coefficient of hydrogen peroxide at 254 nm is 19.6 M^{-1}s^{-1} (Lay 1989), which is very low. Comparatively, the molar extinction coefficients of for example ozone, naphthalene and pentachlorophenol (at pH 7) are 3300, 3300 and 10 000 M^{-1}cm^{-1}, respectively. This means that in order to generate a sufficiently high level of OH radicals in a solution, which contains strong photon absorbers, the concentration of hydrogen peroxide has to be rather high. Low-pressure mercury vapour lamps with a 254 nm peak emission are the most common UV source used in UV/H$_2$O$_2$ systems. The maximum absorbance of hydrogen peroxide occurs at about 220 nm. If low-pressure lamps are used, a high concentration of hydrogen peroxide is needed to generate sufficient hydroxyl radicals because of low absorption coefficient. However, high concentration of H$_2$O$_2$ scavenges the radicals, making the process less effective.

$$OH\bullet + H_2O_2 \rightarrow HO_2\bullet + H_2O \tag{7.2}$$

$$HO_2\bullet + H_2O_2 \rightarrow OH\bullet + H_2O + O_2 \tag{7.3}$$

$$HO_2\bullet + HO_2\bullet \rightarrow H_2O_2 + O_2 \tag{7.4}$$

To overcome the limitation of low absorbtivity of hydrogen peroxide by the wavelength of 254 nm, some researchers/vendors use high intensity, medium-pressure, broadband UV-lamps. The third option is high-intensity, xenon flash lamps whose spectral output can be adjusted to match the absorption characteristics of H$_2$O$_2$.

8.2.2 Photolysis

In an advanced oxidation process (AOP) where irradiation is involved, degradation of the organic compound can take place either by direct or indirect photolysis. An organic molecule can undergo photochemical transformation if and only if both of the following conditions are met (Calvert and Pitts, 1966):

- Light energy is absorbed by the molecule to produce an electronically excited state molecule, and
- chemical transformations of the excited state are competitive with deactivation processes.

The necessity for light absorption is cited as the first law of photochemistry or the Grotthus-Draper Law: Only the light, which is absorbed by a molecule, can be effective in producing photochemical changes in the molecule. The molecule absorbs light in several regions of electromagnetic spectrum, corresponding to different kinds of molecular transitions.

The rate of direct photolysis of a chemical species at concentration [C] is given by Eq. (7.5), which is a combination of Grotthus-Draper Law and Stark-Einstein Law (Leifer, 1988):

$$-\frac{d[C]}{dt} = I_o \phi_C f_C (1 - \exp(-A_t)) \quad (7.5)$$

where Φ_C is the quantum yield of C, i.e., the fraction of absorbed radiation that results in photolysis, and I_o is the incident flux of radiation (254 nm in this case). The factor f_C is the ratio of light absorbed by C to that absorbed by other components of the solution, and A_t is the total absorbance of the solution times a factor of 2.3, defined, respectively, as follows:

$$f_C = \frac{\varepsilon_C [C]}{\Sigma \varepsilon_i [C_i]} \quad (7.6)$$

$$A_t = 2.3 L \Sigma \varepsilon_i [C_i] \quad (7.7)$$

In Equations 6 and 7, the subscript i represents any species present in the solution that absorbs light at the specified wavelength, L is the (effective) path length of the photoreactor, and ε_i is the molar absorpativity or molar extinction coefficient of the ith species at the lamp wavelength.

When the concentration of absorbers is so large that exp($-A_t$) is approximately zero, Equation 2 simplifies to Equation 8 (Leifer, 1988):

$$-\frac{d[C]}{dt} = \phi_C I_o f_C \quad (7.8)$$

UV irradiation for Micropollutant removal from aqueous solutions

If C is a minor absorber, then its rate of decay will be first order. If C is the principal absorber, its rate will be zero order. When the concentration of absorbers is small, $[1 - \exp(A)_t]$ may be expanded in a Taylor series to yield the familiar first order expression (Leifer, 1988):

$$-\frac{d[C]}{dt} = 2.3L\,\phi_C\,I_o\,\varepsilon_C[C] \tag{7.9}$$

Under these conditions, the photolysis of two absorbing substrates will be independent of one another. When the concentration of a single absorber is intermediate in concentration, Equation 5 becomes Equation 10:

$$\frac{d[C]}{dt} = \phi_C\,I_o(1 - \exp(-\alpha[C])) \tag{7.10}$$

where

$$\alpha = 2.3L\,\varepsilon_C \tag{7.11}$$

8.2.3 Mechanisms UV/H$_2$O$_2$ oxidation

The mechanism of oxidation by H$_2$O$_2$/UV has been studied extensively and attributed to the photolysis of hydrogen peroxide (Baxendale and Wilson, 1957) which gives rise to the appearance of hydroxyl radicals (Guittonneau, 1989, Glaze and Lay, 1989). Hence, the elimination rate of organic compounds has at least two contributions: direct photolysis and hydroxyl radical attack. Besides of photolysis the compound C is eliminated by hydroxyl radical attack:

$$-\frac{d[C]}{dt} = I_o\,\varphi_C f_C(1 - \exp(-A_t)) + k_{OH,C}[OH][C] \tag{7.12}$$

Where $k_{OH,C}$ is the second order reaction rate constant of hydroxyl radical with compound C. Pseudo first order rate constant can be used because the concentration of OH radical can be assumed to be constant over the range of reaction and thus it can be incorporated to the rate constant (Stumm and Morgan, 1981):

$$-\ln C/C_o = k't \tag{7.13}$$

where k' = pseudo first order rate constant, s^{-1}.

The effect of light source and intensity, dose of hydrogen peroxide are also listed if provided. However, there is a limit to the beneficial effects of added peroxide, since H$_2$O$_2$ is also an OH radical scavenger ($k_{OH,H2O2} = 2.7 \times 10^7$ M^{-1}s^{-1}, Christensen et al., 1982):

$$-\frac{d[C]}{dt} = I_o\phi_C f_C(1 - \exp(-A_t)) + k_{OH,C}[OH][C] \qquad (7.14)$$

Lot of studies to determine the ratio of hydrogen peroxide and pollutant has been carried out in order to create a tool for the optimisation of treatment process (Ince, 1999).

8.2.4 Ozone/UV

The production of free OH –radicals can take place also by combining UV treatment and ozonation (Glaze *et al.*, 1987; Glaze and Kang, 1990). The first step by O_3/UV process is the formation of H_2O_2 by photolysis of ozone:

$$O_3 + H_2 + UV(<310 \text{ nm}) \rightarrow O_2 + OH\bullet + OH\bullet \rightarrow O_2 + H_2O_2 \qquad (7.15)$$

The ozone/UV process can destroy compounds through direct ozonation, photolysis or reaction with OH-radicals.

8.3 LABORATORY SCALE EXPERIMENTS OF UV/H_2O_2

8.3.1 General

Some of the organic compounds can be decomposed by direct UV radiation. Addition of hydrogen peroxide accelerated the decomposition except for few compounds which are readily photodegradable. The comparison of the purification efficiencies and the economical feasibility of micropollutant removal by UV based techniques is difficult due to various kinds of reactor design and the type and intensity of lamps used. It is especially difficult to compare results gained with low, medium and high pressure mercury lamps because their spectral differences. In generally the decomposition of organic compound are considered as single component system – only the disappearance of model compound has been followed. Usually the formed by-products are neglected.

8.3.2 Treatment of contaminated groundwater

Remediation of contaminated groundwater is probably the most common application of UV/hydrogen peroxide treatment. Lot of studies are available particularly from the laboratory scale experiments with synthetic single compound experimental setup.

Sundstrom *et al.* (1986) studied the destruction of typical halogenated aliphatics by the combination of UV light and hydrogen peroxide. The chlorinated compounds with carbon-carbon double bonds, such as trichloroethylene, degraded much faster than the other compounds studied. The reacted chlorine was converted quantitatively to chloride ion, indicating that the chlorinated structures were effectively destroyed. Sundstrom also studied the degradation rates of typical aromatic compounds. The pseudo-first order rate constants for UV and UV/H_2O_2 experiments are given in the parentheses ($\times 10^{-3}$ s^{-1}): benzene (0.25 and 1.42), chlorobenzene (0.42 and 1.05), toluene (0.38 and 1.57), phenol (0.06 and 1.37), 2-chlorophenol (0.25 and 0.83), 2,4-dichlorophenol (0.67 and 1.36), 2,4,6-trichlorophenol (1.3 and 1.67), dimethyl phthalate (0.02 and 0.78), and diethyl phthalate (0.02 and 0.78).

Weir *et al.* (1987) studied the effects of process operating conditions on the destruction of benzene by UV and hydrogen peroxide. The lamp was low-pressure mercury arc lamp with 5.3 W output at 254 nm. The rates of reaction followed pseudo-first order kinetics and were strongly affected by the contaminant and hydrogen peroxide molar ratio and by the intensity of UV radiation. Many intermediates were formed, but extending the treatment time could eliminate them. The identified intermediates were phenol, catechol and hydroquinone. No polymerisation was observed. They also studied the destruction kinetics of TCE (Weir *et al.*, 1993). The reaction rate of TCE was first order in ultraviolet light intensity and pseudo- first order in TCE concentration. The rate was first order in hydrogen peroxide concentration at low peroxide levels, but independent of peroxide concentration at high peroxide levels.

Ho (1986) studied the synergistic effect of hydrogen peroxide and UV radiation with medium pressure mercury arc lamps (450 W) on the decomposition of 2,4-dinitrotoluene (DNT). The degradation pathways of DNT in aqueous solution were side-chain oxidation, which converted DNT to 1,3-dinitrobenzene, hydroxylation of the benzene ring, which converted 1,3-dinitrobenzene to hydroxynitrobenzene derivatives, benzene ring cleavage of the hydroxynitrobenzenes, which produced lower molecular weight carboxylic acids and aldehydes and further photo oxidation, which eventually converted the lower molecular weight acids and aldehydes to CO_2, H_2O and HNO_3.

Milano *et al.* (1992) studied the intermediate formation from 4-bromodiphenylether degradation by UV/H_2O_2 treatment. The lamp used was 100 W medium pressure mercury arc lamp. They identified several degradation compounds. The complex degradation mechanism was divided roughly into three steps. Firstly, dehalogenatation produced diphenylether and parahydroxydiphenylether, secondly, decomposition of diphenylether and the formation of phenol and benzene, and thirdly, the opening of the aromatic ring produced carboxylic acids leading to the mineralisation. Overall, the mineralisation was much more difficult to obtain than that of aliphatic halogen compounds. During degradation of the aromatic ether, at the same time as the formation of the carboxylic acids, macromolecular compounds were formed, which had a humic acid structure and were more resistant to degradation.

Ishikawa et al. (1992) studied the photochemical degradation of organic phosphate esters by low-pressure mercury arc irradiation (15 W). The degradation followed first order kinetics and the reaction rates varied between $0.2–11.1 \times 10^{-3}$ s^{-1}.

According to Yashura et al. (1977) the mono-chlorophenols also followed first order kinetics in their degradation by UV radiation. The reaction rates were: 4-monochlorophenol 1.4×10^{-3}s^{-1}, 2-chlorophenol 2×10^{-4}s^{-1} and 3-chlorophenol 3×10^{-5}s^{-1}. Addition of hydrogen peroxide highly accelerated the decomposition rates. Degradation of 2-chlorophenol also followed first order kinetics when it was irradiated at a wavelength of 313 nm. The first-order direct and indirect photolysis rate constants were determined in humic solution and in concentrated natural waters. Moza et al. (1988) studied the decomposition of 2-chlorophenol, 2,4-dichlorophenol and 2,4,6-trichlorophenol solutions by UV/H$_2$O$_2$ treatment with 125 W medium pressure lamp. When the initial hydrogen peroxide concentration was 55 ppm the following rate constants were obtained for 2-chlorophenol, 2,4-dichlorophenol and 2,4,6-trichlorophenols respectively in degassed solution: 0.33×10^{-4} s^{-1}, 1.1×10^{-4}s^{-1} and 3.3×10^{-4}s^{-1}. The lower rate constant by UV/H$_2$O$_2$ treatment for 2-chlorophenol compared to the direct photolysis can be explained by the lamp used. The medium pressure lamp produces most of its light at the visible region and less radiation at 254 nm.

2,4-Dichlorophenol and 2,4,5-trichlorophenol were studied by Miller et al. (1988). Both compounds were degraded by 300 nm light. Addition of hydrogen peroxide accelerated the photodecomposition of both chlorophenols. Photodegradation of chlorophenols followed smooth first-order kinetics (pseudo first order in the presence of hydrogen peroxide). The reaction rate constant of DCP increased from 1.2×10^{-3} to 6.7×10^{-3}s^{-1} when the hydrogen peroxide concentration increased from 0 to 0.1 M. The corresponding values for TCP were from 6.7×10^{-3} to 13.4×10^{-3}s^{-1}.

Peterson et al. (1988) studied the efficiency of a high-intensity short-wavelength UV light system to degrade the pesticides carbofuran, fenamiphos sulfoxide and propazine in aqueous solutions. The first order rate constants were 3.33×10^{-3}, 11.20×10^{-3} and 3.07×10^{-3} s^{-1}, respectively. Unlike many other substances, addition of hydrogen peroxide did not accelerate the decomposition. Also captane, chlordane, m-xylene and pentachloronitrobenzene were studied by Peterson et al. (1990) with 5000 W high-pressure mercury vapour lamp. Addition of hydrogen peroxide accelerated only the decomposition of pentachloronitrobenzene.

Methyl tert-butyl ether (MTBE) was removed from aqueous solution by exposure to a low-pressure mercury lamp in the presence of hydrogen peroxide in a recirculating batch reactor. Dark and UV-only tests were conducted to separate the effects of MTBE loss to system components and photolytic processes. Experiments were conducted at initial peroxide:MTBE molar ratios of 4:1, 7:1 and 15:1. The concentrations of MTBE, benzene and by-products were measured over a 120-min period. UV/H_2O_2 treatment resulted in >99.9% removal of MTBE with the major purgable by-product identified as tert-butyl formate (TBF). The second order rate constant for the degradation of MTBE from the hydroxyl radical was estimated to be 3.9×10^9 $M^{-1}s^{-1}$ (Chang and Young, 2000).

More data of laboratory scale UV/H_2O_2 experiments are collected into the Table 8.1.

8.3.3 Drinking water applications

The oxidants commonly used in drinking water treatment are ozone, chlorine, chlorine dioxide and chloramines. Also UV-based techniques have been increasing during the last decade. The formation of unwanted disinfection by-products in the reactions of oxidants with the natural organic matrix of the water is very serious issue and lot of efforts have been used for the modification of drinking water processes. All the above mentioned techniques are able to oxidise also micropollutant compounds in the water. Particularly the effects of UV-disinfection have been studied intensively since there seems to be a formation of unwanted degradation products from the parent compounds. The toxicity of the degradation products can be higher than of the original compounds.

The phototransformation of selected pharmaceuticals has been studied in Milli-Q water and in surface water after sand filtration. Results show that higher depletion of organic compounds with low pressure hg lamps than with medium pressure lamps emitting radiation between 239–334 nm. A typical UV dose for disinfection, 400 Jm^{-2}, lead the degradation of 3.5% of antrazine, 14% of iobromide, 16% of sulfamethazol and 45% of diclofenac irradiated with low pressure Hg lamps. The degradation of 17α-ethylenestradiol was 14% in sand filtered water but just less than 1% in Milli Q water. (Meunier *et al.*, 2005) Therefore, UV-phototransformation can be considered a fairly inefficient process under conditions for typical UV disinfection. According to several studies the removal of micropollutants require UV doses greater than those used to disinfect water indicating that UV inactivation of microorganisms and UV degradation of organic chemicals operate on very different dose levels. Finally, the kinetic results and experimental screening of the can be used for assessing the feasibility of UV based processes for reclaiming water contaminated with pharmaceutical compounds and their intermediates.(Canonica et al., 2008; Lopez et al., 2007)

Table 8.1 Reaction rate constants for different chemicals in UV/ H_2O_2 system

Compound	Light	[H_2O_2] mM	K× 10^{-3} s^{-1}	Reference
Phenol	254 nm + O_3	3.2–12.7	1.74 1.16	Esplugas et al., 2002
o-Nitrotoluene	medium pres. 700 W	7.4 14.7 29.4 88.2	2.1 3.7 5.7 7.9	Ghaly et al., 2001b
p-Hydroxy-benzoic acid	medium pres. 185–436 nm 3.30×10^{-5} Einstein/s			Beltran-Heredia et al., 2001
Isoprene	254 nm 2.25 mW/ m^2	molar ratio 0 1 4 6	k_p+k_{OH} 0.40 0.45 0.63 0.70	Elkanzi and Kheng 2000
Tetrabutyl-azine	125 W high pressure	– 0.2 2.0 2.0	7.40 7.67 9.17 15.83	Sanlaville et al., 1996
Napthalene	1.85×10^{-6} Einstein $l^{-1}s^{-1}$ ”– 1.05×10^{-6}	– 1.0 10 20 40 100 –	0.80 1.6 6.9 5.9 4.0 2.6 0.2	Glaze et al., 1992 Tuhkanen 1994 ”– ”– ”– ”– ”–
PCP	1.85×10^{-6} 1.05×10^{-6} 1.85×10^{-6} 1.85×10^{-6} 1.85×10^{-6} 1.85×10^{-6}	– – 1 10 20 100	1.0 0.2 3.3 9.2 7.2 1.2	”– ”– ”– ”– ”– ”–

Removal of the highly potent Ames mutagen, 3-chloro-4-(dichloromethyl)-5-hydroxy-2(5H)-furanone (MX) from an aqueous solution and a decrease in the mutagenic potency of the solution were possible by direct photolysis with the wavelength of 254 nm as well as with thermal, reductive and different kind of oxidative treatments in which generation of hydroxyl radical was involved. The removal of 1 mole of MX required 2 moles of hydroxyl radicals. Photolytic treatment removed 76% of the MX and 79% of the mutagenic potency within one-hour irradiation. (Fucui *et al.*, 1991).

8.3.4 Municipal waste water

Although modern waste water treatment plants are able to remove over 95% of the BOD, biodegradable organic matter, from the wastewater, there are persistent organic compounds which are not removed completely. Some of them are non-biodegradable and so hydrophilic that they are removed by adsorption into the sludge. With advanced analytical techniques more and more compounds such as PPCPs and their metabolites, endocrine distributing compounds and aromatic amines have been detected. (Ternes *et al.*, 2003)

The UV-base techniques have been used widely for waste water disinfection. The doses required for organic pollution removal are much higher compared to the irradiation required for destroying the microorganisms. Advanced oxidation processes (AOPs) have been shown to be effective at the removal of various organic contaminants from drinking water. Ozone, ozone with hydrogen peroxide (H_2O_2), and UV with H_2O_2 are three of the most commonly studied AOPs. (Wert *et al.*, 2009)

Andreozzi *et al.* (2003) studied the removal of Paracetamol with low pressure mercury arc lamp combined to use of hydrogen peroxide and ozone. Addition of both hydrogen peroxide and ozone enhanced the removal of Paracetamol by destroying the aromatic ring. The formation of degradation products occurred and the mineralization of carbon compounds was 40% with the optimal hydrogen peroxide dose.

Rosario-Ordiz *et al.* (2009) studied the efficiency of low pressure UV/H_2O_2 oxidation was evaluated for the removal of six pharmaceuticals in three different wastewaters under multiple UV and H_2O_2 doses. The compounds evaluated were meprobamate, dilantin, carbamazepine, primidone, atenolol andtrimethoprim. These compounds possess different chemical functionalities that would make them reactive to direct photolysis as well as OH radicals. The role of the organic matter in the effluent, EfOM, was evaluated by utilizing an empirical model which estimated the second order reaction rate constant between OH radicals and EfOM of different waste waters (kEfOM$_{OH}$). The scavenging rates were 4.02×10^5 for E1, 29.3×10^5 for E2 and 12.6×10^5 s-1 for E3. Based on these results, the OH-radical exposure was less for E2 and greater for E1.

This influenced the overall efficiency of UV/H_2O_2 for contaminant removal and that quantity is not the most important factor when evaluating the role of EfOM, but also the intrinsic reactivity. When the H_2O_2 dose was 0 mg/L, dilantin was removed by 0–65%. However, removal of the other five pharmaceuticals was generally < 10% indicating that these compounds are not removed by UV photolysis. During UV/H_2O_2 treatment with 300 mJ/cm^2, pharmaceutical oxidation increased as the H_2O_2 dose was increased, as expected. The role of the chemical composition of waste water was significant since EfOM acts as scavenger and competes with the pharmaceutical molecules. The results indicated that greater UV doses were required to promote greater OH radical exposure needed to overcome the scavenging capacity of the wastewater and produce greater oxidation of pharmaceuticals. The overall removals of studied pharmaceuticals were between 0 and >99%, and were dependent on the overall scavenging rate. These results suggest that the intrinsic reactivity of the EfOM (as opposed to overall concentration alone) is an important parameter in the planning of full scale treatment facilities.

The effectiveness of UV-based processes (UV and UV/H_2O_2) for the removal of pharmaceuticals in real wastewater using bench-scale experiment setup with a treatment capacity of 10m^3/day was investigated in Japan by Kim et al. (2009). Forty-one kinds of pharmaceuticals including 12 antibiotics and 10 analgesics were detected in secondary effluent used for tested water. For UV process a good removal seems to be expected for just a few pharmaceuticals such as ketoprofen, diclofenac and antipyrine. Especially, the removal efficiencies of macrolide antibiotics such as clarithromycin, erythromycin and azithromycin for UV alone process were found to be very low even by the introduction of considerable UV dose of 2768 mJ/cm^2. For UV/H_2O_2 process, a 90% removal efficiency could be accomplished in 39 pharmaceuticals at UV dose of 923 mJ/cm^2, indicating that it will be possible to reduce UV energy required for the effective pharmaceuticals removal by the combination of H_2O_2 with UV process. Finally, the kinetic results and experimental screening of the can be used for assessing the feasibility of UV based processes for reclaiming water contaminated with pharmaceutical compounds and their intermediates. (Canonica et al., 2008; Lopez et al., 2007)

The occurrence of hormione like activity in treated wastewater is a highly acute problem. Numerous studies have investigated degradation of individual endocrine disrupting compounds (EDCs) in lab or natural waters including AOP techniques. Chen et al. (2007) have studied UV/hydrogen peroxide treatment. Four EDCs including estradiol (E2), ethinyl estradiol (EE2), bisphenol-A (BPA) and nonylphenol (NP) were spiked individually or as a mixture at $\mu g\ L^{-1}$-$ng\ L^{-1}$ in laboratory or natural river water. The experimental results suggest that the UV/H_2O_2 reaction exhibited first-order kinetics for removing both *in vitro* and *in vivo* estrogenic activity for a mixture of EDCs at environmentally relevant concentrations (μg L^{-1} to ng L^{-1}) in deionized and d natural water systems. UV/H_2O_2 treatment at H_2O_2 doses (10 ppm) and UV fluence (<1000 mJ cm^{-2}) relevant to water treatment used for chemical oxidation processes was capable of decreasing *in vitro* and *invivo* estrogenic activity for an EDC mixture at low concentrations. In addition, the removal rates of in vitro estrogenic activity of EDC mixtures were lower than that observed with single compounds alone. Also, the removal of in vitro estrogenic activity with UV/H_2O_2 treatments was slightly slower in natural water, compared with that observed in laboratory water. The reduction in the removal rate of estrogenic activity is likely due to a decreased steady-state •OH concentration due to the presence of scavengers and thus a reduced rate of reaction for the degradation of E2 and EE2. Although removal rate of *invivo* estrogenic activity was faster than in vitro for a mixture of EDCs with the treatment of UV+10 ppm of H_2O_2, a UV fluence of 2000 mJ cm^{-2} was not sufficient to completely remove *in vivo* estrogenic activity in natural water.

Roselfeldt *et al.* (2007) studied Endocrine Disrupting Compounds (EDCs), 17-β-estradiol (E2) and 17-α-ethinyl estradiol (EE2) by direct UV photolysis and UV/H_2O_2 compared to direct UV photolysis. The addition of H_2O_2 greatly increases the degradation of ECD compounds in lab water. In natural waters, the degradation was retarded, but UV/H_2O_2 advanced oxidation was still capable of removing greater than 90% of the compound associated estrogenic activity by a UV fluence of 350 mJ cm^{-2}. This UV fluence is much lower than previously reported for UV treatmentof these contaminants.

8.3.5 Paper and pulp industry

Schulte *et al.* (1991) determined the influence of parameters such as the water matrix, OH-radical scavengers, pH, the ratio of hydrogen peroxide to the contaminant and the UV intensity to the treatment of real wastewater, paper mill waste water and landfill leachate. The degradation of COD, AOX and formaldehyde was followed in different hydrogen peroxide based advanced oxidation processes. Both low and medium pressure mercury arc lamps with the output up to 2 kW were used. The activated hydrogen peroxide based techniques were effective but the degree of reduction of parameters and the most cost-effective combination of processes should be determined for each individual case.

8.4 OTHER UV BASED TECHNIQUES

Chaly *et al.* (2001) studied the addition of iron into the UV/H_2O_2 system. The reaction rate constants of p-chlorophenol for UV, UV/H_2O_2, UV/H_2O_2/Fe(II) and UV/H_2O_2/Fe(III). The reaction rate constants increased respectively 1.0×10^{-3}, 2.0×10^{-3}, 1.2×10^{-3} and 1.7×10^{-3}. The energy consumption decreased also significantly from 1105 for UV/H_2O_2 to 184 kWh/kg p-chlorophenol for UV/H_2O_2/Fe(III). The disadvantage of introduction of Fenton –type oxidants were the formation of iron precipitation and requirement of pH adjustment to <4.

Trapido *et al.* (2001) studied the combination of ozone with UV-radiation and hydrogen peroxide for the destruction of nitrophenols. O_3 /UV/ H_2O_2 treatment with the admixture of 8 mM (256 mg/l) hydrogen peroxide at pH 2.5 was the fastest way to degrade NPs. They noticed also almost complete mineralisation of organically bound nitrogen and the elimination of toxicity to *Daphnia mangna*.

Akata and Gurol (1992) studied the removal of nitrobenzene and trihalomethan precursors by UV/H_2O_2, system with and without synthetic polymer Polyethylene Oxide PEO. The hypothesis was that the polymer could enhance the removal of nitrobenzene by polymerisation. The UV/H_2O_2 experiments were conducted in neutral pH (alkalinity 40 and 200 mg/l) and the initial concentration of nitrobenzene was 1.5×10^{-4} M. The experiments were carried our in 2 liters reactor with the 33 W emission power and in 0.1 mM hydrogen peroxide solution. The reaction rate of nitrobenzene was 0.76×10^{-6} M/min. The addition of high concentration (1000 mg/l) of polymer retarded significantly the degradation and the reaction rate was only 1.9×10^{-7} M/min due to the competition of polymer molecules with the nitrobenzene for the hydroxyl radicals. The UV/H_2O_2 treatment was able to remove 40% of the THMFP after 25 minutes irradiation. The low concentration, 10 mg/l, of polymer had no effect on removal of THMFP.

The UV/H_2O_2 is commonly used for the treatment of PCE contaminated groundwater but the efficiency is usually lowered in the precence of radical scavengers and/or UV light absorbers. Alibegic *et al.* (2001) studied the improvement of UV/H_2O_2 photolytic system, by reducing the OH° radical scavengers and UV absorbers in the reacting system. Stripped PCE gas was absorbed into a bubble column reactor equipped with UV light, containing distilled water and hydrogen peroxide. Degradation of PCE in the liquid phase was found to follow pseudo-first-order kinetic and the apparent rate constant was in the order of 0.20 s^{-1}.

Processes involving O_3 and UV dosages in wich the range ozone are 16–24 mgL^{-1} and UV dose 810–1610 mJ/cm^2 has been found more effective for elimination TOC and Disinfection By –Products (DBPs) than UV or ozone alone. The combined O_3-UV AOP led to significant mineralization of organic carbon, as demonstrated by a decrease in concentration of TOC. Approximately 50% of the TOC concentration in the rawwater was removed after 60 min of treatment. The mineralization of TOC followed pseudo-first-order kinetics. The pseudo-first-order rate constant for the mineralization of TOC was $0:04\pm0:02$ min^{-1}. Ozone-UV AOP treatment significantly reduced THM and HAA formation potentials. The chloroformformation potential was reduced by approximately 80% and the HAA3 formation potential was reduced by approximately 70% after 60 min of treatment. The reduction of both of these DBP formation potentials followed pseudo-first-order kinetics. The pseudo-first-order rate constant for chloroform and HAA3 formation potential are $0:12 \pm0:03$ and $0:15 \pm0:04$ min^{-1}, respectively (Chin and Berube, 2005).

8.5 ALTERNATIVE RADIATION SOURCES

Conventional mercury vapor UV lamps are known to present high energy consumption. Research in the past few years has been focusing on lowering this energy charge. For this reason, cheaper sources of UV radiation have been tested, like LED or xenon lamps. Some other experiments have focused in the usage of the sun light as source of radiation, where the reduction of the radiation cost and energy consumption is remarkable.

Light emitting diodes (LED) are semiconductor p-n junction devices which emit light at a very narrow wavelength spectrum. LEDs present a great number of advantages over the conventional UV lamps, apart from their lower energy consumption. They emit only the desirable wavelength and are free of toxic compounds. They transmit a higher amount of energy into light, wasting very little energy as heat (Hu 2006). They are also hard to break and very compact in size. These features make them promising as possible substitutes of the mercury lamps, although very little research has been done in the field. Vilhunen and Sillanpää (2009) worked in the degradation of phenol (100 mg/l) with hydrogen peroxide activated by different wavelengths emitted by LED lamps. The results showed a slower reaction compared to the conventional laboratory scale UV lamps, but a considerable reduction of the energy consumption.

Oxidation of phenol by UV LEDs in presence of hydrogen peroxide was studied by Vilhunen et al. (2009). Photocleaving of hydrogen peroxide via UV LED photolysis resulted in generation of hydroxyl radicals. Phenol emoval was insignificant in the absence of hydrogen peroxide. The degradation of phenol was the most pronounced at 255 nm and hydrogen peroxide:phenol − molar ratio of 50.

The use of solar light as source of radiation for activating the hydroxyl radicals is not a new concept. Sun has been, by far, the most used source of energy since ancient times. Several researches have proved the usage of sunlight to be a good activating agent for the Fenton reaction. Delfín Pazos et al. (2009) successfully degraded EDTA, Cr(III)-EDTA and Cu(II)-EDTA by exposing the Fenton reaction to solar light in Cuernavaca (Mexico). Lucas and Peres (2006) compared the reaction rate of the degradation of *Reactive Black 5* combining Fenton treatment with UV radiation and solar light. They concluded that although UV lamp leads to a faster and higher extent of degradation, results of solar light show over 90% of degradation and thus, is highly recommended for reasons of both price and ecology. Sichel *et al.* (2008) proved the solar light to be a good enhancement of peroxide reaction in water disinfection even under low solar radiation.

8.6 PRACTICAL ISSUES OF UV/H_2O_2 TREATMENT

Practically any organic contaminant that is reactive with the hydroxyl radical can potentially be treated. A wide variety of organic contaminants are susceptible to destruction by UV/oxidation, including petroleum hydrocarbons; chlorinated hydrocarbons used as industrial solvents and cleaners; and ordnance compounds such as TNT, RDX, and HMX. In many cases, chlorinated hydrocarbons that are resistant to biodegradation may be effectively treated by UV/oxidation. Typically, easily oxidized organic compounds, such as those with double bonds (e.g., TCE, PCE, and vinyl chloride), as well as simple aromatic compounds (e.g., toluene, benzene, xylene, and phenol), are rapidly destroyed in UV/oxidation processes.

Limitations of UV/oxidation include:

- The aqueous stream being treated must provide for good transmission of UV light (high turbidity causes interference). This factor can be more critical for UV/H_2O_2 than UV/O_3 (Turbidity does not affect direct chemical oxidation of the contaminant by H_2O_2 or O_3).
- Free radical scavengers can inhibit contaminant destruction efficiency. Excessive dosages of chemical oxidizers may act as a scavenger.
- The aqueous stream to be treated by UV/oxidation should be relatively free of heavy metal ions (less than 10 mg/L) and insoluble oil or grease to minimize the potential for fouling of the quartz sleeves.
- When UV/O_3 is used on volatile organics such as TCA, the contaminants may be volatilized (e.g., "stripped") rather than destroyed. They would then have to be removed from the off-gas by activated carbon adsorption or catalytic oxidation.
- Pretreatment of the aqueous stream may be required to minimize ongoing cleaning and maintenance of UV reactor and quartz sleeves.
- Handling and storage of oxidizers require special safety precautions.

UV irradiation for Micropollutant removal from aqueous solutions 311

Technologies to destroy, detoxify, or recycle hazardous waste materials, and reduce mobility or volume of hazardous wastes is being demonstrated and evaluated by the United States Environmental Protection Agency (EPA). The Superfund Innovative Technology Evaluation (SITE) Program was created in 1986 to provide information such as reliable performance and cost data on alternative and innovative technologies.

A UV/oxidation technology developed by Ultrox International was evaluated in treating ground water contaminated with volatile organic compounds (VOCs) in the field at the Lorenz Barrel and Drum site in San Jose, California under the SITE program. The site was used primarily for drum recycling operations from about 1947 to 1987. The drums contained residual organic wastes, solvents, acids, metal oxides and oils (Lewis *et al.*, 1990).

The Ultrox system consists of UV/oxidation reactor module, an air compressor/ozone generator module, a hydrogen peroxide feed system, and a catalytic ozone decomposition unit. During the operation, contaminated water first comes in contact with hydrogen peroxide as it flows trough the influent line to the reactor. The water then comes in contact with the UV radiation and ozone as it flows through the reactor at a specified rate to achieve the desired hydraulic retention time.

Removal efficiencies for trichloroethylene and total VOCs were as high as 99% and 90%. A few compounds were more difficult to treat (1,1,-dichloroethane 65% and 1,1,1-trichloroethane 85% reduction). Most VOCs were removed through chemical oxidation but, for example, for 1,1,1-trichloroethane and 1,1-dichloroethane, stripping contributed also toward removal. The treated ground water met the applicable discharge standards for discharge into a local waterway at the 95% confidence level (Lewis *et al.*, 1990 and EPA/540/5-89/012).

A full scale UV/H_2O_2 oxidation system that had been designed for 60 to 80 litres per minute flow rate; 200 ppm concentration of H_2O_2 and 180 kW UV power was tested by 13 trial runs. The unit had been in operation for one year. Several parameters were varied: residence time, hydrogen peroxide dose, number of UV bulbs operating and the pH. The parameters were then related to destruction and volatization of the VOCs. The system had been tested previously in a bench-scale for acetone, benzene, chlorobenzene, ethylbenzene, tetrahydrofuran, toluene and total xylenes. At that time the results had been then excellent, with 99.99% removal even for a worst-case scenario. During the test period in the field, the removal of VOCs varied considerably. The most important factor was the precipitation of many chemicals and minerals on the quartz tubes, which deteriorated dramatically the purification efficiencies (Nyer *et al.*, 1991).

A full-scale treatment system was studied for removal of VOCs from contaminated ground water by air stripping towers, which were connected to a $UV/H_2O_2/O_3$ -reactor in order to remove the VOCs prior to release of the air stripper emissions to the atmosphere. Different combinations of UV radiation and addition of ozone and hydrogen peroxide were studied. Of the various AOP combinations the UV radiation alone was the most effective AOP and the most important operation parameter was the dose of UV radiation. When the UV radiation increased from 0 to 12000 watts-s/l the removal of carbon tetrachloride also increased. For tri- and tetrachloroethylene (TCE and PCE), removal increased sharply to 90% with UV dose at UV dose of 1600 watts-s/l and then gradually to 100% at 12000 watts-s/l. The use of ozone or hydrogen peroxide did not improve significantly the VOC removal. (AWWA Research Foundation, 1989)

The commercial scale UV/oxidation systems available for contaminated water treatment include Calgon Carbon Corboration Peroxpure^{TM} and Rayox$^®$ UV/H_2O_2 systems, Magnum Water Technology, Inc. CAV-OX$^®$ UV/H_2O_2 system, WEDECO UV/H_2O_2 and UV/O_3 systems and U.S. Filter/Zimpro, Inc $UV/H_2O_2/O_3$ system (former known as Ultrox).

Calgon peroxy-pure^{TM} and Rayox$^®$ system use 15 kW medium-pressure mercury arc lamps to generate UV-radiation. A typical unit contains portable oxidation units (typically six units in series equipped with a 15 kW lamp each), an H_2O_2 feed module and acid and base feed module, if needed. In a typical application of the Calgon system, contaminated water is dosed with H_2O_2 before the water enters the first reactor. If the treated water contains high concentration of carbonates and bicarbonates, which are radical scavengers, addition of acid is used to lower the pH and shift the carbonic acid-bicarbonate-carbonate equilibrium to carbonic acid. The base is used for neutralization of the treated water if required.

A typical CAV-OX$^®$ system consists of a portable, truck- or skid-mounted module with the following components: a cavitation chamber, a H_2O_2 feed unit, and UV reactors. Depending on the application different kind of UV sources is used. The low energy CAV-OX$^®$ I process uses one UV reactor with six 60 W low-pressure lamps and the reactor is operated at 360 W. The CAV-OX$^®$ II uses two UV reactors, each with high-pressure UV lamp operated at 2.5 or 5 kW. In a CAV-OX system, contaminated water is first pumped in cavitation chamber. The H_2O_2 is added between cavitation chamber and UV reactor or sometimes before cavitation unit.

The U.S. Filter $UV/H_2O_2/O_3$ uses UV radiation, ozone and H_2O_2 to oxidise contaminants in water. The major components are the UV/oxidation reactor, ozone generator; H_2O_2 feed tank and the catalytic ozone decomposition unit. The oxidation unit is divided by vertical baffles into chambers and contains low-pressure mercury vapour lamps (65 W each) in quartz sleeves. The contaminated water comes first contact with H_2O_2 before it is lead into the reactor where it is exposed to UV irradiation and ozone.

The use of combination of different treatment technologies such as UV/oxidation, granular activated carbon filtration (GAC), air stripping, chemical precipitation and biological treatment has proven to be cost-effective in many cases. For instance Calgon Carbon Corporations's Rayox® has coupled UV/oxidation with GAC for the treatment of ground water contaminated with DCE, TCE, PCE and DCM. The role of GAC was to remove easily absorbable background organics. UV/oxidation of ground water contaminated by PCE, TCA, DCM and VC combined to air stripping has reduced the organic load of air stripper and the overall system size and costs. Chemical precipitation has been used to remove the bulk loading of COD and the UV/oxidation has been as a polishing step to destroy the organics to low level. UV/oxiadation has been combined also to membrane technologies. (Lem, 2002)

8.7 COST ESTIMATION & PERFORMANCE

EE/O stands for Electrical Energy consumption per Order of magnitude reduction in contaminant concentration in 1 m^3 of treated water. EE/O gives the number of kilowatt-hours of electricity necessary to reduce concentration of a contaminant in 1 m^3 by one order of magnitude. It allows for easy and accurate scale-up to a full-scale design and costs. EE/O combines light intensity, residence time and percent destruction into a single measurement. The total energy requirement depends on the initial concentration of contaminant and the required purification efficiency. Once the required UV dose is known, the electrical operating cost can be calculated using the equation:

$$\text{Electrical cost (Euro/m}^3) = \text{UV Dose (kWh/m}^3) * \text{Power cost (Euro/kWh)}$$

Another parameter/variable in the operation costs is the consumption of hydrogen peroxide. The optimal ratio of H_2O_2/contaminant depends on the absorptivity of the contaminant(s) and the matrix. The cost of hydrogen peroxide can vary considerably. For calculation purposes a delivered price of 100% hydrogen peroxide can be used. Lamp replacement costs range from 10 to 20% of the electrical costs. (The AOT Handbook, 1997)

According to AOT Handbook the following rules of thumb can be used when performing a cost comparison analysis:

- Unit operating costs for UV/oxidation treatment increase much more slowly with increasing influent concentrations than activated carbon.

- UV/Oxidation treatment costs are almost always less for contaminants that are poor carbon absorber regardless of concentration.
- For average carbon absorbers, the most economical technology must be chosen on a case-by-case basis.
- UV/Oxidation capital costs are typically two to three times that of activated carbon.
- For concentrations below 10 ppm, activated carbon is the most cost effective treatment for good absorbers.

Hirvonen et al. (1998) studied the treatment of PCE contaminated groundwater. The initial concentration was 200 µg/l and the target concentration was 2 µg/l. A hydrogen peroxide dose of 27 mg/l and electrical energy of 0.5 kWh/order/m^3 was determined experimentally. Costs evaluation of UV/H$_2$O$_2$ method was compared with activated carbon adsorption when flow rate was 50 m^3/d and purification period is 360 d/a yielding a total of 18 000 m^3 in annual level (Table 8.2.). The Energy cost was $ 0.07/kWh and hydrogen peroxide cost $0.87/kg as 100% H$_2O_2$. The capital cost of UV/ H$_2$O reactor was calculated as 85 000 $. The cost includes purchase of UV-reactor, sedimentation and equalization tanks, a shelter for equipments and remotely monitored control system.

Table 8.2 Cost estimation for UV/H$_2$O$_2$ vs. activated carbon treatment of TCE contaminated aquifer. (Hirvonen et al., 1998)

Cost category	UV/H$_2$O$_2$	Activated Carbon
Equipment		
Investment × 0.149	12600	8600
Operation and maintenance		
Maintenance (0.02 × Inv)	1700	1200
Consumables		
AC + disposal ($ 0.29/m^3)		5200
H$_2$O$_2$	400	
UV lamps (3/year)	1500	
Utilities		
Electricity	3100	
Labour	6400	6400
Analytical service	10000	10000
Total annual cost	35700	31400
Cost $/m^3	2.00	1.70

Esplugas *et al.* (2002) compared in the laboratory the costs of different kind of UV based AOTs. The cost per kg of degraded phenol was calculated for the 50% (half-life time) and 75% of the initial phenol. The cost for UV treatment was 172.2 and 293.1 $/kg, for UV/$H_2O_2$, 13.1 and 28.7 $/kg and for UV/$H_2O_2$/ O_3 7.12 and 9.51 $/kg, respectively.

Under some conditions, pretreatment may be required. Typical forms of pretreatment that may be used include removal of suspended solids, free phase oil and grease and iron concentration. It is always recommended that a design test or treatability study be conducted for each application. It allows to characterize the water and to optimize treatment performance.

The effect of water matrix, such as other competing organics and alkalinity (bicarbonate ion), and the effect of the initial concentration and the degradation intermediates are useful to know. If we are dealing with known organic chemicals and there are data available on the efficiency of chemical oxidation, studies need be conducted on the toxicity or biodegradability of the oxidation by-products.

Most of the successful applications are related to the treatment of ground water contaminated by VOCs. The most common constituent has been TCE, PCE. Also benzene, chlorobenzene, di- and tri chloro acetic acids, and vinyl chloride have been treated with efficiencies that reach accepted levels (usually MCLs). The cost of treatment depends on the contaminant concentrations: 0.1 ppm (0.13–0.65 $/m^3), 1 ppm (0.45–1.0 $/m^3), 10 ppm (0.5–1.5 $/m^3). (AOT Handbook, 1997).

The main problem of performance has been scaling of the lamps and formation of precipitations. The quartz sleeves of the lamps are equipped with wipers that periodically clean the tubes. Solids, such as precipitated iron and manganese, can also be filtrated.

The highest costs of the purification of contaminated ground water are related to the pumping of the water into the soil surface. A pump and treat technique, such as UV/H_2O_2 treatment, is less attractive solution in a case where high volumes of water with quite low contamination level should to be treated.

The economical feasibility of total treatment of water stream by chemical oxidation has been set questionable. There are lot of research going on related to the integration of chemical oxidation to other treatment techniques such as chemical coagulation, membrane techniques, air stripping, activated carbon filtration and biological processes. This kind of approach might solve the limitation of UV/oxidation processes such as high turbidity, solid particles, heavy metals, high COD or BOD matrix of the treated water stream.

The laboratory scale experiments with synthetic single compound solution are numerous but there is still a lack of multicomponent studies. Just the disappearance of one model compound had been studied. In the real environmental engineering applications the problem is however more complicated. First, multiple organic contaminants are routinely expected from materials such as contaminated ground waters and industrial wastewaters. Second, in the course of mineralisation of any organic contaminant, oxidation will logically involve a series of intermediates on the way to CO_2. Such intermediates make the conversion process multicomponent, even if only a single contaminant existed in the feed.

Usually there has not been any attempt to separate the role of different mechanism in the overall disappearance of chemical or the role of competing species. More information is needed of the interplay of various degradation mechanisms, direct photolysis and radical reactions, in the case of UV/H_2O_2 treatment. Because of the high demand of oxidants and energy, the optimisation of the conditions of a chemical oxidation is essential for economical feasibility.

8.8 REFERENCES

Adams, D. C. and Kuzhikannil, J. J. (2000) Effects of UV/H_2O_2 preoxidation on the aerobic biodegradability of quaternary amine surfactants, *Wat. Res.* **34**, 668–672.

Alibegic, D, Tsuneda, S. and Hirata, A. (2001) Kinetics of tetrachloethyle (PCE) gas degradation and byproducts formation during UV/H_2O_2 treatment in UV-bubble column reactor, *Chem. Eng. Sci.* **56**, 6195–6203.

Andreozzi, R., Caprio, V., Marotta, R. and Vogna, D. (2003) Paracetamol oxidation from aqueous solutions by means of ozonation and H 2O2/UV system, *Water Res.* **37**, 993–1004.

AOT Handbook (1997) Chemiron Carbon Oxidation Technologies (Germany).

Baxendale, J. H. and Willson J. A. (1957) Photolysis of hydrogen peroxide at high light intensities. *Trans. Faraday Soc.* **53**, 344–356.

Belgiorno, V., Rizzo, L., Fatta, D., Della Rocca, C., Lofrano, G., Nikolaou, A., Naddeo, V. and Meric, S. (2007) Review on endocrine disrupting-emerging compounds in urban wastewater: occurrence and removal by photocatalysis and ultrasonic irradiation for wastewater reuse, *Desalination* **215**, 166–176

Beltran, F. J., Encinar., J. M. and Alonso, M. A. (1998) Nitroaromatic hydrocarbon ozonation in water. 2. Combinated ozonation with hydrogen peroxide or UV radiation. *Ind. Eng. Chem. Res.* **37**, 32–40

Beltran, F. J., Rivas. J., Alvarez, P. M. and Alonso, M. A. (1999) A kinetic model for advanced oxidation of artomic hydrocarbons in water: application to phenantrene and nitrobenzene. *Ind. Eng. Chem. Res.* **38**, 4189–4199.

Beltran-Heredia, J., Benitez, F. J., Beltran-Heredia, J., Acero, J. L. and Rubio, F. J. (2001) Oxidation of several chorophenolic derivatives by UV irradiation and hydroxyl. *Water Res.* **35**(4), 1077–85.

Buxton, G. V., Greenstock, C. L., Helman, W. P. and Ross, A. B. (1988) Critical review of data constants for reactions of hydrated electrons, hydrogen atoms and hydroxyl radicals (.OH/.O$^-$) in aqueous solutions. *J. Phys. Chem.* **17**, 513–886.

Canonica, S., Meunier, L. and von Gunten, U. (2008) Phototransformation of selected pharmaceuticals duringUV treatment of drinking water, *Wat. Res.* **42**, 121–128.

Calvert, J. G. and Pitts Jr., J. N. (1966) Photochemistry, John Wiley & Sons, Inc., New York, USA, 899 p.

Chen, P.-J., Rosenfeldt, E., Kullman, S., Hinton, D. and Linden, K. (2007) Biological assessments of a mixture of endocrine disruptorsat environmentally relevant concentrations inwater following UV/H_2O_2 oxidation, *Sci. Tot. Environ.* **376**, 18–26.

Chin, A. and Berube, P. R. (2005) Removal of disinfection by-product precursors with ozone-UV advanced oxidation process, *Water Res.* **10**, 2136–2144.

Chang, P. and Young, T. (2000) Kinetics of methyl-tert-butyl ether degradation and by-product formation during UV/hydrogen peroxide water treatment. *Water Res.* **34**(8), 2233–2244.

Christensen, H. S., Sehested, H. and Corfitzan, H. (1982) Reactions of hydroxyl radicals with hydrogen peroxide at ambient and elevated temperatures. *J. Phys. Chem.* **86**, 15–68.

Delfín Pazos, A., Pineda Arellano, C. A. and Silva Martínez, S. (2009). Degradación del ADTE y los complejos Cu(II)-AEDT y Cr(III)-AEDT mediante los procesos Fenton y foto-Fenton asistido por radiación solar en soluciones acuosas. *Rev. Int. Contam. Ambient.* **25**(4), 239–246.

Adams, C., Scanlan, P. and Secrist, N. (1994) Oxidation and Biodegradability Enhancement of 1,4-Dioxane Using Hydrogen Peroxide and Ozone *Environ. Sci. Technol.* **28**(11), 1812–1818.

EC, Directive of the European Parliament and of the Council 2000/60/EC establishing a framework for community action in the field of water policy, Official Journal, 2000, C513, 23/10/2000.

Eilbeck, W. J. and Mattock, G. (1988) Chemical Oxidation, in Chemical Processes in Waste Water Treatment, Ellis Horwood Limited, Chichester, UK, 331 p.

Elkanzi, E. M. and Kheng, C. B. H_2O_2/UV degradation kinetics of isoprene in aqueous solution, *J. Haz.Mater.* **73**(1), 55–62.

Esplugas, S., Gimenez, J., Contreras, S., Pascual, E. and Rodrigies, M. (2002) Comparison of different advanced oxidation processes for phenol degradation.*Water Res.* **36**, 1034–1024.

Fukui, S., Fucui, Ogawa, S., Motozuka, T. and Hanasaki, Y. (1991) Removal of 3-chloro-4-(dichloromethyl)-5-hydroxy-2(5)-furanone (MX) in water by oxidative, reductive, thermal and photochemical treatments. *Chemosphere* **23**, 761–775. http://www.frtr.gov/matrix2/section4/4-45.html

Ghaly, M. Y., Härtel, G., Mayer, R. and Haseneder, R. (2001) Photochemical oxidation of p-chlorophenol by UV/H_2O_2 and photo-Fenton processes. A comparative study. *Waste Managem.*, **21**, 41–47.

Ghaly, M. Y., Härdel, G., Mayer, R. and Haseneder, R. (2001) Aromatic compounds degradation in water by using ozone and AOPs. A comparative study. O-nitrotoluene as a model substrate. *Ozone. Sci. Eng.* **23**, 127–138.

Glaze, W. H., Kang, J. W. and Chapin, D. H. (1987) The chemistry of processes involving ozone, hydrogen peroxide and ultraviolet radiation. *Ozone Sci & Eng.* **9**, 335–342.

Glaze, W. H., Lay, Y. and Kang, J. W. (1995) Advanced oxidation processes . A kinetic model for the oxidation of 1,2-dibromo-3-chloropropane in water by combination of hydrogen peroxide and UV radiation. *Ind. Eng. Chem. Res.* **34**, 2314–2323.

Glaze, W. H. and Kang, J. W. (1989) Advanced oxidation process. Test of a kinetic model for the oxidation of organic compounds with ozone and hydrogen peroxide in a semi-batch reactor. *Ind. Eng. Chem. Res* **28**, 1580–1587.

Glaze, W. H. and Lay, Y. (1989c) Oxidation of 1,2-dibromo-3-chloropropane (DBCP) using advanced oxidation processes. In Ozone in Water Treatment. New York, USA, IOA, 688–708.

.Guittonneau, S. (1989 or 90) Contribution a l'Etude de la Photooxidation de Quelques Micropolluants Organochlores en Solution Aqueuse en Presence de Peroxide d'Hydrogen- Comparaison des Systemes Oxydants: H_2O_2/UV, O_3/UV et O_3/H_2O_2. These de Docteur. Universite de Poitiers, France.

Guittonneau, S., de Laat, J., Dore, M., Duguet, J. P. and Bonnel, C. (1988) Comparative study of the photodegradation of aromatic compounds in water by UV and $H_2O_2/$UV. *Environ. Technol. Lett.* **9**, 1115–1128.

Heberer, T. (2002) Tracking persistant pharmaceutical residues from municipal sewage to drinking water. *J.Hydrology*, **266**, 175–189.

Hirvonen, A., Tuhkanen, T., Ettala, M. and Kalliokoski, P. (1998) Evaluation of a field – scale UV/H_2O_2 –oxidation system foer purification of groundwater contaminated with PCE. *Environ. Technol.*, **19**, 821–828.

Ho, P. C. (1986) Photooxidation of 2,4-dinitrotoluene in aqueous solution in the presence of hydrogen peroxide. *Environ. Sci. Technol.* **20**, 260–267.

Hu, X., Deng, J., Zhang, J., Lunev, A., Bilenko, Y., Katona, T., Shur, M. Gaska, R., Shatalov, M. and Khan, A. (2006) Deep ultraviolet light-emitting diodes, *Phys. Stat. Sol.* **203**, 1815–1818.

Ince, N.H. (1999) "Chritical" effect of hydrogen peroxide in photochemical dye degradation. *Water Res.* **33**, 1080–1084.

Ishikawa, S., Uchimura, Y., Baba, K., Eguchi, Y. and Kido, K. (1992) Photochemical behavior of organic phosphate esters in aqueous solutions irradiated with a mercury lamp. *Bull. Environ. Contam. Toxicol.* **49**, 368–374.

Kawaguchi, H. (1992) Photooxidation of phenols in aqueous solution in the presence of hydrogen peroxide. *Chemosphere* **24**, 1707–1712.

Kim, I., Yamashita, N. and Tanaka, H. (2009) Performance of UV and UV/H_2O_2 processes for the removal of pharmaceuticals detected in secondary effluent of a sewagetreatment plant in Japan, *J. Hazard. Mater.* **166** ,1134–1140.

Klöpffer, W. (1992) Photochemical degradation of pesticides and other chemicals in the environment: A critical assessment of the state of the art. *Sci Total Environ.*, **123/124**, 145–159.

Lay, Y. S. (1989) Oxidation of 1,2-dibromo-3-chloropropane in ground water using advanced oxidation processes. Ph.D.Thesis. University of California at Los Angeles, USA.

Leifer, A. (1988) The kinetics of environmental photochemistry. Theory and practice. ACS Professional Reference Book, York, PA, 304 p.

Lewis, N., Tapudurti, K., Welshans, G. and Foster, R. (1990) A field demonstration of the UV/oxidation technology to treat ground water contaminated with VOCs. *J. Air Waste Manage. Assoc.* **40**, 540–547.

Lem, W. (2002) Combination of UV/oxidation with other treatment technologies for the remediation of contaminated waters, in the 8th International Conference on

Advanced Oxidation Technologies for Air and Water Remediation, Nov. 17–21, 2002, Toronto, Canada.

Liao, C. H. and Gurol, M. D. (1995) Chemical oxidation by photolytic decomposion of hydrogen peroxide. *Environ. Sci. Technol.* **29**, 3007–3014.

Lopez, A., Bozzi, A., Mascolo, G. and Kiwi, J. (2003) Kinetic investigation on UV and UV/H_2O_2 degradations of pharmaceutical intermediates in aqueous solution, *J. Photochem. Photobiol A- Chem.* **156**, 121–126.

Lucas, M. and Peres, J. 2007 Degradation of *Reactive Black 5* by Fenton/UV-C and ferrioxalate/H_2O_2/solar light processes. Dyes and Pigments **74**,622–629.

Milano, J. C., Yassin-Hussan, S. and Vernet, J. L. (1992) Photochemical degradation of 4-bromodiphenylether: Influence of hydrogen peroxide. *Chemosphere* **25**, 353–360.

Mill, T., Mabey, W. R., Lan, B. Y. and Baraze, A. (1981) Photolysis of polycyclic aromatic hydrocarbons in water. *Chemosphere* **10**, 1281–1290.

Miller, E. M., Singer, G. M., Rosen, J. D. and Bartha, R. (1988) Sequential degradation of chlorophenols by photolytic and microbial treatment. *Environ. Sci. Technol.* **22**, 1215–1219.

Moza, P. N., Fytianos, K., Samanidou, V. and Korte, F. (1988) Photodecomposition of chlorophenols in aqueous medium in presence of hydrogen peroxide. *Bull. Environ. Contam. Toxicol.* **41**, 678–682.

Peterson, D., Watson, D. and Winterlin, W. (1990) Destruction of pesticides and their formulations in water using short wavelength UV light. *Bull. Environ. Contam. Toxicol.* **44**, 744–750.

Peterson, D., Watson, D. and Winterlin, W. (1988) The destruction ground water threatening pesticides using high intensity UV light. *Environ. Sci. Health* **B23**(6), 587–603.

Pinto, D. and Rickabauhg, J. Photocatalytic degradation of chlorinated hydrocarbons. *Hazard Ind Wastes* **23**, 368–373.

Rosario-Ortiz, F. L., Wert, E. C. and Snyder, S. A. (2010) Evaluation of UV/H_2O_2 treatment for the oxidationof pharmaceuticals in wastewater doi:10.1016/j.watres.2009.10.031.

Rosenfeldt, E. Chen, P.-J., Kullman, S. and Linden, K. (2007) Destruction of estrogenic activity in water using UV advanced oxidation, *Sci. Tot. Environ.* **377**,105–113.

Sanlaville, Y., Guittonneau, S., Mansour, M., Feicht, E. A. and Meallier, P. Kettrup (1996) *Chemosphere*, **33**(2), 353–362.

Shulte, P., Volkmer, M. and Kuhn, F. (1991) Aktiviertes Wasserstoffperoxid zur Beseitigung von Schadstoffen im Wasser (H_2O_2/UV). *Wasser, Luft und Boden* **35**, 55–58.

Sichel, C. Fernandez-Ibañez, P, de Cara, M. and Tello, J. (2009) Lethal synergy of solar UV-radiation and H_2O_2 on wild *fusarium solani* spores in distilled and natural well water. *Water Res.* **43**, 1841–1850.

Stumm, W. and Morgan, J. J. (1981) Aquatic Chemistry, 2nd ed. John Wiley & Sons, New York, USA, 780 p.

Sundstrom, D. W., Weir, B. A. and Klei, H. E. (1986) Destruction of aromatic pollutants by UV light catalyzed oxidation with hydrogen peroxide. *Environ. Progress* **8**, 6–11.

Ternes, T., Joss, A. and Sigrist, H. (2003) Scrutinizing Pharmaceuticals and personal care Products in wastewater treatment. *Environ. Sci. Technol.* **38**(20), 392A-399A.

Trapido, M. Veressina, Y. and Kallas, J. (2001) Degradation of aqueous nitrophenols by ozone combined with UV-radiation and hydrogen peroxide. *Ozone Sc.Eng.* **23**, 333–342.
Tuhkanen, T. A., doctoral thesis.
Tuhkanen, T. A. and Beltran, F. J. (1995) Intermediates of the oxidation of naphthalene in water with the combination of hydrogen peroxide and UV radiation. *Chemosphere* **30**, 1463–1475.
USEPA (1998) *Handbook on Advanced Photochemical Oxidation Processes.* EPA/625/R-98/004, USEPA, December 1998.
Vilhunen, S. and Sillanpää, M. (2009) Ultraviolet light emiting diodes and hydrogen peroxide in the degradation of aquesous phenol. *J. Hazard. Mater.* **161** 1530–1534.
Vilhunen, S., Rokhina, E., Virkutyte, J. Evaluation of UV LED performance in photochemical oxidation of phenol in presence of H_2O_2. *J. Environ. Eng.* DOI: 10:1061/ASCE)EE-1943-7870.0000152
Wanga, G.-S., Chena, H.-W., Kangb, F.-S. and Catalyzed (1990) UV oxidation of organic pollutants in biologically treated wastewater effluents. *Sci. Tot. Environ.* **277**, 87–94.
Watts, R. J., Udell, M. D. and Leung, S. W. (1991) Treatment of contaminated soils using catalyzed hydrogen peroxide. In proceedings of the First International Symposium Chemical Oxidation Technologies for the Nineties. 37–50. Eds: Eckenfelder, W. W., Bowers, A. R. and Roth, J. A., Vanderbilt University, Nashville, Tennessee, USA.
Watts, R. J., Udell, M. D. and Laung, S. W. (1990) Treatment of pentachloro-contaminated soil using Fenton's reagent. *Hazardous Waste & Hazardous Material*, **7**, 335–345.
Wert, E. C., Rosario-Ortiz, F. L. and Snyder, S. A. (2009b) Using ultraviolet absorbance and color to assess pharmaceutical oxidation during ozonation of wastewater. *Environ. Sci. Technol.* **43**, 4858–4863.
Weir, B. A., Sundstrom, D. W. and Klie, H. E. (1987) Destruction of benzene by ultraviolet light-catalyzed oxidation with hydrogen peroxide. *Hazardous Waste & Hazardous Materials* **4**, 167–176.
Weir, B. A and Sunstrom, D. W. (1993) Destruction of trichloroethylene by UV light-catalyzed oxidation with hydrogen peroxide. *Chemosphere* **27**, 1279–1291.
Wong, A. S. and Crosby, D. G. (1981) Photodecomposition of pentachlorophenol in water. *J. Agric. Food Chem.* **29**, 125–130.
Yasuhara, A., Otsuki, A. and Fuwa, K. (1977) Photodecomposion of odorous phenols in water. *Chemosphere*, 659–664.

Chapter 9
Hybrid Advanced Oxidation techniques based on cavitation for Micropollutants degradation

Jurate Virkutyte and Ekaterina V. Rokhina

9.1 INTRODUCTION

Generally, chemical reactions (including chemical oxidation as a treatment method) require energy to proceed. The properties of specific energy sources determine the course of the degradation pathway. Ultrasonic irradiation differs from traditional energy source (such as heat, light, or ionizing radiation) in duration, pressure, and energy expenditure per molecule. Ultrasound (US) is unique means of interacting energy and matter (Suslick, 1990).

Since 1990s ultrasound has been widely used in water and wastewater treatment processes as an emerging advanced oxidation processes (AOP) technology, applicable for a wide range of contaminants with various initial concentrations. Ultrasound-based treatment promotes main aims of green chemistry, such as the

©2010 IWA Publishing. *Treatment of Micropollutants in Water and Wastewater.* Edited by Jurate Virkutyte, Veeriah Jegatheesan and Rajender S. Varma. ISBN: 9781843393160. Published by IWA Publishing, London, UK.

replacement or minimization of hazardous chemicals and development of the reaction conditions to increase the selectivity of such reaction (Mason, 2007). Moreover, ultrasound-assisted degradation of contaminants has significant advantages over traditional AOP technologies namely ozonation, ultraviolet irradiation (UV), high voltage corona, etc. such as safety, cleanliness and energy conservation without causing secondary pollution (Chowdhury and Viraraghavan, 2009).

9.2 THEORY OF ULTRASOUND

9.2.1 Cavitation phenomena

Ultrasound are waves at frequencies above those within the hearing range of the average person, i.e., at frequencies above 15 kHz, which are associated with acoustic wavelengths of 10 to 0.01 cm. The chemical effects of ultrasound do not derive from direct interactions with molecular species but are principally based on cavitation. Cavitation is the phenomena of the formation, growth and a subsequent collapse of microbubbles or cavities occurring in extremely small interval of time (milliseconds), releasing large magnitudes of energy (Gogate and Pandit, 2004a).

The release of energy takes place over a very small pocket, however cavitation events occur at multiple locations in the reactor simultaneously and therefore have spectacular overall effects. Cavitation can be classified into four types according to the mode of generation: acoustic, hydrodynamic, optic and particle cavitation. However, only acoustic and hydrodynamic cavitation has been found to be efficient in the treatment processes, whereas optic and particle cavitation fail to induce chemical change in the bulk solution.

9.2.2 The general hypothesis in sonochemical processing

There are several popular theories that explain the ultrasound-induced events: a) hot-spot, b) electrical, c) plasma discharge, and d) supercritical theory. These have led to several modes of reactivity being proposed: pyrolytic decomposition, hydroxyl radical oxidation, plasma chemistry, and supercritical water oxidation (Adewuyi, 2001).

The hot-spot theory suggests the generation of so-called 'hot-spots' with a pressure of thousands of atmosphere (up to 1000 atm) and a temperature of about 5000 K as a result of the violent collapse of the bubble formed due to the irradiation imposed onto the liquid. In addition, Margulis (1992) proposed the 'electrical' theory of local electrification of cavitation bubble to understand the nature of sonolumininescence and sonochemical reactions. The 'electrical' theory hypothesizes that the extreme conditions associated with fragmentative collapse are due to the intense electrical fields. Thus, electric phenomena in cavitational fields are the result of electric double layer formation in any liquid on the surface of the cavitation bubble (Margulis, 1992). During bubble formation and collapse, enormous electrical field gradients are generated that are sufficiently high to cause bond breakage and chemical activity. Furthermore, Lepoint and Mullie (1994) found the analogies between sonochemistry and corona chemistry and introduced plasma theory to explain cavitation. They argued that fragmentation process was due to an intense electrical field rather than a true implosion, indicating the formation of microplasmas inside the bubbles (Lepoint and Mullie, 1994). And finally, the supercritical theory suggests the existence of supercritical water (SCW) as an additional phase available for chemical reactions. This phase of water exists above the critical temperature, of 647 K and the critical pressure of 221 bar and has physical characteristics intermediate between those of a gas and a liquid. The physicochemical properties of water such as viscosity, ion product, density, and heat capacity change dramatically in the supercritical region, and these changes favor substantial increases in most chemical reaction rates (Hua and Hoffmann, 1997). Among these theories, the hot-spot theory is widely accepted in explaining sonochemical reactions in the environmental field in general and degradation of contaminants in aqueous phase in particular (Adewuyi, 2001).

9.2.3 Cavitation effects

According to the hot spot theory, the important effects of cavitation are the generation of conditions of very high temperatures and pressures (few hundred atmospheres pressure and few thousand Kelvin temperature) locally with overall ambient operating conditions, release of highly reactive free radicals and generation of turbulence and liquid circulation (acoustic streaming) enhancing the rates of transport processes (Gogate *et al.*, 2004). All these cavitation effects can improve the degradation of a target compound by two mechanisms: physical and chemical. Among the physical effects of cavitation are the high rates of micromixing, renewal of a solid reactant or catalyst surface by dissolving or pitting the inhibiting layers, and an enhancement of the mass transfer between the reactants. The chemical effects are a consequence of the violent bubble collapses. Thus, under the extreme conditions, oxidizing species are generated by the homolytic cleavage of molecules (gases and solvent). In aqueous media and in the presence of oxygen, radicals such as H·, HO·, HOO·, and ·O are produced according to the well know reactions (Adewuyi, 2005a; Adewuyi, 2005b):

[AQ2]

$$H_2O \xrightarrow{)))} \cdot OH + \cdot H \qquad (9.1)$$

$$\cdot OH + \cdot H \rightarrow H_2O \qquad (9.2)$$

$$2 \cdot OH \rightarrow H_2O + \tfrac{1}{2}O_2 \qquad (9.3)$$

$$2 \cdot OH \rightarrow H_2O_2 \qquad (9.4)$$

$$2 \cdot H \rightarrow H_2 \qquad (9.5)$$

$$\cdot OH + \cdot OH \rightarrow H_2O + \cdot O \qquad (9.6)$$

$$O_2 \rightarrow 2 \cdot O \qquad (9.7)$$

$$\cdot H + O_2 \rightarrow \cdot HO_2 \qquad (9.8)$$

$$2 \cdot HO_2 \rightarrow H_2O_2 + O_2 \qquad (9.9)$$

$$H_2O_2 + \cdot OH \rightarrow H_2O + \cdot HO_2 \qquad (9.10)$$

$$H_2O_2 + \cdot HO_2 \rightarrow H_2O + \cdot OH + O_2 \qquad (9.11)$$

Usually, the mechanism of the micropollutant degradation is governed by pyrolysis and/or free radical reactions (Rokhina et al., 2009). Under critical conditions caused by the cavitation, three well defined reaction zones exist: (1) the cavitation bubble, (2) the supercritical interface and (3) the bulk of the solution (Figure 9.1).

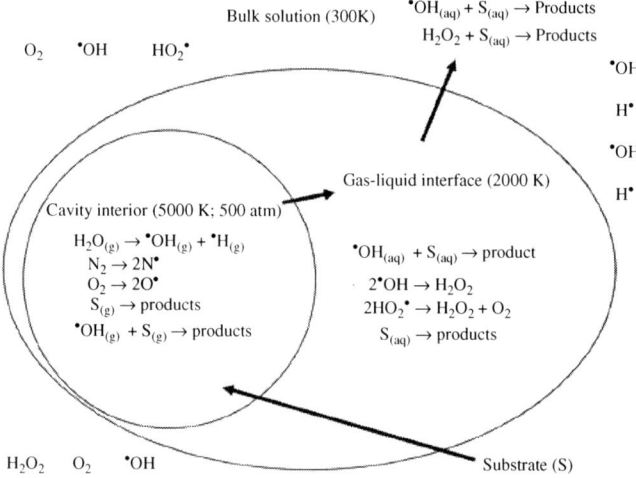

Figure 9.1 Three reaction zones in the cavitation process (Adapted from Adewuyi (2001))

The degradation in the first zone is mainly caused by pyrolytic radical attack that is responsible for the degradation in the third zone and the combined mechanism are responsible for the destruction of a target micropollutant in the cavitation bubble interface. The subsequent mechanism of micropollutant degradation during sonolysis can be described as a two steps process. During the first step, H_2O and O_2 are sonolysed inside the cavitation bubble to produce radicals. During the second step, HO· and HOO· radicals move to the liquid – bubble interface to react with the organic substrate or to recombine with each other to form H_2O_2.

9.2.4 Factors affecting the efficiency of sonochemical degradation

The operational conditions as well as the properties of the reaction system can greatly influence the intensity of cavitation, which directly affects the target contaminant removal rate. These conditions include irradiation frequency, acoustic power, reaction temperature, the presence of additives (e.g., dissolved gases, catalysts, etc.) and the nature of the treated pollutant. It is worthy to note that there is a strong interrelation between all the parameters, and only tuning of all variable parameters will lead to the optimal operational conditions and consequently to the highest removal rates. Each of these factors is described in more details below.

9.2.4.1 Ultrasonic frequency

Ultrasonic frequency is the parameter, responsible for the critical size of the cavitation bubbles and in turn, has a significant effect on the cavitation process. Lower frequency ultrasound produces more violent cavitation, leading to higher localized temperatures and pressures at the cavitation site. Despite less violent cavitation, higher frequencies may actually increase the number of free radicals in the system because there are more cavitational events (e.g., Petrier *et al.* (1996) reported that at 500 kHz, in the absence of a substrate, the yield of hydrogen peroxide generated is 6.2 time higher than obtained at 20 kHz) and thus more opportunities exist for the production of free radicals. In the past, most sonochemical reactions to degrade target contaminants were carried out at frequencies between 20 and 50 kHz (Petrier *et al.*, 1996). However, recent investigations showed that in water, oxidative processes induced by ultrasound (generation of and the reaction with ·OH radicals) are more efficient if performed at high ultrasonic frequencies (200–1000 kHz) in comparison to the reactions conducted at about 20 kHz (Kirpalani and McQuinn, 2006; Chand *et al.*, 2009).

Francony and Mason (1996) determined that optimal frequency highly depended on the physical and chemical properties of the target organic pollutant i.e., its nature, which caused the reaction to follow via radical or pyrolytic reaction pathway. Also, Drijvers *et al.* (1996) determined that the degradation reaction of trichloroethylene (TCE) was energetically more efficient at 520 kHz rather than at 20 kHz. Thus, to obtain the same reaction efficiency at 20 kHz in comparison to 500 kHz, a significant increase in the input electrical power is required, which translates into the excess of the heat dissipated in the reaction medium and may also lead to the deterioration of the ultrasonic horn (Drijvers *et al.*, 1996). Higher frequencies of operation are suitable for effective destruction of pollutants but only until a certain optimum value (Hua and Hoffmann, 1997).

9.2.4.2 Input electrical power

Prior to the input power selection, it should be noted that the actual power dissipated (P_{diss}) in the ultrasonic reactor is not the same as the ultrasonic output power quoted by the manufacturer (P_g), due to the losses in the ultrasonic energy during the transfer processes. One of the most common methods of measuring P_{diss} is based on calorimetry and assumes that all the power entering the process vessel is dissipated as heat (Lorimer and Mason, 1987).

In general, the ultrasound irradiation should not be automatically turned to a maximum power because a power threshold exists beyond which no further increase in pollutant degradation is observed or in some cases, even a decrease in degradation efficiencies may be observed. Thus, the extra power will lead to the significant loss of energy. Two operational parameters are crucial in order to optimize ultrasound-assisted degradation of pollutants: power density (the amount of power dissipated in 1 mL of the reaction mixture, $W\,mL^{-1}$) and power intensity (the amount of power dissipated per square cm of the emitter area, Wcm^{-2}). The intensity of irradiation can be varied either by changing the power input to the system or by changing the transmittance area of the ultrasonic transducers in the equipment. However, if at lower intensities the intensity is changed by changing the transmittance area of ultrasonic equipment, the same power dissipation is taking place over a larger area resulting in uniform dissipation and larger active area of cavitation resulting in higher degradation efficiencies (Gogate, 2008a).

9.2.4.3 Nature of the compound and the reaction pH

A physicochemical data of a target contaminant can be used to predict physical processes, such as sorption, volatilization and dissolution. The important properties to be considered are octanol-water partition coefficient (K_{ow}), aqueous solubility, acid dissociation constant (pK_a) and Henry's Law constant (H_c). Knowledge of chemical partitioning between the aqueous and solid phase can give the insight into pathways of the pollutant transport and transformation (Thompson and Doraiswamy, 1999). The effect of the pollutant nature on the degradation process can be explained by the theory of three different reaction zones formed during sonolysis. The nature of the contaminant determines the zone of the reaction and thus, the mechanism of its degradation. Molecules with relatively large Henry's law constants will be incinerated inside the cavitation bubble, while nonvolatile molecules with low Henry's law constants will be oxidised by the ·OH ejected from the bubble of cavitation (Petrier et al., 2010).

9.2.4.4 The reaction temperature

The effect of the reaction temperatures on the removal rate of target pollutants varies according to the ultrasound frequencies applied. For the low frequencies the reaction rate decreases linearly with an increase in temperature, whereas only a slight change in the degradation efficiency is observed when high ultrasound frequencies are applied. This difference can be explained by four important parameters affected by the temperature. Increasing temperature of the liquid will (1) decrease the energy of cavitation, (2) lower the threshold limit of cavitation, (3) reduce the quantity of the dissolved gas, and (4) increase the vapour pressure (Jiang et al., 2006). At lower frequency (e.g., 20 kHz), some of the bubbles lose their activity due to the increase in the possibility of coalescence among the increasing amount of bubbles. Additionally, there is more likely transient (vaporous) cavitation taking place, which induces a decrease in sound transmission lowering the ultrasonic effect of energy in the liquid. At higher frequency (e.g., 500 kHz), both, the degradation rate of the pollutant and the rate of H_2O_2 formation increase with an increase in solution temperature. Jiang et al. (2006) argued that at low frequency (20 kHz), due to the large number of formed cavitation bubbles, it was expected that an increase in temperature would lead to an increase in the possibility of coalescence among the bubbles, resulting in some of the bubbles losing their activity. Therefore, the contaminant removal rate would decrease with an increase in the solution temperature. However, the presence of the catalyst solid particles can drastically alter the situation (Rokhina et al., 2010).

9.2.4.5 The presence of additives

The additives introduced into the reaction system provide the deformities for the generation of the cavities and thus may intensify the degradation process.

The first group of additives contains inert solids or catalysts, which cause the homolytical dissociation of hydrogen peroxide, formed during the sonolysis. Also, the presence of solid particles can provide additional nuclei for the cavitation phenomena and hence the number of cavitation events occurring in the reactor is enhanced resulting in a subsequent enhancement in the cavitational activity and hence the net chemical effects (Gogate, 2008b).

The second group includes the dissolved salts that can also intensify the cavitational activity by altering the physicochemical properties of the cavitating medium as well as by localizing the reactant species at the cavity implosion site (Gogate and Pandit, 2004b). The presence of salts can increase the hydrophilicity, the surface tension and ionic strength of the aqueous phase and decrease the vapor pressure. For instance, the addition of CCl_4 remarkably enhances the removal rate of the target compound by the formation of reactive chlorine species also capable to degrade the organics (Merouani et al., 2010). Petrier et al. (2010) demonstrated the significant improvement in the degradation efficiency of micropollutant (e.g., BPA) (<0.1 L mol^{-1}) in the presence of bicarbonate ions. Such enhancement may be attributed to the formation of carbonate radical ($CO_3^{\cdot-}$), which is ejected from the cavitation bubble and migrated towards the bulk of the solution. The third group contains various gases. It is hypothesized that the presence of gases leads to the intensified radical recombination and the subsequent formation of new types of radicals available for the oxidation of the pollutants (Petrier et al., 2010). Mixture of sparged gases can improve the sonolytic degradation rates because the chemical reactivity at the cavitation site is determined by the temperature within the bubble during collapse as well as the nature of the oxidizing species produced (Hua et al., 2002). Therefore, high polytropic gas ratio (c) and low conductivity (j) favor higher collapse temperatures. In accordance, the values of c and j for argon, air and oxygen (c = 1.67, j = 177 × 10^{-4} Wm^{-1} K^{-1}; c = 1.40, j = 262 × 10^{-4} Wm^{-1} K^{-1} and c = 1.40, j = 267 × 10^{-4} Wm^{-1} K^{-1}, respectively) suggest that maximum collapse temperature is expected in the presence of argon, and lower/nearly equal temperatures in the presence of air and/or oxygen. However, the reactivity of generated radical species should also be taken into account (e.g., ·HO$_2$ reacts with aromatic compounds is 2–3 orders of magnitude lower than that of ·NO$_x$ (Gultekin and Ince, 2008).

In the presence of oxygen, other pathways different from ·OH recombination can also lead to the formation of hydrogen peroxide, whereas the presence of air influence the pH of the reaction. The presence of N$_2$ promotes the formation of acidic media (HNO$_3$ and HNO$_2$) according to the reactions:

$$N_2 + \cdot O \rightarrow NO + N \tag{9.12}$$

$$N + O_2 \rightarrow NO + O \quad (9.13)$$

$$NO + \cdot OH \rightarrow HNO_2 \quad (9.14)$$

$$HNO_2 \rightarrow H^+ + NO_2^- \quad (9.15)$$

$$HNO_2 + H_2O_2 \rightarrow HNO_3 + H_2O \quad (9.16)$$

The presence of oxygen and nitrogen atoms can promote the generation of additional reactive radical species, which in turn can contribute to the free radical attack on the pollutant:

$$HNO_3 \xrightarrow{\;)))\;} \cdot OH + \cdot NO_2 \quad (9.17)$$

$$HNO_3 \xrightarrow{\;)))\;} \cdot H + \cdot NO_3 \quad (9.18)$$

9.2.4.6 Ultrasonic equipment

The type of transducer, its shape and the dimensions of the reactor have to be taken into account since they determine the ultrasonic energy distribution in the volume (Gondrexon et al., 1998). Four types of laboratory scale ultrasonic equipments are currently used for a great variety of purposes: whistle, bath, probe (horn) and a cup-horn system. The main advantages of bath system are its economy and simplicity. However, the temperature control is not easy in such a system as well as the ultrasound intensity is limited by attenuation by the water and the walls of the reactor. The ultrasonic horn or probe system is characterized by very high intensities but the temperature control is crucial for this type of transducer. The cup-horn system is a combination of bath and horn (probe) system with a much better temperature control device. It allows much higher intensities without any contamination by the dissolution of a horn tip material. The major aim in the selection of a suitable ultrasonic equipment must be to maximize the cavitational effects with maximum possible energy efficiency.

Despite the lowest energy efficiency of the ultrasonic horn (10% in comparison to 18% of ultrasonic bath efficiency, it is the most studied and used laboratory scale equipment for the degradation of pollutants (Gogate et al., 2003). However, the cavitationally active volume in the horn-type reactors is very small and also the circulation flow rates generated due to the acoustic streaming are not capable of causing circulatory motion thorough the entire liquid (Gogate et al., 2003). Greater energy efficiencies were observed for ultrasonic probes with higher irradiating surface, (lower operating intensity of irradiation) which resulted into uniform dissipation of energy (Gogate et al., 2001a). Thus, for the same power density (power input into the system per unit volume of the effluent to be treated), power input to the system should be attained through larger areas of irradiating surfaces.

Moreover, various designs of laboratory and pilot scale sono-reactors are available, differing in terms of the operating and geometric conditions such as multiple frequency flow cells, near field acoustic processors, parallel plate processors, hexagonal flow cells, etc. (Prabhu et al., 2004; Cravotto et al., 2005; Gogate, 2008a; Amin et al., 2010).

9.3 HYBRID CAVITATION-BASED TECHNOLOGIES

Despite some promising laboratory scale results, sonolysis alone applied to the high volumes of real wastewater, fails to completely degrade organic compounds with a complex structure such as micropollutants. Therefore, the use of hybrid configurations is a very promising alternative not only to eliminate the concentrations of recalcitrant micropollutants but also their metabolites and subsequent by-products until well acceptable mineralization levels (Mendez-Arriaga et al., 2008). The combination of two or more advanced oxidation processes such as ultradound (US) and O_3, US and H_2O_2, etc. leads to the enhanced generation of the reactive species, which subsequently increase the degradation rates of target pollutants. The efficacy of the process and the extent of synergism not only depend on the enhancement in the number of free radicals but also on the alteration of the reactor configuration leading to a better contact of the generated free radicals with the pollutant molecules and also better utilization of the oxidants (Gogate, 2008a). Moreover, these hybrid technologies can be used to degrade the complex residue up to a certain level of toxicity beyond which the conventional methods can be successfully used for further degradation if needed.

Herein we report the use of non-catalytic hybrid techniques such as ultrasound coupled with oxidant (US/H_2O_2, O_3, and etc.), ultraviolet irradiation (US/UV), microwave irradiation (US/MW), adsorption (US/A), electrooxidation (US/EO) recently applied for the degradation of organic pollutants.

9.3.1 US/oxidant

9.3.1.1 US/H_2O_2

Hydrogen peroxide is well known source of hydroxyl radicals that is widely utilized for the degradation of organic pollutants. It is attractive because of un-toxic nature of byproducts, relatively low cost and easy handling. However, hydrogen peroxide is unstable and very sensitive to the conditions such as pH, temperature and the presence of metal impurities. The O-O bond in hydrogen peroxide is relatively weak, approx. 213 kJ mol^{-1}, and is susceptible to homolysis by a variety of methods including thermal, photolytic, radiolytic, metal-redox, and sonolytic as well (Jones, 1999). However, the high oxidizing power of formed ·OH radicals correlate to a relative lack of selectivity as an oxidant. Therefore, the hydroxyl radical has the ability to attack not only the contaminant but also hydrogen peroxide itself. This effect is called a 'scavenging effect' and is detrimental for the pollutant degradation process. Thus, the influence of hydrogen peroxide is dual: it acts as a source of free radicals due to cavitation, however also consumes some free radicals during the radical recombination. Therefore, for the most efficient outcome of the process, it is vitally important to initially adjust and monitor the concentration of hydrogen peroxide throughout the pollutants degradation process. Importantly, some amount of hydrogen peroxide is always formed during the radical recombination. This amount strongly depends on the parameters of ultrasound irradiation (frequency, power, etc.) and should be taken into account before selecting the proper H_2O_2 concentration.

In general, several crucial factors should be carefully studied and analyzed before the use of US/H_2O_2 method such as: operating pH, which influences the state of a target pollutant (molecular or ionic), nature of the contaminant (hydrophobic or hydrophilic), composition of treated effluent together with the optimal frequency of ultrasound, power dissipation per volume and reactor configuration.

9.3.1.2 US/O_3

Ozone is an unstable gas, which can be produced at the point of use, known as a strong oxidizing agent that is extremely efficient in degradation of refractory organic pollutants. Its effectiveness is based on the multiple effects produced by the oxidative activity of ozone and ozone-derived oxidizing species such as ·OH radicals. The generation of hydroxyl radicals from ozone in the presence of ultrasound takes place in two steps:

$$O_3 \rightarrow O_2 + O(^3P) \quad (9.19)$$

$$O(^3P) + H_2O \rightarrow 2 \cdot OH \quad (9.20)$$

This process involves the thermolytic destruction of ozone inside the vapour phase of the cavitation bubbles to form atomic oxygen, which subsequently reacts with water vapor yielding gas-phase hydroxyl radicals. Decomposition rate of ozone is significantly enhanced by the application of ultrasound, especially with an increase in power dissipation. However, ozone may also react with atomic oxygen or be scavenged by other reactive species in or near the bubble such as ·OH formed during sonolysis, reducing both, the production efficiency of hydroxyl radicals and ozone available in the solution to react directly with the substrate (Adewuyi, 2005a). On the other hand, the turbulence created by the acoustic streaming induced by the ultrasound enhance the absorption of ozone and eliminates mass transfer resistance in the solution (Gogate, 2008b). Despite the improved performance of such hybrid technique, the reaction rates are still lower than the addition of two individual rates, therefore the synergy of ultrasound irradiation and ozone is minimal. The possible explanation is the intensified recombination of free radicals formed via sonoozonolysis, which decrease the amount of free radicals attacking the contaminant and thus reducing the reaction rates. The increase in ultrasound frequency may also be detrimental due to the bubble dynamics. At higher frequencies the bubble collapse in significantly faster than at low frequencies, therefore, the time available for ozone molecule to diffuse into the cavitation bubble will be significantly lower thereby reducing the amount of hydroxyl radicals produced.

However, if the described drawbacks are overcome, the enhancement factor of concurrent application of ozone and ultrasound over the separate process may be calculated according to Weavers (Weavers and Hoffmann, 1998):

$$\text{Enhancement} = \frac{k_{US/O_3}}{k_{US} + k_{O_3}} \times 100 \tag{9.21}$$

The kinetics of the sonoozonolysis process can be assessed (Weavers et al., 1998; Weavers et al., 2000):

$$-\frac{dS}{dt} = k_{US}[S] + k_{O_3}[S] + k_{US/O_3}[S] \tag{9.22}$$

Yet again, it is important to note that the optimization of the reaction conditions is required. The concentration of ozone strongly depends on the operating frequency and power dissipation as well as the reactor configuration. Taking into account all the parameters, the type of ozone sparging (i.e., continuous bubbling or initial saturation) should be carefully selected in accordance with the optimal sonication parameters to achieve the most desired contaminant degradation results. However, the improvement is limited by: (i) the reactivity of molecular ozone with target micropollutant and its oxidation by-products and (ii) the gas phase decomposition of ozone to more reactive radical species (Kidak and Ince, 2007).

9.3.2 US/UV

The combined application of ultrasound and light has received many attention last decade (Wu *et al.*, 2001; Gogate *et al.*, 2002; Torres *et al.*, 2008; Behnajady *et al.*, 2009). These two physical agents rely on very different phenomena and their simultaneous use should result simply in an addition of their respective characteristics, since the chance of having an interference between purely mechanical, acoustic waves with electromagnetic vibrations is a priori very improbable (Toma *et al.*, 2001). Toma *et al.* (2001) have given an overview of different studies pertaining to the effect of ultrasound on photochemical reactions concentrating on the chemistry aspects, i.e., mechanisms and pathways of different chemical reactions. They anticipated that there is no reason to assume the influence of cavitation on the absorption of light. Absorption of light is very fast (10–15s), and therefore all cavitational events are too slow to interfere or to produce any changes. Therefore, the photosonochemical effect in US/UV system mostly related to the formation of hydrogen peroxide during sonication, which undergoes photolysis. However, a synergism between US and UV cannot be accounted for on the basis of an additive effect. The magnitude of this synergism varies from substrate to substrate, and the process is subject to substantial matrix effects. The addition of hydrogen peroxide at the optimized concentrations will greatly improve US/UV system efficiency (Behnajady *et al.*, 2009).

9.3.3 US/A

Adsorption is highly effective and economical method for the removal of low concentrations of organic pollutants present in the aqueous solutions. Different porous solid materials such as activated carbons prepared from coal, coconut shells, lignite, wood etc, and alternative low cost adsorbents fabricated from various carbon containing wastes such as grape pomace, peanut shells, spent coffee, peat, pine bark, banana pith and peel, rice bran, soybean and cottonseed hulls, hazelnut shells, rice husk, sawdust, wool fibres, orange peel, saffron corm, coirpith, cocoa shells, etc have gained importance in laboratory, pilot and industrial applications of adsorption processes (Gupta and Suhas, 2009). However, this well-established technique has several drawbacks, such as the tendency of the adsorbent to saturate after some time, which makes the further treatment un-feasible. Moreover, spent adsorbent can also cause an additional pollution to the environment. Moreover, adsorption does not lead to the destruction of the pollutant but only to a physical removal from the aqueous phase. Ultrasound provides the economical regeneration of exhausted adsorbents by overcoming the affinity of adsorbates with the adsorbent surface and enhancing the transport between the adsorbent surface and the reaction mixture without actually damaging the adsorbent (Breitbach and Bathen, 2001; Entezari and Sharif Al-Hoseini, 2007). Thus, the beneficial use of US to overcome drawbacks of unassisted adsorption can be divided into three areas: mass transfer enhancement and increase in the surface area of the adsorbent, the adsorbent regeneration both, in-situ and offline, and finally, supply of free radicals formed during sonolysis to improve the degradation of a target contaminant (Hamdaoui and Naffrechoux, 2009; Kuncek and Sener, 2010).

9.3.4 US/EO

Electrochemical oxidation (EO) is considered as an environmentally friendly technology with a wide array of applications because of its excellent oxidation ability without causing secondary pollution. Electrochemical oxidation over anodes made of graphite, Pt, TiO_2, IrO_2, PbO_2, various Ti-based alloys and, more recently, boron-doped diamond (BDD) electrodes in the presence of a suitable electrolyte (typically NaCl) has been employed for the decontamination of various micropollutants. Two mechanisms are responsible for the degradation of target contaminants, namely: (a) direct anodic oxidation where the pollutants are adsorbed on the anode surface and destroyed by the anodic electron transfer reaction and (b) indirect oxidation in the liquid bulk which is mediated by the oxidants that are formed electrochemically (Klavarioti *et al.*, 2009). Such oxidants include chlorine, hypochlorite, hydroxyl radicals, ozone and hydrogen peroxide. Critical operating parameters dictating the performance are the working electrode, the type of supporting electrolyte and the applied current. Other factors include the effluent pH and the initial concentration of the pollutant (Klavarioti *et al.*, 2009).

The coupling of US to EO constitute a thriving field gathering the mechanical effects of ultrasound (activation of the electrodes' surface and enhancement of the solid-liquid mass transfer) and their chemical effects (possible change in the degradation mechanisms) (Trabelsi *et al.*, 1996; Lindermeir *et al.*, 2003; Valcarel *et al.*, 2004; Zhao *et al.*, 2009). The optimization of ultrasound frequency is essential and highly depends on the contaminant to be treated. Low frequency ultrasound (LFU) still yields higher mass transfer (120-fold higher mass transfer by diffusion) in comparison to high frequency ultrasound (HFU) (up to 70-fold higher mass transfer by diffusion) (Trabelsi *et al.*, 1996). However, high frequency ultrasound may considerably affect the reaction pathway. The chemical effects of HFU such as bond cleavage of substrate or water sonolysis (hydroxyl radicals formation) are more pronounced than LFU.

The combination of EO with US significantly enhances degradation of the target micropollutant. In addition, the formation of different intermediate products during US/EO treatment then during the sole application of EO is possible as a result of degradation process intensification (Yasman *et al.*, 2004).

The polymerization of the reaction intermediates during the sole EC of organic compounds can lead to the formation of a layer on the electrode, hindering the electrolysis. US irradiation can prevent or eliminate the impurity layer at the surface of the electrode and increase its activity, as well as increase the reaction rates for the mass transfer (Ai *et al.*, 2010).

9.3.5 US/MW

The microwave region of the electromagnetic spectrum is associated with wavelengths of 1 cm to 1 m (frequencies of 30 GHz to 300 MHz). Microwave dielectric heating is rapidly becoming an alternative method for water depollution. Generally, both, thermal and non-thermal effects of microwave (MW) irradiation are utilized, since MW irradiation vastly increases the rotation, migration and friction for polar molecules (Wu *et al.*, 2008). Moreover, the violent motion of polar substances results in an increase in collisions among the reactants. Combined use of microwave irradiation and sonochemistry is a recent development that significantly accelerates the chemical reactions and other processing applications, especially in the case of heterogeneous systems (Gogate, 2008b). The increase of the target contaminants removal efficiencies may be attributed to a combination of enforced heat transfer due to microwave irradiation and intensive mass transfer at phase interfaces caused by ultrasonic irradiation (Wu *et al.*, 2008). In the majority of heterogeneous reactions, the mass transfer between the two (or more) phases is the rate-determining step. When ultrasound irradiation is applied, the liquid jet caused by cavitation propagates across the bubble towards the phase boundary at a velocity estimated at several hundreds of meter per second, and violently hits the surface. The intense agitate leads to the mutual injection of droplets of one liquid into another, resulting in the formation of fine emulsions. These emulsions are smaller in size and more stable than those obtained conventionally and require little or no surfactant to maintain stability (Gogate, 2008a). Microwave irradiation compliments these effects by the enhanced heating as well as absorption by the molecules resulting in significantly enhanced rates of the reactions. Han *et al.* (2004) reported that MW irradiation significantly accelerated the decomposition of organic compounds in comparison to the conventional heating which may be attributed to polar reactants being activated to a higher-excited-state through an increase in collisions rather than just the thermal effects.

Although MW irradiation can greatly enhance the sono-degradation of organic compounds on the laboratory scale, it requires higher energy expenditure than the ultrasound irradiation alone. Obviously, the energy efficiency of combined irradiation by MW-US might be improved on a large scale, since the relative amounts of energy supplied for cooling, pumping and stirring decrease (Han *et al.*, 2004).

9.4 DEGRADATION OF MICROPOLLUTANTS

This section gives an overview of the representative recent work undertaken in the degradation of the representative groups of the emerging micropollutans such as pharmaceuticals, organic dyes and pesticides by hybrid techniques based on cavitation. The existing disagreement between the micropollutants degradation results from various studies must be attributed to the different frequencies applied and reactor parameters used, which dictate bubble life time and collapse temperatures thus influencing the overall removal efficiencies.

9.4.1 Degradation of pharmaceuticals by hybrid techniques based on cavitation

Water contamination caused by pharmaceuticals discharged from municipal wastewater treatment plants became a serious concern all over the world. The pharmaceutical industry is attaining more effective active pharmaceutical ingredients (APIs) by designing for increased potency, bioavailability and degradation resistance (Khetan and Collins, 2007). Growing concentrations of APIs in the environment are already associated with adverse developmental effects in aquatic organisms therefore it is a matter of time when negative impacts on human health will surface. The most important of these relate to the potential for developmental impairment by trace quantities of pharmaceuticals in drinking water, that is, the subject area of endocrine disruption (Khetan and Collins, 2007). Unfortunately, the current regulatory status of APIs in North America and Europe is that their discharge in wastewater is largely unregulated. There is no water quality or technology based standards for APIs in surface water, groundwater, drinking water, wastewater, soil, sediments, or sludge/biosolids yet. So far most European and North American countries have only determined water quality concentrations for single substances (Chevre *et al.*, 2008). Despite the lack of regulatory actions, it is important to identify the most suitable and efficient sole or hybrid APIs removal and a subsequent toxicity elimination technology that can be used to avoid the adverse effects on humans and environment in the nearest future.

The sonochemical degradation of various estrogen compounds (initial concentrations of 10 µg/L) such as 17ά-estradiol, 17β-estradiol, estrone, estriol, equilin, 17ά-dihydroequilin, 17ά-ethinyl estradiol and norgestrel was investigated by Suri and co-workers (2007). The sonolysis destroyed 80–90% of individual estrogens within 40–60 min. of the reaction time. The degradation process obeyed the first order reaction kinetics. The removal rate increased with an increase in power intensity from 157 to 259 kW m^{-2}. Suri et al. (2007) concluded that from the perspective of the process economics, reactors with high intensities and low power densities would be suitable for the degradation of estrogens. Hence, the 4 kW reactor with the highest power intensity (259 kW m^{-2}) and the lowest power density (0.43 kW L^{-1}) was more energy efficient than the 0.6 and 2 kW reactors for the degradation of estrone. The sonolytic degradation of these pharmaceuticals favors low solution pH and temperature (Suri *et al.*, 2007). Fu *et al.* (2007) explained the sonolytic degradation mechanism of estrogen compounds via their structural properties. Thus, all the estrogens (17ά-estradiol, 17β-estradiol, estrone, estriol, equilin, 17ά-dihydroequilin, 17ά-ethinyl estradiol and norgestrel) have phenolic group in their structure. It is widely accepted that the sonochemical degradation of phenols is more intensive in lower pH ranges (pH 3–4). The *pKa* (acid ionization constant) of estrogens is higher than 10 and therefore, at lower solution pH, more estrogens would exist in nonionic molecular form exhibiting higher hydrophobicity. These conditions could cause easier diffusion into the cavity-liquid interface region where estrogens would undergo both, thermal degradation and concentrated radical oxidation upon cavity implosion. Therefore, an increase in the reaction rates would be expected with a decrease in the solution pH (Fu *et al.*, 2007). Fu *et al.* (2007) determined a straight proportional correlation between the pseudo-first rate constant (k) and molecular weight (MW) of the estrogens. Thus, faster diffusion of smaller estrogen compounds in the bulk solution to the interfacial region occurs, and therefore, faster ultrasound induced degradation takes place. Fu *et al.* (2007) also proposed a linear regression model including the ratio of the rate constant of estrogen to estrone and the ratio between the molecular weight of estrogen with estrone with experimental data of estrone as the most frequently detected compound:

$$\frac{k_e}{k_{estrone}} = a \frac{MW_e}{MW_{estrone}} + b \tag{9.23}$$

where k_e is the rate constant of estrogen, and $k_{estrone}$ is the experimental rate constant of estrone. MW_e and $MW_{estrone}$ are molecular weights of estrogen and estrone, respectively. Based on the linear regression equation, the first-order rate constant of any other estrogen compound during ultrasound-induced degradation could be predicted based on the experimental rate constant of estrone, as following:

$$k_{predict} = \left[a \frac{MW_e}{MW_{estrone}} + b \right] \times k_{estrone} \qquad (9.24)$$

where $k_{predict}$ is the predicted rate constant of estrogen of interest. The values of constants a and b were determined to be -5.04 and 6.08, respectively, at pH 7. The predicted data by Fu et al. (2007) were within 21% of the experimental data.

Moreover, De Bel et al. (2009) studied the sonolysis of ciprofloxacin (CIP) in water. Modification of the physical properties such as the charge of molecules with ionisable functional groups over pH (3, 7, and 10) was found to be the most important factor influencing the degradation process efficiency. De Bel et al. (2009) established the chemical form in which CIP appears in pH range from 2 to 11 (Figure 9.2). They found that degradation was significantly faster (e.g., degradation at pH 3 (k1 = 0.021 ± 0.0002 min^{-1}) was almost four times faster than at pH 7) when the main part of the CIP molecules carried an overall positive charge (De Bel et al., 2009).

Mendez-Arriaga et al. (2008) hypothesized that ibuprofen (IBP) transformation under the applied ultrasound irradiation (300 kHz and 80 W) would mostly occur in the gas–liquid interface and in the bulk of the solution, principally by the free radical attack. They established that there was no direct reaction between the hydrogen peroxide and IBP ion in the solution. Moreover, the pH during the sonolysis of IBP (21 mg L^{-1}) decreased, which can be attributed to the formation of reaction intermediates and terminal products of IBP such as hydroxylated metabolites and aliphatic acids (Mendez-Arriaga et al., 2008). Because of the highly hydrophilic nature of the latter, the aliphatic acids are recalcitrant to the ultrasonic irradiation and usually accumulate in the solution. To study the effect of the dissolved gases on the degradation of IBP, experiments with argon, air and oxygen were carried out in the presence of US irradiation. The lowest degradation rates were achieved using argon. In the presence of argon, the generation of ·OH is due only to the decomposition of vapor water. Up to 98% of IBP was destroyed in 30 min. by sonolysis. Importantly, pH values above IBP *pKa*, reduced the degradation rate of IBP, whereas acidic media (pH 3) provided the most favorable conditions for the degradation.

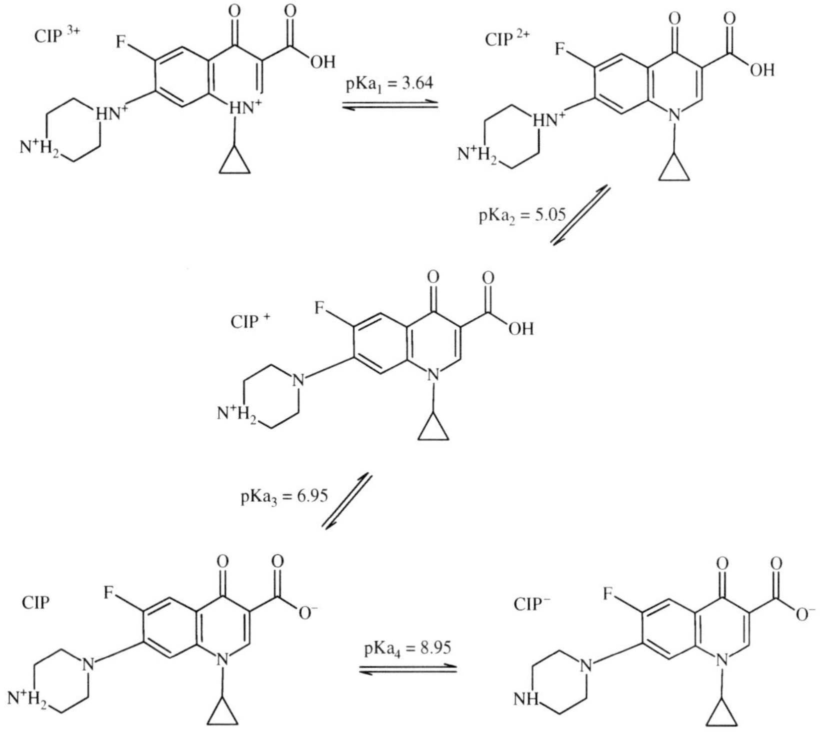

Figure 9.2 (De)protonation equilibria of ciprofloxacin (Adapted from De Bel et al. (2009))

To increase the removal efficiencies of selected pharmaceuticals from water, several groups reported significant findings in coupling ultrasound irradiation with ozonation. Thus, Naddeo et al. (2009) reported a synergistic effect when sonolysis (20 kHz, 400 W L^{-1}) and ozonation (31 g h^{-1}) were coupled for diclofenac (DCF) removal from the aqueous solution. The obtained increase in removal rate (from 0.06 to 0.073 min^{-1}) for US/O$_3$ in comparison with sole ozonation can be attributed to the enhanced O$_3$ decomposition in the collapsing bubbles yielding additional free radicals. When ozone was injected into water simultaneously with applied ultrasonic irradiation, an additional pathway of hydroxyl radical generation occurred upon the decomposition of ozone in the gaseous bubbles during implosive collapse. The advantage of a hybrid technology over sole ozonation was due to enhanced O$_3$ transfer to the solution by mechanical effects of ultrasound. Under the conditions applied

(DCF concentration of 40 mgL^{-1}, temperature 20±3°C), the sole ultrasound irradiation and ozonation led to 36% and 22% diclofenac mineralization, respectively, after 40 min of treatment, in comparison to about 40% for the same treatment duration when hybrid US/O$_3$ technology was used. Furthermore, the rates of TOC degradation ($k_{US/O3}$ = 2.11 × 10^{-1} mg L^{-1}min^{-1}) were found to be larger than those of sole O$_3$ (k_{O3} = 0.190 × 10^{-1} mg L^{-1}min^{-1}) and US (k_{US} = 0.106 × 10^{-1} mg L^{-1}min^{-1}) treatment, but slightly less than their sum ($k_{US/O3}$ = 2.96 × 10^{-1} mg L^{-1} min^{-1}) (Naddeo et al., 2009). One of the most important findings reported by Naddeo et al. (2009) was the fact that sonolysis demonstrated higher efficiency than ozonation for the mineralization of DCF.

Furthermore, Naddeo et al. (2010) also studied the toxicity evolution during sonolysis of DCF with a set of bioassays of Daphnia magna and Artemia salina. Prior to treatment DCF exhibited some toxicity to D. magna, the extent of which depended on its initial concentration. Although 4 mg L^{-1} of DCF was not toxic, a 10-fold increase in concentration led to 15% and 35% immobilization of D. magna at 24 and 48 h of exposure time, respectively (Naddeo et al., 2010). It was found that toxicity increased during initial 45 min of ultrasound irradiation, beyond which it started to decrease and the final sample (60 min. of ultrasound irradiation) was less toxic than the original one. Oxidation of toxic compounds such as DCF usually occurs via the formation of toxic intermediates at the early stages of the treatment, alongside the un-reacted DCF. Various operational parameters (e.g., power density) applied can lead to a different toxicity profile, which is probably caused by the differences in the distribution of the reaction by-products. Throughout the study, the extent of toxicity to D. magna depended on diclofenac concentration and its degradation by-products. It was found that ultrasound irradiation was capable to considerably reduce however not fully eliminate the toxicity. On the other hand, diclofenac and its by-products did not inflict any toxicity to A. salina (Naddeo et al., 2010).

9.4.2 Degradation of organic dyes by hybrid techniques based on cavitation

Dye pollutants from the textile industry are the important source of environmental contamination. About a half of global production of synthetic textile dyes (7 × 10^5 t per year) are classified into azo compounds that have the chromophore of –N=N–unit in their molecular structure. These azo dyes are known to be non-biodegradable in aerobic conditions and can be reduced to more hazardous intermediates in anaerobic conditions (Liu and Sun, 2007). Based on the very high solubility of azo dyes, it is expected that the prime reaction site for their sonolytic decomposition is the bulk solution. During the reaction labile azo (–N=N–) are generally more susceptible to the free radical attack (Liu and Sun, 2007; Abdullah and Ling, 2009). However, it was also reported that in concentrated solutions of highly soluble organic compounds, some of the solutes may migrate to the bubble-liquid interface, where they also undergo pyrolytic or oxidative decomposition (Gultekin et al., 2009). The rate of azo dye bleaching by ultrasound depends on the structural features, such as the largeness and complexity of the molecule, the type and position of the substituents around the azo bond, and the anionicity (Ince and Tezcanli-Guyer, 2004).

Currently, there is a number of studies performed utilizing sole ultrasound to degrade various organic dyes. For instance, Wang et al. (2008) reported sonochemical degradation of reactive brilliant red (K-BP) in initial concentrations ranging from 10 to 100 mgL^{-1}, temperatures from 25 to 55 and pH from 2 to 12. The estimated reaction zone for the low vapor pressure compounds such as K-BP is the interfacial zone between the bubble and the bulk solution where the temperature is lower than that inside the bubble but still high enough for a sonochemical reaction to occur. As the bulk temperature of water gradually increases, the vapor pressure of water and volatile solutes inside the cavitation bubbles also increases. The collapse of cavity is thus cushioned more than that at a lower bulk temperature, resulting in lower sonochemical degradation rates (10.3×10^{-4} and 6.05×10^{-4} min^{-1} at temperature $25 \pm 1°C$ and $55 \pm 1°C$, respectively).

Moreover, Velegraki et al. (2006) studied continuous sonochemical degradation of Acid Orange 7 (AO7) conversion and color removal as a function of time at 24 and 80 kHz, 25 C and 150 W for 240 min. Nearly complete and about 85% AO7 and color, respectively, removal was achieved after 240 min at 80 kHz ultrasound irradiation. However, no conversion was achieved during the experiment carried out at 24 kHz (Velegraki et al., 2006). Degradation significantly decreased with an increase in temperature (from 60% to 10%) and was nearly quenched at 60°C. Velegraki et al. (2006) concluded that low frequency high power ultrasound induced the deep oxidation of azo dyes and its ring intermediates to aliphatic compounds.

Another interesting study by Tezcanli-Guyer and Ince (2003) reported the effectiveness of 520 kHz ultrasound irradiation to destroy and detoxify the selected dyes such as the "reactive" (Reactive Red 141 (RR141) and Reactive Black 5 (RBk5)) and "basic"' (Basic Brown4 (BBr4) and Basic Blue 3 (BBl3)) dyes in aqueous solutions. They determined that RR141 and RBk5 were non-toxic at 20 and 30 mg l^{-1}, while BBl3 and BBr4 were toxic to *V. fisheri* at the same concentrations. Interestingly, a complete detoxification of basic dyes was attained before the color decay. This implied that toxicity causing components in BBr4 were primarily due to the chromophoric characteristics of the dye, which may have been further enhanced by the addition of some impurities in the commercial product (Tezcanli-Guyer and Ince, 2003). The decolorization rate followed first order reaction kinetics in the visible absorption of the dyes and was reported to be slower in azo than in oxazine-origin dye structures. The degradation of aromatic/olefinic content in azo dyes proceeded at a slower rate than that of color due to the increased mass of organic intermediates (Tezcanli-Guyer and Ince, 2003). It was also observed that if the parent dye was toxic, the degradation products were less toxic and the total toxicity elimination was accomplished within shorter contact time than necessary for the total degradation of the target dyes.

Furthermore, Ai *et al.* (2009) studied US/EO degradation of organic dyes. The decolorization of RhB was greatly accelerated by coupling US irradiation with EO and there was an obvious synergetic effect between the US and EO processes, e.g., 0.4% and 24.5% decolorization was achieved during sole US and EO treatment, respectively, in comparison to 91.4% decolorization in 6 min when a hybrid treatment method was applied.

To test the efficiency of ultrasound irradiation coupling to adsorption, Hamdaoui *et al.* (2008) studied dead pine needles as LCA for the Malachite Green (MG) (with initial concentrations ranging from 10 to 125 mgL^{-1}) removal from the model soution. Both the rate and the amount of sorption were appreciably increased and improved in the presence of ultrasound (40 kHz, 125 W). The dye sorption in the conventional method and in the presence of ultrasound alone reached equilibrium in about 330 min while, in the combined process (sonication and mechanical mixing) this time was reduced to about 200 min. (Hamdaoui *et al.*, 2008). Thus, US effects can be expressed as following: (i) the shift the sorption/desorption equilibrium to a new equilibrium, and (ii) the enhancement of the rate of sorption by accelerating the mass transfer by hydrodynamical effects generated by acoustic cavitation.

Another hybrid ultrasound irradiation and adsorption technique was discussed by Küncek and Sener (2010). They studied the removal of Methylene Blue (MB) by ultrasound-assisted adsorption onto sepiolite and found that ultrasound irradiation resulted in a significant increase in the specific surface area of sepiolite from 322 to 487 m^2 g^{-1} after 5 h of sonication. The ultrasound irradiation also improved the adsorption of MB, e.g., the maximum monolayer adsorption capacity increased from 79.37 to 128.21 mg g^{-1} after 5 hours of the irradiation time. Adsorption process of MB onto sepiolite followed the pseudo-second-order reaction kinetics. The thermodynamics parameters, such as entalphy (2.648 kJ mol^{-1} at 25°C), Gibbs energy (-26 kJ mol^{-1} at 25°C) and entrophy (102.47 kJ mol^{-1} at 25°C), indicated that the adsorption process of MB onto sepiolite was spontaneous and endothermic in nature which was favored at higher temperatures and occurred by both, physical adsorption and weak chemical interactions (Kuncek and Sener, 2010).

Several more studies of the dyes removal by hybrid methods based on cavitation are reported in Table 9.1.

9.4.3 Degradation of pesticides by hybrid techniques based on cavitation

Pesticide pollution of environmental waters is a pervasive problem with widespread ecological consequences (Chiron *et al.*, 2000). The main sources of pesticide pollution are wastewaters from agricultural industries and pesticide formulating or manufacturing plants. Nowadays, the use of various pesticides became an inseparable part of a modern agriculture. Due to their recalcitrant nature, pesticides usually do not degrade but accumulate in soil and water, thus posing an enormous risks to the environment. The degradation of pesticides is an extremely difficult task due to the wide array of chemical structures and properties (Matouq *et al.*, 2008). A great majority of currently available pesticides are included in the list of 33 priority hazardous substances or groups of substances of major concern in European Waters to be monitored under the Water Framework Directive [2000/60/EC].

Ultrasound and hybrid ultrasound-assisted techniques are gaining an increased attention for the degradation of pesticides in natural water and the wastewater. For instance, there are numerous works that report successful laboratory scale results in the presence of ultrasound as well as hydrogen peroxide, ozone, etc. in removing organophosphorus, organochlorine carbamate, chlorotriazine and chloroacetanilide pesticides.

Organochlorine (banned in 1970s and 80s in the most of the developed world according to the Stockholm's Convention) and organophosphorus compounds are well known for their insecticidal and herbicidal behavior, therefore they were and still are widely used in agriculture and to combat malaria in many countries around the globe. Organochlorine and organo phosphorus compounds are known to cause various health problems in humans (these compounds are classified as cancerogenic and endocrine disruptors by US EPA (Environmental Protection Agency, 2001) and environment therefore they have to be eliminated by adopting environmentally benign technologies.

Table 9.1 Studies utilizing the cavitation based treatment methods for the removal of organic dyes

Compound	System parameters	Process efficiency	Reference
C. I. Acid Black 210 (AB 210)	US (28 kHz, 500 W) + exfoliated graphite (0.8 gL^{-1})	99.5% at a pH 1 within 120 min. at 51°C	(Li et al., 2008)
C.I. Acid Blue 25 (AB25) (10–150 mg L^{-1})	US (22.5–1700 kHz, 14 W)/ H$_2$O$_2$ (386 to 1928 mg L^{-1})	The rate of degradation was increased 2.4, 1.7 and 1.5 times by the addition of 386, 1157 and 1928 mg L^{-1} of H$_2$O$_2$, respectively.	(Ghodbane and Hamdaoui, 2009a)
C.I. Acid Blue 25 (AB25) (50 mg L^{-1})	US (22.5–1700 kHz, 14 W)/ CCl$_4$ (0–798 mgL^{-1})	k = 0.1467 mg L^{-1} min^{-1} without CCl$_4$ k = 17.325 mg L^{-1} min^{-1} with 798 mg L^{-1} min^{-1} of CCl$_4$	(Ghodbane and Hamdaoui, 2009b)
C.I. Acid Orange7 (AO7) (57 µM)	US (520 kHz, 600 W)	k_{UV254} = (1.75±0.54) × 10^{-3} min^{-1} BOD$_5$ (after 60 min) = 0 mgL^{-1}	(Tezcanli-Guyer and Ince, 2004)
C.I. Acid Orange7 (AO7) (57 µM)	US (520 kHz, 600 W)/UV	BOD$_5$ (after 60 min) = 0.69 mgL^{-1}	(Tezcanli-Guyer and Ince, 2004)
C.I. Acid Orange7 (AO7) (57 µM)	US (520 kHz, 600 W)/ O3(10–60 gm-3)	k_{UV254} = (16.74±1.49) × 10^{-3} min^{-1} BOD$_5$(after 60 min) = 2.40 mgL^{-1}	(Tezcanli-Guyer and Ince, 2004)
C.I. Acid Orange7 (AO7) (57 µM)	US (520 kHz, 600 W)/ O3(10–60 gm-3)/UV	k_{UV254} = (21.71±1.58) × 10^{-3} min^{-1} BOD$_5$ (after 60 min) = 3.18 mgL^{-1}	(Tezcanli-Guyer and Ince, 2004)
C.I. Acid Orange 8 (AO8) (30 µM)	US (300 kHz) + CCl$_4$	85% in 30 min. (decolorization of the dye was enhanced by 27% by CCl$_4$ addition)	(Gultekin et al., 2009)
C.I. Direct Black 168 (200 mg L^{-1})	US (40 kHz, 250 W)/ fly ash (2gL^{-1})/H$_2$O$_2$(2.94 mM)	99.0% after 90 min.	(Song and Li, 2009)

(continued)

Table 9.1 (continued)

Compound	System parameters	Process efficiency	Reference
C.I. Reactive Black 5 (RBk5)(33 µM)	US (640 kHz, 240W)	2.9×10^{-2} min^{-1}	(Vinodgopal et al., 1998)
C.I. Reactive Black 5 (RBk5) (60 mg L^{-1})	US (80 kHz, 135W)	about 10% dye removal after 60 min.	(Kritikos et al., 2007)
C.I. Reactive Blue 19 (RB 19) (83–917 mg dm^{-3})	US (140 W, 850 kHz)/AC (5.8 ç dm^3)	36%, 91% and 99.9% of RB 19 removal for US, AC and US/AC, respectively after 18 min.	(Sayan and Esra Edecan, 2008)
C. I. Reactive Red 22 (RR22)(5–90 µM)	US (200 kHz, 1.25 W cm^{-2})/argon	7.95×10^{-8} M s^{-1}	(Okitsu et al., 2005)
C. I. Reactive Red 22 (RR22)(5–90 µM)	US (200 kHz, 1.25 W cm^{-2})/air	2.78×10^{-8} M s^{-1}	(Okitsu et al., 2005)
C.I. Reactive Brilliant Red (K-BP) (10–100 mg L^{-1})	US (24 kHz, 150W)	10.3×10^{-4}, 8.03×10^{-4}, 6.33×10^{-4} and 4.35×10^{-4} min^{-1} for 10, 20, 50 and 100 mgL^{-1}, respectively	(Wang et al., 2008)
Methyl Orange (MO)	US (130 kHz 150W)/activated aluminas (0.5–5 gL^{-1})	1.6×10^{-4} mol g^{-1}	(Iida et al., 2004)
Methyl Orange (MO) (4–80 µM)	US (200 kHz, 1.25 W cm^{-2}) pH 6.5	8.37×10^{-8} M s^{-1}	(Okitsu et al., 2005)
Methyl Orange (MO) (4–80 µM)	US (200 kHz, 1.25 W cm^{-2}) pH 2.0	7.13×10^{-8} M s^{-1}	(Okitsu et al., 2005)
Methyl Orange (MO)	US/UV/EO	0.0732 min^{-1}	(Zhang and Hua, 2000)

(continued)

Table 9.1 (continued)

Compound	System parameters	Process efficiency	Reference
Methylene Blue (MB) (50 mgL^{-1})	US (20kHz, 33 W cm-2)/ wasted paper (3g)	50% of removal vs 7% during silent adsorption	(Entezari and Sharif Al-Hoseini, 2007)
Orange II (25 mgL^{-1})	US (224, 404, 651 kHz, 11.4, 29.0 and 41.5W)/air	100% after 4 h	(Inoue et al., 2006)
Orange G (OG)	US (213 kHz, 20W)	54% after 75 min.	(Madhavan et al., 2010)
Rifacion Yellow (HE4R) (116–783 ppm)	US (850 kHz, 140W)/AC	57 mg g^{-1}	(Sayan, 2006)
Rhodamine B (RhB) (25 mgL^{-1})	US (224, 404, 651 kHz, 11.4, 29.0 and 41.5W)	100% after 2 h	(Inoue et al., 2006)
Rhodamine B (RhB) (100 mgL^{-1})	HDC reactor (swirling jet) 0.6 MPa	63% after 180 min.	(Wang et al., 2009)

Thangavadivel et al. (2009) studied the degradation of DDT – one of the most widespread organochlorine pesticides, by means of high frequency (1.6 MHz, 150 W/L) ultrasound. The removal efficiency of 10% and 90% after 90 min of the reaction was attained when 19 mgL^{-1} and 8 mgL^{-1} initial DDT concentration, respectively was used. Moreover, the removal efficiency from 32.6 mgL^{-1} of initial DDT concentration in 40 weight% sand slurry was 22% in 90 min. (Thangavadivel et al., 2009). This was remarkably low in comparison to Mason et al. (2004), who utilized 20 kHz ultrasound frequency and 20 min. of the reaction time to sufficiently degrade 75% of DDT with the initial 250 mg L^{-1} concentration in 50 weight% sand slurry. The considerably low efficiency can be explained by the currently existing intensity limitations and higher attenuation of energy in high frequency ultrasound equipment, which subsequently cause low volume coverage and will eventually require multiple sonotrodes, additional circulation, larger sonotrode area and lower slurry densities (Mason et al., 2004). Moreover, Francony and Petrier (1996) determined that sonochemical reaction efficiency is maximal at the frequency range 200–600 kHz and lesser reactive radicals will be produced above 1 MHz compared to 200–600 kHz frequency.

In addition, Zhang et al. (2010) studied the removal of malathion and chlorpyrifos from apple juice by ultrasound (25 kHz, 650 W). The ultrasonic power (100, 300 and 500 W) and treatment time (0–120 min) played a significant role in their degradation efficiency ($p < 0.05$). Both pesticides followed the first-order reaction kinetics at 500 W (reaction constant 0.05 min^{-1} and 0.0138 min^{-1} for malathion and chlorpyrifos, respectively), however chlorpyrifos was found to be much more labile to ultrasonic irradiation than malathion (48.4 and 22.6% after the treatment at 500W for 30 min., respectively). Zhang et al. (2010) hypothesized the degradation pathway and concluded that the destruction took place via the hydroxyl radical attack on the $P = S$ bond of pesticides with malaoxon and chlorpyrifos oxon as the main degradation products (Zhang et al., 2010).

Chloroacetanilide, chlorotriazine and carbamate pesticides are also important micropollutants that can directly and/or indirectly impact freshwater, estuarine, marine ecosystems and humans due to the constantly increasing inputs from industrial activities, sewage discharge, atmospheric deposition, groundwater leaching and run offs.

Torres et al. (2009) studied ultrasound-assisted (20–80 W) degradation of alachlor in deionized and natural water at pH range 3–10 and initial concentrations from 10 to 50 mg L^{-1}. The complete removal was observed in both, deionized and natural water, regardless the presence or absence of potential OH radical scavengers such as bicarbonate, sulphate, chloride and oxalic acid after approximately 75 min. of the reaction time. However, only 20% of the initial

TOC was removed in 240 min. suggesting that alachlor by-products were recalcitrant to the ultrasound irradiation, thus in order to achieve a complete mineralization to carbon dioxide and water, hybrid technique may be necessary (Torres et al., 2009). Importantly, BOD_5/COD tests showed that ultrasound irradiation significantly increased the biodegradability of the initial solution.

Hua and Pfalzer-Thompson (2001) showed that carbofuran decomposition in the presence of ultrasound irradiation (16 and 20 kHz, total applied power of 1600 W and a power-to-area ratio (power intensity of 1.22 W cm^{-2}) could be significantly enhanced with increasing power densities (1.65 W;mL^{-1}–5.55 W mL^{-1}), and with decreasing initial concentrations (25 mM vs. 130 mM). Important findings included the use of a proper ratio off-sparging gas (Ar and O_2) that resulted in optimal decomposition kinetics with pseudo-first order rate constants ranging from 0.00734 to 0.0749 min^{-1} (Hua and Pfalzer-Thompson, 2001).

Examples of various cavitation-based techniques for the removal of pesticides can be seen in Table 9.2.

9.5 SCALE-UP CONSIDERATIONS

Laboratory scale results obtained from a number of studies show an extremely high potential for hybrid ultrasound based techniques to be used for degradation of micropollutants in water and wastewater. However, the applicability and economical feasibility of such hybrid systems or sonolysis alone for the large scale applications remain in question. The major limitation is the energy consumption/the process efficiency ration translated into costs. It is anticipated that the use of alternative ways to produce the cavitational effects such as the use of hydrodynamic cavitation (Gogate and Pandit, 2005), the use of various additives (Gogate, 2008b), multiple frequencies (Gogate, 2008a) or the hybrid methods (Adewuyi, 2005a; Adewuyi, 2005b; Gogate, 2008b) can be useful means in the development of the large scale ultrasound-assisted treatment of target contaminants.

Thompson and Dorayswamy (1999) recommended to first resolve the role of ultrasound in the reacting systems prior to the scaling up and making ultrasound a viable removal rate enhancement technique for many commercial industrial processes. Such evaluation should include the characteristics of the reaction mixture and the kinetics of the reaction as well as the knowledge whether physical or chemical effects of ultrasound are more important for the process. Moreover, the optimization of process parameters during sonication, such as the reaction temperature, pressure, frequency, dissipated power, ultrasound intensity, and their interactions is strongly recommended.

Table 9.2 Studies utilizing the cavitation based treatment methods for the removal of pesticides

Compound	Method	Comments	Reference
2,4,5-T (35 ppm)	US (20 kHz, 150 W)	50% after 10 min.	(Collings and Gwan, 2010)
Alachlor (50 mgL^{-1})	HDC reactor 0.6 Mpa	4.90×10^{-2} min^{-1}	(Wang and Zhang, 2009)
Alachlor (10–50 mg L^{-1})	US (300 kHz 80 W)	100% after 240 min.	(Torres et al., 2009)
Atrazine (0.1 M)	US (20 and 500 kHz, 18.5 W)	8.3×10^9 M s^{-1} (20 kHz) 65×10^9 M s^{-1} (500 kHz)	(Petrier et al., 1996)
Atrazine (58 ppm)	US (20 kHz, 150 W)	70% after 10 min.	(Collings and Gwan, 2010)
Chlordane (715 ppm)	US (20 kHz, 150 W)	70% after 10 min.	(Collings and Gwan, 2010)
DDT (250 mgL^{-1})	US (20 kHz, 375 W L^{-1})	75% after 20 min.	(Mason et al., 2004)
DDT (707 ppm)	US (20 kHz, 150 W)	70% after 10 min.	(Collings and Gwan, 2010)
Diazinon (800–1800 ppm)	US (1.7 MHz 9.5 W)	70% after 300 s 0.01 s^{-1}	(Matouq et al., 2008)
Dichlorvos (5.0×10^{-4} M)	US (500 kHz, 161 W)	0.037–0.002 min.$^{-1}$	(Schramm and Hua, 2001)
Endosulfan (3.3 ppm)	US (20 kHz, 150 W)	70% after 10 min.	(Collings and Gwan, 2010)
Fenitrothion (10 mgL^{-1})	US (150 W, 20 kHz) /UV	100% after 2 h	(Katsumata et al., 2010)
MCPA ((4-chloro-2-methylphenoxy) acetic acid) (100 ppm)	US (500 kHz, 21.4 W)	$k_{MCPA} = 1.10 \times 10^{-4}$ s^{-1}, in pure N$_2$ $k_{MCPA} = 7.78 \times 10^{-4}$ s^{-1}, in pure O$_2$ $k_{MCPA} = 4.58 \times 10^{-4}$ s^{-1}, in pure Ar	(Kojima et al., 2005)

Furthermore, the engineering aspects such as the reactor design should also be taken into account. It is recommended to review a very good recent publication by Gogate (2008a, 2008b), where main aspects of various ultrasonic (for sole and hybrid applications) reactor design aspects are discussed. In addition, increasing commercial interest and multiple researches in the sonochemical field will definitely help to reduce the overall process costs in the near future (Adewuyi, 2001).

The prospects of industrial scale applications will probably be in the focus of attention in the near future due to known detrimental health effects of other relatively low cost water and wastewater treatment technologies such as ozonation and UV processes which make ultrasonic and hydrodynamic cavitation processes even more attractive (Mahamuni and Adewuyi, 2009). One should also understand that, this is a new technology and it is still in the developmental stage. As it matures, the cost of treatment will be significantly reduced especially nowadays, when academics, equipment designers and manufacturers are actively engaged in planning for larger scale sonochemical processing (Mason, 2007). Moreover, several works demonstrated the successful use of US on the large scale (Destaillats *et al.,* 2001; Gogate *et al.,* 2004; Pradhan and Gogate, 2010).

9.6 ECONOMICAL ASPECTS OF CAVITATION BASED TREATMENT

To estimate the operating costs of sole ultrasound irradiation in the laboratory scale application, the 'figure of merit' concept proposed by Bolton *et al.* (2001) is usually used. 'Figure-of-merit' allows the assessment of the technology efficiency with respect to energy consumption associated with the removal of the target compound (Bolton *et al.,* 2001). However, this estimation can only be used for two kinetic regimes: the energy per order of magnitude of removal (E/EO) or energy per mass of removal (E/EM) for first order and zero order reaction kinetics, respectively. To calculate the EE/O and EE/M the following formulas can be used:

$$\text{EE/O} = \frac{P_{elec} \times t \times 1000}{V \times 60 \times \log\left(\frac{C_{A_O}}{C_A}\right)} \qquad (9.25)$$

$$\text{EE/M} = \frac{P_{elec} \times t \times 1000}{V \times M \times 60 \times (C_{A_O} - C_A)} \qquad (9.26)$$

Mahamuni and Adewuy (2009) carried out an excellent cost estimation study of the ultrasound based techniques applicable for water treatment. The factors, which influence the removal rate, in turn influence the cost, e.g., the cost of ultrasound-assisted wastewater treatment containing hydrophobic contaminants is found to be an order of magnitude less than that of wastewater containing hydrophilic contaminants Mahamuni and Adewuyi (2009). Therefore, the combination of ultrasound irradiation with different AOPs is economically more attractive than the use of sole ultrasound for wastewater treatment due to the increased removal efficiency (Adewuyi, 2001). Thus, Figure 9.3 present the comparative results on the dye treatment costs. It is clearly seen in Figure 9.3 that the use of sole ultrasound in degradation of e.g., organic dyes has the lowest cost efficiency.

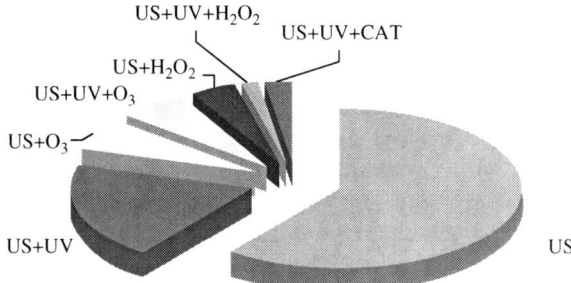

Figure 9.3 Comparative costs of cavitation based methods of treatment for the degradation of organic dyes (Adapted from Mahamuni and Adewuyi, 2009)

Possible reason for such a low energy efficiency is that the conversion of electrical energy into cavitation energy is a very inefficient process utilizing currently available ultrasonic equipment because part of the total energy input is wasted to produce heat and only a fraction is utilized to produce cavitation (Adewuyi, 2001). Important to note, that not all the energy that is utilized for cavitation produces chemical and physical effects. For instance, some energy is consumed in sound re-emission (i.e., harmonics and sub-harmonics). Only 34% of the electrical energy supplied is actually used for the production of the desired final chemical and physical effects (Hoffmann et al., 1996). Another reason for the low energy efficiency is the use of large amounts of energy for treating of very small reaction volumes. For instance, the power densities used in various laboratory scale studies were very high ranging from 0.027 WmL^{-1} to 0.76 WmL^{-1} (Adewuyi, 2001). Mahamuni and Adewuyi (2009) recommended using no more than 0.05 $W\,mL^{-1}$ energy density for ultrasound-assisted processes in order to make the process economically viable at the industrial scale. It is also clearly seen from EE/O equations that values will decrease with a significant increase in the treated volume.

Mahamuni and Adewuyi (2009) proposed the general approach to estimate the economics of large scale applications including capital, operating and maintenance costs. The capital cost estimation is valid for the first and zero order reaction kinetics and can be expressed as:

$$\text{Cost of AOP reactor} = P = N \times C \tag{9.27}$$

where N is the number of treatment units and C is the cost of each unit from the manufacturer. The number of such standard commercial units required:

$$N = \frac{X \times \varepsilon}{E} \tag{9.28}$$

where E is the energy supplied by single unit of AOP taken from the manufacturer quotations, (W), ε is the energy density (W L^{-1}) and X is wastewater treatment reactor capacity (L). The reactor treatment capacity can be expressed as the design flow rate (L min^{-1}) multiplied by time necessary for the 90% of target compound degradation (min). The time (t_{90}) can be calculated from the power law expression according to the reaction rate order. The capital cost is amortized over a span of years at a given amortization rate. Amortized capital costs (A) is given by a following formula:

$$A = \frac{1.2S \times r}{1 - \left(\frac{1}{1+r}\right)^n} \tag{9.29}$$

The O&M (Operating and Maintenance cost) consists of various parts replacement costs (for instance O&M costs for ultrasound irradiation is only 0.5% of the capital costs, whereas for the UV-based processes O&M costs are more than 45% of the annual electrical power consumption costs, and finally for O_3 the annual parts replacement costs are assumed to be 1.5% of the capital costs), labor costs (water sampling, general (system oversight and maintenance such as pressure gauges, control panels, leakages etc.) and specific (inspection, replacement and repair-based on hours of service life) system O&M costs), analytical costs, chemical costs (costs of consumables and chemicals involved in the AOPs) and electrical costs.

As of today, the cost of wastewater treatment using hybrid ultrasonic processes is one to two orders of magnitude higher than conventional AOPs such as sole ozonation, O_3/H_2O_2 and UV/H_2O_2 techniques (Mahamuni and Adewuyi, 2009).

9.7 CONCLUSIONS

The use of sole ultrasound irradiation and ultrasound-based techniques such as ultrasound coupled with oxidant (e.g., H_2O_2, O_3, and etc.), ultraviolet irradiation, microwave irradiation, adsorption, electrooxidation proved to be efficient in the degradation of micropollutants, e.g., pharmaceuticals, organic dyes and pesticides on the laboratory scale. However, there is a lack of studies performed and results reported for the large scale ultrasound-based techniques to eliminate micropollutants from the aqueous phase taking into account economical and technological aspects of the processes. The economical feasibility of ultrasound based technologies is still in question. However, successful attempts are currently surfacing utilizing sole ultrasound for elimination of pharmaceuticals on a larger scale and ultrasound-based techniques to degrade organic dyes and pesticides.

To improve the efficiency of the ultrasound-based techniques it is important to lower energy expenditure using alternative ways of producing the same cavitational effects as observed during ultrasound irradiation with high energy density. The hybrid methods reported in this chapter can also provide the solutions to increase the energy efficiency and reduce the overall costs. The optimization of operational conditions along with a better understanding of chemical oxidation mechanisms is currently one of the most important and relevant issues in order to plan rational chemical treatment of micropollutants in water and wastewater.

9.8 REFERENCES

Abdullah, A. Z. and Ling, P. Y. (2009) Heat treatment effects on the characteristics and sonocatalytic performance of TiO_2 in the degradation of organic dyes in aqueous solution. *J. Hazard. Mater.* **173**, 159–167.

Adewuyi, Y. G. (2001) Sonochemistry: Environmental science and engineering Applications. *Ind. Eng. Chem. Res.* **40**, 4681–4715.

Adewuyi, Y. G. (2005a) Sonochemistry in environmental remediation. 1. Combinative and hybrid sonophotochemical oxidation processes for the treatment of pollutants in Water. *Environ. Sci. Technol.* **39**, 3409–3420.

Adewuyi, Y. G. (2005b) Sonochemistry in environmental remediation. 2. Heterogeneous Sonophotocatalytic oxidation processes for the treatment of pollutants in water. *Environ. Sci. Technol.* **39**, 8557–8570.

Ai, Z., Li, J., Zhang, L. and Lee, S. (2010) Rapid decolorization of azo dyes in aqueous solution by an ultrasound-assisted electrocatalytic oxidation process. *Ultrason. Sonochem.* **17**, 370–375.

Amin, L. P., Gogate, P. R., Burgess, A. E. and Bremner, D. H. (2010) Optimization of a hydrodynamic cavitation reactor using salicylic acid dosimetry. *Chem. Eng. J.* **156**, 165–169.

Behnajady, M. A., Vahid, B., Modirshahla, N. and Shokri, M. (2009) Evaluation of electrical energy per order (EEO) with kinetic modeling on the removal of Malachite Green by US/UV/H2O2 process. *Desalination* **249**.

Bolton, J. R., Bircher, K. G., Tumas, W. and Tolman, C. A. (2001) Figures-of-merit for the technical development and application of advanced oxidation technologies for both electric- and solar-driven systems (IUPAC Technical Report). *Pure Appl. Chem.* **73**, 627–637.

Breitbach, M. and Bathen, D. (2001) Influence of ultrasound on adsorption processes. *Ultrason. Sonochem.* **8**, 277–283.

Chand, R., Ince, N. H., Gogate, P. R. and Bremner, D. H. (2009) Phenol degradation using 20, 300 and 520 kHz ultrasonic reactors with hydrogen peroxide, ozone and zero valent metals. *Sep. Purif. Technol.* **67**, 103–109.

Chevre, N., Maillard, E., Loepfe, C. and Slooten, K. B.-V. (2008) Determination of water quality standards for chemical mixtures: Extension of a methodology developed for herbicides to a group of insecticides and a group of pharmaceuticals. *Ecotoxicology and Environmental Safety* **71**, 740–748.

Chiron, S., Fernandez-Alba, A., Rodriguez, A. and Garcia-Calvo, E. (2000) Pesticide chemical oxidation: State-of-the-art. *Water Res.* **34**, 366–377.

Chowdhury, P. and Viraraghavan, T. (2009) Sonochemical degradation of chlorinated organic compounds, phenolic compounds and organic dyes – a review. *Sci. Total Environ.* **407**, 2474–2492.

Collings, A. F. and Gwan, P. B. (2010) Ultrasonic destruction of pesticide contaminants in slurries. *Ultrason. Sonochem.* **17**, 1–3.

Cravotto, G., Omiccioli, G. and Stevanato, L. (2005) An improved sonochemical reactor. *Ultrason. Sonochem.* **12**, 213–217.

De Bel, E., Dewulf, J., Witte, B. D., Van Langenhove, H. and Janssen, C. (2009) Influence of pH on the sonolysis of ciprofloxacin: Biodegradability, ecotoxicity and antibiotic activity of its degradation products. *Chemosphere* **77**, 291–295.

Destaillats, H., Lesko, T. M., Knowlton, M., Wallace, H. and Hoffmann, M. R. (2001) Scale-up of sonochemical reactors for water treatment. *Ind. Eng. Chem. Res.* **40**, 3855–3860.

Drijvers, D., De Baets, R., De Visscher, A. and Van Langenhove, H. (1996) Sonolysis of trichloroethylene in aqueous solution: volatile organic intermediates. *Ultrason. Sonochem.* **3**, S83–S90.

Entezari, M. H. and Sharif Al-Hoseini, Z. (2007) Sono-sorption as a new method for the removal of methylene blue from aqueous solution. *Ultrason. Sonochem.* **14**, 599–604.

Environmental Protection Agency, U. S. (2001). Preliminary Cumulative Hazard and Dose Response Assessment for Organophosphorus Pesticides: Determination of Relative Potency and Points of Departure for Cholinesterase Inhibition. U. S. E. P. A. Office of Pesticide Programs. Washington, DC.

Fu, H., Suri, R. P. S., Chimchirian, R. F., Helmig, E. and Constable, R. (2007) Ultrasound-induced destruction of low levels of estrogen hormones in aqueous solutions. *Environ. Sci. Technol.* **41**, 5869–5874.

Ghodbane, H. and Hamdaoui, O. (2009a) Degradation of Acid Blue 25 in aqueous media using 1700 kHz ultrasonic irradiation: ultrasound/Fe(II) and ultrasound/H2O2 combinations. *Ultrason. Sonochem.* **16**, 593–598.

Ghodbane, H. and Hamdaoui, O. (2009b) Intensification of sonochemical decolorization of anthraquinonic dye Acid Blue 25 using carbon tetrachloride. *Ultrason. Sonochem.* **16**, 455–461.

Gogate, P. R. (2008a) Cavitational reactors for process intensification of chemical processing applications: A critical review. *Chemical Engineering and Processing: Process Intensification* **47**, 515–527.

Gogate, P. R. (2008b) Treatment of wastewater streams containing phenolic compounds using hybrid techniques based on cavitation: A review of the current status and the way forward. *Ultrason. Sonochem.* **15**, 1–15.

Gogate, P. R., Mujumdar, S. and Pandit, A. B. (2002) A sonophotochemical reactor for the removal of Formic acid from wastewater. *Ind. Eng. Chem. Res.* **41**, 3370–3378.

Gogate, P. R., Mujumdar, S. and Pandit, A. B. (2003) Sonochemical reactors for waste water treatment: Comparison using formic acid degradation as a model reaction. *Adv. Environ. Res.* **7**, 283–299.

Gogate, P. R., Mujumdar, S., Thampi, J., Wilhelm, A. M. and Pandit, A. B. (2004) Destruction of phenol using sonochemical reactors: Scale up aspects and comparison of novel configuration with conventional reactors. *Sep. Purif. Technol.* **34**, 25–34.

Gogate, P. R. and Pandit, A. B. (2004a) A review of imperative technologies for wastewater treatment I: Oxidation technologies at ambient conditions. *Adv. Environ. Res.* **8**, 501–551.

Gogate, P. R. and Pandit, A. B. (2004b) A review of imperative technologies for wastewater treatment II: Hybrid methods. *Adv. Environ. Res.* **8**, 553–597.

Gogate, P. R. and Pandit, A. B. (2005) A review and assessment of hydrodynamic cavitation as a technology for the future. *Ultrason. Sonochem.* **12**, 21–27.

Gondrexon, N., Renaudin, V., Petrier, C., Clement, M., Boldo, P., Gonthier, Y. and Bernis, A. (1998) Experimental study of the hydrodynamic behaviour of a high frequency ultrasonic reactor. *Ultrason. Sonochem.* **5**, 1–6.

Gultekin, I. and Ince, N. H. (2008) Ultrasonic destruction of bisphenol-A: The operating parameters. *Ultrason. Sonochem.* **15**, 524–529.

Gultekin, I., Tezcanli-Guyer, G. and Ince, N. H. (2009) Sonochemical decay of C.I. Acid Orange 8: Effects of CCl4 and t-butyl alcohol. *Ultrason. Sonochem.* **16**, 577–581.

Gupta, V. K. and Suhas (2009) Application of low-cost adsorbents for dye removal – A review. *J. Environ. Manage.* **90**, 2313–2342.

Hamdaoui, O., Chiha, M. and Naffrechoux, E. (2008) Ultrasound-assisted removal of malachite green from aqueous solution by dead pine needles. *Ultrason. Sonochem.* **15**, 799–807.

Hamdaoui, O. and Naffrechoux, E. (2009) Adsorption kinetics of 4-chlorophenol onto granular activated carbon in the presence of high frequency ultrasound. *Ultrason. Sonochem.* **16**, 15–22.

Han, D. H., Cha, S. Y. and Yang, H. Y. (2004) Improvement of oxidative decomposition of aqueous phenol by microwave irradiation in UV/H2O2 process and kinetic study. *Water Res.* **38**, 2782–2790.

Hoffmann, M. R., Hua, I. and Hochemer, R. (1996) Application of ultrasonic irradiation for the degradation of chemical contaminants in water. *Ultrason. Sonochem.* **3**, S163–S172.

Hua, I., Hochemer, R. H. and Hoffmann, M. R. (2002) Sonochemical degradation of p-nitrophenol in a parallel-plate near-field acoustical processor. *Environ. Sci. Technol.* **29**, 2790–2796.

Hua, I. and Hoffmann, M. R. (1997) Optimization of ultrasonic irradiation as an advanced oxidation technology. *Environ. Sci. Technol.* **31**, 2237–2243.

Hua, I. and Pfalzer-Thompson, U. (2001) Ultrasonic irradiation of carbofuran: decomposition kinetics and reactor characterization. *Water Res.* **35**, 1445–1452.

Iida, Y., Kozuka, T., Tuziuti, T. and Yasui, K. (2004) Sonochemically enhanced adsorption and degradation of methyl orange with activated aluminas. *Ultrasonics* **42**, 635–639.

Ince, N. H. and Tezcanli-Guyer, G. (2004) Impacts of pH and molecular structure on ultrasonic degradation of azo dyes. *Ultrasonics* **42**, 591–596.

Inoue, M., Okada, F., Sakurai, A. and Sakakibara, M. (2006) A new development of dyestuffs degradation system using ultrasound. *Ultrason. Sonochem.* **13**, 313–320.

Jiang, Y., Petrier, C. and Waite, T. D. (2006) Sonolysis of 4-chlorophenol in aqueous solution: Effects of substrate concentration, aqueous temperature and ultrasonic frequency. *Ultrason. Sonochem.* **13**, 415–422.

Jones, C. W. (1999) *Applications of Hydrogen Peroxide and Derivatives*. Science Park, Milton Road, The Royal Society of Chemistry: Thomas Graham House.

Katsumata, H., Okada, T., Kaneco, S., Suzuki, T. and Ohta, K. (2010) Degradation of fenitrothion by ultrasound/ferrioxalate/UV system. *Ultrason. Sonochem.* **17**, 200–206.

Khetan, S. K. and Collins, T. J. (2007) Human pharmaceuticals in the aquatic environment: A challenge to green chemistry. *Chem. Rev.* **107**, 2319–2364.

Kidak, R. and Ince, N. H. (2007) Catalysis of advanced oxidation reactions by ultrasound: A case study with phenol. *J. Hazard. Mater.* **146**, 630–635.

Kirpalani, D. M. and McQuinn, K. J. (2006) Experimental quantification of cavitation yield revisited: Focus on high frequency ultrasound reactors,. *Ultrason. Sonochem.* **13**, 1–5.

Klavarioti, M., Mantzavinos, D. and Kassinos, D. (2009) Removal of residual pharmaceuticals from aqueous systems by advanced oxidation processes. *Environ. Int.* **35**, 402–417.

Kojima, Y., Fujita, T., Ona, E. P., Matsuda, H., Koda, S., Tanahashi, N. and Asakura, Y. (2005) Effects of dissolved gas species on ultrasonic degradation of (4-chloro-2-methylphenoxy) acetic acid (MCPA) in aqueous solution. *Ultrason. Sonochem.* **12**, 359–365.

Kritikos, D. E., Xekoukoulotakis, N. P., Psillakis, E. and Mantzavinos, D. (2007) Photocatalytic degradation of reactive black 5 in aqueous solutions: Effect of operating conditions and coupling with ultrasound irradiation. *Water Res.* **41**, 2236–2246.

Kuncek, I. and Sener, S. (2010) Adsorption of methylene blue onto sonicated sepiolite from aqueous solutions. *Ultrason. Sonochem.* **17**, 250–257.

Lepoint, T. and Mullie, F. (1994) What exactly is cavitation chemistry? *Ultrason. Sonochem.* **1**, S13–S22.

Li, M., Li, J.-T. and Sun, H.-W. (2008) Sonochemical decolorization of acid black 210 in the presence of exfoliated graphite. *Ultrason. Sonochem.* **15**, 37–42.

Lindermeir, A., Horst, C. and Hoffmann, U. (2003) Ultrasound assisted electrochemical oxidation of substituted toluenes. *Ultrason. Sonochem.* **10**, 223–229.

Liu, Y. and Sun, D. (2007) Development of Fe2O3-CeO2-TiO2/[gamma]-Al2O3 as catalyst for catalytic wet air oxidation of methyl orange azo dye under room condition. *Appl. Catal. B-Environ.* **72**, 205–211.

Lorimer, J. P. and Mason, T. J. (1987) Sonochemistry part 1 – the physical aspects. *Chem. Soc. Rev.* **16**, 239.

Madhavan, J., Grieser, F. and Ashokkumar, M. (2010) Degradation of orange-G by advanced oxidation processes. *Ultrason. Sonochem.* **17**, 338–343.

Mahamuni, N. N. and Adewuyi, Y. G. (2009) Advanced oxidation processes (AOPs) involving ultrasound for waste water treatment: A review with emphasis on cost estimation. *Ultrason. Sonochem.* In Press, Corrected Proof.

Margulis, M. A. (1992) Fundamental aspects of sonochemistry. *Ultrasonics* **30**, 152–155.

Mason, T. J. (2007) Sonochemistry and the environment – Providing a "green" link between chemistry, physics and engineering. *Ultrason. Sonochem.* **14**, 476–483.

Mason, T. J., Collings, A. and Sumel, A. (2004) Sonic and ultrasonic removal of chemical contaminants from soil in the laboratory and on a large scale. *Ultrason. Sonochem.* **11**, 205–210.

Matouq, M. A., Al-Anber, Z. A., Tagawa, T., Aljbour, S. and Al-Shannag, M. (2008) Degradation of dissolved diazinon pesticide in water using the high frequency of ultrasound wave. *Ultrason. Sonochem.* **15**, 869–874.

Mendez-Arriaga, F., Torres-Palma, R. A., Petrier, C., Esplugas, S., Gimenez, J. and Pulgarin, C. (2008) Ultrasonic treatment of water contaminated with ibuprofen. *Water Res.* **42**, 4243–4248.

Merouani, S., Hamdaoui, O., Saoudi, F., Chiha, M. and Petrier, C. (2010) Influence of bicarbonate and carbonate ions on sonochemical degradation of Rhodamine B in aqueous phase. *J. Hazard. Mater.* **175**, 593–599

Naddeo, V., Belgiorno, V., Kassinos, D., Mantzavinos, D. and Meric, S. (2010) Ultrasonic degradation, mineralization and detoxification of diclofenac in water: Optimization of operating parameters. *Ultrason. Sonochem.* **17**, 179–185.

Naddeo, V., Belgiorno, V., Ricco, D. and Kassinos, D. (2009) Degradation of diclofenac during sonolysis, ozonation and their simultaneous application. *Ultrason. Sonochem.* **16**, 790–794.

Okitsu, K., Iwasaki, K., Yobiko, Y., Bandow, H., Nishimura, R. and Maeda, Y. (2005) Sonochemical degradation of azo dyes in aqueous solution: A new heterogeneous kinetics model taking into account the local concentration of OH radicals and azo dyes. *Ultrason. Sonochem.* **12**, 255–262.

Petrier, C., David, B. and Laguian, S. (1996) Ultrasonic degradation at 20 kHz and 500 kHz of atrazine and pentachlorophenol in aqueous solution: Preliminary results. *Chemosphere* **32**, 1709–1718.

Petrier, C., Torres-Palma, R., Combet, E., Sarantakos, G., Baup, S. and Pulgarin, C. (2010) Enhanced sonochemical degradation of bisphenol-A by bicarbonate ions. *Ultrason. Sonochem.* **17**, 111–115.

Prabhu, A. V., Gogate, P. R. and Pandit, A. B. (2004) Optimization of multiple-frequency sonochemical reactors. *Chem. Eng. Sci.* **59**, 4991–4998.

Pradhan, A. A. and Gogate, P. R. (2010) Removal of p-nitrophenol using hydrodynamic cavitation and Fenton chemistry at pilot scale operation. *Chem. Eng. J.* **156**, 77–82.

Rokhina, E. V., Lahtinen, M., Nolte, M. C. M. and Virkutyte, J. (2009) The influence of ultrasound on the RuI3-catalyzed oxidation of phenol: Catalyst study and experimental design. *Appl. Catal. B-Environ.* **87**, 162–170.

Rokhina, E. V., Repo, E. and Virkutyte, J. (2010) Comparative kinetic analysis of silent and ultrasound-assisted catalytic wet peroxide oxidation of phenol. *Ultrason. Sonochem.* **17**, 541–546.

Sayan, E. (2006) Optimization and modeling of decolorization and COD reduction of reactive dye solutions by ultrasound-assisted adsorption. *Chem. Eng. J.* **119**, 175–181.

Sayan, E. and Esra Edecan, M. (2008) An optimization study using response surface methods on the decolorization of Reactive Blue 19 from aqueous solution by ultrasound. *Ultrason. Sonochem.* **15**, 530–538.

Schramm, J. D. and Hua, I. (2001) Ultrasonic irradiation of dichlorvos: Decomposition mechanism. *Water Res.* **35**, 665–674.

Song, Y.-L. and Li, J.-T. (2009) Degradation of C.I. Direct Black 168 from aqueous solution by fly ash/H2O2 combining ultrasound. *Ultrason. Sonochem.* **16**, 440–444.

Suri, R. P. S., Nayak, M., Devaiah, U. and Helmig, E. (2007) Ultrasound assisted destruction of estrogen hormones in aqueous solution: Effect of power density, power intensity and reactor configuration. *J. Hazard. Mater.* **146**, 472–478.

Suslick, K. S. (1990) Sonochemistry. *Science* **247**, 1439–1445.

Tezcanli-Guyer, G. and Ince, N. H. (2003) Degradation and toxicity reduction of textile dyestuff by ultrasound. *Ultrason. Sonochem.* **10**, 235–240.

Tezcanli-Guyer, G. and Ince, N. H. (2004) Individual and combined effects of ultrasound, ozone and UV irradiation: A case study with textile dyes. *Ultrasonics* **42**, 603–609.

Thangavadivel, K., Megharaj, M., Smart, R. S. C., Lesniewski, P. J. and Naidu, R. (2009) Application of high frequency ultrasound in the destruction of DDT in contaminated sand and water. *J. Hazard. Mater.* **168**, 1380–1386.

Thompson, L. H. and Doraiswamy, L. K. (1999) Sonochemistry: Science and Engineering. *Ind. Eng. Chem. Res.* **38**, 1215–1249.

Toma, S., Gaplovsky, A. and Luche, J.-L. (2001) The effect of ultrasound on photochemical reactions. *Ultrason. Sonochem.* **8**, 201–207.

Torres, R. A., Mosteo, R., Pétrier, C. and Pulgarin, C. (2009) Experimental design approach to the optimization of ultrasonic degradation of alachlor and enhancement of treated water biodegradability. *Ultrason. Sonochem.* **16**, 425–430.

Torres, R. A., Sarantakos, G., Combet, E., Ptrier, C. and Pulgarin, C. (2008) Sequential helio-photo-Fenton and sonication processes for the treatment of bisphenol A. *J. Photochem. Photobiol. A-Chem.* **199**, 197–203.

Trabelsi, F., At-Lyazidi, H., Ratsimba, B., Wilhelm, A. M., Delmas, H., Fabre, P. L. and Berlan, J. (1996) Oxidation of phenol in wastewater by sonoelectrochemistry. *Chem. Eng. Sci.* **51**, 1857–1865.

Valcarel, J. I., Walton, D. J., Fujii, H., Thiemann, T., Tanaka, Y., Mataka, S., Mason, T. J. and Lorimer, J. P. (2004) The sonoelectrooxidation of thiophene S-oxides. *Ultrason. Sonochem.* **11**, 227–232.

Velegraki, T., Poulios, I., Charalabaki, M., Kalogerakis, N., Samaras, P. and Mantzavinos, D. (2006) Photocatalytic and sonolytic oxidation of acid orange 7 in aqueous solution. *Appl. Catal. B-Environ.* **62**, 159–168.

Vinodgopal, K., Peller, J., Makogon, O. and Kamat, P. V. (1998) Ultrasonic mineralization of a reactive textile azo dye, remazol black B. *Water Res.* **32**, 3646–3650.

Wang, X., Wang, J., Guo, P., Guo, W. and Wang, C. (2009) Degradation of rhodamine B in aqueous solution by using swirling jet-induced cavitation combined with H2O2. *J. Hazard. Mater.* **169**, 486–491.

Wang, X., Yao, Z., Wang, J., Guo, W. and Li, G. (2008) Degradation of reactive brilliant red in aqueous solution by ultrasonic cavitation. *Ultrason. Sonochem.* **15**, 43–48.

Wang, X. and Zhang, Y. (2009) Degradation of alachlor in aqueous solution by using hydrodynamic cavitation. *J. Hazard. Mater.* **161**, 202–207.

Weavers, L. K. and Hoffmann, M. R. (1998) Sonolytic decomposition of ozone in aqueous solution: Mass transfer effects. *Environ. Sci. Technol.* **32**, 3941–3947.

Weavers, L. K., Ling, F. H. and Hoffmann, M. R. (1998) Aromatic compound degradation in water using a combination of sonolysis and ozonolysis. *Environ. Sci. Technol.* **32**, 2727–2733.

Weavers, L. K., Malmstadt, N. and Hoffmann, M. R. (2000) Kinetics and mechanism of pentachlorophenol degradation by sonication, ozonation, and sonolytic ozonation. *Environ. Sci. Technol.* **34**, 1280–1285.

Wu, C., Wei, D., Fan, J. and Wang, L. (2001) Photosonochemical degradation of trichloroacetic acid in aqueous solution. *Chemosphere* **44**, 1293–1297.

Wu, Z.-L., Ondruschka, B. and Cravotto, G. (2008) Degradation of phenol under combined Irradiation of microwaves and ultrasound. *Environ. Sci. Technol.* **42**, 8083–8087.

Yasman, Y., Bulatov, V., Gridin, V. V., Agur, S., Galil, N., Armon, R. and Schechter, I. (2004) A new sono-electrochemical method for enhanced detoxification of hydrophilic chloroorganic pollutants in water. *Ultrason. Sonochem.* **11**, 365–372.

Zhang, G. and Hua, I. (2000) Cavitation chemistry of polychlorinated biphenyls: Decomposition mechanisms and rates. *Environ. Sci. Technol.* **34**, 1529–1534.

Zhang, Y., Xiao, Z., Chen, F., Ge, Y., Wu, J. and Hu, X. (2010) Degradation behavior and products of malathion and chlorpyrifos spiked in apple juice by ultrasonic treatment. *Ultrason. Sonochem.* **17**, 72–77.

Zhao, G., Gao, J., Shen, S., Liu, M., Li, D., Wu, M. and Lei, Y. (2009) Ultrasound enhanced electrochemical oxidation of phenol and phthalic acid on boron-doped diamond electrode. *J. Hazard. Mater.* **172**, 1076–1081.

Chapter 10
Advanced catalytic oxidation of emerging Micropollutants

Ekaterina V. Rokhina and Jurate Virkutyte

10.1 INTRODUCTION

Catalysis can be generally described as the action of a catalyst, which is a substance that accelerates the rate of a chemical reaction, while not being consumed during the reaction. Three classes of catalysis exist according to the type of the catalyst: (i) homogeneous catalysis, where the catalyst is in the same phase with the reactants, (ii) heterogeneous catalysis, where the catalyst is present in a different phase from the reactants in the reaction, and (iii) biocatalysis, where the catalysts are natural substances such as protein enzymes.

Several points have to be considered when dealing with the catalysts. First, catalysts do not alter the thermodynamics of the reaction. Generally speaking, the reaction would proceed even without the addition of the catalyst, though in some cases too slowly to be observed or be of any use in a given context.

©2010 IWA Publishing. *Treatment of Micropollutants in Water and Wastewater.* Edited by Jurate Virkutyte, Veeriah Jegatheesan and Rajender S. Varma. ISBN: 9781843393160. Published by IWA Publishing, London, UK.

Second, the use of a catalyst does not change the equilibrium composition because it increases the rates of the forward and reverse reactions by the same extent. The presence of a catalyst may enhance one or more of these reactions or even all of them by various degrees, leading to a different overall selectivity and probably changes in the reaction pathway. They achieve this by providing an alternative pathway of lower activation energy for the reaction to proceed.

Thus, catalysts accelerate reactions by orders of magnitude, enabling them to be carried out under the most favorable thermodynamic regime, and at much lower temperatures and pressures. In this way, an efficient catalyst, in combination with the optimized reactor and plant design, are the key factors in reducing both, the investment and operation costs of chemical processes (Chorkendorff and Niemantsverdriet, 2003).

10.2 HETEROGENEOUS CATALYSIS

Heterogeneous catalysis has a practical advantage over homogeneous such as easy separation of the catalyst from the reaction media. However, it has much more complex mechanism of the reaction, combining several steps: (i) transport of the reactants from the bulk of the fluid to the exterior surface of the catalyst (external mass transfer resistance), (ii) transport of the reactants from the surface to the interior of the catalyst through pores (internal mass transfer resistance), (iii) adsorption of the reactants onto the active sites on the internal surface of the catalyst, (iv) reaction of adsorbed reactants to form adsorbed products, (v) desorption of products, (vi) transport of products out of the pores to the external surface of a particle, (vii) transport of products from the external surface of the catalyst to the main body of the fluid (Inglezakis and Poulopoulos, 2006). As the adsorption is involved in the reaction mechanism, the catalytic activity is closely related to the surface area. Generally, large surface areas are able to enhance the performance of the solid catalyst and the increase in surface area can be provided by the catalyst support or an appropriate carrier. Catalyst supports not only provide the desired surface area of the catalyst but also prevent its deactivation keeping the catalytic phase highly dispersed. Also, these supports may or may not be catalytically active substances with a very high adsorption capacity and huge surface area. Moreover, immobilization of a homogeneous catalyst on the solid support provides its heterogenization even if the catalyst was initially completely homogeneous. Carriers traditionally applied in catalysis can be divided into three groups: (1) porous, when the area mainly results from the porous structure of the support; (2) molecular sieves, when very small pores exist in the support and it is their size that decides which molecules are going to react; and (3) monolithic, when monolith structures are utilized, which allow high surface area to low volume, efficient heat removal, and low pressure drop across the catalyst.

Heterogeneous catalysts include metals, metal oxides, metal alloys and mixtures such as polyoxometalates, heteropolyacids (HPAs- oxo-clusters of early transition metals in their highest oxidation state, namely Mo (VI), W (VI) or V (V), perovskite (mineral species composed of calcium titanate), enzymes, etc. New classes of catalysts are also available: catalytic fibers and clothes, biomimetric catalysts, catalytic membranes, catalysts that are able to operate at supercritical conditions and heterogeneous enantioselective catalysts. Hagen (2006) proposed a generalized classification of the catalysts according to their nature (Figure 10.1).

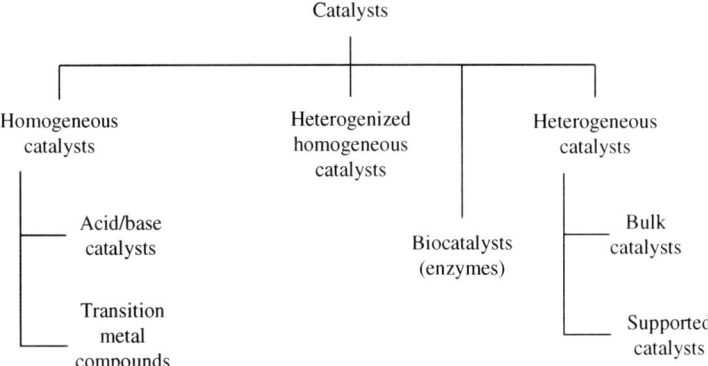

Figure 10.1 Generalized classification of catalysts (Adapted from Hagen, 2006)

10.2.1 Desirable properties of the catalyst

The most fundamental properties of catalysts are their activity, selectivity and stability. The physical properties are also important for its successful application. These properties can be investigated by both, the adsorption methods and various instrumental techniques derived for estimating their porosity and the surface area. The physical characteristics that are really important to the catalyst are its surface area, porosity, particle size distribution, and particle density. However, a number of different processes may contribute to the loss in the catalytic activity over the reaction time. Generally, deactivation of the catalyst can be divided into several groups based on the reasons, which cause changes in reactivity: sintering, leaching of the active component, and poisoning. Loss of catalytic activity during sintering due to the growth of the particle and decrease in surface area (e.g., active catalytic sites) is irreversible process. Poisoning is the formation of carbon or the deposition of impurities or dust onto the catalyst surface. The degree of loss in catalyst activity depends on the operating conditions and the component of the reaction mixture. Catalyst deactivation can be monitored by measuring activity or selectivity as a function of time. After deactivation the catalyst should be regenerated or replaced. Consequently, the desired properties of any catalyst are as follows: high and stable activity, selectivity, controlled surface area and porosity, good resistance to poisons, good resistance to high temperatures and temperature fluctuations and high mechanical strength.

10.3 ENVIRONMENTAL CATALYSIS

Environmental catalysis accounts for (i) waste minimization by providing alternative catalytic synthesis of important compounds without the formation of environmentally unacceptable by-products, and (ii) emission reduction by means of decomposing environmentally unacceptable compounds by using catalysts (Levec and Pintar, 2007). The waste minimization is closely linked with the reaction(s) selectivity and therefore proper choice of the catalyst plays a decisive role. Usually, the reduction of the undesirable emissions or hazardous compounds to the environment does not require the selectivity as the main prerequisite of the efficient catalyst. Transformation of target contaminants without the formation of various by-products is nearly impossible, and therefore the ability of the catalyst to maintain its activity and stability during the by-products decomposition is much more essential than selectivity.

Over the last decade, the increased concern about the deteriorating environment and the subsequent stringent regulations on various emissions provided a basis for the scientific and technological advances in environmental catalysis. Generally, catalysts are called into action to eliminate emissions from mobile (e.g., cars) and stationary (e.g., industry) sources, to take part in liquid and solid waste treatment, as well as to contribute to the efforts to reduce emissions of volatile organic compounds and gases that pose major environmental problems such as photochemical smog and, at a global level, the greenhouse effect (Inglezakis and Poulopoulos, 2006).

There are, however, some distinctive differences between the environmental and the other aspects of catalysis. Importantly, the feed and operation conditions of environmental catalysts cannot be changed in order to increase the conversion or selectivity, as commonly performed for the chemical production catalysts. Also, environmental catalysis has a role to play not only in the industrial processes, but also in emissions control (e.g., auto, ship, and flight emissions), and even in our daily life (e.g., water purifiers). Consequently, the concept of environmental catalysis is vital for a sustainable future. Last but not least, environmental catalysts often operate in much more extreme conditions than catalysts used in the chemical production (Inglezakis and Poulopoulos, 2006).

Interest in innovative methods for wastewater treatment based on catalytic oxidation has been growing rapidly, as these methods have been confirmed the most efficient and powerful to purify wastewater (Levec and Pintar, 2007). In the environmental engineering field, the great interest lies in the determination of the catalytic mechanism for the generation of the reactive species, mostly hydroxyl radical ($\cdot$OH, $E_0 \approx 1.8$ V) by the destruction of some oxidizing compounds (ozone, hydrogen peroxide, etc.) in the presence of the catalyst and/or irradiations such as ultraviolet light (UV), ultrasound (US), microwave (MW), etc.

10.4 ADVANCED CATALYTIC OXIDATION PROCESSES FOR THE REMOVAL OF EMERGING CONTAMINANTS FROM THE AQUEOUS PHASE

10.4.1 Catalytic wet peroxide oxidation processes (CWPO)

Catalytic wet peroxide oxidation reactions are the reactions, which involve hydrogen peroxide as an oxidant (i.e., source of free radicals) and the suitable catalyst to provoke the hemolytic decomposition of hydrogen peroxide to hydroxyl radical as a primary radical.

10.4.1.1 Homogeneous Fenton process

The most popular catalytic wet peroxide oxidation technology applied for the removal of micropollutants is Fenton process, which is characterized by the utilization of a Fenton reagent – iron catalyst and hydrogen peroxide, as inexpensive and abundant chemicals (Catalkaya and Kargi, 2007; Burbano *et al.*, 2008; Li *et al.*, 2009a). Classic Fenton is a homogeneous catalytic process, involving the general set of chain reactions according to the classic interpretation of Haber-Weiss mechanism, which can be divided into three periods:

(1) Initiation (initial reactions):

$$H_2O_2 \rightarrow H_2O + 1/2 O_2 \tag{10.1}$$

$$H_2O_2 + Fe^{2+} \rightarrow Fe^{3+} + \cdot OH + OH^- \tag{10.2}$$

$$H_2O_2 + Fe^{3+} \rightarrow Fe(OOH)_2 + +H^+ \leftrightarrow Fe^{2+} + \cdot HO_2 + H^+ \qquad (10.3)$$

$$OH^- + Fe^{3+} \rightarrow Fe(OH)^{2+} \leftrightarrow Fe^{2+} + OH^- \qquad (10.4)$$

(2) Propagation:

$$HO\cdot + H_2O_2 \rightarrow HO_2\cdot + H_2O \qquad (10.5)$$

$$HO_2\cdot + H_2O \rightarrow HO\cdot + H_2O + O_2 \qquad (10.6)$$

$$HO_2\cdot + HO_2^- \rightarrow HO\cdot + HO^- + O_2 \qquad (10.7)$$

(3) Termination:

$$Fe^{2+} + HO\cdot \rightarrow Fe^{3+} + OH^- \qquad (10.8)$$

$$HO_2\cdot + Fe^{3+} \rightarrow Fe^{2+} + H^+ + O_2 \qquad (10.9)$$

$$HO\cdot + HO_2\cdot \rightarrow H_2O + O_2 \qquad (10.10)$$

$$HO\cdot + HO\cdot \rightarrow H_2O_2 + O_2 \qquad (10.11)$$

However, number of scientists proposed the alternative mechanism, first suggested by Bray and Gorin (1932), in which the ferryl ion, $[Fe_{IV}O]^{2+}$, is assumed to be the active intermediate instead of $\cdot OH$:

$$Fe^{2+} + H_2O_2 \rightarrow [Fe_{IV}O]^{2+} + H_2O \qquad (10.12)$$

Ensing et al. (2002) performed static density functional theory (DFT) calculations to study the active species produced by hydrated Fenton's reagent in vacuo as well as ab initio (DFT) molecular dynamics (AIMD) simulations of Fe^{2+} and H_2O_2 in aqueous solution. They concluded that the ferryl ion is easily formed when hydrogen peroxide coordinates to an iron (II) ion in water, which confirms the reaction mechanism first proposed by Bray and Gorin (X.12) (Bray and Gorin, 1932). Moreover, formation of the ferryl ion from hydrated Fenton's reagent in vacuo was found to be energetically favored over the formation of free hydroxyl radicals (Ensing and Baerends, 2002; Ensing et al., 2003).

Probably in practice, the proposed mechanisms compete with each other depending on the reaction conditions, such as metal ligands present in the reaction, Fe ions to H_2O_2 ratio, pH, the presence of oxygen and the amount of organic substrate to be oxidized. All these parameters are case-specific and should be carefully adjusted for the successful application of Fenton process.

Regardless the lack of a sole concept to explain the mechanism of a Fenton reaction, it is still one of the most effective means to destroy a great variety of micropollutants such as dyes, pharmaceuticals and pesticides in terms of degradation efficiencies and overall treatment costs (Devi et al., 2009; Poerschmann et al., 2009; Zimbron and Reardon, 2009; Fu et al., 2010).

For instance, Hsueh et al. (2005) demonstrated the efficiency of Fenton reaction to degrade selected azo dyes with initial concentrations of (0.1 mM): Red MX-5B, Reactive Black 5 and Orange G at low concentration of Fenton reagent. They reported that even concentrations less than 10 mgL^{-1} of iron and ratio of iron to hydrogen peroxide of 1 to 10 were sufficient for dye removal at the optimum pH range of 2.5–3.0 (Hsueh et al., 2005).

Elmolla and Chaudhuri (2009) reported the effectiveness of Fenton reaction on biodegradability improvement and a complete mineralization of antibiotics amoxicillin (AMX), ampicillin (AMP) and cloxacillin (CLX) in aqueous solution. Initial AMX, AMP and CLX concentrations were 104, 105 and 103 mg L^{-1}, respectively, and the operating conditions were H_2O_2/COD molar ratio of 3, H_2O_2/Fe^{2+} molar ratio of 10 and the initial COD of 520 mg L^{-1} with pH in the range of 2–4. Thus, under the optimum operating conditions (COD/H_2O_2/Fe^{2+} molar ratio 1:3:0.30, and pH 3), a complete degradation of the antibiotics occurred in 2 min. Biodegradability improved from 10 to 0.37 in 10 min., as well as COD and DOC degradation were 81.4% and 54.3%, respectively in 60 min. Maximum biodegradability (BOD$_5$/COD ratio) improvement was achieved in 10, 20 and 40 min. at initial concentrations of 100, 250 and 500 mg L^{-1}, for AMX, AMP and CLX, respectively, in aqueous solution. Also, a complete min.eralization of organic carbon and nitrogen occurred in 60 min. (Elmolla and Chaudhuri, 2009).

Kassinos et al. (2007) demonstrated the feasibility of homogeneous Fenton process for the oxidation of selected pesticides atrazine (ATZ) and fenitrothion (FNT) in aqueous solutions. The maximum TOC removal for ATZ was 50% in the presence of 0.45 mM of the catalyst (FeSO$_4$ × 7H$_2$O, 98.0% purity) and 33.2 mM of the hydrogen peroxide for 24 h of oxidation and 38% were obtained in the presence of 1.79 mM of the catalyst and 33.2 mM of the oxidant during 2 h of the reaction time. For FNT, the maximum decrease in TOC (71%) was obtained in the presence of 0.89 mM of the catalyst and 33.2 mM of the oxidant during 24 h whereas 70% was already obtained during the initial 2 h of the treatment process. For the binary mixture, the maximum of 58% reduction in TOC was achieved in the presence of 0.45 mM of the catalyst and 49.9 mM of the oxidant while similar result was obtained by using the same amount of Fe (II) reagent but different amount of the oxidant during 24 h of the oxidation process. Thus, there was no significant difference in TOC removal regardless the reaction times (2 h and 24 h) applied. This can be attributed to (i) the exhaustion of free hydroxyl radicals, and (ii) to the formation of other radical species (e.g., hydroperoxyl radicals) with significantly lower oxidation potential (Kassinos et al., 2009).

10.4.1.2 Heterogeneous Fenton process

Heterogeneous Fenton has a number of advantages over a traditional homogeneous Fenton reaction: (i) easy catalyst separation makes no need to employ additional extraction units, facilitating all the operations in the treatment of the effluent; (ii) it operates in close to neutral pH, thus eliminating the need for the acidification (pH 3) or further neutralization, preventing the generation of sludge; and (iii) the system can be recycled/regenerated by reducing Fe^{3+} species. Various iron containing solids are utilized for the heterogeneous Fenton reaction, such as Fe_2O_3 and $Fe_2Si_4O_{10}(OH)_2$, Fe(II) sustained on various supports such as zeolites, Al_2O_3 and SiO_2. Various types of mineral carriers such as hematite, saponite, magnetite, ferrhydrite, lepidocrocite, and non-minerals, such as mixed element oxides, steel dust and steel slag were introduced as active supports for the Fenton reagent. Nonetheless, some of these aforementioned systems showed relatively low activity and a tendency to iron leaching at low pH, which resulted in the classical homogeneous Fenton mechanism (Oliveira et al., 2008).

Numerous scientific attempts were demonstrated to utilize heterogeneous Fenton reagent to degrade emerging contaminants on a laboratory scale (Barreiro et al., 2007; Ramirez et al., 2007; Papic et al., 2009). Thus, Kusic et al. (2007) compared traditional Fenton (ferrous and ferric sulphates, $FeSO_4 \times 7H_2O$ and $Fe_2(SO_4)_3 \times 9H_2O$ as catalysts) and Fenton-like (iron powder as a catalyst) processes to decompose two reactive dyes with initial concentrations of 20 mg L^{-1}, C.I. Reactive Blue 49 (RB 49) with anthraquinone chromophore and C.I. Reactive Blue 137 (RB 137) with azo chromophore. In both cases color removal was over 95%, while mineralization of the contaminants ranged between 34 and 72% depending on the dye structure and the type of the reaction. The highest level of mineralization i.e., maximal TOC removal of 72.1% and 47.3%, was obtained for RB49 and absorbable organic halides (AOX), respectively, degraded by the traditional Fenton process (Fe^{2+} to H_2O_2 ratio of 1 to 20, c (Fe^{2+}) 0.5 mM at pH 3). On the contrary, the highest degradation of RB137 (45.3% TOC removal) was achieved utilizing Fenton-like processes (Fe^{3+} to H_2O_2 ratio of 1 to 40, c (Fe^{3+}) 0.5 mM at pH 3). The molecular structure of the treated compound was found to be a very important factor. Both, azo chromophores in RB137 and anthraquinonic chromophore in RB 49 were susceptible to OH radical attack (Kusic et al., 2007). However, it was more likely that the oxidation would preferably occur on the chromophore structure rather than on the dye molecule skeletons (Figure 10.2).

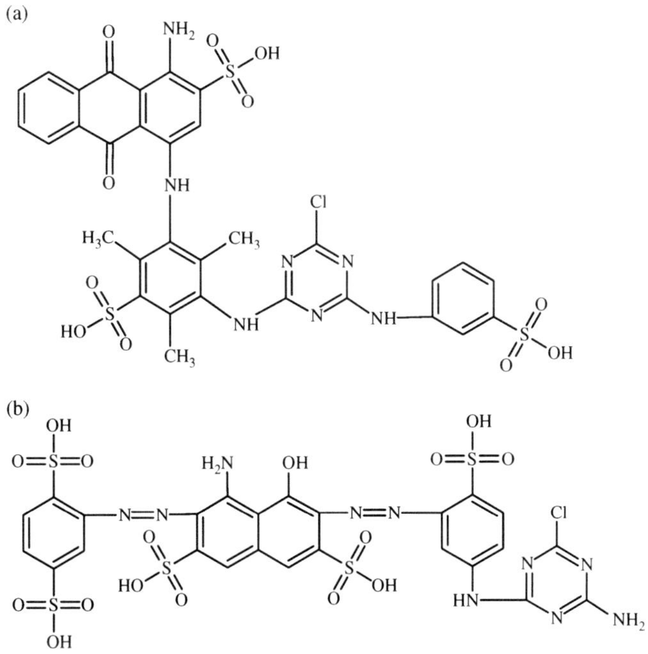

Figure 10.2 Molecular structure of dyes (a) RB49, (b) RB137

10.4.1.3 Heterogenized catalyst for the micropollutants removal

By far, one of the widely used approaches to heterogenize homogeneous catalyst is its immobilization on the active/inert support. Sorokin *et al.* (1995) introduced novel homogeneous iron (III)-tetrasulfophthalocyanine (Fe(III)–TsPc) catalyst (Figure 10.3).

Kim *et al.* (2008) tested both forms, homogeneous iron (III)-tetrasulfophthalocyanine (Fe(III)–TsPc) and immobilized catalyst for the degradation of several micropollutants such as bisphenol-A (BPA), diclofenac (DCF), cefaclor (CFL), and ibuprofen (IBU). They developed a new hybrid system that combines the catalytic oxidation (non-immobilized catalyst) with nanofiltration (NF), in which the catalyst was retained in the system by the NF membrane and was used to continuously oxidize BPA. The catalyst activity varied at different pH, which could be attributed to the various chemical forms of Fe(III)–TsPc, such as µ-oxo dimer (TsPcFe(III)–O–Fe(III)TsPc), the stacked monomer (n[TsPcFe(III)–OH]) and the monomeric (TsPcFe(III)–OH) species (Figure 10.4).

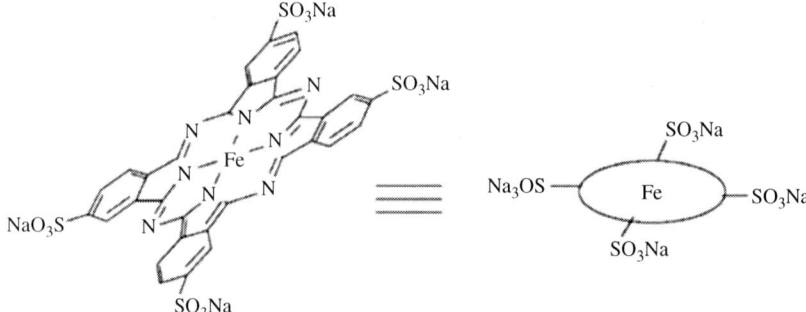

Figure 10.3 Iron(III)–tetrasulfophthalocyanine (Fe(III)–TsPc) (Adapted from Sorokin et al., 2002)

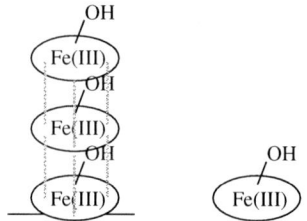

Figure 10.4 Chemical forms of Fe(III)–TsPc (a) μ-oxo dimmer, (b) stacked monomer, and (c) monomer

Monomeric Fe(III)–TsPc with the corresponding pH 3 was found to be the most effective species for BPA decomposition. However, an increase in hydrogen peroxide dose led to the deactivation of the catalyst surface. Essentially, deactivation can be induced by the direct attack of the phtallocianyde by hydrogen peroxide, resulting in the ring opening of the phthalocyanine and degradation of the chromophore and/or by the formation of several byproducts, which leads to the catalyst leaching, e.g., oxalic acid (Kim et al., 2008). Generally, BPA oxidation catalyzed by Fe(III)–TsPc/H_2O_2 resulted in > 90% decomposition of BPA after approx. 40 s under weakly acidic conditions (pH ≤ 4.5).

To overcome rapid deactivation of the homogeneous Fe(III)–TsPc catalyst, Sorokin and Meunier (1994) immobilized it onto amberlite (Amb), which was conventionally used anion exchange resin (Sorokin and Meunier, 1994;

Sorokin et al., 1995). Moreover, the performance of heterogeneous FeTsPc-immobilized amberlite system in the presence of H_2O_2 was investigated for the degradation of BPA, and three pharmaceuticals, DCF, CFL and IBU. In such a heterogeneous system the catalyst demonstrated high stability and reactivity at neutral pH, however, the reaction time increased. Furthermore, Kim et al. (2008) also studied the effect of the hydrophilicity and electrical charge of the pollutants on their removal efficiencies over FeTsPc-immobilized amberlite. Based on the pKa values of each compound and the reaction pH, a negative charge was the most important factor to obtain the highest removal efficiency; hydrophilic DCF and relatively hydrophobic IBU were both readily removed with FeTsPc-Amb owing to their negative charge (Kim et al., 2009). Second important factor was hydrophilicity; electrically neutral and hydrophobic BPA was removed more slowly by adsorption over hydrophobic FeTsPc-Amb than DCF and IBU. Partially charged CFL at the circum neutral reaction pH was removed by an ion-exchange mechanism but the removal efficiency was the lowest because of its hydrophilicity. Supported FeTsPc (300 mg) used in wet peroxide oxidation led to almost a complete removal within 2 h (initial concentration of 2 mg L^{-1}) of BPA, DCF and CFL at circum-neutral pH.

10.4.2 Other metal catalysts in wet peroxide oxidation of micropollutants

Other metals than iron may also be successfully used to produce hydroxyl radical analogues to Fenton reaction. Thus, there are several papers reporting the oxidation of micropollutants utilizing metal-catalyzed wet peroxide oxidation process in the aqueous phase (Kim and Lee, 2004; Aravindhan et al., 2006; Fathima et al., 2008; Rivas et al., 2008; Kondru et al., 2009). However, information on the removal of some important micropollutants such as pharmaceuticals is currently lacking.

Duarte et al. (2009) doped or/and impregnated some metals (Fe, Co, Ni)-on the carbon aerogels catalysts and utilized them for the decolorization of Orange II at atmospheric pressure and 30°C temperature. They found that the surface area of the fabricated catalysts was less that the surface area of the parent aerogel (1032 m^2 g^{-1}) and the catalytic performance highly depended on the metal present in the matrix. For instance, Co and particularly Fe catalysts were highly active in degradation of Orange II, while Ni was almost inactive or only slightly active in the removal of the same compound in aqueous solutions (Duarte et al., 2009).

In addition, Oliveira *et al.* (2008) prepared Nb-substituted goethites and studied their efficiency in the oxidation of methylene blue dye. The characterization of fabricated catalysts revealed that Nb substitution in goethite created surface oxygen vacancies in the Nb-goethite and decreased goethite crystallinity with an increase in BET surface area and the developed pore structure. In the presence of pure goethite, a rather low activity towards methylene blue dye oxidation was observed with only 15% color reduction whereas in the presence of Nb substituted goethite Nb11, catalytic discoloration of up to 85% was observed after 120 min. of the treatment (Oliveira *et al.*, 2008). The reaction products were identified *in situ* using ESI-MS and the reaction scheme was proposed (Figure 10.5).

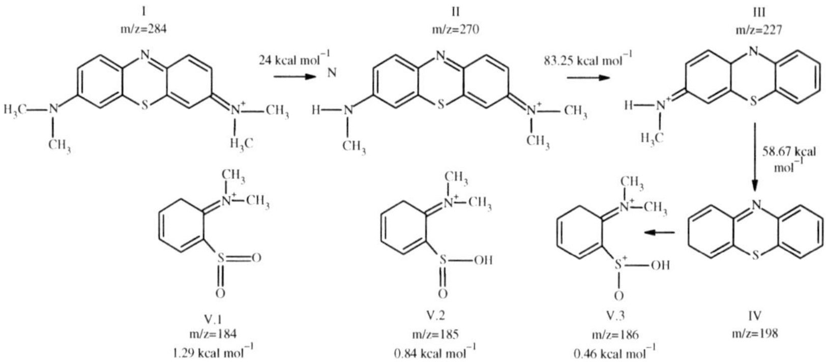

Figure 10.5 The reaction scheme of methylene blue dye (m/z = 284) oxidation with all the identified intermediates by Nb-substituted goethites in the presence of H_2O_2 with the energy barrier for every intermediate (Adapted from Oliveira *et al.*, 2008)

It was determined that the mechanism of the reaction over Nb-goethite catalyst could, therefore, involve oxygen vacancies at the goethite surface, which differed from the heterogeneous iron-based Fenton reaction described in literature (Oliveira *et al.*, 2008). The energy barriers for the formation of each intermediate, calculated by the computational methods suggested two possible mechanisms that could also be applied to other than Nb-based catalysts: (i) reaction at the oxygen vacancy sites in e.g., Nb-substituted goethites, yielding the intense fragments m/z = 270, 227, and 187 (Figure 10.5), which were associated with intermediates II, III, and IV, respectively (Figure 10.5); and (ii) a Fenton mechanism resulting in fragments m/z = 105, 129, and 351.

10.4.3 Catalytic ozonation of micropollutants

The mechanism of micropollutants removal by ozonation can proceed either via direct oxidation in the presence of molecular ozone or via indirect oxidation with ·OH radicals that are formed by the decomposition of ozone in alkaline conditions. Figure 10.6 depicts the principle scheme of ozone reactions at relevant pH.

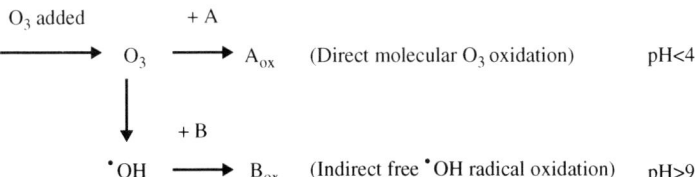

Figure 10.6 Ozonation in aqueous solution at pH < 4 and pH > 9. (Adapted from Chandrasekara Pillai et al., 2009)

The high values of the oxidation rate constants via ·OH radicals (in the range of 10^9) indicate that the removal of organic compounds by these oxidants is very fast but unfortunately non-selective. Molecular O_3 is selective and reacts mainly with certain functional groups such as phenolics. In real (e.g., multi-componential) waters, a large fraction of ·OH radicals (which are much less selective and much more reactive than ozone) are scavenged by the water matrix therefore, most likely the principle pathway of oxidation is direct reaction of ozone with the contaminant (von Gunten, 2003). However, Huber et al. (2003) demonstrated that at circum neutral pH (6–7), the reaction rate constants for O_3 and ·OH radical with micropollutants (e.g., estrogens) are practically the same (7 · 10^9 M^{-1} s^{-1} and 9.8 · 10^9 M^{-1} s^{-1}, respectively). Therefore the pH of the reaction is the crucial parameter to be considered during ozonation in order to select the reaction pathway (radical/non-radical) (Huber et al., 2003). The most widely utilized as catalysts in the catalytic ozonation of organic micropollutants are transition metals such as Fe(II), Mn(II), Ni(II), Co(II), Cd(II), Cu(II), Ag(I), Cr(III), Zn(II), metal oxides (e.g., TiO_2, MnO_2, CeO_2, etc.), activated carbon (AC) and supported metals (e.g., Co/Al_2O_3, Ru/Al_2O_3, Ru/CeO_2 etc.) (Delanoe et al., 2001; Tong et al., 2003; Skoumal et al., 2006; Yunrui et al., 2007; Rosal et al., 2008; Chandrasekara Pillai et al., 2009; Faria et al., 2009; Rosal et al., 2009).

Importantly, the catalyst enhances ozone decomposition with the production of active free radicals or by the acceleration of molecular ozone reactions. Catalytic ozonation process is rather complex and not yet fully understood (Kasprzyk-Hordern et al., 2003). However, it obeys general rules of the catalytic reactions and therefore can be divided into two main groups: homogeneous and heterogeneous. Generally, the oxidation of organic micropollutants by O_3 is an efficient process only for compounds with functional groups such as amino groups, activated aromatic systems (i.e., an electron donor group) or double bonds (von Gunten, 2003).

Chandrasekara Pillai et al. (2009) studied homogeneous catalytic ozonation (O_3/Fe^{2+}) and the combination of several techniques (O_3/H_2O_2/Fe^{2+}/UV) for the terephthalic acid (TPA) degradation. The treatment was performed using 0.089 g L^{-1} of O_3, 0.50 mM of Fe^{2+} and pH 4.5–6.5. The highest COD removal rate constant of $7.28 \pm 0.60 \times 10^{-3}$ min^{-1} was observed at pH 4. When Fe^{2+} was added to the solution, it catalyzed ozone decomposition to produce additional ·OH radicals either through the formation of O_3^- anion or through the formation of $(FeO)^{2+}$:

$$Fe^{2+} + O_3 \rightarrow Fe^{3+} + O_3^- \quad (10.13)$$

$$O_3^- + H^+ \leftrightarrow HO_3 \rightarrow \cdot OH + O_2 \quad (10.14)$$

$$Fe^{2+} + O_3 \rightarrow (FeO)^{2+} + O_2 \quad (10.15)$$

$$(FeO)^{2+} + H_2O \rightarrow Fe^{3+} + \cdot OH + OH^- \quad (10.16)$$

The combination of catalytic (0.50 mM Fe^{2+}) ozonation with simultaneous addition of H_2O_2 (200 mM) and UV (O_3/H_2O_2/Fe^{2+}/UV) was also tested for TPA degradation. During such an unusual treatment, H_2O_2 and Fe^{2+} were added in a 'one-step addition' or using a 'two-step addition' with an initial amendment of 100 mM H_2O_2, and after 60 min., 100 mM H_2O_2 and 0.50 mM of Fe^{2+} for up to 240 min. of the treatment. Thus, the COD removal was significant for the 'two-step addition' approach throughout the treatment and reached 0.38 and 0.10 (62% and 90% degradation efficiency) at 120 and 240 min., respectively, in comparison to 0.55 and 0.27 (45% and 73% degradation efficiency) for the 'one-step addition' method (Chandrasekara Pillai et al., 2009). However, the energy efficiency, estimated as a 'figure-of-merit', (EE/O) was the lowest during catalytic ozonation (21.3 kWh m^{-3} $order^{-1}$), whereas the EE/O of the most effective treatment in terms of the degradation efficiency was almost 1.5-fold higher (32.1 kWh m^{-3} $order^{-1}$).

Rosal et al. (2009) used commercial TiO_2 (80/20 mixture of the anatase/rutile phase, $S_{BET} = 52 \pm 2\, m^2 g^{-1}$) as a catalyst in catalytic oxidation of clofibric acid by ozone. The pseudo-second catalytic rate constant was 2.17×10^{-2} L $mmol^{-1}$ s^{-1} at pH 3 and 6.80×10^{-1} L $mmol^{-1}$ s^{-1} at pH 5, with up to a three-fold increase in comparison to the non-catalytic constants ($8.16 \times 10^{-3} \pm 3.4 \times 10^{-4}$ L $mmol^{-1}$ s^{-1}) using catalyst load of 1g L^{-1}. Thus, clofibric acid was removed in less than 5 min. However, the enhancement of the treatment in the presence of the catalyst was attributed to the adsorption of organics on the catalytic sites rather than to the promotion of ozone decomposition. The mechanism involved in the reaction was probably an interaction of bulk oxidants or oxidized surface sites with the adsorbed organic contaminant.

Faria *et al.* (2009) studied the catalytic ozonation of three commercial dyes from different chemical classes at initial concentrations of 50 mgL^{-1}: acid azo dye – CI Acid Blue 113 (AB113), and the reactive dyes – CI Reactive Yellow 3 (RY3) and CI Reactive Blue 5 (RB5), with azo and anthraquinone chromophores, respectively. Three different catalysts, such as activated carbon (S_{AC0} = 909 m^{2}g^{-1}), cerium oxide (S_{Ce-O} = 72 m^{2}g^{-1}), and a composite of activated carbon and cerium oxide ($S_{AC0-Ce-O}$ = 583 m^{2}g^{-1}), prepared by precipitation were investigated. For all the dyes, the adsorption on AC_0 was negligible, whereas non-catalytic ozonation led to a nearly complete removal of all dyes in 10 min. regardless their composition, however, the complete mineralization, monitored by changes in TOC was not attained. The acid dye AB113 was found to be the least refractory compound among the studied dyes. Farria and co-workers (2009) also investigated the proposed catalytic system for the real wastewater treatment. Ozonation showed excellent decolorization potential (nearly 100% after 5 min.) with negligible mineralization (less than 4% of TOC removal). In the presence of activated carbon, the removal of TOC after 30 min. was approximately 57%, which was significantly higher in comparison to 30% TOC removal during the non-catalytic ozonation. The Orange II ozonation mechanism catalyzed by the ceria-activated carbon composite was assumed to comprise both, the surface reactions, similar to what occurs in activated carbon promoted ozonation, and the liquid bulk reactions involving ·OH radicals, which resulted from the catalytic decomposition of ozone on the surface of the catalyst.

10.4.4 Photocatalytic degradation of micropollutants

Photocatalysis is based on the double aptitude of the photocatalyst (semi-conductor material) to simultaneously adsorb both, the reactants and efficient photons. The popularity of the technique is based on the fact that semiconductors are inexpensive and can easily mineralize various organic compounds.

A semiconductor has a manifold of electron energy levels filled with electrons – the valence band (VB) and also many higher energy levels that are largely vacant – the conduction band (CB). The energy difference between these two bands is called the band gap energy (E_{bg}). A general photocatalytical reaction can be summarized as following:

$$\text{Catalyst} + h\upsilon \rightarrow \text{Catalyst} \ (e_{cb}^- + h_{vb}^+) \quad (10.17)$$

where, e_{cb}^- and h_{vb}^+ are the electrons in the conduction band and the electron vacancy in the valence band, respectively. The change in the Gibbs free energy for this reaction is usually negative during photocatalysis and positive during photosynthesis.

Both these electron–hole pair entities can migrate to the catalyst surface, where they can enter in a redox reaction with other species present on the surface. In most cases, h_{vb}^+ can easily react with surface-bound H_2O to produce ·OH radicals, whereas, e_{cb}^- can react with O_2 to produce superoxide radical anion of oxygen (Rauf and Ashraf, 2009):

$$H_2O + h^+{}_{vb} \rightarrow \cdot OH + H^+ \quad (10.18)$$

$$O_2 + e^-{}_{cb} \rightarrow \cdot O^-{}_2 \quad (10.19)$$

$$\cdot O^-{}_2 + H^+ \rightarrow \cdot OOH \quad (10.20)$$

This set of the reactions prevents the combination of the electron and the hole which are produced in the first step. Subsequently, the OH, ·OOH and $\cdot O_2^-$ produced in the above manner can then react with the dye to form other species and is thus responsible for the elimination of micropollutants, while the electron/hole pairs may also directly react with organic substrate on the surface:

$$\cdot O^-{}_2 + H_2O \rightarrow H_2O_2 \quad (10.21)$$

$$H_2O_2 \rightarrow 2 \cdot OH \quad (10.22)$$

$$\cdot OH + \text{micropollutant} \rightarrow \text{micropollutant}_{ox} \quad (k = 10^9 - 10^{10} M^{-1} s^{-1}) \quad (10.23)$$

$$\text{micropollutant} + e^-{}_{cb}/h^+{}_{vb} \rightarrow \text{micropollutant}_{red} \quad (10.24)$$

A schematic presentation of oxidative species generation in a photocatalytic study is shown in Figure 10.7.

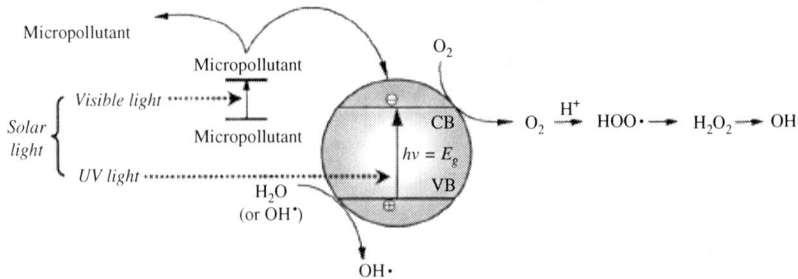

Figure 10.7 Mechanism of generation of oxidative species in photocatalytic oxidation of micropollutant (Adapted from Rauf and Ashraf, 2009)

The standard light used in the photocatalytic degradation of organic compounds can be divided into three groups according to the wave length ranges: UVA (λ 315–400 nm), UVB (λ 285–315 nm), and UVC ($\lambda < 285$ nm). The intensity and wavelength of radiation have a significant influence on the destruction rates of organic pollutants. Recent work demonstrated the feasibility of sunlight or simulated sunlight (visible light with $\lambda > 300$ nm) as a free and renewable energy source in the so-called solar photocatalytic process that may be efficiently utilized for wastewater treatment (Kaur and Singh, 2007; Song et al., 2007; Rafqah et al., 2008; Ji et al., 2009; Song et al., 2009; Zapata et al., 2009). The main advantage of such a process is prevention of lamps by using an environmentally friendly, cheap and widely available energy source, i.e., the sun. In fact, the determination of a parameter called the "environmental-economic index" (EEI) showed that solar driven photocatalysis is superior to the lamp-driven photocatalysis to provide a balance between both, the environmental and economic aspects. However, the band edges of traditional photocatalysts (e.g., TiO_2, ZnO, etc.) lie in the UV region which drives them inactive under visible light irradiation. Therefore, the modification of electronic and optical properties of the semiconductor materials for their efficient use under visible light irradiation is required. Rehman et al. (2009) reviewed the existing strategies of making TiO_2 and ZnO visible light active. They reported that visible light activity can be induced in TiO_2 and ZnO by surface modification via organic materials/semiconductor coupling (conjugated polymer sensitization) and band gap modification by doping with metals (e.g., Cu, Co, Mn, Ni, etc.) and nonmetals (B, C, N, S, etc.), co-doping with nonmetals, creation of oxygen vacancies and oxygen sub-stoichiometry. However, such an effective mean as the surface complexation (creation of oxygen vacancies) can be detrimental due to that surface complexes can undergo self-degradation with the course of photocatalytic reaction. Although oxygen vacancies can activate TiO_2 under visible light they can also promote electron–hole recombination process (Rehman et al., 2009). Also, doping with transition metals can be beneficial to nonmetals doping if the amount of dopants and the reaction conditions are properly optimized.

10.4.4.1 Titanium dioxide catalyzed degradation of micropollutants

The highest photocatalytic performance and maximum quantum yields are always associated with titanium dioxide (TiO_2). Titanium dioxide occurs in nature as well-known minerals rutile, anatase and brookite. The most common and stable form is rutile, whereas anatase is the most widely used in photocatalytic processes. Anatase and brookite both convert to rutile upon heating. Various forms of TiO_2 such as P25 Degussa TiO_2, TiO_2 Tiona PC 100, TiO_2 Tiona PC 500 and etc. are commercially available and extensively used for wastewater treatment processes. TiO_2 may be applied as an aqueous suspension, immobilized thin films, doped with various chemical elements or supported onto a solid support. The surface properties of TiO_2 such as the surface charge, the size of the catalyst particle during aggregation, and the band edge position of TiO_2 are affected by pH of the reaction due to the ability of the photocatalyst surface to be protonated in acidic and de-protonated in alkaline medium (Harir *et al.*, 2008).

Akpan and Hameed (2009) reviewed 125 papers from 1991–2008 and summarized the effects of operating parameters on the photocatalytic degradation of textile dyes using TiO_2-based photocatalysts. They also gave an insight on the TiO_2 preparation methods and the effects of the initial pH of the solution, various oxidizing agents, temperature at which the catalysts must be calcined, dopant(s) content and the catalyst loading on the photocatalytic degradation of any dye in wastewater. The most common preparation method reported was sol–gel with several modifications such as sol–gel, ultrasonic assisted sol–gel, aerogel, method similar to sol–gel, sol–gel and photo-reductive decomposition, precipitation, two-step wet chemical, and extremely low temperature precipitation methods (Akpan and Hameed, 2009). Also, Akpan and Hammed (2009) anticipated that the nature of the dye was an essential factor due to the structural properties and difference in functional groups that could favor different process conditions such as pH, temperature, etc.

There were successful attempts demonstrated to not only remove organic dyes but also pesticides from wastewater. For instance, Rafqah *et al.* (2006) used different types (Degussa P25, PC50 and PC500) of TiO_2 suspensions (S_{BET} = 55 m^2 g^{-1}) under the wavelength range of 300–450 nm for the photocatalytic degradation of the antimicrobial triclosan (5-chloro-2-(2,4-dichlorophenoxy)-phenol). Triclozan exhibited a band with a maximum at 280 nm and a shoulder at 232 nm with the molar extinction coefficient of 4200 mol^{-1} L cm^{-1} at the maximum absorption. Under the experimental conditions applied (initial triclozan concentration of 3.3×10^{-5} mol L^{-1}, catalyst concentration 1.0 g L^{-1}, pH 5) the direct photolysis as well as the adsorption on TiO_2 powder were clearly negligible (less than 8% after 60 min. of irradiation time). However, in the presence of TiO_2 and UV light the triclozan disappeared following the first order reaction kinetics with a half-lifetime of 10 min. Rafgah and co-workers (2006) also studied the degradation of triclosan in natural waters (from the river Allier near Clermont Ferrand in France (DOC 5.3 mg L^{-1}). They found that the addition of TiO_2 exhibited two times slower initial degradation rates when compared to those obtained in pure water due to the presence of humic substances. Moreover, 90% of mineralization was achieved within 25 h of the irradiation time (Rafqah et al., 2006). The main products of triclozan photodegradation were chlorocatechol and 2,4-dichlorophenol. It was hypothesized that these products appeared as a result of three primary processes applicable to any micropollutant with the phenolic group: (i) homolytic scission of C–O bond leading to the formation of 2,4-dichlorophenol and chlorocatechol, (ii) hydroxylation of the phenolic group and (iii) dechlorination process leading to the formation of 5-chloro-2-(4-chlorophenoxyl)phenol. No dioxin derivatives were detected under the applied experimental conditions proving that dioxins were exclusively formed only if triclosan (in its anionic form) absorbed light at $\lambda < 300$ nm (Rafqah et al., 2006).

Harir et al. (2008) studied the photocatalytic degradation of imazamox in the presence of hydrogen peroxide. The degradation of such herbicide can be efficiently described by a pseudo first-order rate expression according to the Langmuir–Hinshelwood kinetic model, commonly used to explain most of the photocatalytic oxidation processes. The optimum values for the abatement of imazamox were determined as pH 5.0, $1 gL^{-1}$ (TiO_2) and 10 mM of H_2O_2. The point of zero charge (pzc) of TiO_2 P25 is at pH 6.25. Thus, the TiO_2 surface would remain positively charged in acidic medium (pH $<$ 6.25) and negatively charged in alkaline medium (pH $>$ 6.25). This behavior affected the adsorption and desorption properties of TiO_2. Furthermore, the structures of contaminants change with changes in pH, depending on the availability of ionisable organic functional groups, which can be protonated or deprotonated depending on pH of the solution.

10.4.4.2 Photo-Fenton process for the degradation of micropollutants

The addition of UV irradiation to a conventional Fenton reaction (reaction 10.2) not only leads to the formation of additional hydroxyl radicals but also to a recycling of the ferrous catalyst (Maletzky and Bauer, 1998). Alternative source of hydroxyl radicals is UV photolysis of hydrogen peroxide, but its contribution to the photo-Fenton method is negligible, because hydrogen peroxide and its dissociated form poorly absorb light (Krutzler and Bauer, 1999). Main photo-Fenton reaction can be rewritten as (10.27), since ferric ions usually exist as hexaquo complexes that dissociate in the pH values of 2 to 4 according to Eqs. (10.25) and (10.26). The photo-Fenton-reaction is mainly metal to ligand charge transfer (LMCT) that produces the hydroxyl radical. In addition, it can be compared to reactions like Eq. (10.28) that are also of great importance for the contaminant degradation (Safarzadeh-Amiri et al., 1996). As indicated by the Eq. (10.28), no hydroxyl radicals are produced, but organic ligands like carboxylates are oxidized directly via a LMCT reaction that is usually faster than the reaction of free radicals with a target pollutant (Krutzler and Bauer, 1999).

$$[Fe(H_2O)_6]^{3+} \rightarrow [Fe(H_2O)_5(OH)]^{2+} + H^+ \quad (10.25)$$

$$[Fe(H_2O)_5(OH)]^{2+} \rightarrow [Fe(H_2O)_4(OH)_2]^+ + H^+ \quad (10.26)$$

$$[Fe(OH)]^{2+} + h\upsilon \rightarrow Fe^{2+} + \cdot OH \quad (10.27)$$

$$[Fe(OOC - R)]^{2+} + h\upsilon \rightarrow Fe^{2+} + CO_2 + \cdot R \quad (10.28)$$

Arslan et al. (2000) studied the combination of UV-A (300 nm $> \lambda >$ 400nm) irradiation, homogeneous Fenton (ferrioxalate (Fe$(C_2O_4)_3^{3-}$) and hydrogen peroxide) as well as heterogeneous TiO$_2$ as a photocatalyst for the degradation of various mono- and bi-functional aminochlorotriazine reactive dyes under O_2^- saturated conditions at pH 2.6 and 7.0, respectively. Complete decolorization and partial mineralization of 17–23% and 73–86% in terms of TOC at 280 nm were achieved by the ferrioxalate-Fenton/UV-A and TiO$_2$/UV-A processes, respectively, within a 1 h treatment time. Dye decomposition was successfully fitted to the empirical Langmuir-Hinshelwood kinetic model. The dye decomposition over Fe(III)/H$_2$O$_2$/UV-A light was explained by the fact that ferrioxalate absorbed light more effectively in the 250±500 nm region, whereas H$_2$O$_2$ only photolysed below 300 nm. Thus, only dyes would normally compete with Fe(III)-ions for the UV-light (Arslan et al., 2000).

In addition, Du et al. (2009) synthesized natural clay-supported iron oxide for the degradation of cationic (malachite green (MG) and fuchsin basic (FB) and anionic (orange II (OII) and X3B) dyes in water at pH 6.5. They found that under visible light and in the presence of H$_2$O$_2$, clay-supported iron oxide catalyst was highly active in comparison to the sole iron oxide or the clay-supported iron oxide sintered at 350°C. The prepared catalysts had a larger (in a range of 86–106 m^{2}g^{-1}) surface area than the parent clay (72 m^{2}g^{-1}) or bare Fe$_2$O$_3$ prepared in parallel (18 m^{2}g^{-1}) and the band-gap energies estimated for these samples were in the range of 2.12–2.17 eV being slightly higher than that of a pure ά-Fe$_2$O$_3$ ($E_g = 2.10$ eV). The cationic dyes (MG and FB) was adsorbed on all the sorbents, but the anionic dyes (OII and X3B) only displayed a notable adsorption on the sample with the surface area of 102 m^{2}g^{-1}due to that Na$^+$ ions were only exchangeable with cationic dyes in the clay interlayer. The photodegradation of MG (0.36 mM) in water over clay-supported iron oxide (0.50 m^{2}g^{-1}) in the presence of H$_2$O$_2$ (2.0 mM) under visible light irradiation led to the degradation of 0.44 mM MG, 0.20 mM OII with apparent reaction rate constants of 0.49 and 1.09 min^{-1}, respectively. One of the main advantages of the clay-supported iron oxide catalyst was the recyclability and excellent stability over six consecutive runs with only 18 μM (1 ppm) of detected iron leaching into the solution. The outstanding performance of the catalyst was correlated with its high sorption capacity toward both types of dyes, thus resulting in enhanced dye degradation via a photosensitization pathway (Du et al., 2009).

The photo-Fenton process that utilizes solar energy offers a very valuable alternative for the degradation and mineralization of the organic dyes. Chacon et al. (2006) showed that Fe^{2+} and H_2O_2 at concentrations of 1.43×10^{-4} M and 5.2×10^{-3} M, respectively, were optimal for the photocatalytic degradation of Acid orange 24 (AO24) with 95% removal efficiency and mineralization of the solution (88% and 84% reduction of COD and TOC, respectively) using 105 kJ L^{-1} of accumulated energy. In general, solar photocatalysis is currently considered the most successful application of solar photons, mostly because it is non-selective and can be applied to complex mixtures of contaminants (Chacon et al., 2006).

10.4.4.3 Other photocatalysts in the degradation of micropollutants

The semiconductor materials, which can act as sensitizers for light-induced redox-processes due to the electronic structure of metal atoms in chemical combination are characterized by a filled valence band, and an empty conduction band. The wide range of such materials, e.g., metal oxides ZnO and CeO_2, metal sulfides (CdS and ZnS), $SrTiO_3/CeO_2$, Al and Fe modified silicates, $CuO-SnO_2$ and etc. are known as photocatalysts for the degradation of micropollutants due to their exceptional properties (Wu et al., 2007; Rafqah et al., 2008; Fu et al., 2009; Song et al., 2009; Elmolla and Chaudhuri, 2010). For instance, ZnO absorbs over a larger fraction of solar spectrum than TiO_2. However, one of the major limitation of the semiconductors like ZnO and CdS is fast fouling of the catalyst leading to the blockage of the catalyst active sites and therefore impairing the overall photocatalytic activity.

Chakrabarti and Dutta (2004) used ZnO in the photocatalytic degradation of Methylene Blue (MB) and Eosin Y. It is important to note that MB and Eosin Y differ in their molecular structure and functional groups, as well as in the extent of ionization in aqueous solution. Thus, MB is a cationic/basic dye, whereas, Eosin Y is an anionic/acidic dye and they are expected to behave differently during the photocatalytic process. Chakrabarti and Dutta (2004) treated both dyes (initial concentration of 50 mg L^{-1}), at the same experimental conditions (catalyst loading 1g L^{-1}, UV irradiation of 16 W, temperature 30°C, airflow rate 6.13 L min^{-1}, and pH 7.0) for 2 h. Both dyes were only partially decomposed with 24% and 8.1% reduction in COD for Methylene Blue and Eosin Y, respectively. d ZnO was regenerated first by treating with boiling distilled water till a colorless wash liquid was obtained and then by drying it in a hot air oven at a temperature of 90 to 100°C. The regenerated product was marked as RC-1. Afterwards it was heated in a muffle furnace at about 600°C to yield the second regenerated catalyst named RC-2. The Eosin Y photodegradation efficiencies of the two recycled catalysts RC-1 and RC-2 were also examined and were 21% and 23% for RC-1 and RC-2, respectively, in comparison to 39% obtained with the fresh catalyst under the same experimental conditions. Authors attributed the decrease in the efficiency of the recycled catalyst to the deposition of photoinsensitive hydroxides (fouling) on the photocatalysts surface blocking its active sites (Chakrabarti and Dutta, 2004).

Also, Chen et al. (2006) used the pelagite as photocatalyst for Methyl Orange (MO) degradation in the presence of air (400 mL min^{-1}) and hydrogen peroxide (0–12 mmol L^{-1}). Pelagite is an autogenic manganese ore found in a deep sea bed, also referred to as manganese nodule, manganese ball and ocean polymetallic nodule, etc. It contains not only oxides and hydroxides of manganese and iron, but also a number of metallic elements such as copper, nickel and cobalt. XRD analysis of pelagite revealed that e.g., manganese was mainly amorphous MnO_2 and MnOOH featuring low degree of crystallization. The photocatalytic decomposition of MO was completed in 120 min. in air and in 60 min. in the presence of hydrogen peroxide (pelagite load of $2gL^{-1}$ and 6 mmol L^{-1} of hydrogen peroxide). Pelagite demonstrated high surface activity, strong absorption capability and oxidation-reduction activity in photodegradation of MO (Chen et al., 2006).

Xia et al. (2008) synthesized cobalt doped mesoporous silica (Co-SBA-15) by the templating method and tested it for the photodegradation of methyl violet (MV) and methylthionine chloride (MC) under solar light irradiation. Un-doped SBA-15 exhibited high adsorption capacities but was photocatalytically inactive under solar light irradiation. Due to a coalescence process favored by the presence of dopant, the surface area, pore volume, and average pore diameter of Co-SBA-15 (690 m^{2}g^{-1}, 0.51 cc g^{-1}, and 32A°, respectively) slightly decreased in comparison with un-doped SBA-15 (701 m^{2}g^{-1}, 0.64 cc g^{-1}, and 45A°, respectively). Photodegradation rate of MV (initial concentration of 50 mg L^{-1}) was 7% and 69% in the presence of un-doped and doped SBA-15, and 6% as well as 76% for MC (initial concentration of 50 mg L^{-1}), respectively. Thus, Co-SBA-15 exhibited very good photoactivity under solar light due to the presence of Co^{2+}/Co^{3+} redox couple in the catalyst. Therefore, well dispersed transition metal ions can be used as efficient dopants for the fabrication of visible and UV light photocatalysts for the photodegradation of dyes (Xia et al., 2008).

10.4.5 Sonocatalytic degradation of micropollutants

The enhancement in the sonolytic degradation rates in the presence of particulates such as the catalyst is due to an increase in the US cavitational activity in the presence of an additional phase (Suslick, 1990; Gogate, 2008). Although, the possible effects due to the capacity of some catalysts (e.g., titanium dioxide) to absorb the light produced by sonoluminescence could not in principle be excluded (Mrowetz *et al.*, 2003). The recombination of ·OH to yield H_2O_2 both, in the gas phase within the bubbles and in the solution are two of the major processes that limit the amount of reactive radicals accessible for the target molecules. In most cases, the sonochemically generated H_2O_2 is not able to react with the target molecules and eventually decomposes to water and oxygen or scavenge hydroxyl radicals. In the presence of a suitable catalyst, hydrogen peroxide becomes a secondary source of ·OH, recovering part of its chemical activity otherwise lost in the production of relatively large amounts of H_2O_2 during sonication. Moreover, ultrasound enhances the catalyst performance via several effects. There are two modes of cavitation bubbles collapse that can affect the surface of the solids. First, cavitation bubbles collapsing directly on a surface may cause direct damage by shock waves produced upon implosion. These bubbles are formed on nuclei such as surface defects, entrapped gases or impurities on the surface of the material. Second, cavitational collapse near the solid surface in the liquid phase would cause microjets to hit the surface and produce a nonsymmetrical shock wave. This phenomenon causes the cleaning action of ultrasound. Thus, the enhanced overall destruction of micropollutants during sonocatalysis is the complex phenomena with simultaneous adsorption, desorption and degradation by the ultrasound in the presence of the catalyst (Pandit *et al.*, 2001). The most popular catalysts utilized so far in the sonochemical treatment of micropollutants are Fenton, Fenton-like and photocatalysts (mostly TiO_2) (Beckett and Hua, 2003; Bejarano-Perez and Suarez-Herrera, 2007; Bahena *et al.*, 2008; Torres *et al.*, 2008a; Abdullah and Ling, 2009). Thus, Abdullah and Ling (2009) argued that rutile phase is more active for the sonocatalytic degradation of several types of micropollutants (e.g., Basic Blue 41 (BB41) dye and methyl parathion) than anatase. The transformation of anatase to rutile and possibly brookite usually takes place at high temperatures ($>700°C$) with corresponding changes in the physical properties. So, Abdullah and Ling (2009) conducted the study, mainly dedicated to the effects of various heat treatments on the characteristics of the TiO_2 sonocatalysts and their consequent effects on the catalytic activity towards Congo Red (CR), Methyl orange (MO) and Methylene blue (MB) degradation. They treated 15 g of TiO_2 powders at various temperatures (400–1000°C) and durations (2 and 4 h) in a muffle furnace to induce phase transformations. The characterization of the resulting catalysts revealed that the heat treatment at temperatures below 700°C did not lead to the phase transformation even at prolonged treatment, whereas TiO_2 sample treated at 1000°C for 4 h, significantly (27.7%) transformed to rutile with an increase in the crystallite size from 56.0 to 62.2 nm. It was also found that anatase-TiO_2 with minimal amount rutile phase demonstrated nearly 5–10% increase in the sonocatalytic activity than pure anatase. The degradation efficiencies for three types of dyes with initial concentration of 20 mg L^{-1}, after 75 min. of ultrasonic irradiation at circum neutral pH with 2.0 g L^{-1} of the catalyst load (calcinated at 800°C for 2h) was in the order of: CR (23%) > MB (19%) > MO (16%). The high degradability of the CR was attributed to the presence of 2 azo (–N N–) labile bonds that were generally more susceptible to the attack by the free radicals formed during the reaction (Figure 10.8).

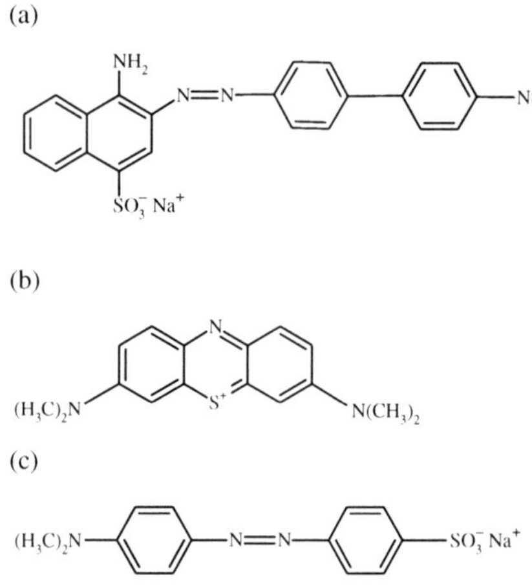

Figure 10.8 Molecular structures of (a) Congo Red, (b) Methylene Blue and (c) Methyl Orange dyes

Meanwhile, MB and MO were nearly identical in molecular size. However, MB showed lower stability due to the presence of one charged site in the aromatic ring while that of MO was located at the external of the ring (Abdullah and Ling, 2009). The first order reaction rates constants drastically increased with the addition of TiO_2 and H_2O_2 from 0.001 to 0.01 min^{-1}. The influence of ultrasound on the catalyst was studied through particle size distribution.

Interestingly, the mode values of the particle sizes of heat-treated TiO_2 catalysts before and after ultrasonic irradiation occurred at the range between 0.3–1.1 μm and 0.2–0.7 μm, respectively, after 2h of the treatment. It was also observed that the cumulative values of the higher end (near 1.5 μm) of the distribution significantly reduced while that of the lower end (near 0.04 μm) significantly increased (Abdullah and Ling, 2009). These observations indicated that partial disintegration of the sonocatalyst particles after the ultrasonic irradiation occured. This disintegration could reduce the micro and mesopores at the interstices of the particles leading to the lower surface area available for the contact with the ultrasonic wave. However, as the ultrasonic wave was also expected to 'grind' the sonocatalyst particles, some new surface was expected to be available for the reaction (Abdullah and Ling, 2009).

Minero et al. (2008) studied the sonochemical degradation (frequency 354.5 kHz and power 35 W) of Methylene Blue (MB) as a nonvolatile and charged substrate, which is unlikely to be found in the gas phase and therefore is unlikely to undergo gas-phase pyrolysis. Its rate constant for the reaction between MB and hydroxyl radical is high and is widely reported in literature (2.1×10^{10} M^{-1} s^{-1}). The sonication experiments were carried out at pH 2 to keep the homogeneous iron catalyst in its dissolved monomeric form and thus to avoid the formation of Fe(III) polynuclear species and colloids. The presence of the catalyst alone provided the substantial increase in the reaction rates (from 3.48×10^{-9} mol L^{-1} s^{-1} during sonolysis to 8.91×10^{-9} mol L^{-1} s^{-1} in the presence of 1×10^{-3} M of Fe (III). Two hypotheses accounted for the enhanced MB sonochemical degradation in the presence of Fe(III): 1) Fe(III) acted as a scavenger of OH radicals, preventing the recombination of hydrogen and hydroxyl radicals to hydrogen peroxide and 2) the degradation of MB occurred due to Fenton reaction during the sonolysis between the catalyst and formed hydrogen peroxide. The initial degradation rate of MB was a function of the initial catalyst concentration, without taking into account the amount of H_2O_2. Moreover, the addition of hydrogen peroxide enhanced the reaction rate constants of MB degradation to 1.68×10^{-8} mol L^{-1} s^{-1} in the presence of 8×10^{-3} M H_2O_2. It was also determined that H_2O_2 acted both, as a source and as a sink of hydroxyl, radicals however it gave an almost negligible contribution to the ·OH budget (either production or scavenging) below 1×10^{-3} M of the initial concentration. Above this concentration, the role of H_2O_2 became significant, with both, the production and scavenging of hydroxyl radicals. A moderate prevalence of the scavenging effect was observed in the presence of 1.0×10^{-2} M H_2O_2 (Minero et al., 2008).

The synergy between the photocatalysis and sonolysis can be quantified as the normalised difference between the rate constants obtained during sonophotocatalysis and the sum of both processes obtained during separate photocatalysis and sonolysis processes in the presence of a suitable catalyst:

$$\text{Synergy} = \frac{k_{US+UV+TiO_2} - (k_{US+TiO_2} + k_{UV+TiO_2})}{k_{US+UV+TiO_2}} \quad (10.29)$$

The main purpose of US in sonophotocatalysis is to contribute to the scission of H_2O_2 produced by both, photocatalysis and sonolysis through cavitation and to promote the performance of the catalyst. It is vitally important to address some factors while designing the combined sonophoto-reactor: (i) it is important to apply US and UV irradiation simultaneously, (ii) the stability of the photocatalyst should be carefully monitored, (iii) the operational conditions must be optimized and (iv) the rate of sonophotochemical degradation can be further enhanced by a Fenton reagent (Gogate and Pandit, 2004). Generally, ultrasound affects the rate of the photocatalytic degradation of micropollutants as a result of three different phenomena: (1) an increase in the photocatalyst surface area, (2) decrease in its aggregation, and (3) the enhancement of the mass transfer and the scission of H_2O_2 produced by both, the photocatalysis and sonolysis processes. Moreover, the sonophotocatalytic treatment presents the best cost-effective ratio for mineralization of the micropollutants. Torres et al. (2006) estimated figure-of-merit (EE/O) values for BPA mineralization under different experimental conditions presented in Table 10.1. The following assumptions were made: (i) only energy cost was estimated, without taking into account the investment cost for chemicals, apparatus, or buildings, and (ii) only processes with more than 60% of TOC removal were considered (Torres et al., 2006).

Table 10.1 Electric energy cost estimates for Bisphenol A degradation by various AOPs (Adopted from Torres et al., 2006)

Process	BPA (μmol L^{-1})	Power (W)	Volume (mL)	Time (min)	% TOC removed	EE/O (kWh m^{-3})
UV	118	25	300	600	Less than 60%	NR
US	118	80	300	600	Less than 60%	NR
US/Fe(II)	118	80	300	600	64	6010
UV/US	118	105	300	300	66	3735
US/UV/Fe(II)	118	105	300	120	79	1033

As it is clearly seen from Table 10.1, the EE/O value for the US/UV/Fe(II) system, 1033 kWh m^{-3}, is four and six times lower than the US/UV and the

US/Fe(II) systems, respectively. This effect is closely related to the process efficiency, which is one of the most important factors in the calculation of EE/O.

Mrowetz et al. (2003) reported the decomposition of two azo dyes, i.e., Acid Orange 8 (AO8) and Acid Red 1 (AR1) in aqueous solution at 20 kHz of US irradiation and in the presence of titanium dioxide particles (Degussa P25 titanium dioxide (80/20% anatase to rutile, surface area of 35 m^2 g^{-1}). After 6 h of ultrasound irradiation treatment, the surface area of the catalyst increased by 30%. The synergy between photocatalysis and sonolysis processes was 0.68 for AO8 (6×10^{-5} M) and 0.56 for AR1 (2.56×10^{-5} M) in the presence of 0.10 g l^{-1} of TiO_2 load.

In addition to the removal of dyes, Bertelli and Selli (2004) investigated the feasibility of sonophotocatalytic treatment to decompose methyl tert-butyl ether (MTBE) in water under atmospheric conditions. Commercially available Degussa P25 titanium dioxide was employed as the photocatalyst. Commonly, the main degradation pathway in H_2O_2 photolysis is ·OH radicals attack, which initiate the degradation of MTBE with a rate constant of 1.6×10^9 L mol^{-1} s^{-1}. Under the photocatalytic treatment ·OH radicals are mainly produced through the interface oxidation of hydroxide anions or water molecules adsorbed on the semiconductor surface by the holes photo-produced in the semiconductor valence band. Whereas under the impact of ultrasound irradiation, volatile organic compounds, such as MTBE, undergo degradation through the direct pyrolysis in the vapor phase of cavitation bubbles, within the hot interfacial region between the vapor and the surrounding liquid phases. The photocatalytic degradation of MTBE (5×10^{-3} M) in the presence of 0.1 g L^{-1} of TiO_2 proceeded at a higher rate than under low frequency ultrasound with the first-order reaction constant $(21 \pm 2) \times 10^5$ s^{-1} while in the presence of US or UV irradiations, the highest reaction rate constants were nearly 8 times lower ($3.71 \pm 0.09) \times 10^5$ s^{-1} and $(13.2 \pm 0.2) \times 10^5$ s^{-1}, respectively) (Bertelli and Selli, 2004).

Torres et al. (2008) demonstrated the effective removal of bisphenol A (BPA) by the coupled ultrasound (300 kHz, 80W) and Fe (II) enhanced solar light-assisted process carried out at pH 5 in the presence of air or oxygen. They investigated a number of possible combinations of US, UV, Fe^{2+} and H_2O_2 and found that after 60 min. of the treatment, the efficiency of BPA degradation followed the order of Fe^{2+}/solar light (8%) ≤ Fe^{2+}/solar light → ultrasound (8%) < Fe^{2+}/solar light/H_2O_2 (73%) < ultrasound (85%) = ultrasound → Fe^{2+}/solar light (85%) < ultrasound/Fe^{2+} (89%) < ultrasound/Fe^{2+}/solar light (92%). However, the complete mineralization of the solution was not achieved reaching 4–70% after 240 min. of the treatment. Thus, the highest efficiency in terms of both, BPA degradation and mineralization of the solution was obtained for the sonophoto-Fenton process even at initial pH around 5 in the presence of the solar light. It can be seen in X.9, that ·OH radicals generated in the bubbles oxidize BPA to the more hydrophilic intermediates at the interface bubble-solution. The hydrophilic BPA intermediates are degraded mainly in the so-called third reaction zone (the solution bulk) by ·OH radicals coming from the photo-Fenton reaction (H_2O_2/Fe^{2+}). This reaction is possible due to the residual H_2O_2 that is formed in the system during the sonication. Moreover, Fe^{3+} can form complexes with aliphatic acids, which suffer a rapid decarboxylation in the presence of light. In such a reaction, Fe^{2+} that is used for the photo-Fenton process, is regenerated. Fe^{2+} can also be regenerated by the light, with the concomitant formation of additional ·OH radicals (Torres *et al.*, 2008b).

10.4.6 Microwave-assisted catalytic degradation of miropollutants

Microwave (MW) irradiation is an energy that forms electromagnetic waves with the wavelengths of 1 mm–1 m and is composed of an electric and magnetic fields. In general, microwave irradiation is associated with any electromagnetic radiation in the microwave frequency ranging from 300 MHz to 300 GHz, ($\lambda = 1$ mm to100 cm), which is located in the segment between the infrared and radio wave in the region of the electromagnetic spectrum (Banik *et al.*, 2003). The application of MW as an AOP is promoted by the fact that MW is able to produce hydroxyl radicals in aqueous solution according to the following reactions:

$$O_2 \rightarrow \cdot O + \cdot O \quad (10.30)$$

$$\cdot O + H_2O \rightarrow 2 \cdot OH \quad (10.31)$$

However, there are some limiting factors making its sole application unfeasible. First of all, the presence of oxygen is critical and second of all, the energy of microwave irradiation (E = 0.4–40 kJ mol^{-1} at $v = 1-100$ GHz) is insufficient to disrupt the bonds of common organic molecules (Muller *et al.*, 2003). However, MW is known as an effective means to promote a wide range of AOP technologies in general, and catalytic oxidation reactions for the degradation of micropollutants in particular (Gromboni *et al.*, 2007; Zhang *et al.*, 2007; Zhanqi *et al.*, 2007; Yang *et al.*, 2009).

The improvement in the catalytic oxidation is attributed to the way the catalyst is heated by the microwaves. The fundamentally different method of transferring energy from the source to the sample is the main benefit of utilizing microwave energy when compared to the conventional thermal processing. By directly delivering energy to the microwave-absorbing materials, complications such as long heating up periods, thermal gradients, and energy loss to the system can often be avoided. Furthermore, the penetrating capacity of microwave allows volumetric heating of samples (Mutyala et al., 2010). During microwave irradiation, the catalyst is heated directly by the action of the microwaves, and so it is at a higher temperature than the surrounding atmosphere. Although, it should be remembered that the sample heating induced by the microwave fields is strictly influenced by the ability of the material to absorb microwave energy. Therefore, allowing specimens to be selectively heated, which can improve the efficiency of introducing energy into the system (Mutyala et al., 2010).

For instance, the improvement in the photocatalytic activity of TiO_2 by MW irradiation was attributed to the polarization effect of the highly defected catalysts in a microwave field, which increased the transition probability of photon-generated electrons and decreased the electron–hole recombination on semiconductor surface (Horikoshi et al., 2009). Horikoshi et al. (2002) anticipated that whatever the precise nature of microwave non-thermal factors, they nonetheless do cause changes in the overall degradation process that cannot be ascribed to the simple conventional heating or microwave generated heating of the dispersions. They proposed the theory of the formation of localized hot micro-/nano-scale domains, so-called "hot spots" on the catalyst surface by microwave irradiation, or the selective heating of catalytic active sites on the catalyst, which may lead to the microwave effects similar to those attributed to the microwave non-thermal factors (Horikoshi et al., 2004a). They also presumed that such localized hot domains had plasma-like properties in certain parts of the catalyst surface. In addition, the deposits on the catalyst surface may also generate polarized domains on the particle surface under the microwave irradiation field. Therefore, the degradation of a target compound may be enhanced at or near hot localized or polarized domains during the MW irradiation (Horikoshi et al., 2002).

MW is a versatile technique, which can substantially lower the reaction time, improve the catalyst performance (e.g., enhance its regeneration), restructure the catalyst surface to increase the number of catalytically active sites, and to produce more hydroxyl radicals (Horikoshi et al., 2002). Moreover, it can also be easily coupled to several technologies simultaneously, e.g., electron paramagnetic resonance (EPR) showed that a greater number of ·OH radicals are produced in the microwave-assisted photocatalytic process (PD/MW) in aqueous TiO_2 dispersions (Horikoshi et al., 2004b).

Quan et al. (2007) studied the combination of commercially available activated carbon (particle size of 1.0–2.0 mm) as a catalyst with MW (800 W) irradiation to degrade hydroxybenzoic acid (salicylic acid SA) and pentachlorophenol (PCP). The rate of ·OH generation was 0.036 µmol s^{-1}, and the ·OH yield reached 3.2 µmol in less than 3 min. under the applied conditions. PCP degradation was nearly 72–100% (corresponding to 40% and 82% TOC removal) in 60 min. for two initial PCP concentrations of 500 and 2000 mg L^{-1}, which implied that continuous production of ·OH during MW irradiation supported free radical reactions that benefited PCP degradation. Simultaneous presence of activated carbon and oxygen proved to be critical in ·OH generation. Temperature of water and AC increased at the different rates under MW irradiation due to the internal heating by MW energy, which could cause locally higher temperatures in some micro- surfaces of the AC particles than water bulk (Quan *et al.*, 2007).

Also, Bi *et al.* (2009) studied perovskite type catalyst (CuO$_n$–La$_2$O$_3$/γ-Al$_2$O$_3$) in the microwave enhanced ClO$_2$ catalytic oxidation process to eliminate remazol golden yellow RNL at the optimized reaction conditions (ClO$_2$ concentration 80 mg L^{-1}; microwave power 400W; contact time 1.5 min., catalyst dosage 70 g L^{-1}). In such a system, the microwave enhanced ClO$_2$ catalytic oxidation process, ·OH could be generated by a free radical chain auto-oxidation process. It was found that microwave irradiation substantially shortened the reaction time from 90 min. to 1.5 min. The general mechanism of the ClO$_2$ catalytic oxidation was not determined, but microwave irradiation had a synergistic effect when applied in the presence of the catalyst to initiate the additional formation of free radicals in the presence of ClO$_2$ (Bi *et al.*, 2009).

Finally, Yang *et al.* (2009) investigated the applicability of microwave-assisted Fenton-like reaction (MW-Fenton-like) for the treatment of pharmaceutical wastewater and compared its performance with un-assisted Fenton reaction and the conventional heating assisted Fenton-like (CHFL) reaction. They reported that MW (300 W power) improved the degradation efficiency of Fenton-like reaction in comparison to the Fenton-like reaction performed at ambient temperature and conventional heating only up to 5%, however was beneficial to enhance the settling quality of sludge, reduce the production of sludge and improve the biodegradability of effluent (Yang *et al.*, 2009). Also, Yang *et al.* (2009) estimated the treatment costs in MW-Fenton-like system in terms of the electricity consumption per gram of COD removed (0.0209 kWh g COD^{-1}), which was considered as applicable to treat biorefractory wastewater contaminated with the residual pharmaceuticals.

10.4.7 Electrocatalytic oxidation

Electrochemistry is the surface science that studies the physicochemical phenomena at the interface between an electrode and the electrolyte. Electrocatalysis is used to accelerate the electrochemical reactions by structural or chemical modifications of the electrode surface and by additives to the electrolyte. Structural modifications include changes in the surface geometry (crystal planes, clusters, adatoms), and variations in the electronic state of the catalyst material.

The main driving force in electrocatalysis is the oxidation-reduction reactions, which occur in several steps. Redox reactions at the electrodes where only electrons pass the interface and the electrode surface remains unaffected after the reaction, usually reaches a steady state similar to the heterogeneous catalysis. However, there is a fundamental difference between the catalysis and electrocatalysis: in the case of an electrode reaction, the catalyst contributes one of the reactants to the process, e.g., the electrons, which are either consumed or generated in the net reaction. Charge transfer rates and electrosorption equilibrium highly depends on the electrode potential. Therefore, the driving force of an electrode reaction is not only controlled by the chemical forces, which depend on temperature, pressure, reactant concentrations etc., but also by the electrical forces which affect the rate of electron transfer through the interface (Hagen, 2006). The most important parameter for the characterization of these electric forces is the electrode potential relative to a suitable reference electrode, which can be altered in an electrolytic cell by an external voltage applied to such cell. One of the great advantages is that the rate of the reaction can be followed with high sensitivity in the form of the electric current passing the electrode interface.

In practice, there are two kinds of electrochemical reactions of technical importance: (i) direct and (ii) indirect reactions. In the direct anodic oxidation, the adsorbed pollutants are oxidized on the anode surface without the involvement of any substances other than the electron. Moreover, during a direct electrochemical reaction, the substrate undergoes a heterogeneous redox reaction at the electrode surface within the Helmholtz layer.

On the other hand, in indirect electrochemical reactions, the heterogeneous reaction between the substrate and the electrode is replaced by a homogeneous redox reaction in the solution between the substrate and an electrochemically activated species from water discharge at the anode such as physically adsorbed "active oxygen" (physisorbed hydroxyl radical ($\cdot$OH)) or chemisorbed "active oxygen" (oxygen in the lattice of a metal oxide (MO) anode) (Panizza and Cerisola, 2007). The action of these oxidizing species leads to the total or partial decontamination, respectively. Importantly, if electrode passivation occurs and impairs direct electrolyses, an indirect pathway could be an advantage. By researching the indirect or mediated oxidation with different heterogeneous species formed from water discharge, it is possible to propose two main approaches for the pollution abatement in wastewaters by EO process (Martinez-Huitle and Brillas, 2009):

(i) The electrochemical conversion method, in which refractory organics are selectively transformed into biodegradable compounds, usually carboxylic acids, with chemisorbed "active oxygen";

(ii) The electrochemical combustion (or electrochemical incineration) method, where organics are completely mineralized, i.e., oxidized to CO_2 and inorganic ions, with physisorbed OH radical. This radical is the second strongest oxidant known after fluorine, with a high standard potential ($E_0 = 2.80$ V vs. SHE) that ensures its fast reaction with most organic contaminants producing dehydrogenated or hydroxylated derivatives up to the conversion to CO_2.

Brillas et al. (2009) wrote an excellent review, where they summarized efficient elctro-Fenton and related electrochemical technologies based on the Fenton chemistry developed for the degradation of various micropolluttants such as pesticides, dyes, pharmaceuticals and personal care products, industrial chemicals and etc. with all fundamental aspects of the technology. These electrochemical processes are an effective, a simple, and a versatile alternative to conventional methods. Despite of electro-Fenton and related processes are perhaps the most ecological and effective electro-AOPs, they present some drawbacks that complicate their setup such as the need to operate at pH ca. 3.0 and the continuous supply of O_2 (Brillas et al., 2009).

Also, Martınez-Huitle and Brillas (2009) summarized efficient electrochemical technologies developed to decolorize and/or degrade dyeing effluents for environmental protection. According to the Comninellis model, the most important factor determining the extent of dyes mineralization in EO process is the anode material. The electrode materials that can be effectively utilized for the degradation of various dyes during EO process are polypyrrole, granular activated carbon, ACF, perovskite-like structures, glassy carbon, graphite, Ti/Pt and Pt, and doped and undoped PbO_2 and mixed metal oxides of Ti, Ru, Ir, Sn and Sb. Despite the huge variety of electrode materials available, synthetic boron doped diamond electrode (BDD) thin films are currently preferred as anodes by their better oxidative performance (Brillas et al., 2005).

10.4.7.1 Degradation of micropollutants with electrocatalytic and coupled electrocatalytic methods

Despite the successful attempts to utilize sole electrochemical treatment to degrade various organic contaminants, coupling of electrocatalytic methods with Fenton, Fenton-like, ultrasound and MW irradiation processes may provide an increased activities and selectivities towards the destruction of micropollutants in the aqueous phase (Panizza and Cerisola, 2007; Sires et al., 2008; Martinez and Bahena, 2009; Zhao et al., 2009; Ai et al., 2010). Thus, the removal of chlorbromuron herbicide (1.19×10^{-4} M) by the electro-Fenton was investigated by Martınez and Bahena (2009). The electrolyses were carried out in a three-electrode divided and undivided cell and also in a two-electrode undivided cell in which Fe(II) formed from the anode (Martinez and Bahena, 2009). The three-electrode divided cell was comprised of a Pt gauze anode separated by a cation permeable membrane (Nafion 117) using a (25 mm × 25 mm × 10 mm) reticulated vitreous carbon (RVC) cathode (60 ppi, Electrolytic Inc., NY). The reference electrode was a saturated calomel electrode (SCE, Orion). The undivided two-electrode cell comprised of a stainless steel (SS) anode (316) and a RVC cathode.

Theoretically, the stoichiometric conversion of 1 mol of chlorbromuron to CO_2 requires 25 mol of H_2O_2 according to the following reaction:

$$C_9H_{10}BrClN_2O_2 + 25H_2O_2 \rightarrow 9CO_2 + Br^- + Cl^- \\ + 2NO_3^- + 2H^+ + 33H_2O \quad (10.32)$$

However, it was observed that a molar ratio of H_2O_2 to $C_9H_{10}BrClN_2O_2$ of 25 to 1 led to 80% TOC decrease, whereas, a molar ratio of 37.5 to 1 (with a 50% of H_2O_2 more than the stoichiometric amount) led to 97% TOC decrease at the end of the treatment. Obviously, higher amount of H_2O_2 was necessary for the nearly complete mineralization of the initial wastewater to further degrade the by-products formed during the chlorbromuron degradation by the Fenton reaction. Electro-Fenton reaction led to 89.6% of TOC removal after the passage of 237.8 C of electric current, whereas a theoretical charge of 109.1 C would be required to completely oxidize 1.19×10^{-4} M of chlorbromuron, calculated by the Farraday law. This behavior can be explained by the low current efficiency, the catalyst deactivation and/or by the formation of iron complexes (such as $FeH_2O_2^{2+}$, $Fe(OH)^{2+}$, $FeOOH^{2+}$, etc.). A 92% of TOC was removed at pH 2, followed by TOC removal of 87% and 80% at pH 1 and pH 5, respectively. TOC decayed rapidly during the first 30 min. of the reaction that corresponded to a 150 C charge passed with high current efficiency that was able to oxidize chlorbromuron reaching as high as 60% TOC decrease for all the concentration tested. As the treatment proceeded, the current efficiency decreased and, thus, TOC removal rate also decreased. At the end of the treatment (75 min.), the pH of the solutions increased with a subsequent formation of iron precipitates. Authors also estimated the electrical energy expenditure (0.04 U.S.$ kWh^{-1}) to assess the economic feasibility of Electro-Fenton treatment. The rough calculation showed that electro-Fenton process was rather economical, with 0.08 U.S.$ m^{-3} to eliminate 90% of TOC from the solution (Martinez and Bahena, 2009).

Combination of ultrasound and electro-Fenton (sonoelectro-Fenton (SEF)) was successfully used by Yasman et al. (2004) to degrade the common herbicide

2,4-dichlorophenoxyacetic acid (2,4-D) with initial concentrations of 0.25–1.5 mM and its derivative 2,4-dichlorophenol (2,4-DCP) with the initial concentrations of 0.35–1.5 mM in aqueous solutions in the presence of 0.5 to 50 mM of Fe^{2+}, 30 mM of H_2O_2, pH 3 and ultrasound irradiation of 20 kHz and ultrasound power of 75 W. Both, the cathode and the anode were made of nickel foil (thickness of 0.125 mm) in the form of cylindrical segments of 11 mm in radius and 20 mm in height. Both electrodes were placed around the ultrasonic horn (11 mm radius) in the reactor filled with Na_2SO_4 (0.5 g L^{-1}) as electrolyte. The typical reaction time to fully degrade 2 mM of 2,4-D or its toxic metabolite – 2,4-DCP was less than 600 s. Yasman *et al.* (2004) also conducted a comparative study employing the classic Fenton reaction for the degradation of 2,4-D (initial concentration of 1,2 mM) with 3.0 mM of Fe^{2+} and 3.0 mM of H_2O_2 under violent mechanical stirring. The insignificant degradation of a target contaminant was achieved only after 7 h of the reaction and was attributed to the Fenton oxidation, since the half lifetime of 2,4-D in ambient water was much longer (6–170 days, depending on the environmental conditions it was kept under) (Yasman *et al.*, 2004). It was determined that the actual efficiency of the SEF process was much higher than all the reference degradation processes namely Fenton and sono-Fenton processes. The fast and complete degradation of 2,4-D and 2,4-DCP during SF and SEF processes was attributed to the higher efficiency of ·OH radicals production in concurrent process (SEF) as well as to the ultrasonic cleaning of the active surfaces of the electrodes during the treatment processes.

Sires *et al.* (2007) performed electro- and photoelectroFenton for the removal of clofibric acid (179 mg L^{-1}, which is equivalent to 100 mg L^{-1} of TOC). All electrolyses were conducted in an open, undivided and thermostated cylindrical cell containing 100 mL of solution stirred with a magnetic bar with a 3-cm^2 Pt anode and 3-cm^2 carbon-PTFE cathode (Sires *et al.*, 2007). The application of the electro-Fenton method only resulted in 80% of decontamination due to the formation of products hardly oxidizable with ·OH such as benzenediols, benzenetriols, and etc. On the contrary, the photoelectro-Fenton method with UVA light demonstrated the mineralization of clofibric acid higher than 96% in aqueous medium of pH 3.0. Moreover, its efficiency increased with increasing metabolite content and with decreasing in the current density from 150 mA cm^{-2} to 33 mA cm^{-2}. This tendency corroborates the enhancement of parallel non-oxidizing reactions of ·OH (e.g., its anodic oxidation to O_2 and its recombination into H_2O_2) when current density raises, yielding a smaller proportion of this oxidant able to destroy pollutants (Sires *et al.*, 2007).

Gao *et al.* (2009) anticipated that the application of microwave (125 W) irradiation could also improve the oxidation ability of the electrode during the electrochemical oxidation of micropollutants, such as herbicide 2,4-dichlorophenoxyacetic acid (2,4-D) in a continuous flow system under atmospheric pressure. It was found that 88.5% of 2,4-D (100 mg L^{-1}) was efficiently removed at low current density in less than 3 hours using BDD electrode activated by the microwave irradiation. However, in the absence of microwave irradiation, only 54.3% of 2,4-D was removed due to the blockage of the electrode. The primary products of 2,4-D degradation included 2,4-dichlorophenol, hydroquinone, fumaric acid and oxalic acid. In addition, the concentrations of all the intermediates were much lower in MW-EC in comparison to the EC method (Gao *et al.*, 2009).

10.4.8 Biocatalytic oxidation of micropollutants

Enzymes are globular proteins and range from just 62 amino acid residues in size, for the monomer of 4-oxalocrotonate tautomerase, to over 2,500 residues in the animal fatty acid synthase (Karam and Nicell, 1997). The enzymatic activity is determined by the three-dimensional structure. Although structure does determine the function, it is extremely difficult to predict the enzymatic activity only based on its structure. Nomenclature Committee of the International Union of Biochemistry and Molecular Biology (NC-IUBMB) developed a special nomenclature for enzymes, based on the EC numbers – each enzyme was described by a sequence of four numbers preceded by "EC". The first number broadly classified the enzyme based on its mechanism: EC1 oxidoreductases, EC2 transferases, EC3 hydrolases, EC4 lyases, EC5 isomerases, and EC6 ligases (http://www.chem.qmul.ac.uk/iubmb/enzyme/). Oxidoreductases, especially peroxidazes (e.g., HPR), polyphenol oxidaze (e.g., tyrozinase, laccase), which are able to catalyze the oxidation/reduction reactions, are widely used for the laboratory-scale biocatalytic treatment of the aqueous organic contaminants. Generally, peroxidases are oxidoreductases produced by a number of microorganisms and plants. Horseradish peroxidase (HRP, EC 1.11.1.7) is undoubtedly one of the most studied enzymes in the relatively new area of enzymatic waste treatment (Duran and Esposito, 2000). Once activated by HRP, H_2O_2 is able to eliminate a wide array of toxic aromatic micropollutants including biphenols, anilines, benzidines and related heteroaromatic compounds.

The replacement of the existing chemical oxidation with enzymatic technologies can be a very attractive alternative from the ecological point of view (Bozic and Kokol, 2008). The main advantages of enzyme-catalyzed reactions are (i) mild temperatures of processing, (ii) absence of by-products, and (iii) environmentally benign nature of the processes. Moreover, enzymatic oxidation exhibits several advantages over conventional biological methods including (i) high selectivity, (ii) removal of toxic pollutants to low levels, and (iii) easier handling and storage of enzymes in comparison to microorganisms (Entezari and Petrier, 2004). However, the enzyme-assisted process also includes several disadvantages such as (i) the dependence of the enzymatic activity on the temperature and pH, (ii) high costs and (iii) the disposal of solid products that may limit the applicability of the method.

Taking into account all the advantages and disadvantages of the enzyme-catalyzed processes, several research groups utilized enzymes in combination with ultrasound and hydrogen peroxide addition to remove various estrogens and azo dyes from aqueous solutions (Tamagawa *et al.*, 2006; Vilaplana *et al.*, 2008; Marco-Urrea *et al.*, 2010; Rodriguez-Rodriguez *et al.*, 2010). For instance, to overcome the existing drawbacks, to prolong the enzymatic life time and activity and to improve the enzyme-catalyzed removal of estrogens it is advisable to use a protective additive (e.g., polyethylene glycol (PEG), polyvinyl alcohol (PVA), violuric acid and 2,2-azinobis-3-ethylbenzthiazoline-6-sulfonic acid (ABTS), etc.) to create a protective hydrophobic layer around the enzyme in the reaction solution or/and to immobilize the enzyme on various surfaces including nano (nanostructured enzymes) or use a plug-flow reactor instead of a batch reactor (Rokhina *et al.*, 2009).

Among all the enzymes, horseradish peroxidase (HRP) is effectively used in wet peroxide oxidation of aqueous micropollutants, especially for the destruction of estrogenic compounds with aromatic structures. The by-products of such HRP-assisted treatment are polymerized through a non-enzymatic process, which leads to the formation of high molecular weight polymers of low solubility that can be easily removed from wastewater by co-precipitation, sorption to solids, sedimentation or filtration (Karam and Nicell, 1997).

In addition, Auriol *et al.* (2008) studied HRP and laccase-catalyzed treatment of natural and synthetic estrogens such as estrone (E1), 17β-estradiol (E2), estriol (E3), and 17ά-ethinylestradiol (EE2). They found that the initial HRP activity of 0.032 U mL^{-1} was sufficient to completely remove estrogenic compounds from a model solution, whereas 8–10 U mL^{-1} activities were necessary when a real wastewater was introduced. Moreover, laccase activities of 20 U mL^{-1} were effective in the complete removal of each steroid estrogen at pH 7.0 and 25 ± 1°C within 1h of the model and real wastewater treatment. Besides, laccase and its activity was not affected, whereas HRP was significantly affected by the wastewater constituents. However, when laccase was introduced, the residual estrogenic activity was slightly lower (3%) in comparison to the same experimental conditions when HRP was utilized (12%). In addition, laccase required oxygen as an oxidant, which was relatively less expensive than the hydrogen peroxide required by peroxidase enzymes (Auriol *et al.*, 2008).

One of the recent advances in enzymatic catalysis is the use of ultrasound to enhance the enzymatic performance. However, the increase in the removal efficiency is accompanied by the fast deactivation of the enzyme with the subsequent decrease in its life-time. The overall effect of ultrasound irradiation on enzymes would therefore depend on parameters, such as energy input and irradiation duration. The optimization of these parameters would allow to determine the threshold conditions for ultrasound irradiation of enzymes that do not adversely affect their activity (Rokhina *et al.*, 2009).

In addition to the estrogens, azo dyes may also be removed by the combination of ultrasound and enzymatic treatment. Indeed, Rehorek *et al.* (2004) found that the use of laccase is viable to degrade several industrial azo dyes such as Acid Orange 5 and 52, Direct Blue 71, Reactive Black 5 and Reactive Orange 16 and 107. Therefore, a long-term (up to 12 h) enzymatic treatment in the presence of ultrasound irradiation was able to degrade azo dyes and detoxify occurring intermediates. Degradation measured in terms of the absolute quantities (L mol h^{-1}) implied a linear correlation between the radical formation and the amount of pollutants (Rehorek *et al.*, 2004). The combination of ultrasound irradiation with laccase indicated that ultrasound irradiation at 90 W did not affect the activity of the enzyme. However, increasing the power input to 120 W, significantly deactivated the enzyme resulting in 4 times shorter life time (5 h) in comparison to 90 W power applied. The ultrasound-assissted decolorization of various textile dyes with laccase from *Trametes villosa* was also successfully attempted in laboratory and pilot scale experiments (Rehorek *et al.*, 2004; Basto *et al.*, 2007; Tauber *et al.*, 2008). For instance, Rehorek *et al.* (2004) and Tauber *et al.* (2008) demonstrated that in the presence of 60–120 W ultrasound irradiation at 850 kHz, a complete mineralization of azo dyes could be attained within 1–9 hours. Basto *et al.* (2007) achieved 65 to 77% decolorization of indigo carmine when 47–72 W with an optimum ultrasound irradiation frequency of 150 kHz was applied for 60 min.

All these aforementioned studies provide evidence that a concurrent application of enzymes and ultrasound can considerably be more effective in terms of reduced reaction time and treatment efficiency than either ultrasound-assisted or enzyme-catalyzed reactions alone, regardless of the enzyme or process applied (Rokhina *et al.*, 2009).

10.4.9 Catalytic wet air oxidation of micropollutants

Wet air oxidation (WAO) or so called 'hydrothermal treatment', in which the generation of active oxygen species, such as hydroxyl radicals, takes place at high temperatures and pressures, is known to show a great potential for the treatment of effluents containing a high content of organic matter (chemical oxygen demand (COD) 10–100 g l^{-1}), or toxic contaminants for which the direct biological purification is unfeasible. Typical operational conditions for WAO range from 100°C to 372°C temperatures and 20 to 200 bar of pressures. Residence times of the liquid-phase in a three-phase reactor may range from 15 to 120 min., and the extent of COD removal may typically be about 75–90% (Luck, 1996; Luck, 1999). The WAO process is widely used for breaking down biologically refractory compounds to simpler, easily treated materials before they can be released into the environment. However, one of the main drawbacks of the WAO process is its inability to achieve a complete mineralization of organics, since some low molecular weight oxygenated compounds, originally present in a wastewater or accumulated in the liquid-phase during the oxidation process, are resistant to further transformation to carbon dioxide (Levec and Pintar, 2007). Compared to the conventional wet-air oxidation, catalytic wet air oxidation (CWAO) has lower energy requirements and provides much higher oxidation efficiencies at shorter times, which in turn allows reducing COD to the same degree as during non-catalytic process, however, at less severe conditions. Nonetheless, even at these milder conditions the rapid deactivaton of the catalyst surface may occur.

Commercial catalytic wet oxidation processes rely on both, homogeneous such as Fe^{2+} or Cu^{2+} and heterogeneous catalysts such as noble metals: iridium (Ir) gold (Au), platinum (Pt), palladium (Pd), rhodium (Rh) and rhenium (Re), and ruthenium (Ru); metal oxides: $FeO(OH)$, CuO-ZnO; and mixed metallic catalysts Fe-Cu-Mn, Cu-Zn, $CoAlPO_4$-5 and CeO_2, Fe-CeO_2, perovskite-type oxide $LaFeO_3$, and heteropolyacids (HPAs) or the supported precious metal and/or base metal oxide catalysts (Lee *et al.*, 2004; Lei *et al.*, 2007; Liu and Sun, 2007; Mikulova *et al.*, 2007; Yang *et al.*, 2007; Wang *et al.*, 2008; Carrier *et al.*, 2009; Li *et al.*, 2009b; Zhang *et al.*, 2009b). The support materials that can be employed to enhance the CWAO are CeO_2, TiO_2–CeO_2, TiO_2, ZrO_2 and graphite, activated carbon, Al_2O_3, and carbon nanofibres (Oliviero *et al.*, 2000; Chang *et al.*, 2003; Milone *et al.*, 2006; Quintanilla *et al.*, 2007; Rodriguez *et al.*, 2008).

Therefore, Arslan-Alatona and Ferry (2002) studied the catalytic effect of two Keggin-type polyoxotungstates, namely $H_4SiW_{12}O_{40}$ and $Na_2HPW_{12}O_{40}$, on the wet air oxidation of azo dye Acid Orange 7 (AO7; C_{dye} = 248 mM) at varying reaction temperatures (T = 160–290°C) at pH 2 and oxygen atmosphere of 0.6–3.0 MPa. The activation energy in terms of TOC decreased nearly two-fold from 40 kJ mol^{-1} in the absence of the catalyst to 28 kJ mol^{-1} (PW_{12}^{3-}) and 22 kJ mol^{-1} (SiW_{12}^{4-}) at the end of the treatment process when both catalysts were employed. The catalytic processes were less sensitive to the presence of ·OH scavengers (an organic (isopropanol; IsOH) and an inorganic (bromide in the form of KBr) than un-catalyzed WAO indicating that polyoxotungstate-catalyzed wet air oxidation did not solely depend upon a free radical type reaction mechanism. It was also determined that upon the use of polyoxotungstates, the oxidation mechanism shifts from a free radical (·OH) chain reaction to a charge (an electron) transfer controlled reaction which is more selective and hence preferable process in the presence of SiW_{12}^{4-} and PW_{12}^{3-} (Arslan-Alaton and Ferry, 2002).

Moreover, Liu and Sun (2007) synthesized Fe_2O_3-CeO_2-TiO_2/g-Al_2O_3 catalyst, which contributed to nearly complete (up to 98%) degradation of Methyl Orange (MO) at room temperature and atmospheric pressure within 2.5 h of the reaction time. The leaching of the catalyst was negligible due to the preventive effect of Ti addition into the catalyst structure where Ce and Ti formed stable oxides. However, the deactivation of the catalyst after several consecutive runs was caused by the adsorption of reaction intermediates on the catalyst surface, therefore covering the active sites and impairing the catalytic activity. Nonetheless, the catalytic activity of the deactivated catalyst can be generally restored by rinsing it with hydrochloric acid followed by calcination at 350°C for 3 h.

In the recent study, Zhang et al. (2009) used as-prepared stable and insoluble in water $Zn_{1.5}PMo_{12}O_{40}$ catalyst (calcined at 280°C) in the presence of native cellulose fiber templates for CWAO of organic dye Safranin-T (ST) at room temperature and atmospheric pressure. The catalyst demonstrated excellent catalytic activity in the CWAO process towards a complete mineralization of ST within 40 min. of the reaction. Also, the catalyst exhibited outstanding stability without a significantly leaching observed for at least six consecutive runs.

In addition to the destruction of dyes, CWAO may also be utilized to degrade various pesticides and other micropollutants in aqueous solutions. However, Carrier et al. (2009) investigated supported ruthenium (3% Ru/TiO_2) catalyst for the abatement of diuron and concluded that despite the thermal degradation of diuron was observed during the process, CWAO was not a favorable method to be used to degrade such pesticide in the wastewater. The main drawbacks, besides the heating of a very diluted aqueous solutions, were the incomplete mineralization even at higher temperatures, the leaching of metal catalyst induced by amines formed during the CWAO process (Carrier et al., 2009).

10.5 ADVANCED NANOCATALYTIC OXIDATION OF MICROPOLLUTANTS

The recent rapid development in the field of nanotechnology has driven a considerable volume of research into the use of metal nanoparticles as efficient catalysts for water and wastewater treatment. The use of nanocatalysts in the treatment processes benefits from the enhanced reactivity, extremely large surface area, and/or enhanced mobility of nanometre (10^{-9}m)-scale particles, which produces more rapid or cost-effective clean-up of wastes compared to the conventional catalytic technologies (Pradeep and Anshup, 2009). The current section will cover only the application of nanocatalysts in AOP in the presence of various oxidants and irradiations in order to evaluate the potential of nanocatalytic advanced oxidation processes for the removal of micropollutants in the aqueous phase.

One of the most important nanomaterials studied for aqueous phase decontamination is zero-valent iron (ZVI), which is usually used in wet peroxide oxidation as a primary source of Fe^{2+} to initiate the ·OH radicals formation. In acidic conditions, the surface of the ZVI corrodes and generates ferrous ions *in situ*, which leads to Fenton reactions, especially in the presence of hydrogen peroxide. It is hypothesized that Fe^0 is first oxidized to Fe^{2+} via the following reaction:

$$Fe^0 + H_2O_2 \rightarrow Fe^{2+} + 2OH^- \qquad (10.33)$$

And then the aforementioned traditional Fenton reactions according to the Eqs. (10.2) and (10.3) occur. This process is called advanced Fenton process (AFP). AFP has several advantages over the conventional Fenton's process such as: (i) the use of ZVI instead of iron salts reduces or even eliminates the unnecessary loading of aquatic system with counter anions, (ii) the concentration of ferrous and ferric ions in wastewater treated by AFP is significantly lower in comparison to the classical Fenton's process that utilizes iron salts, (iii) faster recycling of ferric iron at iron surface occurs.

AFP shows better efficiency than conventional Fenton. Therefore, some recent works in AFP for the degradation of micropollutants with all the process parameters and relevant data are summarized in Table 10.2.

ZVI catalysts (fabricated by research groups or commercially available) in the presence of hydrogen peroxide or ultrasound, UV as well as microwave irradiations are widely applied for the destruction of various micropollutants (Table 10.2). For instance, Bergendahl and Thies (2004) synthesized highly reactive ZVI in the presence of sodium borohydride and tested it in wet peroxide oxidation of methyl tert-butyl ether (MTBE). They observed a nearly complete degradation of MTBE in 125 min. of the reaction at pH 4 and 7 with H_2O_2 to MTBE molar ratio of 220:1. Importantly, the final concentration of acetone produced by this process was approximately 400 µg L^{-1} and no by-products were detected after 24 h of ZVI/hydrogen peroxide treatment. The degradation of MTBE followed the second-order reaction kinetics with the reaction rate constants of 1.9×10^8 M^{-1} s^{-1} at pH 7.0 and 4.4×10^8 M^{-1} s^{-1} at pH 4.0.

Table 10.2 Advanced nanocatalytic oxidation of micropollutants

Nanocatalyst	Synthesis of nanocatalyst	Treated micropollutant	Process efficiency	Reference
ZVI with average particle size 10–30 nm (250 mg L^{-1}) Fe^0:H$_2$O$_2$ (molar ratio 1.8:1) H$_2$O$_2$:MTBE (molar ratio of 220:1) pH 4.0 and 7.0	Fe^0 was synthesized with sodium borohydride (NaBH$_4$) in the presence of aqueous iron salt (FeSO$_4 \times$7H$_2$O). The iron solids were repeatedly washed (5 times) by centrifuging for 10 min at 7000 rpm, decanted, and refilled with water. The Fe^0 was stored at 4°C and sonicated before use.	Methyl tert-butyl ether (MTBE), (1 mg L^{-1})	99% within 10 min.	(Bergendahl and Thies, 2004)
ZVI (0.12 g) H$_2$O$_2$ (30%; 1.7 mL) US (acoustic 20 kHz, 45 W and hydrodynamic cavitator maximum discharge pressure of 4500 psi)	Commercially available ZVI	2,4-dichlorophen-oxyacetic acid (2,4-D), (0.235 g L^{-1})	60% TOC removal by acoustic cavitation 70% TOC removal by hydrodynamic cavitation in 20 min.	(Bremner et al., 2008)
UV (photon flux of the light source is 7.75mW/cm^2) ZVI (50 mgL^{-1}) H$_2$O$_2$ (100 mgL^{-1}) ammonium persulfate (APS)(200 mgl^{-1}) pH3	Commercially available ZVI (95% purity, 300-mesh size, electrolytic)	Alizarin Red S (ARS), (200 mg L^{-1})	100% after 3h The rate constant calculated for the process Fe^0/APS/UV is approx. 1.5 times higher than Fe^0/H$_2$O$_2$/UV process.	(Devi et al., 2009)

(continued)

Table 10.2 (continued)

Nanocatalyst	Synthesis of nanocatalyst	Treated micropollutant	Process efficiency	Reference
ZVI (1 gL^{-1}) US (20 kHz, 385W) Air or Ar (1.0 L min^{-1})	Commercially available ZVI (S_{BET} = 0.0786 ± 0. m^{2}g^{-1})	4-chlorophenol (4CP), (100mgL^{-1}) Ethylenediaminetetraacetic acid (EDTA), (0.32mM)	k_{obs}(EDTA) = 0.41 h^{-1} k_{obs}(4CP) = 0.32 h^{-1}	(Zhou et al., 2010)
UV (photon flux of the light source is 7.75mW/cm^2) ZVI (10 ppm) H$_2$O$_2$ (10ppm) ammonium persulfate (APS)(40 ppm) pH3	Commercially available ZVI (95% purity, 300-mesh size, electrolytic)	Methyl Orange (MO), (10 ppm)	Fe^0–UV–APS k = 0.0297 min^{-1} Fe^0–UV–H$_2$O$_2$ k = 0.1025 min^{-1}	(Gomathi Devi et al., 2009)
ZVI (0.5 gl^{-1}) H$_2$O$_2$ (15mM) US (20 kHz; power density is 201 WL^{-1}) pH 3	Commercially available ZVI (200-mesh size)	C.I. Acid Orange 7 (AR7), (200 mgL^{-1})	90% within 2 min. 56% COD over 60 min.	(Zhang et al., 2009a)
ZVI (0.3 gL^{-1}) H$_2$O$_2$ (2mM) pH 3.0	Commercially available ZVI (analytical grade, 99% purity, 200 meshes)	C.I. Acid Red 73 (AR 73), (200 mg L^{-1})	96.8% after 30 min.	(Fu et al., 2010)
Fe$_3$O$_4$ magnetic nanoparticles (MNPs) as a peroxidase mimetic (0.5 g L^{-1}) H$_2$O$_2$ (40 mmol L^{-1}) pH 5.0 US (20kHz, 6W)	FeCl$_3$×6H$_2$O (2.22 g) and FeSO$_4$×7H$_2$O (2.28 g) were dissolved in 30 mL 0.01 mol L^{-1} HCl aqueous solution and heated to 80°C.	Rhodamine B (RhB), (0.02 mmol L^{-1})	k = 0.15 min^{-1}	(Wang et al., 2010)

(continued)

Advanced catalytic oxidation of emerging Micropollutants 403

Table 10.2 (continued)

Nanocatalyst	Synthesis of nanocatalyst	Treated micropollutant	Process efficiency	Reference
	The warmed Fe(II)/Fe(III) solution was added drop-wise into 40 mL of 3.0 mol L^{-1} ammonia solution at 80°C under magnetic stirring. After 3 h reaction, the generated black Fe$_3$O$_4$ nanoparticles were collected by magnetic separation, washed with water to neutral pH, and then re-dispersed to 100 mL of water and stored for use (referred to as the Fe$_3$O$_4$ MNPs stock solution with a concentration of 12.5 g L^{-1}).			
Composite Tin and Zinc oxide nanocrystalline (ZnO (600 nm) /SnO$_2$ (10–15 nm) (2 g l^{-1}) UV (intensity 200 Wm^{-2})	Colloidal tin (IV) oxide aqueous dispersion (0.3 mL, crystallite size 10–15 nm, Alfa Chemicals) was mixed with few drops of HCl and the mixture was thoroughly ground with 60 mg of ZnO (particle size 600 nm, BDH).	Eosin Y, (1.59 × 10^{-4} M)	ZnO 1.08 × 10^{-4} mol L^{-1} h^{-1} SnO$_2$ 0.82 × 10^{-4} mol L^{-1} h^{-1} ZnO/SnO$_2$ 2.12 × 10^{-4} mol L^{-1} h^{-1} TiO$_2$ 1.31 10^{-4} mol L^{-1} h^{-1}	(Bandara et al., 2002)

(continued)

Table 10.2 (continued)

Nanocatalyst	Synthesis of nanocatalyst	Treated micropollutant	Process efficiency	Reference
	The mixture was diluted to 100 ml by addition of water, ultrasonically agitated, washed with water by centrifuging. The ZnO content of the catalyst prepared in the above manner was 54% by wt. when fully dried. For certain experiments sintering of the catalyst was done at 500°C for 30 min. under atmospheric pressure.			
Nano TiO_2 (1000 mg L^{-1}) US (40 kHz, power 50 W) pH = 10.0 t 40°C	Commercially available TiO_2 powders were heat treated at 450°C for two hours for activation. The X-ray diffraction (XRD) patterns were recorded on a Siemens (D-5005) diffractometer using Cu Kα radiation with a scan rate of 2.0 μmin^{-1}. The size of nanometer and anatase TiO_2 particles are 30–50 nm and 90–150 nm, respectively.	Methyl Parathion (MP), (50 mg L^{-1})	Up to 97% of removal with nano TiO_2, 75% using ordinary TiO_2 after 50 min. 22% with US only	(Wang et al., 2006)

(continued)

Table 10.2 (continued)

Nanocatalyst	Synthesis of nanocatalyst	Treated micropollutant	Process efficiency	Reference
Ni–Fe bimetallic particles (average particle size 30 nm) (0.06 g 40 mL^{-1} (Ni 30%) US (20 kHz, 0–250W) $pH_0 = 1.70$	At pH 12, 6 M NaOH, the aqueous solution of NaHB4 (0.5 M) was added drop-wise into an aqueous mixture of FeSO$_4 \times$ 7H$_2$O (0.2 M) and NiSO$_4 \times$ 7H$_2$O (0.02 M) at room temperature. The process is shown as below: $2Fe^{2-}(Ni^{2+}) + 2H_2O + BH_4^- \rightarrow 2Fe(Ni)\downarrow + BO_2^- + 4H^+ + 2H_2\uparrow$ The addition of NaBH$_4$ was excessive to ensure the complete reduction of the metallic ions in the solution. During the addition of NaBH$_4$, the solution was stirred vigorously. No special precautions were taken to eliminate oxygen from the reaction vessel. The mixture was stirred for 5 min and then filtered through a 0.45 μm membrane with vacuum pump. To get rid of the excess borohydride, the	Pentachlorphenol (PCP), (0.19 mM)	98% of PCP conversion 96% of dechlorination efficiency after 30 min	(Zhang et al., 2006)

(continued)

Table 10.2 (continued)

Nanocatalyst	Synthesis of nanocatalyst	Treated micropollutant	Process efficiency	Reference
	particles were washed with deionized water for five times and then rinsed with absolute ethanol to eliminate water and followed by filtration as dry as possible. At last, the black powder was spread in a thin layer on a filter paper and dried at room temperature in argon atmosphere. Then it was collected and stored in refrigerator for the subsequent use.			
Nano Ti (0.2 g L^{-1}) H$_2$O$_2$ US (35 kHz using ultrasonic power of 160W) t 25±1°C	Commercially available Ti nanoparticle (Degussa P25) (particle size appr. 100 nm, about 97% purity and 80:20 anatase to rutile ratio)	Basic Blue 41(BB41) (15 mgL^{-1})	For different concentrations of H$_2$O$_2$ 0 to 1000 mg l^{-1} the first order reaction rates were 9×10^{-4} to 9.9×10^{-3} min^{-1}, respectively	(Abbasi and Asl, 2008)

(continued)

Table 10.2 (continued)

Nanocatalyst	Synthesis of nanocatalyst	Treated micropollutant	Process efficiency				Reference
Au–TiO$_2$ nanoparticles US (358 kHz, 17W) UV	Sample 1 (S1): Au–TiO$_2$ nanoparticles were prepared by stirring 2 g of TiO$_2$ (Deguzza P-25) in the HAuCL$_4$×3H$_2$O (0.2 mM) containing polyvinyl-pyrrolidone (0.1 wt%) and 1-propanol (0.1 M) solution containing sonochemically synthesized gold nano-particles for 66 h (the longer time was needed in order to uniformly distribute the gold nanoparticles on TiO2 particles). This slurry was then evaporated in an air oven at 80°C. The dried samples were grounded to the fine powder, loaded onto a silica boat and introduced into a tubular furnace for sintering at 450°C for 4 h.	nonylphenol ethoxylate (NPE), (1.5×10^{-4} M) (Teric GN9), (1.5×10^{-4} M)		Rate constants, ×10^{-4} s^{-1}			(Anandan and Ashokkumar, 2009)
					US	UV	US/UV
				S1	1.7	5.6	4.2
				S2	1.6	5.3	5.4
				S3	0.5	1.1	0.69

(continued)

Table 10.2 (continued)

Nanocatalyst	Synthesis of nanocatalyst	Treated micropollutant	Process efficiency	Reference
	Sample 2 (S2): Au–TiO$_2$ nanoparticles were prepared by sonicating (20 kHz) an aqueous solution (70 ml) of HAuCl$_4 \times$3H$_2$O (0.2 mM) containing polyvinyl-pyrrolidone (0.1 wt%), 1-propanol (0.1 M) and 2 g of TiO$_2$ (Degussa P-25) at room temperature under nitrogen atmosphere. In this case higher ultrasonic power (input power = 120 W) was required for the reduction of Au(III). The purple powder formed was dried in a constant temperature oven at 80°C and sintered at 450°C.			
	Sample 3 (S3): titanium tetraisopropoxide Au–TiO$_2$ nanocolloids were prepared by sonicating (20 kHz) a mixture of solutions A and B (described below) at room			

(continued)

Table 10.2 (continued)

Nanocatalyst	Synthesis of nanocatalyst	Treated micropollutant	Process efficiency	Reference
	temperature under nitrogen atmosphere. Solution A: 70 ml of an aqueous solution containing $HAuCl_4 \times 3H_2O$ (0.2 mM), polyvinylpyrrolidone (0.1 wt%) and 1-propanol (0.1 M). Solution B: 0.2 g of titanium tetraisopropoxide dissolved in 2 ml of iscpropanol was injected into 20 ml of acidified water, pH adjusted to 1.5. In this case a higher power (input power = 160 W) of irradiation was required for the reduction of Au(III).			
Nano-nickel dioxide (0.04 g) (S_{BET} 105 m^2 g^{-1} and the average particle size with 3 nm) MW (750 W) pH 9	Precipitation–oxidation method in the presence of microwave irradiation was employed. Nickel oxide was obtained from nickel chloride with sodium hypochlorite in the sodium hydroxide solution. The black precipitate was then filtered,	Crystal violet CV (100 mg L^{-1})	97% in 5 min.	(He et al., 2010)

(continued)

Table 10.2 (continued)

Nanocatalyst	Synthesis of nanocatalyst	Treated micropollutant	Process efficiency	Reference
	washed with deionized water and irradiated with 2450 MHz microwave for 10 min. at 100 W, dried in a muffle furnace at 110°C for 24 h. The dried product was grounded to fine powders and stored in a desiccator. The as-prepared sample was separately calcined at different temperatures for 3 h.			

To enhance the effectiveness of ZVI/hydrogen peroxide process, Bremner and co-workers (2008) reported the concurrent application of AFP with acoustic and hydrodynamic cavitation for the elimination of 2,4-dichlorophenoxyacetic acid (2,4-D) from aqueous solutions. The successful removal of 2,4-D was enhanced by the ultrasound irradiation inducing a two-fold increase in the hydrogen peroxide consumption during 20 min. in comparison to 40 min. of the reaction in the absence of ultrasound. After 20 min. the total removal of the contaminant was observed with the residual 20% of TOC. However, the more efficient removal was obtained by using hydrocavitator (HC), and the advantages of such process included the possibility for the unit to operate in a continuous mode where larger volumes of contaminated water can be cost-effectively treated at equivalent energy dissipation levels.

However, iron is not the only metal, which can be utilized in nanocatalytic oxidation of micropollutants. Indeed, Bandara *et al.* (2002) studied composite ZnO/SnO_2 nanocrystalline particles for the decomposition of Eosin Y (20; 40; 50; 70- tetrabromofluorescein disodium salt). Catalytic activity of the newly fabricated nanocatalyst was superior to individual ZnO, SnO_2 or TiO_2 particles and its higher activity was assigned to the wider charge separation ability of the composite ZnO/SnO_2 system. Bandara and co-workers (2002) reported that sintering of the catalyst enhanced the catalytic properties of ZnO/SnO_2 composite system because it improved the contacts between both ZnO and SnO_2 particles, which adversely facilitated the charge transfer ability of ZnO/SnO_2 coupled system. In addition, interconnected SnO_2 crystallites structure may further facilitate the charge separation (Bandara *et al.*, 2002). Futhermore, the nanophotocatalytic decomposition of micropollutants was reported by several researchers. Thus, Arabatzis *et al.* (2003) studied nanocrystalline titania thin-film photocatalysts by gold deposition via electron beam evaporation for the degradation of representative azo dye – Methyl Orange (MO). They found that surface deposition of gold particles improved the photocatalytic efficiency of the titania films by the synergistic action on the charge separation process onto the semiconductor surface. The most beneficial surface concentration of gold particles in the Au/TiO_2 photocatalyst was 0.8 µg cm^{-2}, leading to a two times faster degradation and decolorization (from 5 h to 2.5 h) of MO in comparison to the reaction rates obtained with the parental TiO_2 material (Arabatzis *et al.*, 2003). Further increase in Au load resulted in the considerable efficiency decrease; however, it is important to note that the efficiency of the gold-modified materials was always higher than those with the non-modified catalysts. The enhanced performance of such loaded catalyst may be attributed to the ability of the catalyst to act both, as photons capturing sponge and gold-promoted substrate. Also, the surface coverage and the diameter of the nanoparticles are essential parameters to take into account because small metal islands deposited on the TiO_2 surface can provide a favorable geometry for facilitating the interfacial charge transfer under UV irradiation (Arabatzis et al., 2003). Yet again, enhancement of the catalytic activity is attributed to the action of Au particles, which play a key role in attracting conduction band photoelectrons and preventing the electron-hole recombination.

Moreover, Wang et al. (2006) studied the applicability of nano-TiO_2 prepared from the ordinary anatase TiO_2 for methyl parathion elimination. They determined an essential difference between the performance of nanometer and ordinary anatase TiO_2 catalysts resulting in the different degradation process of methyl parathion. Small particle size of the nanometer anatase TiO_2 catalysts can be regarded as series of microreactors in the solution with only one microreactor controlling one step of the whole degradation process (Wang et al., 2006). However, ordinary anatase TiO_2 catalysts with relatively large particle size, can be described as an integrative reactor. In such a reactor, entire degradation reaction takes place as the chain of adsorption and degradation reactions on the large surface of all the particles. The treatment parameters are represented in Table 10.2.

Aarthi et al. (2007) synthesized nano-sized anatase TiO_2 by solution combustion method and investigated its photocatalytic activity on the destruction of dyes with similar structure but different functional groups: Azure (A and B) and Sudan (III and IV). The degradation rates were directly proportional to the initial dye concentration of 10–20 mg L^{-1} with nearly the same reaction rate constants for Sudan III and Sudan IV, whereas the rate constants of Azure B was 2.1 times higher than that of Azure A. Moreover, Aarthi et al. (2007) also compared the performance of combustion synthesized (CS) TiO_2 with traditional Degussa P25, at the similar initial (21 mg L^{-1}) concentrations of the dyes. In the case of Sudan III degradation, the initial rates were higher for CS TiO_2 compared to that of Degussa P25, with the initial rates of 1 and 1.7 mg L^{-1} min^{-1} for Degussa P-25 and CS TiO_2, respectively. However, in the case of Azure A and B, degradation process was faster in the presence of Degussa P25 compared to that of CS TiO_2. The Langmuir–Hinshelwood kinetic model applied to describe the degradation of dyes in CS TiO_2 photocatalytic process, proved that the indirect pathway of electrons producing ·OH radicals is the main pathway for degradation of the substrate (Aarthi et al., 2007). In addition, the direct oxidation of the dyes by holes and the indirect oxidation by holes via OH· radicals had a negligible contribution.

To evaluate the effect of irradiation on the enhancement of pesticide degradation rates, Zhanqi et al. (2007) studied TiO_2 nanotubes in the presence of microwave irradiation to degrade atrazine (ATZ) in aqueous solution. They used two types of materials: TiO_2 nanotubes prepared by hydrothermal processing and TiO_2 particles prepared by the ultrasound assisted hydrolysis. The microwave-assisted photodegradation of atrazine on TiO_2 nanoparticles was much faster (5 min. with 20 mg L^{-1} of initial ATZ concentration reaching 98.5% of mineralization) than previous photocatalytic degradation reported by Bianchi et al. (2006) (4 h for 21.5 mg L^{-1} ATZ) and Parra et al. (2004) (45 min. for 20 mg L^{-1} ATZ), comparatively (Parra et al., 2004; Bianchi et al., 2006). TiO_2 nanotubes were more effective for atrazine degradation than TiO_2 nanoparticles, which could probably be due to the larger specific surface area, which would absorb more organic molecules to degrade. These findings can be explained as the cooperative effect between the MW irradiation and photocatalysis process. The surface of TiO_2 nanotubes became more hydrophobic under microwave irradiation and UV–vis light, which can increase the population of OH^- groups or O_2 that can be further oxidized to ·OH (Zhanqi et al., 2007). Moreover, it also caused the desorption of water molecules on the surface of TiO_2 that provided more active sites of the reactants to attach for oxidation (Horikoshi et al., 2002). Also, microwave irradiation can generate additional defect sites on TiO_2, to increase the transition probability of e^-–h^+ and decrease the e^-–h^+ recombination on TiO_2 surface Ai et al. (2005).

Furthermore, Abassi and Asl (2008) studied ultrasound-assisted degradation of Basic Blue 41 (BB41) by wet peroxide oxidation with commercially available nano TiO_2 (Degussa P25, average primary particle size around 100 nm, 97% purity and 80:20 anatase to rutile ratio). Efficiency of the nanocatalyst was higher than the efficiency reported for the same type of the catalysts with larger particle size (Table 10.2). The reason can be attributed to an increased surface area of nanoparticles which provides more active sites to produce radicals (Abbasi and Asl, 2008). Authors comprehensively studied the influence of the process parameters such as the reactant concentrations and pH. BB 41 degradation was significantly affected by pH. Thus, when pH was 4.5, the decolorization efficiency was 51% after 180 min. of ultrasound irradiation while an increase in pH to 8, increased the decolorization efficiency to 89.5%. This effect is related to the properties of TiO_2 itself, such as the point of zero charge (pzc) of the TiO_2 particles (pH_{pzc} = 6.8). It means that when the pH of the reaction increases, the number of negatively charged sites, which favor the adsorption of dye cations due to the electrostatic attraction also increases. The increase in H_2O_2 concentration resulted in the increase in the degradation efficiency (Table 10.2). The color decay also fitted the pseudo-first order reaction kinetics. The degradation products and the reaction intermediates were identified as benzaldehyde, formamide, 2-propyn-1-ol, indene, 2,2,3,3-tetramethylbutane, cyclopentanol,

1,3-benzodioxol-2-one, 2-methoxy-2-methyl, pentadecanal, 2-ethoxy-2,3-dihydro-3,3-dimethyl-methanesulfonate, acetic acid, oxalic acid, formic acid.

Nanocatalysts have different properties often determined by their synthesis protocol. For instance, Anandan and Ashokkumar (2009) prepared three types of Au–TiO$_2$ nanoparticles by several treatment methods: (i) precipitation, (ii) sonicating (20 kHz) commercially available Degussa P25 at a room temperature under nitrogen atmosphere, and by (iii) sonicating Au–TiO$_2$ nanocolloids using titanium tetraisopropoxide as a parental compound. The prepared nanocatalysts were used in the sonophotocatalytic degradation of nonylphenol ethoxylate (NPE). Characterization of the catalysts revealed the slight agglomeration of Au particles in the samples fabricated by precipitation and sonication methods, whereas Au–TiO$_2$ nanoparticles prepared by the simultaneous ultrasound irradiation of titanium tetraisopropoxide and gold chloride ions clearly showed that 2–3 nm gold nanoparticles were immobilized on the surface of TiO$_2$ without any visible aggregation. However, the latter sample was less effective in the degradation of NPE (Table 10.2). Lower sonocatalytic activity can be attributed to the presence of alcohol in the Au-TiO$_2$ sample, which retard the OH radical-mediated oxidation by scavenging the free radicals (Anandan and Ashokkumar, 2009).

Besides, He *et al.* (2010) determined that several semiconductors, ferromagnetic metals, transition metal oxides and especially transition metal oxide NiO$_2$ appeared to be especially efficient catalyst in catalytic oxidation of organic contaminants in the present of microwave irradiation. They synthesized nanosized nickel oxide through precipitation–oxidation in microwave and utilized it for the MW-assisted catalytic degradation (MICD) of crystal violet (CV) organic dye. Nanosized NiO$_2$ with OH group and active oxygen in its structure had a large specific area of 105 m^2 g^{-1} and the average particle size of 3 nm. Its high catalytic activity was attributed to its strong microwave absorbing capacity and the role of active oxygen and OH group, which finally transformed into ·OH radicals under microwave irradiation. MICD with nanosized nickel dioxide led to the deep oxidation of CV via three processes such as N-de-methylation, destruction of conjugated structure and the opening-benzene ring (He *et al.*, 2010).

10.6 CONCLUSIONS

The development of advanced catalytic processes for the removal of micropollutants from water and wastewater is mainly driven by the necessity for clean, sustainable and low–cost green process combined with the high efficiency. The great advantage of advanced catalytic oxidation process is the flexibility and numerous possibilities for the further development and innovation. Different types of catalysts can be used for the treatment of micropollutants. However, the tailor-made synthesis of the catalyst is still a challenging task due to inability of a single catalyst to fulfill all the needs and requirements of a target process. Apparently, future development in manufacturing of the catalysts should be directed towards the ability of the catalysts to carry out the selective oxidation reaction at desirable conditions.

Future examination of catalytic oxidation process applicability to a variety of industrial effluents for the full scale application is needed in order to demonstrate the capability of these processes to be utilized in industrial wastewater treatment. Some of them such as CWAO already found their application, but most of the advanced catalytic treatment processes demonstrated high removal efficiency only on the laboratory scale accompanied with a questionable feasibility (e.g., sonophotocatalytic oxidation). Therefore, further developments of the advanced catalytic technology should not only include the development of high durability/low cost catalysts but also solutions to overcome the energy consumption issues by, for example, replacing the UV-A,B,C irradiation with the solar light. If successfully implemented, catalytic oxidation would thus provide a cost-effective environmentally attractive option to manage the growing and toxic wastewater treatment problems.

10.7 REFERENCES

Aarthi, T., Narahari, P. and Madras, G. (2007) Photocatalytic degradation of Azure and Sudan dyes using nano TiO2. *J. Hazard. Mater.* **149**, 725–734.

Abbasi, M. and Asl, N. R. (2008) Sonochemical degradation of Basic Blue 41 dye assisted by nanoTiO2 and H2O2. *J. Hazard. Mater.* **153**, 942–947.

Abdullah, A. Z. and Ling, P. Y. (2009) Heat treatment effects on the characteristics and sonocatalytic performance of TiO2 in the degradation of organic dyes in aqueous solution. *J. Hazard. Mater.* **173**, 159–167.

Ai, Z., Li, J., Zhang, L. and Lee, S. (2010) Rapid decolorization of azo dyes in aqueous solution by an ultrasound-assisted electrocatalytic oxidation process. *Ultrason. Sonochem.* **17**, 370–375.

Akpan, U. G. and Hameed, B. H. (2009) Parameters affecting the photocatalytic degradation of dyes using TiO2-based photocatalysts: A review. *J. Hazard. Mater.* **170**, 520–529.

Anandan, S. and Ashokkumar, M. (2009) Sonochemical synthesis of Au-TiO2 nanoparticles for the sonophotocatalytic degradation of organic pollutants in aqueous environment. *Ultrason. Sonochem.* **16**, 316–320.

Arabatzis, I. M., Stergiopoulos, T., Andreeva, D., Kitova, S., Neophytides, S. G. and Falaras, P. (2003) Characterization and photocatalytic activity of Au/TiO2 thin films for azo-dye degradation. *J. Catal.* **220**, 127–135.

Aravindhan, R., Fathima, N. N., Rao, J. R. and Nair, B. U. (2006) Wet oxidation of acid brown dye by hydrogen peroxide using heterogeneous catalyst Mn-salen-Y zeolite: A potential catalyst. *J. Hazard. Mater.* **138**, 152–159.

Arslan-Alaton, I. and Ferry, J. L. (2002) Application of polyoxotungstates as environmental catalysts: wet air oxidation of acid dye Orange II. *Dyes Pigment.* **54**, 25–36.

Arslan, I., Balcioglu, I. A. and Bahnemann, D. W. (2000) Advanced chemical oxidation of reactive dyes in simulated dyehouse effluents by ferrioxalate-Fenton/UV-A and TiO2/UV-A processes. *Dyes Pigment.* **47**, 207–218.

Auriol, M., Filali-Meknassi, Y., Adams, C. D., Tyagi, R. D., Noguerol, T.-N. and Pica, B. (2008) Removal of estrogenic activity of natural and synthetic hormones from a municipal wastewater: Efficiency of horseradish peroxidase and laccase from Trametes versicolor. *Chemosphere* **70**, 445–452.

Bahena, C. L., Martinez, S. S., Guzman, D. M. and del Refugio Trejo Hernandez, M. (2008) Sonophotocatalytic degradation of alazine and gesaprim commercial herbicides in TiO2 slurry. *Chemosphere* **71**, 982–989.

Bandara, J., Tennakone, K. and Jayatilaka, P. P. B. (2002) Composite Tin and Zinc oxide nanocrystalline particles for enhanced charge separation in sensitized degradation of dyes. *Chemosphere* **49**, 439–445.

Banik, S., Bandyopadhyay, S. and Ganguly, S. (2003) Bioeffects of microwave–a brief review. *Bioresource Technol.* **87**, 155–159.

Barreiro, J. C., Capelato, M. D., Martin-Neto, L. and Bruun Hansen, H. C. (2007) Oxidative decomposition of atrazine by a Fenton-like reaction in a H2O2/ferrihydrite system. *Water. Res.* **41**, 55–62.

Basto, C., Tzanov, T. and Cavaco-Paulo, A. (2007) Combined ultrasound-laccase assisted bleaching of cotton. *Ultrason. Sonochem.* **14**, 350–354.

Beckett, M. A. and Hua, I. (2003) Enhanced sonochemical decomposition of 1,4-dioxane by ferrous iron. *Water. Res.* **37**, 2372–2376.

Bejarano-Perez, N. J. and Suarez-Herrera, M. F. (2007) Sonophotocatalytic degradation of congo red and methyl orange in the presence of TiO2 as a catalyst. *Ultrason. Sonochem.* **14**, 589–595.

Bergendahl, J. A. and Thies, T. P. (2004) Fenton's oxidation of MTBE with zero-valent iron. *Water. Res.* **38**, 327–334.

Bertelli, M. and Selli, E. (2004) Kinetic analysis on the combined use of photocatalysis, H2O2 photolysis, and sonolysis in the degradation of methyl tert-butyl ether. *Appl. Catal. B-Environ.* **52**, 205–212.

Bi, X., Wang, P., Jiao, C. and Cao, H. (2009) Degradation of remazol golden yellow dye wastewater in microwave enhanced ClO2 catalytic oxidation process. *J. Hazard. Mater.* **168**, 895–900.

Bianchi, C. L., Pirola, C., Ragaini, V. and Selli, E. (2006) Mechanism and efficiency of atrazine degradation under combined oxidation processes. *Appl. Catal. B-Environ.* **64**, 131–138.

Bozic, M. and Kokol, V. (2008) Ecological alternatives to the reduction and oxidation processes in dyeing with vat and sulphur dyes. *Dyes Pigment.* **76**, 299–309.

Bray, W. C. and Gorin, M. H. (1932) Ferryl ion, a compound of tetravalent iron. *J. Am. Chem. Soc.* **54**, 2124–2125.

Bremner, D. H., Carlo, S. D., Chakinala, A. G. and Cravotto, G. (2008) Mineralisation of 2,4-dichlorophenoxyacetic acid by acoustic or hydrodynamic cavitation in conjunction with the advanced Fenton process. *Ultrason. Sonochem.* **15**, 416–419.

Brillas, E., Sires, I., Arias, C., Cabot, P. L., Centellas, F., Rodriguez, R. M. and Garrido, J. A. (2005) Mineralization of paracetamol in aqueous medium by anodic oxidation with a boron-doped diamond electrode. *Chemosphere* **58**, 399–406.

Brillas, E., Sires, I. and Oturan, M. A. (2009) Electro-Fenton Process and Related Electrochemical Technologies Based on Fenton's Reaction Chemistry. *Chem. Rev.* **109**, 6570–6631.

Burbano, A. A., Dionysiou, D. D. and Suidan, M. T. (2008) Effect of oxidant-to-substrate ratios on the degradation of MTBE with Fenton reagent. *Water. Res.* **42**, 3225–3239.

Carrier, M., Besson, M., Guillard, C. and Gonze, E. (2009) Removal of herbicide diuron and thermal degradation products under Catalytic Wet Air Oxidation conditions. *Appl. Catal. B-Environ.* **91**, 275–283.

Catalkaya, E. C. and Kargi, F. (2007) Effects of operating parameters on advanced oxidation of diuron by the Fenton's reagent: A statistical design approach. *Chemosphere* **69**, 485–492.

Chacon, J. M., Teresa Leal, M., Sanchez, M. and Bandala, E. R. (2006) Solar photocatalytic degradation of azo-dyes by photo-Fenton process. *Dyes Pigment.* **69**, 144–150.

Chakrabarti, S. and Dutta, B. K. (2004) Photocatalytic degradation of model textile dyes in wastewater using ZnO as semiconductor catalyst. *J. Hazard. Mater.* **112**, 269–278.

Chandrasekara Pillai, K., Kwon, T. O. and Moon, I. S. (2009) Degradation of wastewater from terephthalic acid manufacturing process by ozonation catalyzed with Fe2+, H2O2 and UV light: Direct versus indirect ozonation reactions. *Appl. Catal. B-Environ.* **91**, 319–328.

Chang, D.-J., Chen, I. P., Chen, M.-T. and Lin, S.-S. (2003) Wet air oxidation of a reactive dye solution using CoAlPO4-5 and CeO2 catalysts. *Chemosphere* **52**, 943–949.

Chen, J.-Q., Wang, D., Zhu, M.-X. and Gao, C.-J. (2006) Study on degradation of methyl orange using pelagite as photocatalyst. *J. Hazard. Mater.* **138**, 182–186.

Chorkendorff, I. and Niemantsverdriet, J. W. (2003) *Concepts of Modern Catalysis and Kinetics*. Weinheim, Germany, Wiley-VCH.

Delanoe, F., Acedo, B., Karpel Vel Leitner, N. and Legube, B. (2001) Relationship between the structure of Ru/CeO2 catalysts and their activity in the catalytic ozonation of succinic acid aqueous solutions. *Appl. Catal. B-Environ.* **29**, 315–325.

Devi, L. G., Rajashekhar, K. E., Raju, K. S. A. and Kumar, S. G. (2009) Kinetic modeling based on the non-linear regression analysis for the degradation of Alizarin Red S by advanced photo Fenton process using zero valent metallic iron as the catalyst. *J. Mol. Catal. A: Chem.* **314**, 88–94.

Du, W., Sun, Q., Lv, X. and Xu, Y. (2009) Enhanced activity of iron oxide dispersed on bentonite for the catalytic degradation of organic dye under visible light. *Catal. Commun.* **10**, 1854–1858.

Duarte, F., Maldonado-Hydar, F. J., Perez-Cadenas, A. F. and Madeira, L. M. (2009) Fenton-like degradation of azo-dye Orange II catalyzed by transition metals on carbon aerogels. *Appl. Catal. B-Environ.* **85**, 139–147.

Duran, N. and Esposito, E. (2000) Potential applications of oxidative enzymes and phenoloxidase-like compounds in wastewater and soil treatment: a review. *Appl. Catal. B-Environ.* **28**, 83–99.

Elmolla, E. and Chaudhuri, M. (2009) Optimization of Fenton process for treatment of amoxicillin, ampicillin and cloxacillin antibiotics in aqueous solution. *J. Hazard. Mater.* **170**, 666–672.

Elmolla, E. S. and Chaudhuri, M. (2010) Degradation of amoxicillin, ampicillin and cloxacillin antibiotics in aqueous solution by the UV/ZnO photocatalytic process. *J. Hazard. Mater.* **173**, 445–449.

Ensing, B. and Baerends, E. J. (2002) Reaction Path Sampling of the Reaction between Iron(II) and Hydrogen Peroxide in Aqueous Solution. *J. Phys. Chem. A* **106**, 7902–7910.

Ensing, B., Buda, F. and Baerends, E. J. (2003) Fenton-like Chemistry in Water: Oxidation Catalysis by Fe(III) and H2O2. *J. Phys. Chem. A* **107**, 5722–5731.

Entezari, M. H. and Petrier, C. (2004) A combination of ultrasound and oxidative enzyme: sono-biodegradation of phenol. *Appl. Catal. B-Environ.* **53**, 257–263.

Faria, P. C. C., Orfao, J. J. M. and Pereira, M. F. R. (2009) Activated carbon and ceria catalysts applied to the catalytic ozonation of dyes and textile effluents. *Appl. Catal. B-Environ.* **88**, 341–350.

Fathima, N. N., Aravindhan, R., Rao, J. R. and Nair, B. U. (2008) Dye house wastewater treatment through advanced oxidation process using Cu-exchanged Y zeolite: A heterogeneous catalytic approach. *Chemosphere* **70**, 1146–1151.

Fu, F., Wang, Q. and Tang, B. (2010) Effective degradation of C.I. Acid Red 73 by advanced Fenton process. *J. Hazard. Mater.* **174**, 17–22.

Fu, X., Wang, X., Long, J., Ding, Z., Yan, T., Zhang, G., Zhang, Z., Lin, H. and Fu, X. (2009) Hydrothermal synthesis, characterization, and photocatalytic properties of Zn2SnO4. *J. Solid State Chem.* **182**, 517–524.

Gao, J., Zhao, G., Shi, W. and Li, D. (2009) Microwave activated electrochemical degradation of 2,4-dichlorophenoxyacetic acid at boron-doped diamond electrode. *Chemosphere* **75**, 519–525.

Gogate, P. R. (2008) Treatment of wastewater streams containing phenolic compounds using hybrid techniques based on cavitation: A review of the current status and the way forward. *Ultrason. Sonochem.* **15**, 1–15.

Gogate, P. R. and Pandit, A. B. (2004) A review of imperative technologies for wastewater treatment II: hybrid methods. *Adv. Environ. Res.* **8**, 553–597.

Gomathi Devi, L., Girish Kumar, S., Mohan Reddy, K. and Munikrishnappa, C. (2009) Photo degradation of Methyl Orange an azo dye by Advanced Fenton Process using zero valent metallic iron: Influence of various reaction parameters and its degradation mechanism. *J. Hazard. Mater.* **164**, 459–467.

Gromboni, C. F., Kamogawa, M. Y., Ferreira, A. G., Nobrega, J. A. and Nogueira, A. R. A. (2007) Microwave-assisted photo-Fenton decomposition of chlorfenvinphos and cypermethrin in residual water. *J. Photochem. Photobiol. A-Chem.* **185**, 32–37.

Hagen, J. (2006) *Industrial Catalysis*. Weinheim, Germany, Wiley-VCH.

Harir, M., Gaspar, A., Kanawati, B., Fekete, A., Frommberger, M., Martens, D., Kettrup, A., El Azzouzi, M. and Schmitt-Kopplin, P. (2008) Photocatalytic reactions of imazamox at TiO2, H2O2 and TiO2/H2O2 in water interfaces: Kinetic and photoproducts study. *Appl. Catal. B-Environ.* **84**, 524–532.

He, H., Yang, S., Yu, K., Ju, Y., Sun, C. and Wang, L. (2010) Microwave induced catalytic degradation of crystal violet in nano-nickel dioxide suspensions. *J. Hazard. Mater.* **173**, 393–400.

Horikoshi, S., Hidaka, H. and Serpone, N. (2002) Environmental remediation by an integrated microwave/UV-illumination method II. Characteristics of a novel UV-VIS-microwave integrated irradiation device in photodegradation processes. *J. Photochem. Photobiol. A-Chem.* **153**, 185–189.

Horikoshi, S., Hidaka, H. and Serpone, N. (2004a) Environmental remediation by an integrated microwave/UV illumination technique: VI. A simple modified domestic microwave oven integrating an electrodeless UV-Vis lamp to photodegrade environmental pollutants in aqueous media. *J. Photochem. Photobiol. A-Chem.* **161**, 221–225.

Horikoshi, S., Matsubara, A., Takayama, S., Sato, M., Sakai, F., Kajitani, M., Abe, M. and Serpone, N. (2009) Characterization of microwave effects on metal-oxide materials: Zinc oxide and titanium dioxide. *Appl. Catal. B-Environ.* **91**, 362–367.

Horikoshi, S., Tokunaga, A., Hidaka, H. and Serpone, N. (2004b) Environmental remediation by an integrated microwave/UV illumination method: VII. Thermal/non-thermal effects in the microwave-assisted photocatalyzed mineralization of bisphenol-A. *J. Photochem. Photobiol. A-Chem.* **162**, 33–40.

Hsueh, C. L., Huang, Y. H., Wang, C. C. and Chen, C. Y. (2005) Degradation of azo dyes using low iron concentration of Fenton and Fenton-like system. *Chemosphere* **58**, 1409–1414.

Huber, M. M., Canonica, S., Park, G.-Y. and von Gunten, U. (2003) Oxidation of Pharmaceuticals during Ozonation and Advanced Oxidation Processes. *Environ. Sci. Technol.* **37**, 1016–1024.

Inglezakis, V. J. and Poulopoulos, S. G. (2006) *Adsorption, Ion Exchange and Catalysis* New York, Elsevier Publishing Co.

Ji, P., Zhang, J., Chen, F. and Anpo, M. (2009) Study of adsorption and degradation of acid orange 7 on the surface of CeO2 under visible light irradiation. *Appl. Catal. B-Environ.* **85**, 148–154.

Karam, J. and Nicell, J. A. (1997) Potential applications of enzymes in waste treatment. *J. Chem. Tech. Biotechnol.* **69**, 141–153.

Kasprzyk-Hordern, B., Zi?lek, M. and Nawrocki, J. (2003) Catalytic ozonation and methods of enhancing molecular ozone reactions in water treatment. *Appl. Catal. B-Environ.* **46**, 639–669.

Kassinos, D., Varnava, N., Michael, C. and Piera, P. (2009) Homogeneous oxidation of aqueous solutions of atrazine and fenitrothion through dark and photo-Fenton reactions. *Chemosphere* **74**, 866–872.

Kaur, S. and Singh, V. (2007) Visible light induced sonophotocatalytic degradation of Reactive Red dye 198 using dye sensitized TiO2. *Ultrason. Sonochem.* **14**, 531–537.

Kim, J.-H., Kim, S.-J., Lee, C.-H. and Kwon, H.-H. (2009) Removal of Toxic Organic Micropollutants with FeTsPc-Immobilized Amberlite/H2O2: Effect of Physico-chemical Properties of Toxic Chemicals. *Ind. Eng. Chem. Res.* **48**, 1586–1592.

Kim, J.-H., Park, P.-K., Lee, C.-H., Kwon, H.-H. and Lee, S. (2008) A novel hybrid system for the removal of endocrine disrupting chemicals: Nanofiltration and homogeneous catalytic oxidation. *J. Membr. Sci.* **312**, 66–75.

Kim, S.-C. and Lee, D.-K. (2004) Preparation of Al-Cu pillared clay catalysts for the catalytic wet oxidation of reactive dyes. *Catal. Today* **97**, 153–158.

Kondru, A. K., Kumar, P. and Chand, S. (2009) Catalytic wet peroxide oxidation of azo dye (Congo red) using modified Y zeolite as catalyst. *J. Hazard. Mater.* **166**, 342–347.

Krutzler, T. and Bauer, R. (1999) Optimization of a photo-fenton prototype reactor. *Chemosphere* **38**, 2517–2532.

Kusic, H., Loncaric Bozic, A., Koprivanac, N. and Papic, S. (2007) Fenton type processes for minimization of organic content in coloured wastewaters. Part II: Combination with zeolites. *Dyes Pigment.* **74**, 388–395.

Lee, D.-K., Cho, I.-C., Lee, G.-S., Kim, S.-C., Kim, D.-S. and Yang, Y.-K. (2004) Catalytic wet oxidation of reactive dyes with H2/O2 mixture on Pd-Pt/Al2O3 catalysts. *Sep. Purif. Technol.* **34**, 43–50.

Lei, L., Dai, Q., Zhou, M. and Zhang, X. (2007) Decolorization of cationic red X-GRL by wet air oxidation: Performance optimization and degradation mechanism. *Chemosphere* **68**, 1135–1142.

Levec, J. and Pintar, A. (2007) Catalytic wet-air oxidation processes: A review. *Catal. Today* **124**, 172–184.

Li, R., Yang, C., Chen, H., Zeng, G., Yu, G. and Guo, J. (2009a) Removal of triazophos pesticide from wastewater with Fenton reagent. *J. Hazard. Mater.* **167**, 1028–1032.

Li, W., Zhao, S., Qi, B., Du, Y., Wang, X. and Huo, M. (2009b) Fast catalytic degradation of organic dye with air and MoO3:Ce nanofibers under room condition. *Appl. Catal. B-Environ.* **92**, 333–340.

Liu, Y. and Sun, D. (2007) Development of Fe2O3-CeO2-TiO2/[gamma]-Al2O3 as catalyst for catalytic wet air oxidation of methyl orange azo dye under room condition. *Appl. Catal. B-Environ.* **72**, 205–211.

Luck, F. (1996) A review of industrial catalytic wet air oxidation processes. *Catal. Today* **27**, 195–202.

Luck, F. (1999) Wet air oxidation: past, present and future. *Catal. Today* **53**, 81–91.

Maletzky, P. and Bauer, R. (1998) The Photo-Fenton method – Degradation of nitrogen containing organic compounds. *Chemosphere* **37**, 899–909.

Marco-Urrea, E., Perez-Trujillo, M., Cruz-Morató, C., Caminal, G. and Vicent, T. (2010) White-rot fungus-mediated degradation of the analgesic ketoprofen and identification of intermediates by HPLC-DAD-MS and NMR. *Chemosphere* **78**, 474–481.

Martinez-Huitle, C. A. and Brillas, E. (2009) Decontamination of wastewaters containing synthetic organic dyes by electrochemical methods: A general review. *Appl. Catal. B-Environ.* **87**, 105–145.

Martinez, S. S. and Bahena, C. L. (2009) Chlorbromuron urea herbicide removal by electro-Fenton reaction in aqueous effluents. *Water. Res.* **43**, 33–40.

Mikulova, J., Rossignol, S., Barbier Jr, J., Mesnard, D., Kappenstein, C. and Duprez, D. (2007) Ruthenium and platinum catalysts supported on Ce, Zr, Pr-O mixed oxides prepared by soft chemistry for acetic acid wet air oxidation. *Appl. Catal. B-Environ.* **72**, 1–10.

Milone, C., Fazio, M., Pistone, A. and Galvagno, S. (2006) Catalytic wet air oxidation of p-coumaric acid on CeO2, platinum and gold supported on CeO2 catalysts. *Appl. Catal. B-Environ.* **68**, 28–37.

Minero, C., Pellizzari, P., Maurino, V., Pelizzetti, E. and Vione, D. (2008) Enhancement of dye sonochemical degradation by some inorganic anions present in natural waters. *Appl. Catal. B-Environ.* **77**, 308–316.

Mrowetz, M., Pirola, C. and Selli, E. (2003) Degradation of organic water pollutants through sonophotocatalysis in the presence of TiO2. *Ultrason. Sonochem.* **10**, 247–254.

Muller, P., Klan, P. and Cirkva, V. (2003) The electrodeless discharge lamp: a prospective tool for photochemistry: Part 4. Temperature- and envelope material-dependent emission characteristics. *J. Photochem. Photobiol. A-Chem.* **158**, 1–5.

Mutyala, S., Fairbridge, C., Pare, J. R. J., Belanger, J. M. R., Ng, S. and Hawkins, R. (2010) Microwave applications to oil sands and petroleum: A review. *Fuel Process. Technol.* **91**, 127–135.

Oliveira, L. C. A., Ramalho, T. C., Souza, E. F., Gonzalves, M., Oliveira, D. Q. L., Pereira, M. C. and Fabris, J. D. (2008) Catalytic properties of goethite prepared in the presence of Nb on oxidation reactions in water: Computational and experimental studies. *Appl. Catal. B-Environ.* **83**, 169–176.

Oliviero, L., Barbier, J., Duprez, D., Guerrero-Ruiz, A., Bachiller-Baeza, B. and Rodriguez-Ramos, I. (2000) Catalytic wet air oxidation of phenol and acrylic acid over Ru/C and Ru-CeO2/C catalysts. *Appl. Catal. B-Environ.* **25**, 267–275.

Pandit, A. B., Gogate, P. R. and Mujumdar, S. (2001) Ultrasonic degradation of 2:4:6 trichlorophenol in presence of TiO2 catalyst. *Ultrason. Sonochem.* **8**, 227–231.

Panizza, M. and Cerisola, G. (2007) Electrocatalytic materials for the electrochemical oxidation of synthetic dyes. *Appl. Catal. B-Environ.* **75**, 95–101.

Papic, S., Vujevic, D., Koprivanac, N. and Sinko, D. (2009) Decolourization and mineralization of commercial reactive dyes by using homogeneous and heterogeneous Fenton and UV/Fenton processes. *J. Hazard. Mater.* **164**, 1137–1145.

Parra, S., Elena Stanca, S., Guasaquillo, I. and Ravindranathan Thampi, K. (2004) Photocatalytic degradation of atrazine using suspended and supported TiO2. *Appl. Catal. B-Environ.* **51**, 107–116.

Poerschmann, J., Trommler, U., Gorecki, T. and Kopinke, F.-D. (2009) Formation of chlorinated biphenyls, diphenyl ethers and benzofurans as a result of Fenton-driven oxidation of 2-chlorophenol. *Chemosphere* **75**, 772–780.

Pradeep, T. and Anshup (2009) Noble metal nanoparticles for water purification: A critical review. *Thin Solid Films* **517**, 6441–6478.

Quan, X., Zhang, Y., Chen, S., Zhao, Y. and Yang, F. (2007) Generation of hydroxyl radical in aqueous solution by microwave energy using activated carbon as catalyst and its potential in removal of persistent organic substances. *J. Mol. Catal. A: Chem.* **263**, 216–222.

Quintanilla, A., Casas, J. A. and Rodriguez, J. J. (2007) Catalytic wet air oxidation of phenol with modified activated carbons and Fe/activated carbon catalysts. *Appl. Catal. B-Environ.* **76**, 135–145.

Rafqah, S., Chung, P. W.-W., Forano, C. and Sarakha, M. (2008) Photocatalytic degradation of metsulfuron methyl in aqueous solution by decatungstate anions. *J. Photochem. Photobiol. A-Chem.* **199**, 297–302.

Rafqah, S., Wong-Wah-Chung, P., Nelieu, S., Einhorn, J. and Sarakha, M. (2006) Phototransformation of triclosan in the presence of TiO2 in aqueous suspension: Mechanistic approach. *Appl. Catal. B-Environ.* **66**, 119–125.

Ramirez, J. H., Maldonado-Hydar, F. J., Perez-Cadenas, A. F., Moreno-Castilla, C., Costa, C. A. and Madeira, L. M. (2007) Azo-dye Orange II degradation by heterogeneous Fenton-like reaction using carbon-Fe catalysts. *Appl. Catal. B-Environ.* **75**, 312–323.

Rauf, M. A. and Ashraf, S. S. (2009) Fundamental principles and application of heterogeneous photocatalytic degradation of dyes in solution. *Chem. Eng. J.* **151**, 10–18.

Rehman, S., Ullah, R., Butt, A. M. and Gohar, N. D. (2009) Strategies of making TiO2 and ZnO visible light active. *J. Hazard. Mater.* **170**, 560–569.

Rehorek, A., Tauber, M. and Gubitz, G. (2004) Application of power ultrasound for azo dye degradation. *Ultrason. Sonochem.* **11**, 177–182.

Rivas, F. J., Carbajo, M., Beltran, F., Gimeno, O. and Frades, J. (2008) Comparison of different advanced oxidation processes (AOPs) in the presence of perovskites. *J. Hazard. Mater.* **155**, 407–414.

Rodriguez-Rodriguez, C. E., Marco-Urrea, E. and Caminal, G. (2010) Degradation of naproxen and carbamazepine in spiked sludge by slurry and solid-phase Trametes versicolor systems. *Bioresource Technol.* **101**, 2259–2266.

Rodriguez, A., Ovejero, G., Romero, M. D., Diaz, C., Barreiro, M. and Garcia, J. (2008) Catalytic wet air oxidation of textile industrial wastewater using metal supported on carbon nanofibers. *J. Supercrit. Fluids* **46**, 163–172.

Rokhina, E. V., Lens, P. and Virkutyte, J. (2009) Low-frequency ultrasound in biotechnology: state of the art. *Trends Biotechnol.* **27**, 298–306.

Rosal, R., Gonzalo, M. S., Rodriguez, A. and Garcia-Calvo, E. (2009) Ozonation of clofibric acid catalyzed by titanium dioxide. *J. Hazard. Mater.* **169**, 411–418.

Rosal, R., Rodriguez, A., Gonzalo, M. S. and Garcia-Calvo, E. (2008) Catalytic ozonation of naproxen and carbamazepine on titanium dioxide. *Appl. Catal. B-Environ.* **84**, 48–57.

Safarzadeh-Amiri, A., Bolton, J. R. and Cater, S. R. (1996) Ferrioxalate-mediated solar degradation of organic contaminants in water. *Sol. Energy* **56**, 439–443.

Sires, I., Arias, C., Cabot, P. L., Centellas, F., Garrido, J. A., Rodriguez, R. M. and Brillas, E. (2007) Degradation of clofibric acid in acidic aqueous medium by electro-Fenton and photoelectro-Fenton. *Chemosphere* **66**, 1660–1669.

Sires, I., Guivarch, E., Oturan, N. and Oturan, M. A. (2008) Efficient removal of triphenylmethane dyes from aqueous medium by *in situ* electrogenerated Fenton's reagent at carbon-felt cathode. *Chemosphere* **72**, 592–600.

Skoumal, M., Cabot, P.-L., Centellas, F., Arias, C., Rodriguez, R. M., Garrido, J. A. and Brillas, E. (2006) Mineralization of paracetamol by ozonation catalyzed with Fe2+, Cu2+ and UVA light. *Appl. Catal. B-Environ.* **66**, 228–240.

Song, L., Qiu, R., Mo, Y., Zhang, D., Wei, H. and Xiong, Y. (2007) Photodegradation of phenol in a polymer-modified TiO2 semiconductor particulate system under the irradiation of visible light. *Catal. Commun.* **8**, 429–433.

Song, L., Zhang, S. and Chen, B. (2009) A novel visible-light-sensitive strontium carbonate photocatalyst with high photocatalytic activity. *Catal. Commun.* **10**, 1565–1568.

Sorokin, A. and Meunier, B. (1994) Efficient H2O2 oxidation of chlorinated phenols catalysed by supported iron phthalocyanines. *J. Chem. Soc., Chem. Com.*, 1799–1800.

Sorokin, A., Meunier, B. and Séris, J.-L. (1995) Efficient oxidative dechlorination and aromatic ring cleavage of chlorinated phenols catalyzed by iron sulfophthalocyanine *Science*, 1163–1166

Sorokin, A. B., Mangematin, S. and Pergrale, C. (2002) Selective oxidation of aromatic compounds with dioxygen and peroxides catalyzed by phthalocyanine supported catalysts. *J. Mol. Catal. A: Chem.* **182–183**, 267–281.

Suslick, K. S. (1990) Sonochemistry. *Science* **247**, 1439–1445.

Tamagawa, Y., Yamaki, R., Hirai, H., Kawai, S. and Nishida, T. (2006) Removal of estrogenic activity of natural steroidal hormone estrone by ligninolytic enzymes from white rot fungi. *Chemosphere* **65**, 97–101.

Tauber, M. M., Gubitz, G. M. and Rehorek, A. (2008) Degradation of azo dyes by oxidative processes – Laccase and ultrasound treatment. *Bioresource Technol.* **99**, 4213–4220.

Tong, S.-P., Liu, W.-P., Leng, W.-H. and Zhang, Q.-Q. (2003) Characteristics of MnO2 catalytic ozonation of sulfosalicylic acid and propionic acid in water. *Chemosphere* **50**, 1359–1364.

Torres, R. A., Nieto, J. I., Combet, E., Petrier, C. and Pulgarin, C. (2008a) Influence of TiO2 concentration on the synergistic effect between photocatalysis and high-frequency ultrasound for organic pollutant mineralization in water. *Appl. Catal. B-Environ.* **80**, 168–175.

Torres, R. A., Petrier, C., Combet, E., Moulet, F. and Pulgarin, C. (2006) Bisphenol A Mineralization by Integrated Ultrasound-UV-Iron (II) Treatment. *Environ. Sci. Technol.* **41**, 297–302.

Torres, R. A., Sarantakos, G., Combet, E., Petrier, C. and Pulgarin, C. (2008b) Sequential helio-photo-Fenton and sonication processes for the treatment of bisphenol A. *J. Photochem. Photobiol. A-Chem.* **199**, 197–203.

Vilaplana, M., Marco-Urrea, E., Gabarrell, X., Sarra, M. and Caminal, G. (2008) Required equilibrium studies for designing a three-phase bioreactor to degrade trichloroethylene (TCE) and tetrachloroethylene (PCE) by Trametes versicolor. *Chem. Eng. J.* **144**, 21–27.

von Gunten, U. (2003) Ozonation of drinking water: Part I. Oxidation kinetics and product formation. *Water. Res.* **37**, 1443–1467.

Wang, J., Pan, Z., Zhang, Z., Zhang, X., Wen, F., Ma, T., Jiang, Y., Wang, L., Xu, L. and Kang, P. (2006) Sonocatalytic degradation of methyl parathion in the presence of nanometer and ordinary anatase titanium dioxide catalysts and comparison of their sonocatalytic abilities. *Ultrason. Sonochem.* **13**, 493–500.

Wang, J., Zhu, W., He, X. and Yang, S. (2008) Catalytic wet air oxidation of acetic acid over different ruthenium catalysts. *Catal. Commun.* **9**, 2163–2167.

Wang, N., Zhu, L., Wang, M., Wang, D. and Tang, H. (2010) Sono-enhanced degradation of dye pollutants with the use of H2O2 activated by Fe3O4 magnetic nanoparticles as peroxidase mimetic. *Ultrason. Sonochem.* **17**, 78–83.

Wu, L., Li, A., Gao, G., Fei, Z., Xu, S. and Zhang, Q. (2007) Efficient photodegradation of 2,4-dichlorophenol in aqueous solution catalyzed by polydivinylbenzene-supported zinc phthalocyanine. *J. Mol. Catal. A: Chem.* **269**, 183–189.

Xia, F., Ou, E., Wang, L. and Wang, J. (2008) Photocatalytic degradation of dyes over cobalt doped mesoporous SBA-15 under sunlight. *Dyes Pigment.* **76**, 76–81.

Yang, M., Xu, A., Du, H., Sun, C. and Li, C. (2007) Removal of salicylic acid on perovskite-type oxide LaFeO3 catalyst in catalytic wet air oxidation process. *J. Hazard. Mater.* **139**, 86–92.

Yang, Y., Wang, P., Shi, S. and Liu, Y. (2009) Microwave enhanced Fenton-like process for the treatment of high concentration pharmaceutical wastewater. *J. Hazard. Mater.* **168**, 238–245.

Yasman, Y., Bulatov, V., Gridin, V. V., Agur, S., Galil, N., Armon, R. and Schechter, I. (2004) A new sono-electrochemical method for enhanced detoxification of hydrophilic chloroorganic pollutants in water. *Ultrason. Sonochem.* **11**, 365–372.

Yunrui, Z., Wanpeng, Z., Fudong, L., Jianbing, W. and Shaoxia, Y. (2007) Catalytic activity of Ru/Al2O3 for ozonation of dimethyl phthalate in aqueous solution. *Chemosphere* **66**, 145–150.

Zapata, A., Velegraki, T., Sanchez-Perez, J. A., Mantzavinos, D., Maldonado, M. I. and Malato, S. (2009) Solar photo-Fenton treatment of pesticides in water: Effect of iron concentration on degradation and assessment of ecotoxicity and biodegradability. *Appl. Catal. B-Environ.* **88**, 448–454.

Zhang, H., Zhang, J., Zhang, C., Liu, F. and Zhang, D. (2009a) Degradation of C.I. Acid Orange 7 by the advanced Fenton process in combination with ultrasonic irradiation. *Ultrason. Sonochem.* **16**, 325–330.

Zhang, W., Quan, X., Wang, J., Zhang, Z. and Chen, S. (2006) Rapid and complete dechlorination of PCP in aqueous solution using Ni-Fe nanoparticles under assistance of ultrasound. *Chemosphere* **65**, 58–64.

Zhang, X., Li, G. and Wang, Y. (2007) Microwave assisted photocatalytic degradation of high concentration azo dye Reactive Brilliant Red X-3B with microwave electrodeless lamp as light source. *Dyes Pigment.* **74**, 536–544.

Zhang, Y., Li, D., Chen, Y., Wang, X. and Wang, S. (2009b) Catalytic wet air oxidation of dye pollutants by polyoxomolybdate nanotubes under room condition. *Appl. Catal. B-Environ.* **86**, 182–189.

Zhanqi, G., Shaogui, Y., Na, T. and Cheng, S. (2007) Microwave assisted rapid and complete degradation of atrazine using TiO2 nanotube photocatalyst suspensions. *J. Hazard. Mater.* **145**, 424–430.

Zhao, G., Gao, J., Shi, W., Liu, M. and Li, D. (2009) Electrochemical incineration of high concentration azo dye wastewater on the *in situ* activated platinum electrode with sustained microwave radiation. *Chemosphere* **77**, 188–193.

Zhou, T., Lim, T.-T., Li, Y., Lu, X. and Wong, F.-S. (2010) The role and fate of EDTA in ultrasound-enhanced zero-valent iron/air system. *Chemosphere* **78**, 576–582

Zimbron, J. A. and Reardon, K. F. (2009) Fenton's oxidation of pentachlorophenol. *Water. Res.* **43**, 1831–1840.

Chapter 11
Existence, Impacts, Transport and Treatments of Herbicides in Great Barrier Reef Catchments in Australia

Dimuth Navaratna, Li Shu and Veeriah Jegatheesan

11.1 INTRODUCTION

This chapter is focused on the present and future impacts due to the discharge of large quantities of herbicides and pesticides, which are mostly persistent, bio-accumulative and toxic, to the Great Barrier Reef (GBR) annually. As herbicides and pesticides are considered as Persistent Organic Pollutants (POPs), initially the chapter describes the POPs including their classifications, common characteristics and impacts to the humans and the environment in general. The second generation herbicides (Photosystem II herbicides), which are currently used in the farmlands located in the GBR catchment are discussed in this chapter as most of the pesticides listed as POPs by the Stockholm Convention in 2004 are banned or strictly controlled in many countries including Australia. Then the chapter describes the background of the GBR catchment, reasons for the existence of herbicides in the waterways and their sources and modes of discharge. This chapter also provides evidence to show the persistence of these herbicides and their impacts, to the ecosystem and marine life. Finally, the chapter discusses possible treatment methods to reduce the quantities of pesticides, herbicides and other POPs including trace organic compounds such as Pharmaceuticals, Endocrine Disrupting Compounds (EDC), Disinfection By-Products (DBPs) and other micro-pollutants.

11.2 PERSISTENT ORGANIC POLLUTANTS (POPS)

Persistent Organic Pollutants (POPs) are carbon-based chemical substances that persist in the environment, bio-accumulate through the food web, capable of long-range transport and pose a risk of causing adverse effects to human health and to the environment at large. There are only a very few natural sources of POPs, but mostly they are generated by human beings through industrial processes, either intentionally or as by-products. Most of the POPs are the pesticides used in the past and the present and others are used or generated in industrial processes and manufacturing of products such as solvents, polyvinyl chloride and pharmaceuticals substances. This group of priority pollutants consists of pesticides (such as DDT), industrial chemicals (such as Polychlorinated Biphenyls-PCBs) and unintentional by-products of industrial processes (such as dioxins and furans). After their usage for the intended purpose, a large fraction of these substances will be discharged to the environment. In addition to this, as most of the existing conventional wastewater treatment plants in the world are not designed for the removal of these persistent organic compounds, a significant quantity of persistent and toxic matter is discharged to the environment unintentionally.

With the evidence of long-range transport of these POP substances (semi-volatile) to regions where they have never been used or produced and the consequent threats they pose to the environment of the whole globe, the societies and organizations that are concerned about the global environmental issues have now at several occasions called for urgent global actions to reduce and eliminate releases of these chemicals (United Nations Environment Program – UNEP). According to *Northern Perspectives published by the Canadian Arctic Resources Committee (vol. 26, No. 1, Fall/Winter 2000)* most of the POPs generated in the other parts of the world have been transported to the Arctic by wind and water, and tends to stay and accumulate due to the low evaporation rates in the region devastating the environment and living being.

In general POPs resist photolytic, biological and chemical degradation to a varying degree and they are often halogenated and demonstrated low water solubility, resulting bioaccumulation in fatty tissues. On the other hand, POPs are highly toxic, causing a wide array of adverse effects and diseases to the human and other life forms. They can cause dangerous diseases such as cancer, chronic allergies and hypersensitivity; damage to the central and peripheral nervous systems; reproductive disorders; and disruption to the immune system. Most of these POPs have the ability to transmit from present generation to the next generation via human or animal body and therefore, the consequences of these POPs will not be known for another 50 to 100 years.

The Stockholm Convention on POPs (*managed by UNEP*), which was adopted in 2001 and entered into action in 2004, is a global treaty whose purpose is to safeguard human health and the environment from highly harmful chemicals that are already persisting in the environment and generated by human activities intentionally and unintentionally. This convention on POPs initially identified 12 dangerous chemicals (Table 11.1) and considered that these chemicals could damage the health and life of humans and wildlife mostly. Because of this reason, most of the countries in the world have already banned or strictly limited the usage and production of these chemicals (*Australia has banned the usage and production of all pesticides except the insecticide Mirex and other industrial chemicals listed by the Stockholm Convention in 2004*) and the Convention began adding new additional nine chemical substances in May 2009 (Table 11.2) and committed to adding chemicals that cause adverse impact to the human life and the environment globally in an on-going basis.

POPs are broadly categorised in to two groups: (i) intentionally produced chemicals and (ii) unintentionally generated chemicals. However, according to the Stockholm Convention, the POPs are divided in to three groups: (i) pesticides, (ii) industrial chemicals and (iii) by-products. POPs also can be classified as Endocrine Disrupting Compounds (EDCs), dioxins and furans (see Tables 11.3 & 11.4). As per *World Health Organisation/ International Programme on Chemical safety* (WHO/IPCS) 2002, EDC is an exogenous substance or mixture that alters function(s) of the endocrine system and consequently causes adverse health effects in an intact organism, its progeny or (sub) population. EDCs are subdivided into two broad categories and they are Pesticides/Insecticides and Pharmaceutical and Personal Care Products (PPCPs). According to the United states Environmental Protection Agency (US EPA), the term dioxin is commonly used to refer to a family of toxic chemicals that share similar chemical structure and a common mechanism of toxic action. This family includes seven of the polychlorinated dibenzo dioxins (PCDDs), ten of the polychlorinated dibenzo furans (PCDFs) and twelve of the polychlorinated biphenyls (PCBs). PCDDs and

Table 11.1 The POPs that have been recognized as the "Dirty Dozen"

Category	POP Name	Global Historical Use/Source	Adverse Impact to Humans and Wildlife
Pesticides	Aldrin	Applied to soil to kill termites, grasshoppers, corn rootworms and other insect pests	Lack of quantitative information as Aldrin is readily metabolized to Dieldrin in both plants and animals. Overdoses can kill birds, fish and humans. Signs and symptoms of Aldrin intoxication include headaches, dizziness, nausea, vomiting, etc.
	Chlordane	Used on agricultural crops such as vegetable, grains, maize, potatoes, sugarcane, nuts, cotton, etc., as a insecticide to control termites.	Significant changes in the immune system, a possible human carcinogen, acute toxicity to pink shrimp, rats, monkeys, etc.
	DDT	Used excessively during World War II to control spreading of malaria, typhus and other vector borne diseases. Also used for agricultural crops to control certain diseases.	A possible human carcinogen, highly toxic to fish, shrimp, rainbow trout, birds (adverse impact on reproduction), very persistent in the environment and can be transported long distances.
	Dieldrin	Used principally to control termites and textile pests and control insect-borne diseases and insects living in agricultural soils.	Highly toxic to most species of fish, frogs, birds and most of other animals. High bio-accumulating and log range transport properties and deposited heavily in Arctic.

(continued)

Table 11.1 (continued)

Category	POP Name	Global Historical Use/Source	Adverse Impact to Humans and Wildlife
	Endrin	Sprayed as an insecticide on the leaves of crops such as cotton and grains. Also used to control rodents (mice and voles).	Highly toxic to fish and very high potential to bio-concentrate in organism. Long range transport properties and detected in Arctic freshwater.
	Hexachloro-benzene (HBC)	Introduced in 1945 to treat seeds and kills fungi in crops and control wheat bunt. By-product of manufacturing certain industrial chemicals.	Potential to have symptoms of photosensitive skin lesions, hyperpigmentation, hirsutism, colic, severe weakness, prophyrinuria and debilitation. Can develop a metabolic disorder called porphyria turcica and 14% could die.
	Heptachlor	Used to kill soil insects and termites and cotton insects, grasshoppers, other crop pests and malaria carrying mosquitoes.	A possible human carcinogen and Heptachlor can affect the immune responses. Impacted severely to cause declination in the population of several bird species. Bio-concentrates in organisms.
	Mirex	Used as an insecticide mainly to combat fire ants. Also used as a fire retardant in plastics, rubber, and electrical goods.	A possible human carcinogen and it is toxic to several plant and fish species. One of the most stable pesticides having a half-life of 10 years.

(continued)

Table 11.1 (continued)

Category	POP Name	Global Historical Use/Source	Adverse Impact to Humans and Wildlife
	Toxaphene	Used as an insecticide on cotton, cereal grains, fruits, nuts and vegetables. Also used to control ticks and mites in livestock.	A possible human carcinogen and 50% of a toxaphene release can persist in soil up to 12 years. It is highly toxic and it has long range transport properties.
Industrial Chemicals	Polychlorinated Biphenyls (PCBs)	Used in variety of industrial processes including in electrical transformers and capacitors, heat exchange fluids, as plant additives, in carbonless copy papers, in paint additives and in plastics. Also produced unintentionally during combustion.	PCBs are toxic to fish, killing them at high doses. Affect reproductive and immune systems in various wild animals. Humans are exposed to PCBs via food contamination. Have evidence for disorders in kids whose mothers are contaminated with PCBs. PCBs also suppress the human immune system and are listed as probable human carcinogens.
By-Products	Dioxins	Unintentionally produced as by-products mainly in the production of pesticides and other chlorinated substances, and sometimes found as trace contaminants in certain herbicides, wood preservatives and in PCB mixtures. Not used for any purpose.	Cause adverse effects in humans including immune and enzyme disorders and chloracne. Also recognized as a possible human carcinogen.
	Furans	Furans are a major by-product during production of PCBs. Also detected in emissions of waste incinerators and automobiles.	Impacts on human and other species are similar to Dioxins.

Table 11.2 Newly listed chemicals as POPs by Stockholm Convention in May 2009

Category	New additional Chemicals
Pesticides	Chlordecone, Alpha Hexachlorocyclohexane, Beta hexachlorocyclohexane, Lindane, Pentachlorobenzene
Industrial Chemicals	Hexabromobiphenyl, Hexabromodiphenyl ether and Heptabromodiphenyl ether, Pentachlorobenzene, Perfluorooctane sulfonic acid, its salts and Perfluorooctane Sulfonyl fluoride, Tetrabromodiphenyl ether and Pentabromodiphenyl ether
By-products	Alpha hexachlorocyclohexane, Beta hexachlorocyclohexane and Pentachlorobenzene

Table 11.3 Classification of POPs as EDCs

Category	Sub Category	POP Chemicals
EDCs	Pesticides	2,4-D, Atrazine, Benomyl, Carbaryl, Cypermethrin, Chlordane (γ-HCH), DDT and its metabolites, Dicofol, Dieldrin/Aldrin, Endosulfan, Endrin, Heptachlor, Hexachlorobenzene (HCB), Iprodione, Kepone (Chlordecone), Lindane, Malathion, Mancozeb, Methomyl, Methoxychlor, Mirex, Parathion, Pentachlorophenol, Permethrin, Simazine, Toxaphene, Trifluralin and Vinclozolin
	Organohalogens	Dioxins and furans, PCBs, PBBs and PBDEs, 2,4-Dichlorophenol
	Alkylphenols	Nonylphenols, Octylphenols, Pentaphenols, Nonylphenol ethoxylates, Octylphenol ethoxylates and Butylphenols
	Heavy Metals	Cadmium, Lead, Mercury and Arsenic
	Organotins	Tributyltin (TBT), Triphenyltin (TPhT)
	Phthalates	Di-ethylhexyl phthalate, Butyl benzyl phthalate, Di-n-butyl phthalate, Di-n-pentyl phthalate, Di-hexyl phthalate, Di-propyl phthalate, Dicyclohexyl phthalate, Diethyl phthalate
	Natural Hormones	17β-Estradiol, Estrone, Estriol and Testosterone

(continued)

Table 11.3 (*continued*)

Category	Sub Category	POP Chemicals
	Pharmaceuticals	Ethinyl estradiol, Mestranol, Tamoxifen and Diethylstilbestrol (DES)
	Phytoestrogens	Isoflavonoids, Coumestans, Lignans, Zearalenone and β-sitosterol
	Phenols	Bisphenol A and Bisphenol F
	Aromatic Hydrocarbons	Benzo(a)pyrene, Benzo(a)anthracene, Benzo(b/h) fluoranthene, 6-hydroxy-chrysene, Anthracene, Pyrene, Phenanthrene and n-Butyl benzene

Table 11.4 Classification of POPs as Dioxins and Furans (Source: Jones and Sewart, 1997)

Category	Homolog Name and Abbreviation	Possible compounds of PCDDs and PCDFs
Dioxins	MonochloroDD (MCDD)	2
	DichloroDD (DCDD)	10
	TrichloroDD (TrCDD)	14
	TetrachloroDD (TCDD)	22
	PentachloroDD (PeCDD)	14
	HexachloroDD (HxCDD)	10
	HeptachloroDD (HpCDD)	2
	OctachloroDD (OCDD)	1
Furans	MonochloroDF (MCDF)	4
	DichloroDF (DCDF)	16
	TrichloroDF (TrCDF)	28
	TetrachloroDF (TCDF)	38
	PentachloroDF (PeCDF)	28
	HexachloroDF (HxCDF)	16
	HeptachloroDF (HpCDF)	4
	OctachloroDF (OCDF)	1

PCDFs are not commercial chemical products but are trace level unintentional by-products from most forms of combustion and several industrial chemical processes. As explained by Jones and Sewart (1997), PCDDs and PCDFs have two basic chemical structures (Figure 11.1), however, two benzene rings of

PCDDs are connected by two oxygen atoms while two benzene rings of the PCDFs are bonded by C-O-C and C-C chains. Both groups of chemicals could have up to eight chlorine atoms and their toxicity would vary depending on the number of atoms and their position. All dioxins and furans that have same number of chlorine atoms are in the same "homologous" series and depending on their position they are called as different compounds or "congeners". The most toxic congener is 2,3,7,8-TCDD (tetrachlorinated dibenzo-p-dioxin).

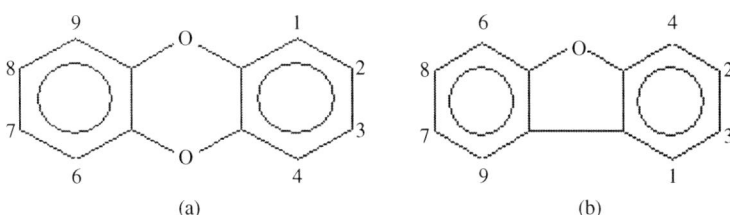

Figure 11.1 Basic chemical structures of (a) PCDDs and (b) PCDFs

11.3 HERBICIDES AND PESTICIDES

Pesticides are generally used as a chemical substance against any pest. On the other hand herbicides are used to kill unwanted weeds and plants. Selective herbicides kill specific targeted weeds and plants while leaving the desired crop relatively unharmed. Some of these act by interfering with the growth of the weed and are often synthetic "imitations" of plant hormones.

Generally, herbicide and pesticide contaminated surface water is mainly discharged from the agricultural lands during wet season, and at the same time a significant amount of herbicide/pesticide residues are discharged in to the environment through the existing wastewater treatment plants all over the world unintentionally. According to Gerecke *et al.* (2002), in Switzerland, 75% of the herbicide/pesticide load that is entering the surface waters is through the existing wastewater treatment plants. In addition to the above, a large amount of pesticide/ herbicide residues can be discharged in to the soil and water ways via the industries related to manufacturing, packaging, transporting, storing and delivering & distributing pesticides and herbicides.

The use of pesticides to protect crops has become current practice which enhances the crop yield significantly. Although this benefits the agricultural industry, the risk to the environment by polluting the soil and ground and surface waters must be considered seriously. The quality of soil and water bodies deserves particular attention in order for the survival of ecosystems and water supplies.

Similar to other POPs, herbicides and pesticides also undergo a number of degradation processes during storage, and both during and after the application. These reactions require certain time to reach equilibrium and so far it is not known exactly the specific percentages or effects of degradation products resulting from breakdown of pesticides and herbicides. However, it is a fact that they are persistent in the environment as active pesticides/herbicides or as their metabolites with high ecotoxicity.

As mentioned above, nine out of twelve priority POPs identified by the United Nations Environment Programme (UNEP) belong to the pesticide group and they are: DDT, Mirex, Hexachlorobenzene (HCB), Aldrin, Dieldrin, Toxaphene, Heptachlor, Endrin and Chlordane and they all are organic chlorinated compounds. Herbicides that are designated as high priority POPs by the UNEP include, 2,4-dichlorophenoxy acetic acid (2,4-D), 4-chloro-2-methylphenoxya-cetic acid (MCPA), 3-chlorobenzonic acid (3-CBA) and 2,4,5-trichlorophenox-yacetic acid (2,4,5-T). However, most of these pesticides and herbicides are now banned in many countries as it was identified that these POPs have serious adverse influence on human health in addition to the damage they cause to the environment.

Therefore, this chapter will mainly focus on the new generation herbicides whose main properties are listed in Table 11.5. At present, these herbicides are widely used in the farmlands located in the GBR catchment areas. Their IUPAC names are: (a) Diuron -3-(3,4-dichlorophenyl)-3, 3-dimethylurea, (b) Atrazine – 6-chloro-N2-ethyl-N4-isopropyl-1,3,5-triazine-2,4-diamine, (c) Ametryn – N2-ethyl-N4-isopropyl-6-methylthio-1,3,5-triazine-2,4-diamine, (d) Hexazinone – 3-Cyclohexyl-6-dimethylamino-1-methyl-1,3,5-triazine-2,4-dione, (e) Simazine – 6-chloro-N,N'-diethyl-1,3,5-triazine-2,4-diamine, (f) Tebuthiuron – 1-(5-tert-butyl-1,3,4-thiadiazol-2-yl)-1,3-dimethylurea. All these six herbicides fall into Photosystem II herbicide group, which is broadly divided into two distinct groups called Phenylurea and Triazines (Jones et al., 2005).

As stated earlier, herbicides/pesticides are persistent and accumulate in the fatty tissues of living organisms and are harmful to human and wildlife. A large number of research studies have been carried out in this area and those studies have revealed that residues of these pesticides and herbicides and other POPs in human body could have caused many common diseases such as cancer, immunological and reproductive disorders and blocking of hormones. Low levels of pesticide residues in potable water generally may not pose acute toxicity problems, but could cause chronic effects (Ahmed et al., 2008).

Table 11.5 Properties of photosystem II herbicides used in the farmlands of the GBR catchments

Properties	Diuron	Atrazine	Ametryn	Hexazinone	Simazine	Tebuthiuron
Molecular Weight/ (g)	233.10	215.69	227.33	252.31	201.66	228.3
Molecular Formula	$C_9H_{10}Cl_2N_2O$	$C_8H_{14}ClN_5$	$C_9H_{17}N_5S$	$C_{12}H_{20}N_4O_2$	$C_7H_{12}ClN_5$	$C_9H_{16}N_4OS$
Melting Point (°C)	158–159	173–175	84–85	115–117	225–227	161.5–164
Appearance	White Crystalline Solids	Colourless Crystals	White Powder	White Crystalline solids	White Crystalline Powder	Colourless Crystalline Powder
Solubility	36–42 mg/L in water (25°C)	34.7 mg/L (water 22°C) and 31 g/L (acetone 25°C)	185 mg/L (water 20°C) and readily dissolves in solvents (acetone)	Soluble, 33,000 mg/L at 25°C	Only 5 mg/L in water (insoluble)	2.3 g/L at 25°C
Purpose	phenyl-urea herbicide to enhance grass killing	chloro-triazine herbicide to control broad leaf weeds	methyl-thio-triazine herbicide to control grass	Triazine class herbicide to control broadleaf grass and woody plants	Triazine class herbicide to control broadleaf weeds and annual grasses	phenyl-urea herbicide to control mimosa

(continued)

Table 11.5 (continued)

Properties	Diuron	Atrazine	Ametryn	Hexazinone	Simazine	Tebuthiuron
IUPAC Name	3-(3,4-dichloro-phenyl)-1,1-dimethylurea	6-chloro-N2-ethyl-N4-isopropyl-1,3,5-triazine-2,4-diamine	N2-ethyl-N4-isopropyl-6-methylthio-1,3,5-triazine-2,4-diamine	3-cyclohexyl-6-dimethyla-mino-1-methyl-1,3,5-triazine-2,4(1H,3H)-dione	6-chloro-N2,N4-diethyl-1,3,5-triazine-2,4-diamine	1-(5-tert-butyl-1,3,4-thiadiazol-2-yl)-1,3-dimethylurea
Stability	Sunlight degrades					Sunlight degrades
Hydrolysis half life (days)	1490 (pH 5), 1240–2180 (pH 7) & 2020 (pH 9)					
Aqueous photolysis half-life (days)	43.1–2180 (pH 7, 25°C)					12.9 months
Aerobic/anaerobic soil degradation (days)	372/995	103				
Field dissipation half-life (days) – average	99.9–134	41	60	90	45–100	79

(continued)

Table 11.5 (continued)

Properties	Diuron	Atrazine	Ametryn	Hexazinone	Simazine	Tebuthiuron
Octanol-water coefficient (Log K_{ow})	2.81–2.87	2.60–2.71	2.83	−4.40	2.18	1.80
Soil adsorption coefficient (Log K_{oc})	2.62–2.75	1.96–2.98	2.88	1.73	2.13	1.90
Density		1.23 g/cm^3 (22°C)				
Chemical Structure						

As it was found that there are many adverse impacts to human life by consumption of pesticide/herbicide and POP contaminated water for a long time, most of the major drinking water treatment plants have been upgraded in developed countries with suitable advanced treatment methods such as Reverse Osmosis (RO) or nano-filtration (NF). However, the rapid deterioration to the global ecosystem and to the marine life due to the deposition of these POPs including pesticide and herbicide residues has been recognized as a major problem but ignored for a long period of time.

11.4 GREAT BARRIER REEF (GBR)

11.4.1 Background

The GBR World Heritage Area (GBRWHA) is jointly managed by the Australian Federal Government, through the Great Barrier Reef Marine Park Authority, and the Queensland State Government, through the Environmental Protection Agency (Queensland Parks and Wildlife Service). In addition to this, the land management in the catchment areas, where the sources of major pollutants are generated and discharged to the GBR lagoon are managed by the Queensland State Department of Natural Resources, Mines and Energy (Hutchings *et al.*, 2005).

The GBR, which was designated as a world heritage area in 1981 is the largest coral reef ecosystem in the world, spreading over an area of 350,000 km^2 in the North/East and spans almost 2000 km of the East coast of Queensland, Australia (Johnson and Ebert, 2000). Australia's GBR is precious to the entire nation due to its ecological and biological processes, significant habitats for biodiversity and its exceptional natural beauty.

The coastal region adjoining GBR World Heritage Area (GBRWHA) is divided in to a number of wet and dry tropical catchments and most of them are less than 10,000 km^2 in area (Brodie *et al.*, 2001). However, the Burdekin (133,000 km^2) and Fitzroy River catchments (143,000 km^2) are the largest catchments in Australia.

According to Moss *et al.* (2005), GBR catchment is primarily used for cattle grazing for beef production (77%). In addition to this, about 1% of the land in the river valleys and the floodplains are used for cropping sugarcane and 0.2% of the land is used for cropping each horticulture and cotton.

The Queensland sugarcane industry established in the GBR catchments is the most intensive agricultural industry and it generates approximately AUD 1.75 billion annually to the Australian economy. In addition to the above, there are thousands of small-scale cropping lands located in GBR catchments and growing various types of agricultural products.

Australia is willing and able to finance towards the protection of environment and the control of water pollution and it has expert knowledge and resources in order to achieve those tasks. The Government of Australia has put forward the Reef Water Quality Protection Plan (Anon, 2003) in order to control and eliminate the deterioration of the Australian Reef areas including GBR World Heritage Area due to the discharge of sediments, nutrients and persistent pollutants such as herbicides. The efforts taken towards the prevention of decline in the water quality in the GBR lagoon have been mainly focused on analysing the extent of damage to the ecosystem and identifying the root causes. These studies and results would be very valuable to design practical, sustainable and economical strategies to control the discharge of pollutants that damages in the GBR World Heritage Area.

11.4.2 Transport of Herbicides and Pesticides into the GBR

It is a known fact that the GBR catchment has been extensively modified and changed since the European settlement by forestry, urbanization and agriculture. As shown in Figure 11.2, the largest land use in the GBR catchment is cropping, mainly with sugarcane and this industry has increased steadily over the last 100 years with a total area reaching 390,000 ha in 1997 (Brodie *et al.*, 2001). During this transition time significant areas of freshwater wetlands of the major rivers in the GBR catchments have been reclaimed for agricultural and urban use. Most of the sugarcane cultivation areas are mainly located near the coast (lowland areas) of the catchments and due its rapid increase in application of herbicides and pesticides, sugarcane industry is considered to be the most influencing industry on the sustainability of the GBR ecosystem. Cotton, horticulture and bananas are the other mostly influencing agricultural sectors to the GBR ecosystem. However, major proportion of the GBR catchment is still occupied by cattle grazing (Table 11.6).

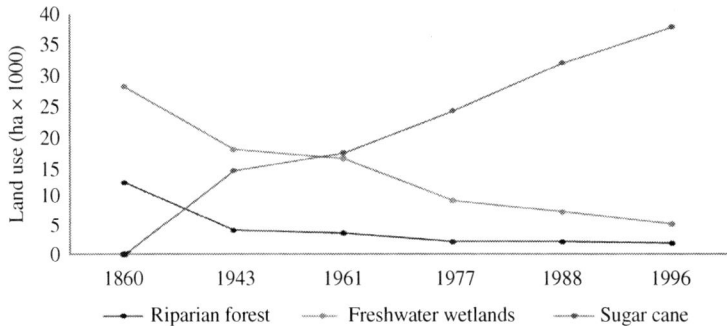

Figure 11.2 Change of land use (ha × 1000) in the lower Herbert River catchment over 140 years (Brodie *et al.*, 2001)

Table 11.6 Land uses in selected Queensland Catchments adjoining GBR Marine Park

Catchment	Total area (ha)	% of Catchment				
		Timber	Pristine	Grazing	Sugar	Other
Daintree	213	37.7	31.7	26.7	1.8	2.1
Mossman	49	30.4	11.0	44.6	10	4.0
Barron	218	36.4	2.0	47.7	2.1	11.8
Mulgrave/Russel	202	16.9	25.1	38.9	13.1	6.0
Johnston	233	25.3	12.8	41.6	14.8	5.5
Tully	169	62.5	2.1	20.7	9.6	5.1
Murray	114	32.9	27.3	29.6	6.1	4.1
Herbert	1,013	9.5	9.7	71.1	6.6	3.1
Black	108	18.0	9.3	67.4	0.7	4.6
Houghton	365	0.8	10.8	74	10.4	4.0
Burdekin	12,986	1.0	1.3	94.8	0.2	2.7
Don	389	0.2	2.6	91.3	1.1	4.8
Proserpine	249	9.6	4.0	74.6	7.5	4.3
O'Conelle	244	7.6	4.4	70.5	11.1	6.4
Pioneer	149	22.7	6.1	48.5	17.9	4.8
Plane	267	4.3	2.9	67.4	21	4.4
Fitzroy	15,264	6.7	2.3	87.5	0	3.5
Baffle	386	12.2	4.4	75.9	0.4	7.1
Kolan	298	12.5	0.0	79	4.5	4.0
Burnett	3315	12.9	0.4	79.9	0.8	6.0
Burrum	334	26.9	6.3	53.4	8.8	4.6
Mary	960	28.3	0.6	64.5	1.2	5.4

Other – Banana/fruits, vegetables, grain, cotton, sunflower, peanuts, irrigated forage crops, urban areas (roads, railways, dwellings, etc.)
Pristine – National Parks, Conservation Parks and Resource Reserves
Source – Report submitted to Productivity Commission (September 2002) by Queensland Cane Growers Organization Limited

The recent actions taken by the Australian sugar industry towards more sustainable practices, such as minimum cropping-land preparation, have resulted in an increased reliance on herbicides (Johnson and Ebert, 2000), particularly for the control of weeds in ratoon crops. This continuous rapid expansion of the farming industry in the GBR catchments as well as the increase in the usage of herbicides, insecticides and fungicides have contributed to high rate of discharge of these toxic wastes to the GBR lagoon during wet seasons (Table 11.7). It has been found that there is a 3 to 7 fold increase in the usage of herbicides (e.g., Atrazine, Diuron and 2,4-D) over the last 30 to 40 years (Johnson and Ebert, 2000). It is a fact that organochlorine insecticides such as DDT, Aldrin, Heptachlor, Chlordane, Lindane and Dieldrin were used in the sugar and horticultural industries since the 1950s and have been banned in the 1980s and 90s (Cavanagh et al., 1999). However, large quantities of those chemicals are still deposited in the farmlands of the GBR catchments and are being transported to the GBR lagoon with the agricultural run-offs.

Table 11.7 Annual herbicide loads discharged to the GBR lagoon and its catchments (Source Lewis et al., 2009 and Davis et al., 2009)

Location	Quantities of Herbicides discharged to the GBR Lagoon Annually					
	2007/2008		2006/2007		2005/2006	
	Diuron	Atrazine	Diuron	Atrazine	Diuron	Atrazine
West Bararatta Creek	44	70	79	116	46	80
Houghton River	16	25	39	26	63	72
Pioneer River	RNA	RNA	470	310	RNA	RNA
Sandy Creek	RNA	RNA	200	66	RNA	RNA
O'Connell River	RNA	RNA	31	20	30	6.6
Upper Barratta Creek	53	77	45	100	37	57
East Barratta Creek	28	44	53	108	RNA	RNA

There are many evidences available to state that Trazine (Atrazine), Organochlorine and Phenylurea herbicides (Diuron and 2, 4-D) and Organophotphate pesticides (Chlorpyrifos) are still being hezavily used in the agricultural areas of the GBR catchments (McMahon et al., 2005; Mitchell et al., 2005; Negri et al., 2009; Haynes et al., 2000 a, b; Duke et al., 2005; Moss et al., 2005; Shaw and Müller 2005; Davis et al., 2009; Lewis et al., 2009; Johnson, et al., 2000; Cavanagh et al., 1999). Table 11.7 illustrates annual flow of Diuron and Atrazine loads to the GBR lagoon via some of the selected waterways. Although the fertilizers and pesticides applied to the land are taken by the crops, a significant portion is collected in recycling ponds which exist only in a few large farmlands. However overflows from those ponds during wet seasons to neighbouring creeks, rivers, etc., end up in the GBR lagoon (Figure 11.3).

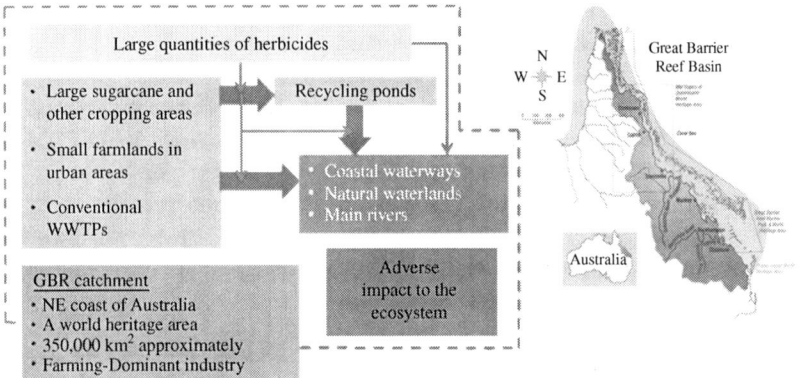

Figure 11.3 Transport routes of herbicides to the GBR lagoon

11.5 PERSISTENCE OF HERBICIDES AND PESTICIDES IN THE GBR CATCHMENTS AND LAGOON

Until the recent past, the impacts of herbicides and pesticides were not considered as a serious issue due to lack of research work in this area. However, now it has been found that there is a significant amount of pesticide and herbicide residues are in the GBR lagoon.

Lewis et al. (2009) showed a comprehensive dataset that examines the sources, transport and distribution of pesticide residues from selected GBR catchments to the GBR lagoon. They also showed that elevated concentrations of herbicide residues persist in the GBR lagoon even several weeks after the floods have reached the lagoon. They detected several pesticides (mainly herbicides) in both freshwater and coastal marine waters which were attributed to specific land uses in those catchments and found that elevated herbicide concentrations were particularly associated with sugar cane cultivation in adjacent catchments. Hence, the management of agricultural runoff is a key goal in improving the water quality in the coastal GBR lagoon (Anon et al., 2003). Herbicide residues have been detected in waterways of the GBR catchments (Davis et al., 2009; Ham et al., 2007; McMahon et al., 2005; Mitchell et al., 2005; Stork et al., 2008) as well as in intertidal/sub-tidal sediments (Duke et al., 2005; Haynes et al., 2000a), mangroves (Duke et al., 2005), sea-grass (Haynes et al., 2000a) and waters surrounding inshore coral reefs (Shaw and Müller,

2005). However, pesticide runoff has not previously been traced from those catchments to the GBR lagoon. River water plumes form in the GBR lagoon following wet season rains (December to April) that lead to large water volumes being discharged from the rivers of the GBR catchments. These event flows supply virtually all land-based materials (suspended sediment, nutrients and pesticides) transported annually to the GBR lagoon (Devlin and Brodie, 2005). The herbicides such as Diuron, Atrazine, Hexazinone and Ametryn were detected frequently and in relatively high concentrations (Table 11.8), while other pesticides were detected only infrequently. These herbicides were frequently detected at the highest concentrations at sites draining sugarcane, and the former three compounds also detected at sites in the urban land use category (Lewis et al., 2009).

According to the results revealed by Lewis et al. (2009), the highest Diuron concentrations were 19 $\mu g L^{-1}$ in the Tully-Murray region, 3.8 $\mu g L^{-1}$ in the Burdekin-Townsville region and 22 $\mu g L^{-1}$ in the Mackay Whitsunday region; all associated with more than 10% sugarcane cultivation as the main land use. They noted that Diuron residues were consistently above the Australian and New Zealand Environment Conservation Council (ANZECC) and Agriculture and Resource Management Council of Australia and New Zealand (ARMCANZ) ecological trigger value (0.2 $\mu g L^{-1}$) at the sites draining sugarcane farm run-offs monitored in three regions.

Davis et al. (2009) found Atrazine (<0.01 to 0.08 µg/L: 13 out of 14 samples collected) and Diuron (<0.01 to 0.08 µg/L: 12 out of 14 samples collected) in flood plume produced from the Haughton River and Barratta Creek in 2007. It was also found that Diuron in certain seagrass (Haynes et al., 2000a) and coral species (0.10 and 0.30 µg/L respectively) (Jones and Kerswell, 2003).

On the other hand, Lewis et al. (2009) found that the peak concentrations of Atrazine residues were 1.0 $\mu g L^{-1}$ in the Tully-Murray region, 6.5 $\mu g L^{-1}$ in the Burdekin-Townsville region and 7.6 $\mu g L^{-1}$ in the Mackay Whitsunday region; all peak concentrations in these regions were associated with sugarcane farms sites draining more than 10% runoff.

The summary of herbicides found in the GBR lagoon and its catchments, together with their maximum concentrations by several researchers are shown in Table 11.8 below.

Table 11.8 Herbicide concentrations in the GBR lagoon and its catchments

No	Type of Herbicide	Maximum Concentration (μg/L)	Location	Reference
1	Diuron	8.50	Rivers flowing to GBR	White et al. (2002)
2	Diuron	0.1–1.00	Coastal waters in North QLD	Haynes et al. (2000)
3	Sum of 8 herbicides	0.070	Hervey Bay (water)	McMahon K. et al. (2005)
	Diuron	0.050	Hervey Bay (water)	
	Diuron/ Atrazine	1.1 ng/g	Hervey Bay (sediments)	
	Sum of 8 herbicides	4.26	Mary river (water)	
4	Atrazine	1.20	Pioneer river catchment, Gooseponds Creek, Sandy Creek and Carmila Creek in Mackay Witsunday Region in QLD	Mitchell et al. (2005)
	Diuron	8.50		
	2,4-D	0.40		
	Hexazinone	0.30		
	Ametryn	0.30		
5	Diuron	19.00	Tully Murray	Lewis et al. (2009)
	Diuron	3.80	Burdekin-Townsville	
	Diuron	22.00	Mackay Whitsunday	
	Atrazine	1.00	Tully Murray	
	Atrazine	6.50	Burdekin-Townsville	
	Atrazine	7.60	Mackay Whitsunday	
6	Diuron	1.2–6.0	McCreadys Creek	Duke et al. (2005)
		1.0–8.2	Pioneer river	
		2.4–6.2	Bakers Creek	

11.6 IMPACT TO THE GBR ECOSYSTEM DUE TO THE PERSISTENCE OF HERBICIDES AND PESTICIDES

Laboratory-based ecotoxicological tests show that marine photosynthetic organisms are vulnerable to the exposure of herbicides, including macroalgae (Magnusson *et al.*, 2008; Seery *et al.*, 2006), mangroves (Bell and Duke, 2005), seagrass (Haynes *et al.*, 2000a) and corals (Cantin *et al.*, 2007; Jones 2005; Jones and Kerswell, 2003; Jones *et al.*, 2003; Negri *et al.*, 2005; Owen *et al.*, 2003) with certain species are more sensitive than the others.

During the investigations carried out by Haynes *et al.*, 2000b to check the impact of Diuron on seagrass, it was revealed that the lowest observable effect concentrations of Diuron exposure can be up to two orders of magnitude different for the species *Halophila ovalis* and *Zostera capricorni* (both 0.1 µgL^{-1}) compared to that for *Cymodocea serrulata* (10 µgL^{-1}). They concluded that the impact from these herbicides and pesticides depends on the type of marine species.

The assessment of risk of herbicide exposure in the GBR marine life is further complicated by the different toxicity of various herbicides. Studies on the same species of marine plants have shown that Diuron affects photosynthesis at lower doses than those of Atrazine, Hexazinone or Tebuthiuron (Bell and Duke, 2005; Jones, 2005; Jones and Kerswell, 2003; Jones *et al.*, 2003; Magnusson *et al.*, 2008; Owen *et al.*, 2003; Seery *et al.*, 2006). In addition, the toxicities of degradation products of herbicide residues in the GBR lagoon are largely unstudied and may be equal or greater than the toxicities of the parent compounds (Giacomazzi and Cochet, 2004; Graymore *et al.*, 2001; Stork *et al.*, 2008). The majority of eco-toxicological studies have quantified short-term effects of herbicide exposure (exposure times of hours to days) using pulse amplitude modulation chlorophyll fluorescence techniques as a measure of effective quantum yield of the photosystem of the target plant (Bell and Duke, 2005; Haynes *et al.*, 2000b; Jones, 2005; Magnusson *et al.*, 2008). Lowest observable effect concentrations (decline in quantum yield) in these experiments have been recorded within hours of exposure at levels as low as 0.1 µgL^{-1}, although most species recovered after the exposure ceased (Haynes *et al.*, 2000b; Jones, 2005; Jones and Kerswell, 2003; Jones *et al.*, 2003; Negri *et al.*, 2005). The results (Lewis *et al.*, 2009) show that herbicide residues can persist in the GBR lagoon over longer timescales (weeks) than the exposure times applied in most eco-toxicological studies. However, chronic effects of long-term herbicide exposure to GBR plant communities would develop over a longer timeframe. A decline in the reproductive output of corals was reported following Diuron exposure over a period of 50 days (Cantin *et al.*, 2007). In addition, chronic exposure to Diuron (and possibly Ametryn) residues have been implicated as the cause of severe mangrove dieback of *A. marina* in the Mackay Whitsunday region which has been developed progressively over a period of 10 years (Duke *et al.*, 2005).

Based on Hayes *et al.* (2002) and Hays *et al.* (2003), the endocrine disrupting effects of Atrazine on some amphibian fauna, such as inducement of hermaphroditism at concentrations as low as 0.1µg/L highlight their sensitivity to pesticide concentrations far below traditional toxicological methodologies. Some of the impacts to the GBR marine species due to the existence of common herbicides are tabulated in Table 11.9. The Table implies that the key task is to manage the agricultural runoff towards the GBR lagoon in order to achieve a sustainable solution to eliminate further deterioration of the ecosystem in the GBR. Hence, improvement in the quality of discharge that enters the GBR lagoon is the best solution to save the ecosystem of the lagoon.

Table 11.9 Impacts of herbicides on marine species

No	Description of Impacts	Reference
1	Diuron up to 1000 µg/L – *Acropora millepora* and *Montipora aequituberculata oocytes* (not inhibited)	Negri *et al.* (2004)
	Diuron 10 µg/L exposure 96 hrs. – Bleaching of two weeks old *P. damicornis*	
	Diuron 1 µg/L exposure 2 hrs. – reduction of photosynthetic efficiency in *P. damicornis*	
2	Atrazine 0.1 µg/L – Disrupt Steroidogenesis in amphibians	Hayes *et al.* (2002)
3	Diuron 10–100 µg/L exposure 2 hrs. – Decline quantum yield in *Cymodocea Serrulata*, and *Zostera Capricorni*	Haynes *et al.* (2000)
	Diuron 0.1–1 µg/L exposure 1 hr. – Decline quantum yield in *Halophila Ovalis*	
4	Diuron, Atrazine and Simazine 10–50 ng/L – impact to Seagrass health	McMahon *et al.* (2005)
5	Diuron 0.5–2 µg/L – 10 reduction of microalgae photosynthetic efficiency	Mitchell *et al.* (2005)
6	Diuron 1 µg/L or 3 µg/L exposure of 10 hrs *Acropora Formosa* (coral) – reduce photosynthetic efficiency	Jones *et al.* (2003)
7	When Diuron concentration is more than 2 µg/kg in sediments, it was noticed that *A. marina* was either absent, unhealthy or dead in all estuaries including in Mackay region	Duke *et al.* (2005)

Although there have been many research work carried out in order to identify and quantify the herbicide and pesticide residues in the GBR lagoon and the water bodies in the GBR catchments, very little effort has been taken to find a suitable economical solution to improve the quality of effluent discharged to GBR lagoon and to the water plumes in the catchments.

11.7 REMOVAL OF HERBICIDES BY DIFFERENT WATER TREATMENT PROCESSES

Most of the community water supply schemes, which are located in the GBR catchment areas, consist of conventional treatment methods such as coagulation-flocculation, sedimentation and conventional filtration. As it is a fact that the waterways in the GBR catchment are contaminated with pesticides and other POPs and micropollutants, these conventional treatment strategies are not effective in removing such pollutants. The inefficiency in the removal of pesticides from such conventional water treatment methods is evident from the results obtained by Miltner *et al.* (1989) (Tables 11.10, 11.11 and 11.12).

Table 11.10 Removal of pesticides in surface water sources by coagulation (Miltner *et al.*, 1989)

Pesticide	Coagulant (Dose, mg/L)	Initial Concentration (g/L)	% Removal
Atrazine	Alum (20)	65.7	0
Simazine	Alum (20)	61.8	0
Metribuzin	Alum (30)	45.8	0
Alachlor	Alum (150)	43.6	4
Metolachlor	Alum (30)	34.3	11
Linuron	Alum (30)	51.8	0
Carbofuran	Alum (30)	93.2	0

On the other hand, the type of disinfectant used and the length of contact time are important to assess the level of water treatment. Generally the disinfection process in water treatment is carried out to kill the pathogens such as bacteria, viruses, amoebic cysts, algae and spores from the treated water. According to a study carried out by Miltner *et al.* (1987), different oxidants (ozone, chlorine dioxide, chlorine, hydrogen peroxide and potassium permanganate) were tested for their ability to remove Alachlor in water and only ozone was found to remove Alachlor with removal efficiencies ranging from 75–97% when Alachlor was present in distilled, ground and surface water.

Table 11.11 Removal of pesticides by softening and clarification at full scale treatment plants (Miltner et al., 1989)

Pesticide	Initial Concentration (g/L)	% Removal or Transformation
Atrazine	7.24	0
Cyanazine	2.00	0
Metribuzin	0.53–1.34	0
Simazine	0.34	0
Alachlor	3.62	0
Metolachlor	4.64	0
Carbofuran	0.13–0.79	100∗

∗this study could not distinguish the removal and transformation to another metabolite/s. In the case of Carbofuran, author assumed that it was transformed to Carbofuran-Phenol and Hydroxy-Carbofuran.

Table 11.12 Removal of pesticides in surface water by chlorination process in full scale plants (Miltner et al., 1989)

Pesticide	Initial Concentration (g/L)	% Removal or Transformation
Atrazine	1.59–15.5	0
Cyanazine	0.66–4.38	0
Metribuzin	0.10–4.88	24–98*
Simazine	0.17–0.62	0–7
Alachlor	0.94–7.52	0–9
Metolachlor	0.98–14.1	0–3
Linuron	0.47	4
Carbofuran	0.13	24

*this study could not distinguish the removal and transformation to another metabolite/s. In the case of Metribuzin, author assumed that it may be the result of sample oxidant quenching.

Among the advanced treatment processes, powdered activated carbon (PAC) filtration, Granulated activated carbon (GAC) filtration and high pressure membrane processes such as Reverse Osmosis (RO) are considered as efficient treatment processes to remove organic chemicals including pesticides. During the recent past many of the water treatment plants have been upgraded in Australia to meet the higher treatment water quality standards and the removal of these micropollutants.

Another study carried out by Miltner *et al.* (1989) found that removal efficiencies of Atrazine and Alachlor using PAC during full scale water treatment were between 28%–87% and 33%–94% respectively. On the other hand, they found that GAC adsorption could remove Atrazine (47%), Cyanazine (67%), Metribuzin (57%), Simazine (62%), Alachlor (72%–98%) and Metolachlor (56%) to varying degrees.

According to the report on *"The Incorporation of Water treatment effects on Pesticide Removal and transformations in food Quality Protection Act (FQPA) Drinking Water Assessments",* submitted to the office of pesticide programs in United States Environmental Protection Agency in October 2001, water treatment by reverse osmosis shows superior performance in removing pesticides (Table 11.13). Thin-Film Composite membranes provide better efficiencies in removing pesticides.

Table 11.13 Removal efficiencies of RO membranes for different pesticide classes (Source: US EAP, 2001)

Pesticide Class	Cellulose Acetate (CA)	Polyamide	Thin Film Composite
Triazine	23–59	65–85	80–100
Acetanilide	70–80	57–100	98.5–100
Organochlorine	99.9–100	–	100
Organophosphorus	97.8–99.9	–	98.5–100
Urea derivatives	0	57–100	99–100
Carbamate	85.7	79.6–93	>92.9

Further, 100% removal of organochlorines (Chlordane, heptachlor and Methoxychlor) and Alachlor could be achieved using ultra-filtration. However, ultra-filtration was not effective in removing Dibromochloropropane and Ethylene dibromide. Nanofiltration, gives better results and it removes Atrazine (80–98%), simazine (63–93%), Diuron (43–87%) and Bentazone (96–99%). Integrated membrane /absorbent systems, aeration and air stripping systems are also used in present water treatment facilities to eliminate these pesticides and other micro-pollutants that cause health problems to humans.

11.8 POSSIBLE METHODS OF TREATMENT OF POPS INCLUDING HERBICIDES AND PESTICIDES FROM CATCHMENT DISCHARGES

Herbicides and pesticides and are generally removed by biological, adsorption, wetland and membrane processes. Some of the researches illustrating the performance of those processes are briefly mentioned below.

11.8.1 Biological Processes

Mangat and Elefsiniotis (1999) used laboratory-scale Sequencing Batch Reactors (SBRs) in order to study the efficiency of biodegradation of the herbicide 2,4-Dichlorophenoxyacitic Acid (2,4-D) and found that over 99% removal efficiency could be achieved with a hydraulic retention time (HRT) of 48 hours. It was revealed that the removal rate of 2,4-D was affected by the substrate (phenol or dextrose) and was significantly lower (30%–50%) in the case of dextrose. The study also found that the main mechanism of 2,4-D disappearance was biodegradation as adsorption onto the biomass and volatilization were insignificant.

Stasinakis *et al.* (2009) investigated activated sludge reactors and impacts of aerobic and anoxic conditions during the biodegradation of Diuron. It was found that almost 60% of Diuron was biodegraded under aerobic conditions (major metabolite was 3,4-dichloroaniline(DCA)) and over 95% of Diuron was biodegraded under anoxic conditions while the major metabolite was 1-3,4-dichlorophenylurea (DCPU). DCA and DCPU were biodegraded much faster than the parent compound under aerobic conditions and therefore, anoxic followed by aerobic biological treatment could provide efficient removal of Diuron and its metabolites from wastewater.

Ghosh *et al.* (2004) studied the degradation of Atrazine by anaerobic mixed culture microorganisms in co-metabolic process and in the absence of external carbon and nitrogen sources and revealed that in the presence of 2000 mg/L dextrose, the degradation of Atrazine was between 8 and 15%. Pure culture bacteria used Atrazine as sole source of carbon and/or nitrogen and the degradation depended on the type of bacterial culture present in the reactor and level of absence of various external carbon and nitrogen sources, carbon to nitrogen (C/N) ratio, pH and moisture content. It was found that the degradation of Atrazine by anaerobic mixed culture microorganisms was better in co-metabolic process than in the absence of external carbon and nitrogen sources. There was no significant inhibition effect on mixed anaerobic microbial consortia even at a concentration of 15 mg/L of Atrazine. However, it was observed that the rate of degradation of Atrazine declined at high organic contents in the reactor.

Znad et al. (2006) researched the performance of an air-lift bioreactor using the biodegradation of an herbicide, S-ethyl dipropylthiocarbamate (EPTC) in batch experiments and found that the rate of biodegradation of EPTC was decreased at high substrate concentration with free suspended activated sludge. On the other hand, the biodegradation of EPTC was more effective when it had immobilized activated sludge in the bioreactor. The results of this research noted that the rate of biodegradation of herbicide could be doubled by immobilizing the acclimated activated sludge inside the riser using non-woven textile.

During another research study, Gisi et al. (1997) used fixed-bed column reactors for measuring the rate of biodegradation of the pesticide 4,6-dinitro-ortho-cresol by introducing microbial cultures in batch and found that a rate of biodegradation of 30 mmol/day could be achieved for the above pesticide.

11.8.2 Adsorption Processes

Ratola et al. (2003) used pine bark as a natural adsorbent to remove persistent organic pollutants such as pesticides. The removal of Lindane and Heptachlor were found to be 80.6% and 93.6% respectively. On the other hand, Sannino et al. (2008) used sorption technique in order to investigate the removal efficiency of ionic herbicides (Paraquat and 2,4-D). They used Polymerin as the sorption media and achieved a rate of removal of about 44% for 2,4-D.

Removal of herbicide/pesticide using activated carbon (either PAC or GAC) is considered as very effective. Fontecha-Cámara et al. (2008) studied the activated carbon adsorption kinetics of the herbicides, Diuron and Amitrole in aqueous solution. Despite its lower driving force for adsorption, Amitrole showed faster adsorption kinetics compared to Diuron because of its smaller molecular size compared to that of Diuron. During another study carried out by Namasivayam and Kavitha (2003) found that coir pith carbon is an effective absorbent for the removal of 2-cholorophenol, which is a metabolite of pesticides, herbicides, pharmaceuticals, etc., from aqueous solution.

Jones et al. (1998) carried out another study on GAC filters inoculated with bacterial culture and found that the rate of degradation of Atrazine was about 70%. PAC was found to be a very suitable adsorbent compared to bentonite and chitosan in the removal of isoproturon pesticide (98–99%) from spiked distilled water (Sarkar et al., 2007a,b).

11.8.3 Wetland Processes

Sub-surface flow wetlands remove chemicals from the run-offs by microbial degradation, plant uptake, sorption, chemical reactions and volatilization.

Stearman *et al.* (2003) studied the efficiency of constructed sub-surface flow wetlands for the removal of herbicides (Simazine and metolachor) and found that vegetated cells with 5.1 day of HRT could remove around 82% of these herbicides. Another study carried out by Heather *et al.* (2003) found that there is a 21% removal of Atrazine from a surface flow wetland. Sorption was the main mechanism for the removal of Atrazine. On the other hand Moore *et al.* (2000) found that average of 52% of measured Atrazine were transferred to or transformed in the wetland system. In addition to that, Kristen *et al.* (2002) carried out another study regarding Atrazine mineralization (measured by $^{14}CO_2$ evolution from U-ring-^{14}C) using two wetlands (one was a constructed wetland and the other was a natural fen – Cedar bog) and revealed that the constructed wetland achieved 70 to 80% mineralization of Atrazine while the natural fen – Cedar bog achieved less than 13%. Marsh plant systems have also been found to remove herbicide (Atrazine) from wastewater effectively (Mackinlay and Kasperek, 1999). Matamoros *et al.* (2007), investigated a total of eight European priority pollutants listed in the Water Framework Directive including a variety of chemical classes such as organochlorine, organophosphorus, phenols, chloroacetanilides, triazine, phenoxycaboxylic acid and phenylurea pesticides. They evaluated the performance of wetlands in removing pesticides after 21 days of operations and categorized the pesticides into four groups depending on their degradation: (a) highly efficiently removed (more than 90% removal) – lindane, pentachlorophenol, endosulfan and pentachlorobenzene; (b) efficiently removed (between 80 and 90% removal) – alachlor and chlorpyriphos; (c) poorly removed (20% removal) – mecoprop and simazine and (d) resistant to elimination – clofibric acid and diuron.

11.8.4 Pressure Driven Membrane Filtration Processes

Since some of the above systems are very inefficient and difficult to operate in large scale, pressure driven membrane processes in treating pesticide contaminated water are considered as promising alternatives. When reviewing the past research work it is clear that high-pressure membrane processes (reverse osmosis and nanofiltration) are very effective in removing pesticides from ground and surface waters (Majewska-Nowak *et al.*, 2002). Boussahel *et al.* (2000, 2002) studied the performance of two types of nano-filtration membranes (Deasal DK and NF200 having molecular weight cut-off 150–300 Dalton and 300 respectively) for the removal of some selected pesticides (Simazine, Cyanazine, Atrazine, Isoproturon, Diuron and desethyl-atrazine – DEA) and found that all pesticides were rejected by Desal DK membrane (over 90%) except Diuron (less than 70%). It was also found that the presence of organic matter (humic acid) and inorganic matter (sulphates and chlorides) improve the elimination of pesticides except Duiron either by forming macromolecules with the pestcides or by reducing the pore size of the membrane. A similar study conducted by Plakas *et al.* (2006) in order to identify the role of organic matter and calcium concentration on herbicide retention from a nanofilter revealed that the presence of humic acid as well as calcium significantly improved the retention of herbicides by increasing the fouling of organic membrane. Nano-filtration membranes made of polyamide and cellulose with same molecular weight cut-off gave 60 to 95% and 25% herbicide/pesticide rejection respectively (Causserand *et al.,* 2005). Another research carried out by Van der Bruggen *et al.* (1998) found that two stage nano-filtration system could give over 99% removal of pesticide from water. Ahmad *et al.* (2008) found that increasing the pH of wastewater enhanced the rejection of Atrazine and dimethoate but reduced the permeate flux.

When using Reverse Osmosis (RO) for desalination, hardness removal, disinfection and removal of herbicide/pesticide and other micro-pollutants, adequate level of pre-treatment (equivalent to ozonation, biological activated carbon filtration (BACF), and slow sand filtration) should be carried out (Bonné *et al.*, 2000). RO process with ozonation and BACF showed over 99.5% removal of pesticide from water (Bonné *et al.*, 2000). Another study carried out by Majewska-Nowak *et al.* (2002) found that 80% of Atrazine was rejected from a low pressure driven ultra-filtration membrane when the concentration of humic substances was equal to 20 g/m^3 at a pH of 7.

11.8.5 Hybrid Systems

Hybrid wastewater treatment systems are defined as combination of two or more individual treatment processes (different biological, adsorption, wetland, or membrane processes). These hybrid systems perform better than a single treatment process. Recent research studies have found that these hybrid systems could improve the treatment of micropollutants. The following studies are examples of such hybrid systems: A study carried out by Tomaszewska *et al.* (2004) investigated the removal efficiency of humic acid and phenol by coagulation (A PAX XL-69 polyyaluminum chloride) and adsorption (PAC) and revealed that the integrated adsorption-coagulation system is effective in removing organic matter than coagulation alone. A different study carried out by Areerachakul *et al.* (2007) showed that a combined granular activated carbon (GAC) fixed bed and a continuous photo-catalysis system could remove 90% of the herbicide metasulfuron-methyl.

However, Membrane Bioreactor (MBR) technology, which is a combination of biological and membrane filtration processes, is an ideal example for a popular hybrid wastewater treatment system. Recently, many researchers have studied MBR to improve its performance and to reduce its drawbacks in industrial applications. It is a known fact that MBR is a better treatment process than Activated Sludge Process (ASP) for the treatment of micropollutants and POPs.

11.8.6 Hybrid Systems – Membrane Bioreactors (MBR)

Although there were not many MBR research work have been carried out for highly persistent organic pollutants such as herbicides and pesticides, significant number of research work have been carried out related to the treatment and removal of moderately persistent trace organic compounds such as pharmaceutically active compounds, surfactants, industrial chemicals and micro-pollutants from wastewater.

The results obtained by Petrović *et al.* (2003, 2007) on MBR has shown a significantly improved removal of pharmaceutically active lipid regulators and cholesterol lowering statin drugs (gemfibrozil, bezafibrate, clofibric acid and pravastatin), β-blockers (atenolol and metoprolol), antibiotics (ofloxacin and erythromycin), anti-ulcer agent (ranitidine) and some analgesics and anti-inflammatory drugs (propyphenazone, mefenamic acid and diclofenac). Petrović *et al.* (2003, 2007) also have also found that surfactants such as alkylehpenol ethoxylates (APEOs) are removed at higher efficiency.

González *et al.* (2006) performed a comparative study on the removal of acidic pesticides (MCPP, MCPA, 2,4-D, 2,4-DP and Bentazone) and the acidic pharmaceutical diclofenec by a MBR and a fixed bed bioreactor (FBBR) and found that the MBR is more efficient (44 to 85%) in treating all these pesticides and diclofenec except bentanone. They also confirmed that the microorganisms that were present in the MBR were capable of degrading ubiquitous pollutants present in wastewater treatment plants (WWTPs) such as MCPP, MCPA, 2,4-D and 2,4-D.

Kim *et al.* (2007) studied the performance of a MBR using 14 pharmaceutical substances, 6 hormones, 2 antibiotics, 3 personal care compounds and 1 flame retardant and found that a MBR process could provide effective removal of hormone and some pharmaceutical compounds such as acetaminophen, ibuprofen and caffeine. However, combining MBR with NF and RO could provide excellent removal (more than 95%) of all toxic trace organic compounds mentioned above.

Yuzir and Sallis (2007) measured the performance of an anaerobic membrane bioreactor (AMBR) for the treatment of synthetic ((RS)-2-(2-methyl-4-chlorophenoxy)-propionic acid (contains MCPA, 2,4-D and MCPB), which is an herbicide widely used for agriculture and horticulture including domestic gardening. AMBR was operated under methanogenic conditions and only 15% removal efficiency was achieved at a HRT of 3.3 days. There was no significant impact on COD reduction and methane yield of the reactor due to the addition of above herbicide.

Yiping *et al.* (2008) used anaerobic-membrane bioreactor in order to remove organic micro-pollutants that were present in the landfill leachate effluent. In this study, 17 organo-chlorine pesticides (OCPs), 16 polycyclic aromatic hydrocarbons (PAHs) and technical 4-nonylphenol (4-NP) were investigated and found that 4-NP compound were removed from the MBR and OCPs and PAHs were mainly removed from the anaerobic process. Finally, an overall removal of 94% of OCPs, 77% of 4-NPs and 59% of PAHs were achieved.

Grimberg *et al.* (2000) used a hollow fibre membrane bio-film-reactor (bioreactor) in order to study the removal efficiency of 2,4,6-trinitrophenol (TNP). TNP is a common nitro-aromatic compound, which is generally used for the production of pesticides, herbicides, pharmaceuticals and explosives. TNP has been shown to be biodegradable by four strains of a close relative of Nocardioides Simplex and MBRs with these organisms removed 85% of TNP. It was also found that the microorganisms used NP as their sole carbon and energy source. Table 11.14 summarises the findings on the removal of POPs using the MBR technology.

Table 11.14 Summary of past research results on the removal of POPs using MBR technology

No	Trace Organic/ POP or Micro-pollutant	MBR Performance/ Observation or Findings	Reference
1	**Pharmaceutically active** lipid regulators cholesterol lowering statin drugs-gemfibrozil, bezafibrate, β-blockers -atenolol and metoprolol, antibiotics-ofloxacin & erythromycin, anti-ulcer agent (ranitidine) and some analgesics and anti-inflammatory drugs (propyphenazone, mefenamic acid and diclofenac)	Removal percentages; gemfibrozil (89.6%), bezafibrate (95.8%), atenolol (65.5%), mctoprolol (58.7%), ofloxacin (94%), erythromycin (67.3%), r anitidine (95%), propyphenazone (64.6%), mefenamic acid (74.8%) and diclofenac (87.4%)	Radjenović *et al.* (2006) and Petrović *et al.* (2003, 2007)
2	**Acidic Pesticides** (MCPP, MCPA, 2.4-D, 2,4-DP and Bentazone – compared the MBR treatment efficiency with a fixed bed bioreactor	MBR is more efficient in the treatment of acidic pesticides (**44% – 85%**). Microorganisms could degrade these pollutants	González *et al.* (2006)

(*continued*)

Table 11.14 (*continued*)

No	Trace Organic/ POP or Micro-pollutant	MBR Performance/ Observation or Findings	Reference
3	**Dissolved organic carbon and Trihalomethane** precursors using Powdered Activated Carbon (**PAC**)	Removal efficiency varied between **20% – 60%** depending on the carbon dose	Williams et al. (2007)
4	**Acidic pharmaceuticals** using sludge as inoculum under **aerobic** conditions	Removal Percentages are Diclofenac (25%), Ketoprofen (60%), Bezafibrate (90%), Naproxen (75%) and Ibuprofen (98%)	Quintana et al. (2005)
5	Two different radio-labelled 17α-ethinylesstradiols (EE2), which is a EDC and a synthetic estrogen used as an active agent of contraceptive pills.	Satisfactory removal rate of 80% (about 5% withdraw from the sludge removal and about 16% found in the MBR effluent).	Cirja et al. (2007)
6	An EDC (Bisphenol A) and a Phamaceutical compound (Sulfamethoxazole)	90% removal of Bisphenol A and 50% removal of Sulfamethoxazole	Nghiem et al. (2009)
7	Pentachlorophenol (PCP), which is used for formulation of pesticides, herbicides etc.,	PCP removal rate of 99% at a loading rate of 12–40 mg/m^3/d. Found that bio-sorption plays an important role in addition to biodegradation	Visvanathan et al. (2005)
8	1,2-dichloroethane and 2,4-D-acetic acid (component of commercial herbicide) using a suitable microbial culture	99% removal of 1,2-dichloroethane and successful removal of 2,4-D-acetic acid	Livingston (1994) and Buenrostro-Zagal et al. (2000)
9	17 organochlorine pesticides (OCPs), 16 polycyclic aromatic hydrocarbons (PAHs) and technical 4-nonylphenol (4-NP) under anaerobic conditions (AMBR)	overall removal 94% of OCPs, 77% of 4-NPs and 59% of PAHs	Yiping et al. (2008)
10	Atrazine using bio-augmented genetically engineered microorganisms (GEM)	Over 90% removal efficiency and MBR start up time reduced to 2–12 days (under different operating conditions)	Liu et al. (2008)

11.8.7 Other Processes

Treatment processes such as photocatalytic degradative oxidation, dielectric barrier discharge – DBD, solar photo-Fenton technologies and phyto-remediation techniques could also be used to remove POPs and other micropollutants from wastewater. However, those processes are in their early research stages and therefore this chapter does not cover the details of such treatment processes.

11.9 CONCLUSIONS

The ecosystem of the Great Barrier Reef is one of the best managed ecosystems in the world. The authorities responsible ensure that the discharge from the catchments cause minimal adverse impacts to the GBR lagoon. Although the agricultural industry located in the GBR catchments implements best management practices, the release of herbicides and pesticides to the GBR lagoon is unavoidable especially during the wet season. Thus, integrated treatment systems are necessary to minimise the herbicides and pesticides loads that are being released to GBR lagoon with agricultural discharges. Research studies indicate that the membrane processes are the best in reducing those pollutants from those discharges followed by adsorption, biological and/or wetland processes.

11.10 REFERENCES

Abegglen, C., Joss, A., McArdell, C., Fink, G., Schusener, M. P., Ternes, T. A. and Hansruedi, S. (2009) The fate of selected micropollutants in a single-house MBR. *Water Res.* **43**, 2036–2046.

Ahmed, A. L., Tan, L. S. and Shukor, S. R. A. (2008) The role of pH in nanofiltration of atrazine and dimethoate from aqueous solution. *J. Hazard. Mater.* **154**, 633–638.

Anon (2003) The state of Queensland and Commonwealth of Australia. Reef Water quality protection plan for catchments adjacent to the Great Barrier Reef world heritage area. Queensland Department of Premier and Cabinet, Brisbane, http://www.the premier.qld.gov.au/library/pdf/rwqpp.pdf

Areerachakul, N., Vigneswaran, S., Ngo, H. H. and Kandasamy, J. (2007) Granular activated carbon (GAC) adsorption-photocatalysis hybrid system in the removal of herbicide from water. *Sep. Pur. Technol.* **55**, 206–211.

Bell, A. M. and Duke, N. C. (2005) Effects of photosystem II inhibiting herbicides on mangroves – preliminary toxicology trials. *Mar. Pollut. Bull.* **51**, 297–307.

Bernhard, M., Müller, J. and Knepper, T. P. (2006) Biodegradation of persistent polar pollutants in wastewater: comparison of an optimised lab-scale membrane bioreactor and activated sludge treatment. *Water Res.* **40**, 3419–3428.

Bonné, P. A. C., Beerendonk, E. F., van der Hoek, J. P. and Hofman, J. A. M. H. (2000) Retention of herbicides and pesticides in relation to aging of RO membranes. *Desalination* **132**, 189–193.

Bouju, H., Buttiglieri, G. and Malpei, F. (2008) Perspectives of persistent organic pollutants (POPS) removal in an MBR pilot plant. *Desalination* **224**, 1–6.

Boussahel, R., Bouland, S., Moussaoui, K. M. and Montiel, A. (2000) Removal of pesticide residues in water using the nanofiltration process. *Desalination* **132**, 205–2009.

Boussahel, R., Montiel, A. and Baudu, M. (2002) Effects of organic and inorganic matter on pesticide rejection by nanofiltration, *Desalination* **145**, 109–114.

Brodie, J., Christine, C., Devlin, M., Morris, S., Ramsay, M., Waterhouse, J. and Yorkston, H. (2001) Catchment management and the Great Barrier Reef. *Water Sci. Technol.* **43**, 203–211.

Buenrostro-Zagal, J. F., Ramirez-Oliva, A., Caffarel-Mendez, S., Schettino-Bermudez, B. and Poggi-Varaldo, H. M. (2000) Treatment of 2,4-dichloroacetic acid (2,4-D) contaminated wastewater in a membrane bioreactor. *Water Sci. Technol.* **42**(5–6), 185–192.

Canadian Arctic Resources Committee (2000) Persistent Organic Pollutants: Are we close to a solution? *Northern Perspectives*, Fall/Winter **26**,(1).

Calabro, V., Curcio, S., De Paola, M. G. and Iorio, G. (2009) Optimization of membrane bioreactor performances during enzymatic oxidation of waste bio-polyphenols. *Desalination* **236**, 30–38.

Cantin, N. E., Negri, A. P., and Willis, B. L. (2007) Photoinhibition from chronic herbicide exposure reduces reproductive output of reef-building corals. *Marine Ecol. Progress Series* **344**, 81–93.

Causserand, C., Aimar, P., Carvedi, J. P. and Singlande, E. (2005) Dichloroaniline retention by nanofiltration membranes.*Water Res.* **39**, 1594–1600.

Cavanagh, J. E., Burns, K. A., Brunskill, G. J. and Coventry, R. J. (1999) Organochlorine Pesticide Residues in soils and sediments of the Herbert and Burdekin river regions, North Queensland – implication for contamination of the Great Barrier Reef. *Marine Poll. Bull.* **39**, 367–375, 1999.

Chang, C., Chang, J., Vigneswaran, S. and Kandasamy, J. (2008) Pharmaceutical wastewater treatment by membrane bioreactor process – a case study in southern Taiwan. *Desalination* **234**, 393–401.

Cicek, N. (2003) A review of membrane bioreactors and their potential application in the treatment of agricultural wastewater. *Canadian Biosystems Engineering* **45**.

Cirja, M., Zuehlke, S., Ivashechkin, P., Hollender, J., Schaffer, A. and Corvini, P. F. X. (2007) Behaviour of two differently radiolabelled 17α-ethinylestradiols continuously applied to a laboratory-scale membrane bioreactor with adapted industrial activated sludge. *Water Res.* **41**, 4403–4412.

Davis, A., Lewis, S., Bainbridge, Z. and Brodie, J. (2009) Pesticide residues in waterways of the lower Burdekin region: Challenges in ecotoxicological interpretation of monitoring data (submitted to *Australasian Journal of Ecotoxicology*).

De Wever, H., Weiss, S., Reemtsma, T., Wereecken, J., Müller, J., Knepper, T., Rörden, O., Gonzales, S., Barcelo, D. and Hernando, M. D. (2007) Comparison of sulfonated and other micropollutants removal in membrane bioreactor and conventional wastewater treatment. *Water Res.* **41**, 935–945.

Devlin, M. J. and Brodie, J. (2005) Terrestrial discharge in to the Great Barrier Reef lagoon: nutrient behaviour in coastal waters. *Marine Poll. Bull.* **51**, 9–22.

Duke, N. C., Bell, A. M., Pederson, D. K., Roelfsema, C. M. and Nash, S. B. (2005) Herbicides implicated as the cause of sever mangrove dieback in the Mackay region, NE Australia: consequences for marine plant habitats of the GBR World Heritage Area. *Marine Poll. Bull.* **51**, 308–324.

Fontecha-Cámara, M. A., Lópeza-Ramón, M. V., Pastrana-Martines, L. M. and Moreno-Castilla, C. (2008) Kinetics of diuron and amitrole adsorption from aqueous solution on activated carbons. *J. Haz. Mater.* **156**, 472–477.

Frietas dos Santos, L. M. and Lo Biundo, G. (1999) Treatment of pharmecuitical industry process wastewater using the extractive membrane bioreactor. *Environ. Prog.* **18**, 34–39.

Gerecke, A. C., Schärer, M., Singer, H. P., Müller, S. R., Schwarzenbach, R. P., Sägesser, M., Ochsenbein, U. and Popow, G. (2002) Sources of pesticides in surface waters in Switzerland: pesticide load through waste water treatment plants-current situation and reduction potential. *Chemosphere* **48**, 307–315.

Ghosh, P. K. and Philip, L. (2004) Atrazine degradation in anaerobic environment by a mixed microbial consortium, *Water Res.* **38**, 2277–2284.

Giacomazzi, M. and Cochet, N. (2004) Environmental impact of diuron transformation: a review. *Chemosphere* **56**, 1021–1032.

Grimberg, S. J., Rury, M. J., Jimenez, K. M. and Zander, A. K. (2000) Trinitrophenol treatment in a hollow fibre membrane biofilm reactor. *Water Sci. Technol.* **41**, 235–238.

Gisi, D., Stucki, G. and Hanselmann, K. W. (1997) Biodegradation of the pesticide 4,6-dinitro-ortho-cresol by microorganisms in batch cultures and in fixed-bed column reactors. *Appl. Microbial Biotechnol.* **48**, 441–448.

Gonzáles, S., Müller, J., Petrovic, M., Barceló, D. and Knepper, T. P. (2006) Biodegradations studies of selected priority acidic pesticides and diclofenac in different bioreactors. *Environ. Poll.* **144**, 926–932.

Graymore, M., Stagnitti, F. and Allinson, G. (2001) Impacts of atrazine in aquatic ecosystems. *Environ. Int.* **26**, 4823–495.

Ham, G. (2007) Water quality of the inflows/ outflows of the Barratta Creek system. *Proc. Aust. Soc. Sugar Cane Technol.* **29**, 149–166.

Hayes, T, Collins, A, Lee., M, Mendoza, M., Noriega, N. and Stuart A. A. (2002) Hermaphroditic, demasculinized frogs after exposure to the herbicide atrazine at low ecologically relevant doses. *Proc Natl Acad Sci.* **99**:5476–5480.

Haynes, D., Müller, J. and Carter, S. (2000a) Pesticide and herbicide residues in sediments and seagrasses from the great barrier reef world heritage area and Queensland coast. *Marine Poll. Bull.* **41**, 279–287.

Haynes, D., Ralph, P., Prang, J. and Dennison, B. (2000b) The Impact of the herbicide diuron on photosynthesis in three species of tropical seagrass. *Marine Poll. Bull.* **41**, 288–293.

Hays, T., Haston, K., Ysui, M., Hoang, A., Haeffele, C. and Vonk, A. (2003) Atrazine-Induced Hermaphroditism at 0.1 ppb in American Leopard Frogs (Rana pipiens): Laboratory and Field Evidence. *Environ. Health Persp.* **111**, 4.

Heather B. R., Jenkins, J. J., Moore, J. A., Bottomley, P. J. and Wilson, B. D. (2003) Treatment of atrazine in nursery irrigation runoff by a constructed wetland. *Water Res.* **37**, 539–550.

Hutchings, P., Haynes, D., Goudkamp, K. and McCook, L. (2005) Catchment to Reef: Water quality issues in the Great Barrier Reef Region – An overview of papers. *Marine Poll. Bull.* **51**, 3–8.

Johnson, A. K. L. and Ebert, S. P. (2000) Quantifying inputs of pesticides to the Great Barrier Reef marine park – A case study in the Herbert River Catchment of North-East Queensland. *Marine Poll. Bull.* **41**, 302–309.

Jones, K. C. and Stewart, A. P. (1997) Dioxins and Furans in sewage Sludges: A review of their occurrences and sources in sludge and their environmental fate, behaviour, and significance in sludge-amended agricultural systems. *Crit. Rev. Environ. Sci. Technol.* **27**, 1–85.

Jones, L. R., Owen, S. A., Horrell, P. and Burns, R. G. (1998) Bacterial inoculation of Granular Activated Carbon Filters for the removal of Atrazine from surface water. *Water Res.* **32**, 2542–2549.

Jones, R. (2005). The ecotoxicological effects of Photosystem II herbicides on corals. *Marine Poll. Bull.* **51**, 495–506.

Jones, R. J. and Kerswell, A. P. (2003) Photo-toxicity of photosystem II (PSII) herbicides to coral. *Mar. Ecological Progress Series* **261**, 149–159.

Jones, R. J., Muller, J., Haynes, D. and Schreiber, U. (2003) Effects of herbicides diuron and atrazine on corals on the Great Barrier Reef, Australia. *Mar. Ecol. Prog. Series* **251**, 153–167.

Kim, K., Ahmed, Z., Ahn, K. and Paeng, K. (2009) Biodegradation of two model estrogenic compounds in a pre-anoxic/anaerobic nutrient removing membrane bioreactor. *Desalination* **243**, 265–272.

Kim, S. D., Cho, J., Kim, I. S., Vanderford, B. J. and Snyder, S. A. (2007) Occurrence and removal of pharmaceuticals and endocrine disruptors in South Korean surface, drinking, and waste waters. *Water Res.* **41**, 1013–1021.

Kimura, K., Hara, H. and Watanabe, Y., Removal of pharmaceutical compounds by submerged membrane bioreactors (MBRs) *Desalination* **178**, 135–140.

Kristen, A. L., Wheeler, K. A. and Robinson, J. B. (2002) Atrazine mineralization potential in two wetlands, *Water Res.* **36**.

Lewis, S. E., Brodie, J. E., Bainbridge, Z. T., Rohde, K. W., Davis, A. M., Masters, B. L., Maughan, M., Devlin, M. J., Muller, J. F. and Schaffelke, B. (2009) Herbicides: A new threat to the Great Barrier Reef. *Environ. Poll.* 1–15.

Liu, C., Huang, X. and Wang, H. (2008) Start-up a membrane bioreactor bio-augmented with genetically engineered microorganism for enhanced treatment of atrazine containing wastewater. *Desalination* **231**, 12–19.

Livingston A. G. (1994) Extractive membrane bioreactors: A new process technology for detoxifying chemical industry wastewater. *J. Chem. Technol. Biotechnol.* **60**, 117–124.

Mack, C., Burgess, J. E. and Duncan, J. R. (2004) Membrane bioreactors for metal recovery from wastewater: A review. *Water SA* **30**.

Magnusson, M., Keimann, K. and Negri, A. (2008) Comparative effects of herbicides on photosynthesis and growth of tropical estuarine microalgae. *Marine Poll. Bull.* **56**, 545–1552.

Majewska-Nowak, K., Kabsch-Korbutowicz, M. and Dodz, M. (2002) Effects of natural organic matter on atrazine rejection by pressure driven membrane processes. *Desalination* **145**, 281–286.

Manem, J. and Sanderson, R. (eds.) (1996) Membrane bioreactors in water treatment processes, Ch17 AWWARF/Lyonnaise des Eaux/ WRC. McGraw Hill.

Mangat, S. S. and Elefsiniotis, P. (1999) Biodegradation of the herbicide 2,4-dichlorophenosyacetic acid (2,4-D) in sequencing batch reactors. *Water Res.* **33**, 861–867.

Matamoros, V., Puigagut, J., Garcia, J. and Bayona. (2007) Behaviour of selected priority organic pollutants in horizontal subsurface flow constructed wetlands: A preliminary screening. *Chemosphere* **69**, 1374–1380.

Mckinlay, R. G. and Kasperek, K. (1999) Observations on decontamination of herbicide polluted water by marsh plant systems. *Water Res.* **33**, 505–511.

McMahon, K., Nash, S. B., Raglesham, G., Müller, J. F., Duke, N. C. and Winderlich, S. (2005) Herbicide contamination and the potential impact to seagrass meadows in Hervey Bay, Queensland, Australia. *Marine Poll. Bull.* **51**, 325–334.

Miltner, R. J., Baker, D. B., Speth, T. F. and Fronk, C. A. (1989) Removal of Alachlor from drinking water. Proc. National conference on Environmental Engineering. ASCE. Orlando, FL (July 1987).

Miltner, R. J., Fronk, C. A. and Speth, T. F. (1987) Treatment of seasonal Pesticides in Surface Waters. *Journal AWWA*. **81**: 43–52.

Mitchell, C., Brodie, J. and White, I. (2005) Sediments, nutrients and pesticide residues in event flow conditions in streams of the Mackay Whitsunday Region, Australia. *Marine Poll. Bull.* **51**, 23–36.

Moore, M. T., Rodgers Jr, J. H., Cooper, C. M. and Smith, Jr, S. (2000) Constructed wetlands for mitigation of atrazine-associated agricultural runoff. *Environ. Poll.* **110**, 393–399.

Moss, A., Bordie, J., Furnas, M. (2005) Water quality guidance for the Great Barrier Reef World Heritage Area: a basis for development and preliminary values. *Marine Poll. Bull.* **51**, 76–88.

Namasivayam, C. and Kavitha, D. (2003) Adsorptive removal of 2-chlorophenol by low-cost coir pith carbon. *J. Haz. Mater.* **B98**, 257–274.

Negri, A. P., Mortimer, M., Carter, S. and Müller, J. F. (2009) Persistent organochlorines and metals in estuarine mud crabs of the Great Barrier Reef, Baseline. *Marine Poll. Bull.* **58**, 765–786.

Nghiem, L. D., Tadkaew, N. and Sivakumar, M. (2009) Removal of trace organic contaminants by submerged membrane bioreactors. *Desalination* **236**, 127–134.

Owen, R., Knap, A., Ostrander, N. and Carbery, K. (2003) Comparative acute toxicity of herbicides to photosynthesis of coral zooxanthellae. *Bull. Environ. Contam. Toxicol.* **70**, 541–548.

Peter-Varnamets, M., Zurbrügg, C., Swartz, C. and Pronk, W. (2009) Decentralized systems for potable water and the potential of membrane technology. *Water Res.* **43**, 245–265.

Petrović, M., Gonzales, S. and Barceló, D. (2003) Analysis and removal of emerging contaminants in wastewater and drinking water. *Trends in Anal. Chem.* **22**.

Petrović, M., Radjenović, J. and Barceló, D. (2007) Elimination of emerging contaminants by membrane bioreactor (MBR).

Phattaranawik, J., Fane, A., Pasquier, A. C. S. and Bing, W. (2008) A novel membrane bioreactor based on membrane distillation. *Desalination* **223**, 386–395.

Plakas, K. V., Karabelas, A. J., Wintgens, T. and Melin, T. (2006) A study of selected herbicides retention by nanofiltration membranes – the role of organic fouling. *J. Mem. Sci.* **284**, 291–300.

Quintana, J. B., Weiss, S. and Reemtsma, T. (2005) Pathways and metabolites of microbial degradation of selected acidic pharmaceutical and their occurrence in municipal wastewater treated by a membrane bioreactor. *Water Res.* **39**, 2654–2664.

Radjenovic, J., Matosic, M., Mijatovic, I., Petrovic, M. and Barcelo, D. (2008) Membrane Bioreactor (MBR) as an Advanced Wastewater Treatment Technology. *Hdb. Env. Chem.* **5**, 37–101.

Radjenovic, J., Petrovic, M., M. and Barceló (2006) Analysis of pharmaceuticals in waste water and removals using a membrane bioreactor..

Ratola, N., Botelho, C. and Alves, A. (2003) The use of pine bark as a natural adsorbent for persistent organic pollutants – study of lindane and heptachlor adsorption. *J. Chem. Technol. Biotechnol.* **78**, 347–351.

Sannino, F., Iorio, M., De Martino, A., Pucci, M., Brown, C. D. and Capasso, R. (2008) Remediation of waters contaminated with ionic herbicides by sorption on polymerin. *Water Res.* **42**, 643–652.

Sarkar, B., Venkateshwarlu, N., Rao, R. N., Bhattacharjee, C. and Kale, V. (2007a) Potable water production from pesticide contaminated surface water – A membrane based approach. *Desalination* **204**, 368–373.

Sarkar, B., Venkateswralu, N., Rao, R. N., Bhattacharjee, C. and Kale, V. (2007b) Treatment of pesticide contaminated surface water for production of potable water by a coagulation-adsorption-nanofiltration approach. *Desalination* **212**, 29–140.

Seery, C. R., Gunthorpe, L. and Ralph, P. J. (2006) Herbicide impact on Hormosira banksii gametes measured by fluorescence and germination bioassays. *Environ. Poll.* **140**, 43–51.

Shaw, M. and Müller, J. F. (2005) Preliminary evaluation of the occurrence of herbicides and PAHs in the wet tropics regions of the Great Barrier Reef, Australia, using passive samplers. *Marine Poll. Bull.* **51**, 876–881.

Spring, A. J., Bagley, D. M., Andrews, R. C., Lemanik, S. and Yang, P. (2007) Removal of endocrine disrupting compounds using a membrane bioreactor and disinfection. *J. Environ. Eng. Sci.* 131–137.

Stasinakis, A. S., Kotsifa, S., Gatidou, G. and Mamais, D. (2009) Diuron biodegradation in activated sludge batch reactors under aerobic and anoxic conditions. *Water Res.* **43**, 1471–1479.

Stearman, G. K., George, D. B., Carlson, K. and Lansford, S. (2003) Pesticide removal from container nursery runoff in constructed wetland cells. *J. Environ. Qual.* **32**, 1548–1556.

Stork, P. R., Bennett, F. R. and Bell, M. J. (2008) The environmental fate of diuron under a conventional production regime in a sugarcane farm during the plant cane phase. *Pest Manag. Sci.* **64**, 954–963.

Tang, H. L., Regan, J. M. and Eix, Y. F. (2007) DBP precursors removal by membrane bioreactors http://www.hbg.psu.edu/etc/spwstac/research/hxt154/paper.pdf.

Tomaszewska, M., Mozia, S. and Morawski, A. W. (2004) Removal of organic matter by coagulation enhanced with adsorption on PAC. *Desalination* **161**, 79–87.

United States Environmental Protection Agency (October 2001) The Incorporation of Water treatment effects on Pesticide Removal and transformations in food Quality Protection Act (FQPA) Drinking Water Assessments.

Van der Bruggen, B., Schaep, J., Maes, W., Wilms, D. and Vandecasteele, C. (1998) Nanofiltration as a treatment method for the removal of pesticides from ground waters. *Desalination* **117**, 139–147.

Visvanathan, C., Thu, L.N., Jagatheesan, V. and Anotai, J. (2005) Biodegradation of pentachlorophenol in a membrane bioreactor. *Desalination* **183**, 455–464.

Weiss, S. and Reemtsma, T. (2008) Membrane bioreactors for municipal wastewater treatment – A viable option to reduce the amount of polar pollutants discharged into surface waters. *Water Res.* **42**, 3837–3847.

White, I., Brodie, J. and Mitchell, C. (2002) Pioneer river catchments event based water quality sampling. Healthy waterways programme, Mackay Whitsunday Regional Strategy Group, Mackay.

Williams, M. D. and Pirbazari. M. (2007) Membrane bioreactor process for removing biodegradable organic matter from water. *Water Res.* **41**, 3880–3893.

World Health Organization and International Programme on Chemical Safety -WHO/IPCS (2002) Global Assessment of the State-of-the-Science of Endocrine Disruptors. (eds). Damstra, T. Barlow, S. Bergman, A, Kavlock, R. Van Der Kraak, G. WHO/PCS/EDC/02.2 World Health Organization, Geneva, Switzerland. http://ehp.niehs.nih.gov/who/

Wintgens, T., Gallenkemper, M. and Melin, T. (2002) Endocrine disrupter removal from wastewater using membrane bioreactor and nanofiltration technology. *Desalination* **146**, 387–391.

Yiping, X., Yiqi, Z., Donghong, W., Shaohua, C., Junxin, L. and Zijian, W. (2008) Occurrence and removal of organic micropollutants in the treatment of landfill leachate by combined anaerobic-membrane bioreactor technology. *J. Environ. Sci.* **20**, 1281–1287.

Yuzir, A. and Sallis, P. J. (2007) Performance of anaerobic membrane bioreactor (AMBr) in the treatment of a synthetic (RS)-MCPP wastewater, IWA 8th National UK Young Water Professionals Conference, University of Surrey April 1997.

Znad, H., Kasahara, N. and Kawase, Y. (2006) Biological decomposition of herbicides (EPTC) by activated sludge in a slurry bioreactor. *Process Biochem.* **41**, 1124–1128.

Index

A
abiotic degradation 257–258
activated carbon
 catalysts 389–390
 granular activated carbon (GAC) 170–172, 313, 451
 powdered activated carbon (PAC) 219, 451
activated sludge systems
 bioavailability 261
 biodegradation in 258–259
 biological treatment in 268–272, 450
 conventional 270
 estrogen removal 256–257, 269–271
 nonylphenol (NP) formation 271–272
 oxidation ditches 270–271
 and pH 262
 photolysis in 257
 removal efficiency 264, 268–271, 272, 450
 sorption to sludge 256–257, 262
 volatization 258
 see also hydraulic retention time (HRT); sludge retention time (SRT)
active pharmaceutical ingredients (APIs) 336
adsorbents *see* adsorptive materials
adsorption and ion exchange 165–168, 171, 173–187
 in activated sludge systems 256–257
 biosorption 187–190, 193–194
 in catalysis 362
 definitions 166
 ferrocyanides 177–178
 functional groups 173–175

©2010 IWA Publishing. *Treatment of Micropollutants in Water and Wastewater.* Edited by Jurate Virkutyte, Veeriah Jegatheesan and Rajender S. Varma. ISBN: 9781843393160. Published by IWA Publishing, London, UK.

future development 192–194
herbicide/pesticide removal 451
hybrid materials 191–192
of hydrogen peroxide 297
hydrous metal oxides 179–184
ion exchange 168, 171, 175
ion exchange resins 173–176
mechanism 184
membrane processes 138–139, 216, 218
of pharmaceuticals 16
removal efficiency 451
and temperature variations 268
ultrafiltration (UF) 218–219
and ultrasound treatment (US/A) 333–334
water purification 174
water softening 174
zeolites 172–173
see also ion exchangers
adsorptive materials 168–169
activated alumina 181–182
biosorbents 187–191
hybrid-composite 191–192
hydrotalcite type 183
hydrous metal oxides 179–182
hydrous oxides 182–183, 184, 186–187
manganese oxide 180–181
metal sulfides 183–184
zeolites 172–173
see also activated carbon; carbon adsorbents
advanced catalytic oxidation 361–415
biocatalytic oxidation 395–397
catalytic ozonation 373–375
catalytic wet air oxidation (CWAO) 397–399
catalytic wet peroxide oxidation processes (CWPO) 365–372, 396, 413
of dyes 371–372, 380–381, 383–385, 390, 413
electrocatalytic oxidation 390–395
and hydrogen peroxide 365–368
microwave (MW) assisted 388–390
nanocatalytic oxidation 399–414
and nanofiltration (NF) 369

photocatalysis 375–382
reaction mechanisms 365–366
removal efficiency 369–371
sonocatalytic degradation 383–388
titanium dioxide photocatalysis 377–379
see also catalysis; catalysts; Fenton process
advanced Fenton process (AFP) 400, 411
advanced nanocatalytic oxidation 399–414
advanced oxidation processes (AOP) see advanced catalytic oxidation; ultrasound (US) treatment; UV/H_2O_2 oxidation
AFP see advanced Fenton process (AFP)
agricultural runoff 441, 442–443, 446
alachlor 347–348, 447, 449
aldrin 3, 6, 13, 428, 434, 441
alkylphenol polyethoxylates 252, 266, 277
ametryn 434
ammonia oxidizing bacteria (AOB) 269, 279
amoxicillin (AMX) 367
amperometric biosensors 99
ampicillin (AMP) 367
anaerobic biological treatment 275–277
antibiotics 14, 272
biodegradability 367
and Fenton process 367
removal efficiency 210–211, 220–221, 276–277, 306
resistance 81
sorption 255
antimicrobial activity 81
API see active pharmaceutical ingredients (APIs)
aquatic environment
biodegradation 19
perfluorinated compounds (PFCs) 30–33
personal care products (PCPs) 25–30
pesticides 4–13
pharmaceuticals 13–20
steroid hormones 21–25
surfactants 25–30

aquatic organisms 4, 5–6, 8, 10
    and estrogens 21, 24–25
    Great Barrier Reef (GBR) 445–446
    nonylphenol (NP) toxicity 29
    pharmaceutical toxicity 19–20
arsenic
    adsorption 186–187
    in drinking water 181–182
    and human health 181
    nanofiltration (NF) 143–144
    removal 181–182
artificial neural networks (ANN) 66
atrazine 9–10, 13, 434
    adsorption 219, 451
    advanced nanocatalytic oxidation
        412–413
    biodegradation 450
    endocrine disrupting effects 446
    and the GBR 441, 443, 446
    nanofiltration (NF) 145, 146, 449, 453
    oxidation 367
    removal efficiency 218, 219, 225,
        449, 450
    toxicity 446
    ultrafiltration (UF) 218–219, 219
    wetland processes 452
azo dyes 340–341
    degradation 340–341
    removal 368, 397, 398–399, 411
    wet air oxidation 398–399

B

bacteria removal, nanofiltration (NF)
    146–148
bentazone 449
benzene 301
benzothiazoles 220, 265
    removal efficiency 220, 265
bezafibrate 220, 267, 275
    removal efficiency 220, 267
bilinear least squares (BLLS) 65
binding assays 97–99
bioaccumulation 20, 24
    in humans 4–5, 32–33
    of organochlorines 4–5
    personal care products (PCPs) 29–30
    of surfactants 29

bioassays 79, 80, 252–253
    for estrogens 252–254
biocatalytic oxidation 395–397
bioconcentration factors (BCFs) 4, 24, 33
biodegradation
    in activated sludge systems 258–259
    in the aquatic environment 19
    and compound structure 259–260
    endocrine disrupting chemicals
        (EDCs) 261
    of estrogens 24, 249
    fire retardants 272
    of herbicides 8, 450
    nonylphenol (NP) 27, 278–279
    pharmaceuticals 16, 19, 259, 265,
        267, 272–275
    and sludge retention time (SRT)
        264–265
    and temperature 266–267
    of urine 247–249
biofouling 140, 216–217
biological treatment 249–281
    in activated sludge systems 268–272,
        450
    anaerobic 275–277
    biodegradation 247–249
    bioreactor combinations 277
    by-products 280–281
    herbicide/pesticide removal 450–451
    microorganisms 278–280
    removal efficiency 259–268, 268–269
    removal mechanisms 254–259
    of sludge 278
    in trickling filters 277
    of urine 247–249
    in wetlands 272–274
    see also membrane bioreactors (MBRs)
bioluminescent bacteria (BLB) 79–80
biomagnification 4–5, 33
biosensors 81–82, 96–100
    nanotechnology 104
biosorbents 187–190
biosorption 187–190, 193–194
bisphenol A (BPA) 98, 262, 280, 386
    in activated sludge systems 271–272
    catalytic oxidation 369–371
    immunosensor for 100

removal efficiency 213, 220, 221, 226, 274
sonophotocatalytic treatment 387–388
UV/H_2O_2 oxidation 306–307
black water 243–244, 245
BPA *see* bisphenol A (BPA)
4-bromodiphenylether 301–302

C

cancer 10, 25, 98, 181, 427, 434
  and estrogens 25, 98
capillary electrophoresis (CE) 69, 70, 72–73, 94
carbamazepine (CBZ) 243, 278
  metabolites 14, 16
  removal efficiency 221, 265, 273
  in riverwater 16, 19
  in seawater 19
  in soils 18
  sorption 255
  in wetlands 273
carbofuran 348
carbon adsorbents 167, 169–172
  functional groups 170–172
  granular activated carbon (GAC) 170–172, 313, 451
  ion exchange ability 171
  powdered activated carbon (PAC) 219, 451
carbon nanotube field-effect transistor (CNTFET) 98
carbon nanotubes (CNTs) 98, 107–108
  nanofiltration membranes 134–136
catalysis 361–365
  adsorption during 362
  environmental 364–365
  heterogeneous 362–364
  homogeneous 365–368
  and ozone 373–375
  photocatalysis 375–382
  solar photocatalytic process 377, 381
  see also advanced catalytic oxidation; catalysts; Fenton process
catalysts 363, 398
  activated carbon 389–390
  cavitation 328
  deactivation of 363–364, 370, 399

enzymes 395–397
heterogenized 369–371
iron 307–308, 365–371
metals 371–372
nanocatalysts 399–414
niobium based 372
photocatalysts 377–382
properties 363–364
catalytic ozonation 373–375
catalytic wet air oxidation (CWAO) 397–399
catalytic wet peroxide oxidation processes (CWPO) 365–372, 396, 413
cation exchangers 180–181
cavitation 322–330
  catalysts 328
  effects 323–325
  electrical theory 322–323
  hot spot theory 322, 323
  intensity 325–330
  nucleation sites 327–328
  plasma theory 323
  supercritical theory 323
  transient 327
  see also ultrasound (US) treatment
CE *see* capillary electrophoresis (CE)
cefaclor (CFL) 369, 371
celestolide 221
ceramic membranes 133
cesium removal 177–178
charcoals *see* carbon adsorbents
Chemical Analysis Working Group (CAWG) 83
chemometrics 63–66
chlordane 428, 434
chloroform 218–219, 226
chlorpyrifos 6–8, 13, 98, 347, 441
chromatographic separation 69
ciprofloxacin (CIP) 255, 338
cloxacillin (CLX) 367
CNT *see* carbon nanotubes (CNTs)
CNTFET (carbon nanotube field-effect transistor) 98
coagulation 205–215
  coagulant dosage 210–211
  coagulant type 210–211
  enhanced 206–213

oxidation-coagulation 213–215
  removal efficiency 447
combined sewer overflow (CSO) 247
composite adsorbents 191–192
composite ion exchangers 191–192
composite membranes 135–136
compound structure 259–260
computational chemistry 60–63
conducting polymers 112–113
cork biomass 189, 190
costs
  dye removal 351
  groundwater treatment 315–316
  ultrasound (US) treatment 348, 350–352
  UV/H_2O_2 oxidation 313–315
CWAO (catalytic wet air oxidation) 397–399
CWPO (catalytic wet peroxide oxidation processes) 365–372, 396, 413
cyanazine 449

D

2,4-D 394, 395, 411, 441
  removal efficiency 450, 451
DBP *see* disinfection by-product (DBP)
DCA (3,4-dichloroaniline) 11, 12, 450
DCF *see* diclofenac (DCF)
2,4-DCP (2,4-dichlorophenol) 302, 394
DCPU (1-3,4-dichlorophenylurea) 11, 450
DDT 3–6, 13, 428, 434
  at the GBR 441
  bioaccumulation 4–5
  removal efficiency 347
degradation
  abiotic 257–258
  anaerobic 276–277
  mechanism 323–325
  photocatalytic 375–382, 386
  photodegradation 18–19, 28, 257, 303, 378–379
  products 78–82, 280
  rates 300–303, 412
  sonocatalytic 383–388
  sonochemical 325–330, 337–348, 340–343

see also biodegradation
density functional theory (DFT) 61, 63
desalination pretreatment 148–149
detection and quantification
  endocrine disrupting chemicals (EDCs) 251–254
  estrogens 251
  transformation products 71–78
diazepam 210, 221, 278
3,4-dichloroaniline (DCA) 11, 12, 450
2,4-dichlorophenol (2,4-DCP) 302, 394
2,4-dichlorophenoxyacetic acid *see* 2,4-D
1-3,4-dichlorophenylurea (DCPU) 11, 450
diclofenac (DCF) 14, 16
  biodegradation 19, 248
  coagulation removal 206, 208, 210, 211, 213
  MBR removal 221 265, 267, 275
  metabolites 14, 19
  photodegradation 18–19, 257, 303
  removal efficiency 278, 371
  in soils 18
  toxicity 20, 340
  ultrasound (US) treatment 339–340
  US/O_3 treatment 339–340
  in wastewater 240, 242
dieldrin 3–4, 13, 428, 434, 441
  toxicity 6
dilantin 306
2,4-dinitrotoluene (DNT) 301
diode array detector (DAD) 251
dioxins 427, 432 433
dirty dozen 13, 427, 428–430
disinfection 146–148
disinfection by-product (DBP) 145
  enhanced coagulation 206, 208–209, 212–213
  removal efficiency 208–209
  UV/H_2O_2 oxidation treatment 308–309
diuron
  adsorption 451
  biodegradation 450
  catalytic wet air oxidation (CWAO) 399
  Great Barrier Reef (GBR) 434, 441, 443, 445–446

legislation 13
metabolites 11, 12, 450
nanofiltration (NF) 145, 449, 452
properties 434
removal efficiency 449
toxicity 11–12, 445–446
DNT (2,4-dinitrotoluene) 301
drinking water
  arsenic in 181–182
  carbon sorbents 169
  legislation 12–13, 181, 206
  nonylphenol (NP) in 27
  steroid hormones in 22
  treatment 206, 303–305
dyes
  advanced catalytic oxidation 371–372, 380–381, 390
  catalytic ozonation 375
  degradation rates 412
  electrocatalytic oxidation 392
  Fenton process 367, 368
  organic 340–343
  photocatalytic degradation 378, 381–382
  removal 150, 340–343
  sonocatalytic degradation 383–385
  sonophotocatalytic treatment 387
  toxicity 341–342
  treatment costs 351
  ultrasound (US) treatment 340–343
  see also azo dyes

E
E-screen assay 252, 253
ecotoxicological assessment 79–80, 445–447
EDCs *see* endocrine disrupting chemicals (EDCs)
EDTA 217–218
effluent wastewater *see* WWTP effluent
electrical theory of cavitation 322–323
electro-Fenton process 392–393
electrocatalytic oxidation 390–395
electrochemical oxidation (EO), and ultrasound treatment (US/EO) 334–335
electrochemical sensors 95

electrodialysis 215, 227
ELISA *see* enzyme-linked immunosorbent assays (ELISA)
endocrine disrupting chemicals (EDCs) 6, 10, 93–94
  biodegradation 261
  detection and quantification 251–254
  estrogens 24–25
  and human health 250, 427
  persistent organic pollutants (POPs) 427
  personal care products 30
  removal efficiency 220–221
  in riverwater 253–254
  sensors and biosensors 94–105
  surfactants 29
  UV/H_2O_2 oxidation 306–307
  wastewater analysis 252–254
endrin 429, 434
energy efficiency
  of catalytic ozonation 374
  ultrasound (US) treatment 329–330, 336, 351, 353
enhanced coagulation 206–213
  physico-chemical properties of micropollutants 206–213
  removal efficiency 207–209, 210–211, 213–214
  removal mechanism 206
environmental catalysis 364–365
environmental monitoring 96–97, 104, 107, 114
enzyme-catalyzed reactions 395–397
enzyme-linked immunosorbent assays (ELISA) 94–95
EO *see* electrochemical oxidation (EO)
EPTC (S-ethyl dipropylthiocarbamate) 451
erythromycin 14, 220
estrogenicity
  assessment 80
  of black water 249
  personal care products (PCPs) 99
estrogens 21–25
  in activated sludge systems 256–257, 269–271
  anaerobic biological treatment 275–276

bioassays for 252–254
biocatalytic oxidation 396
biodegradation 24, 249
biosensors for 97–99
  degradation products 280
  detection and quantification 251
  photolysis 257–258
  physico-chemical properties 208–209
  removal efficiency 227, 263–265, 267, 269–271, 278, 279
  removal mechanisms 255–258, 269
  in riverwater 253–254
  sorption 208, 209, 255–257, 262
  ultrasound (US) treatment 337–338
  in urine 21
  UV/H_2O_2 oxidation 306–307
  in wastewater treatment plants (WWTPs) 253–254, 266
17α-ethylenestradiol 303
excretion pathways of pharmaceuticals 244–245
excretion rates
  musk fragrances 242
  of pharmaceuticals 14, 242
  steroid hormones 21, 242

F
fate
  of micropollutants in ultrafiltration (UF) 218–219
  of micropollutants in urine source separation 247–249
  of organochlorines 3–5
  of organophosphates 7–8
  of perfluorinated compounds (PFCs) 31–32
  of personal care products (PCPs) 27–29
  of pharmaceuticals 15–19
  of steroid hormones 22–24
  of substituted ureas 11
  of surfactants 27–29
  of triazine herbicides 9
fenitrothion 367
Fenton process 365–368
  advanced Fenton (AFP) 400, 411
  for dye removal 367, 368
  electro-Fenton 392–393
  heterogeneous 368
  homogeneous 365–368
  microwave (MW) assisted 390
  and pH 368, 370
  photo-Fenton 379–381
  photoelectro-Fenton 394
  sonoelectro-Fenton (SEF) 393–394
Fenton reagent 365–368
ferrate (FeVI) 213–215
ferrocyanides 177–178
fertility 10, 250
figure-of-merit 350
fire retardants 272
flocculation 209, 210–211, 213
fluorescence detector 251
fluorophores 103–104
fouling
  of filtration membranes 132–133, 139–141, 148
  membrane processes 218
  in reverse osmosis (RO) 227
Fourier transform (FT) 73, 74
fragmentation pattern analysis (MS2) 71
fragrances 221, 242, 255
Frontier electron density analysis 61–62
Fukui function 63
fullerene 193
functional groups 373, 374
  of biosorbents 189
  of carbon adsorbents 170–172
  of ion exchange resins 173–175
  pharmaceuticals 255
furans 427, 432–433

G
GAC *see* granular activated carbon (GAC)
galaxolide 210, 221, 278
gas chromatography (GC) 69–70, 72, 94
  with mass spectrometry (MS) 72, 94, 251–252
gemfibrozil (GEM) 14, 16, 20
granular activated carbon (GAC) 170–172, 451
  and UV/oxidation 313
Great Barrier Reef (GBR) 438–457
  agricultural runoff 441, 442–443, 446

catchment area 438, 439–442
ecosystem 445–446
ecotoxicological studies 445–447
herbicide/pesticide 434, 439–449
  concentrations 443
  persistence 442–447
  removal 447–457
  transport 439–441
lagoon water quality 441, 442–443
land use 439, 442
marine organisms 445–446
possible treatment methods 450–457
sugar industry 439, 440, 442–443
water treatment methods 447–449
Great Barrier Reef World Heritage Area (GBRWHA) 438, 439
groundwater
  PCE contamination 308, 313, 314
  pharmaceuticals in 16, 18
  radium removal 181
  softening 142–143
  treatment 311–313, 314, 315
  treatment costs 315–316
  UV/H_2O_2 oxidation treatment 300–303, 308
  volatile organic compound (VOC) removal 311–312, 315

H

Haber-Weiss mechanism 365–366
HBC (hexachlorobenzene) 429, 434
HCH (hexachlorocyclohexanes) 3–5
heavy metal removal 188
heptachlor 3, 13, 429, 434, 441
  removal 449, 451
  toxicity 6
herbicides 433–438
  adsorption 451
  biodegradation 8, 450
  biosensors for 99–100
  concentrations 443
  Great Barrier Reef (GBR) 434, 439–449
  and membrane bioreactors (MBRs) 454–455
  nanofiltration (NF) 146
  persistence 442–447
  photosystem II (PSII) 9, 434

removal 447–457
substituted ureas 10–13
toxicity 445–446
triazine 8–10
  see also atrazine
heterogeneous catalysis 362–364
hexachlorobenzene (HBC) 429, 434
hexachlorocyclohexanes (HCH) 3–5
hexazinone 434
high performance liquid chromatography (HPLC) 94, 251
hormones see steroid hormones
horseradish peroxidase (HRP) 396–397
hot spot theory of cavitation 322, 323
HPLC (high performance liquid chromatography) 94, 251
HRT see hydraulic retention time (HRT)
human health 4–5
  and arsenic 181
  cancer 10, 25, 98, 181, 427, 434
  endocrine disrupting effect 250, 427
  fertility 10, 250
  and organochlorines 3, 343
  and organophosphates 8, 343
  and persistent organic pollutants (POPs) 426–427, 434, 438
  and pharmaceuticals 336–337
hybrid adsorption materials 191–192
hybrid wastewater treatment systems 330–336, 336–348, 453–455
hydraulic retention time (HRT) 270–271
  and removal efficiency 263–264
hydrogen peroxide 385, 386
  adsorption 297
  advanced catalytic oxidation 365–368
  and catalyst deactivation 370
  photolysis 297, 299
  and ultrasound treatment (US/H_2O_2) 330–331
  see also UV/H_2O_2 oxidation
hydrophobicity 254–255
hydrotalcites 183
hydrous metal oxides 179–184
hydrous oxides 182–183, 184, 186–187
hydroxyl radicals 297, 299, 331–332, 373, 379
hyperfiltration see reverse osmosis (RO)

Index

I
ibuprofen 14, 214–215
  advanced catalytic oxidation 369, 371
  bioaccumulation 20
  biodegradation 19, 267, 272, 273, 275
  MBR treatment 220, 221, 265
  metabolites 14, 16, 19
  in municipal sewage 254, 255, 260, 262
  photodegradation 19
  removal efficiency 211, 218, 220, 267, 272, 273, 371
  removal mechanism in WWTPs 254, 255
  in riverwater 16, 19
  in seawater 16
  ultrafiltration (UF) 218, 220
  of ultrafiltration (UF) 218, 220
  ultrasound (US) treatment 338
  wetland treatment 273
identification levels 82–84
immunosensors 100–104
industrial waste products 188
industrial wastewater 415
  pH value 262
  textile wastewater 150, 340–343, 378
insecticides *see* organochlorines; organophosphates
instrumental analysis methods 66–82
  capillary electrophoresis (CE) 69, 70, 72–73, 94
  chromatographic separation 69
  NMR spectroscopy 76–78
  sample extraction 67–69
  transformation product detection 71–78
  UV-Visible spectroscopy 71, 76, 251
  see also mass spectrometry (MS)
integrated membrane system (IMS) 150–151
iobromide 303
ion exchangers
  adsorptive properties 184
  cation exchangers 180–181
  composite 191–192
  ferrocyanides 177–178
  functional groups 173–175
  hybrid 191–192
  hydrous metal oxides 179–184
  inorganic 176–187

  ion exchange resins 173–176
  materials 168–169
  properties 184
  synthesis of 184–187
  zeolites 173
  see also adsorption and ion exchange
isoproturon 11, 13
  removal 145, 146, 451, 452
  toxicity 12

K
ketoprofen 272

L
layered mixed hydrous oxides 183, 186–187
LC *see* liquid chromatography (LC)
legislation
  arsenic in drinking water 181
  drinking water 12–13, 206
  perfluorinated compounds (PFCs) 33
  personal care products 30
  pesticides 12–13
  pharmaceuticals 20–21
  steroid hormones 25
  surfactants 30
life cycle analysis (LCA) 143
light absorption 298
lindane 3–4, 441, 452
  legislation 13, 21, 431
  removal 451, 452
  toxicity 6
liquid chromatography (LC) 69–70, 71
  high performance liquid chromatography (HPLC) 94, 251
  with mass spectrometry (MS) 72, 77–78, 251
  ultra-performance liquid chromatography (UPLC) 69, 72, 74
lithium removal from seawater 180–181

M
malathion 347
manganese oxide adsorbents 180–181
mass spectrometry (MS) 71, 72–76, 78, 251, 252
  EDC detection 94

with gas chromatography (GC) 72, 94, 251–252
with liquid chromatography (LC) 72, 77–78, 251
time-of-flight (TOF) mass spectrometry 73, 74, 251
MBR *see* membrane bioreactors (MBRs)
membrane bioreactors (MBRs) 220–221, 453–455
   bisphenol A (BPA) 274
   herbicide removal 454–455
   ibuprofen removal 220, 221, 265
   metal removal 221
   pesticide removal 454–455
   and pH 262
   pharmaceutical removal 265, 267, 275, 454
   removal efficiency 220–225, 259–269
   sludge retention time (SRT) 274–275
   and ultrafiltration (UF) 220–221
membrane processes 215–227
   adsorption 138–139, 216, 218
   biofouling 140, 216–217
   electrodialysis 215, 227
   in membrane bioreactors (MBRs) 220–221
   microfiltration (MF) 217–218
   and natural organic matter (NOM) 216–217, 218
   pesticide removal 452–453
   powdered activated carbon (PAC) 219
   reverse osmosis (RO) 225–227
   solute rejection 216–217
   ultrafiltration (UF) 218–221
metabolites
   of carbamazepine (CBZ) 14, 16, 243
   of diclofenac (DCF) 14, 19
   of diuron 11, 12, 450
   of erythromycin 14
   of gemfibrozil (GEM) 14
   of ibuprofen 14, 16, 19
   identification levels 82–83
   methyl parathion 8
   of trimethoprim (TMP) 14
metal sulfides 183–184
metals 188
   catalysts 371–372

ion detection 112
removal efficiencies 221, 226
methyl parathion 6–8, 412
methyl tert-butyl ether (MTBE) 303, 387, 400
metolachlor 449
metribuzin 449
MF *see* microfiltration (MF)
MFI (modified fouling index) 141
MIC (minimal inhibitory concentration) 81
microbiotests 80
microcontact printing 111
microfiltration (MF) 129–131, 217–218
microorganisms
   biological treatment 278–280
   biosorbents 188–189
   and temperature variations 266–267
micropollutants
   advanced catalytic oxidation 361–415
   analysis 53–85, 250–254
   and aquatic environment 1–33
   biological treatment 249–281
   compound structure 259–260
   emerging contaminants 59
   exposure to 250
   hybrid advanced oxidation 321–353
   identification 53–58
   physico-chemical treatment 165–194, 205–227
   in sewage 239–243
   transformation products 71–85
   treatability studies 250–251
   UV/H_2O_2 oxidation 295–316
microwave (MW) irradiation
   and advanced catalytic oxidation 388–390
   and ultrasound (US/MW) 335–336
minimal inhibitory concentration (MIC) 81
MIPs (molecular imprinted polymers) 108–111
mirex 429, 434
mixed hydrous oxides 182–183
modified fouling index (MFI) 141
molecular imprinted polymers (MIPs) 108–111

molecular sieves 173
see also zeolites
MS *see* mass spectrometry (MS)
MS2 (fragmentation pattern analysis) 71
MTBE *see* methyl tert-butyl ether (MTBE)
multi-walled carbon nanotubes (MWNTs) 134–135, 144
multistage flash desalination (MSF) 148–149
multivariate curve resolution (MCR) 64–65
municipal sewage
 biological treatment 240–249
 ibuprofen in 254, 255, 260, 262
municipal wastewater treatment, UV/ H_2O_2 oxidation 305–307
musk fragrances 221
 excretion rates 242
MWNT (multi-walled carbon nanotubes) 134–135, 144

N

nanocatalysts 399–414
nanocatalytic oxidation 399–414
nanofiltration (NF) 129–152, 453
 and advanced catalytic oxidation 369
 arsenic removal 143–144
 bacteria removal 146–148
 desalination pretreatment 148–149
 disinfection 146–148
 disinfection by-product (DBP) 145
 fouling 132–133, 139–141
 of herbicides 145, 146, 449, 452
 metal ion removal 149–150
 of micropollutants in water 142–152
 natural organic matter (NOM) 145–146
 nitrate rejection 149
 organic acids 149
 of pesticides 145–146
 polymer membranes 132–133
 removal efficiency 449
 separation 136–139
 surfactants 149
 textile wastewater treatment 150
 virus removal 146–148
 water softening 142–143

nanofiltration (NF) membranes 131
 biofouling 140
 carbon nanotubes (CNTs) 134–136
 ceramic 133
 composite 135–136
 fouling 132–133, 139–141, 148
 high temperature resistant 150
 membrane materials 132–136
 negative surface charge 144–145
 organic fouling 139–140
 scaling 139, 140, 148
 solute characteristics and membrane performance 137–138
 surface morphology 137
 zeolite 133–134, 144
nanotechnology
 advanced nanocatalytic oxidation 399–414
 sensors and biosensors 104, 106–108, 110
naproxen
 removal efficiency 208, 210, 211, 220, 267, 272, 278
 in surface waters 257
napthalene sulfonates 220
 removal efficiency 220
natural organic matter (NOM)
 and membrane processes 216–217, 218
 nanofiltration (NF) 145–146
NF *see* nanofiltration (NF)
niobium based catalysts 372
nitrification 269–271, 272
nitrobenzene 308
nitrophenols 308
NMR spectroscopy 76–78
NOM *see* natural organic matter (NOM)
non-steroidal anti-inflammatory drugs (NSAIDs) 14, 16, 18
 toxicity 19–20
 see also diclofenac (DCF); ibuprofen; naproxen
nonylphenol ethoxylates (NPEs) 25, 276–277
nonylphenol (NP) 25, 27, 280
 in activated sludge systems 271–272
 biodegradation by microorganisms 278–279

legislation 30
toxicity 29
NP *see* nonylphenol (NP)
NPEs *see* nonylphenol ethoxylates (NPEs)
NSAIDs *see* non-steroidal anti-inflammatory drugs (NSAIDs)
nuclear magnetic resonance (NMR) spectroscopy 76–78
nutrient recycling 246

O
organic acids 149
organic dyes 340–343
organic fouling 139–140
organic load rate 266
organochlorines 3–6, 13, 434, 441
   bioaccumulation 4–5
   bioconcentration factors (BCFs) 4
   biomagnification 4–5
   fate 3–5
   and human health 3, 4–5, 343
   persistent organic pollutants (POPs) 427–430, 434
   toxicity 5–6
   ultrafiltration (UF) removal 449
   ultrasound (US) treatment 343, 347
   wetland treatment 452
organophosphates 6–8
   biosensors for 97, 98, 99, 106
   degradation 7–8, 343
   fate 7–8
   and human health 8, 343
   metabolites 8
   removal 347
   toxicity 8
oxidation ditches 270–271
oxidation processes *see* advanced catalytic oxidation; ultrasound (US) treatment; UV/H_2O_2 oxidation
oxidation-coagulation 213–215
oxidizing agents 213–215
ozonation
   catalytic 373–375
   with ultrasound treatment (US/O_3) 331–332, 339–340
   with UV irradiation (UV/O_3) 300, 308–309

P
p-chlorophenol 307–308
p-Nitrophenol 219
PAC (powdered activated carbon) 219, 451
PAC1 coagulation 211–212
PAH (polycyclic aromatic hydrocarbons) 221, 261
paper and pulp industry 307
parabens 26, 27, 29
paracetamol 305
parallel factor analysis (PARAFAC) 64
PCDD *see* polychlorinated dibenzo dioxins (PCDDs)
PCDF *see* polychlorinated dibenzo furans (PCDFs)
PCP *see* personal care products (PCPs)
pentachlorophenol (PCP) 389–390
perfluorinated compounds (PFCs) 30–35
   physico-chemical characteristics 30
   toxicity 32–33
perfluorooctane sulfonate (PFOS) 31–33
perfluorooctanoate (PFAO) 31–33
persistence
   of herbicides 9, 442–447, 445
   organochlorines 4
persistent organic pollutants (POPs) 13, 426–433
   dioxins 427, 432–433
   dirty dozen 13, 427, 428–430
   endocrine disrupting chemicals (EDCs) 427
   furans 427, 432–433
   and human health 426–427, 434, 438
   newly listed 427
   pesticides 434
   removal 450–457
   transport 426
personal care products (PCPs)
   in the aquatic environment 25–30
   bioaccumulation 29–30
   in effluent wastewater 27
   estrogenicity 99
   fate 27–29
   legislation 30
   parabens 26
   removal from wastewater 27

Index

toxicity 29–30
triclosan (TCS) 26
see also pharmaceuticals and personal care products (PPCPs)
pesticides 2–13, 433–438
  in the aquatic environment 4–13
  biosensors 97, 98, 99
  concentrations 443
  Great Barrier Reef (GBR) 434, 439–449
  immunosensors 101–102
  legislation 12–13
  membrane bioreactors (MBRs) 454–455
  nanocatalytic oxidation 412–413
  nanofiltration (NF) 145–146
  persistence 434, 442–447
  persistent organic pollutants (POPs) 434
  photocatalytic removal from wastewater 378–379
  removal 367–368, 447–457, 451–453
  substituted ureas 10–13
  triazines 8–10
  ultrasound (US) treatment 343–348, 347–348
  UV/H_2O_2 oxidation 302
  see also organochlorines; organophosphates
PFAO (perfluorooctanoate) 31–33
PFCs *see* perfluorinated compounds (PFCs)
PFOS (perfluorooctane sulfonate) 31–33
pH
  and activated sludge systems 262
  and coagulation 211–213
  and Fenton process 368, 370
  of industrial wastewater 262
  in membrane bioreactors (MBRs) 262
  and removal efficiency 261–262
  and sorption 262
pharmaceuticals 13–21, 211
  adsorption 16
  in the aquatic environment 13–20
  at WWTPs 240–243, 306
  bioconcentration 20
  biodegradation 16, 19, 265, 267, 272–275

compound structure 259
in effluent wastewater 16
environment, introduction into 15, 16
excretion pathways 244–245
excretion rates 14, 242
fate 15–19
functional groups 255
in groundwater 16, 18
and human health 336–337
legislation 20–21
in membrane bioreactors (MBRs) 265, 267, 275, 454
metabolites 16
photodegradation 18–19
photolysis 257–258
removal efficiency 16, 240, 265
removal mechanisms 255
in riverwater 19
sonochemical degradation 337–340
sorption 18, 255
in surface water 16
and temperature 267
toxicity 19–20
ultrasound (US) treatment 336–340
in urine 14, 243, 244–246
UV/H_2O_2 oxidation 305–306, 306
pharmaceuticals and personal care products (PPCPs)
  detection and quantification 251, 252
  enhanced coagulation removal 210
  fragrances 221, 242, 255
  hormones 255
  and hydraulic retention time (HRT) 263
  removal efficiency 210–211, 220–221, 260
  and sludge retention time (SRT) 263
  sorption 255
  wetland treatment 273–274
phosphate removal 213–214
photo-Fenton process 379–381
photocatalysis 375–382
photocatalysts 377–382
photocatalytic degradation 375–382, 386
photodegradation
  pharmaceuticals 18–19
  of triclosan (TCS) 28, 378–379
photoelectro-Fenton process 394

photolysis 257–258
  activated sludge systems 257
  advanced oxidation processes (AOP) 298–299
  of estrogens 257–258
  of hydrogen peroxide 297, 299
  UV/H_2O_2 oxidation 298–299
photosystem II (PSII) based biosensors 99–100
photosystem II (PSII) herbicides 9, 434
photosystem II (PSII) inhibitors 9, 10
physico-chemical properties
  and enhanced coagulation 206–213
  of estrogens 208–209
  hydrophobicity 254–255
  of micropollutants 252, 254–255, 326–327
  perfluorinated compounds (PFCs) 30
  and sample preparation techniques 67–69
  steroid hormones 23
  and ultrasound (US) treatment 326–327
physico-chemical treatment *see* adsorption and ion exchange; coagulation; membrane processes
plasma theory of cavitation 323
plasmon resonance (SPR) 98
polychlorinated dibenzo dioxins (PCDDs) 427, 432–433
polychlorinated dibenzo furans (PCDFs) 427, 432–433
polycyclic aromatic hydrocarbons (PAHs) 221, 261
polymer membranes 132–133
POPs *see* persistent organic pollutants (POPs)
powdered activated carbon (PAC) 219, 451
PPCPs *see* pharmaceuticals and personal care products (PPCPs)
pressure driven membrane processes 452–453
pretreatment of biosorptive biomasses 189–190

Q
quantum dots (QDs) 107, 108

R
radium removal 181
Reef Water Quality Protection Plan 439
removal efficiency 259–268
  of activated sludge systems 264, 268–271, 272, 450
  and adsorption 451
  advanced catalytic oxidation 369–371
  anaerobic treatment 276, 278
  antibiotics 210–211, 220–221, 276–277, 306
  and bioavailability 261
  biological treatment 259–268, 268–269, 450
  in bioreactor combinations 277–278
  of coagulation 447
  and compound structure 259–260
  disinfection by-product (DBP) 208–209
  and dissolved oxygen (DO) 261
  endocrine disrupting chemicals (EDCs) 220–221
  of enhanced coagulation 207–209, 210–211, 213–214
  estrogens 227, 263–264, 265, 269–271, 279
  herbicides 347–348, 447, 450
  and hydraulic retention time (HRT) 263–264
  in membrane bioreactors (MBRs) 220–225, 260
  metals 221, 226
  of microfiltration (MF) 217
  of nanofiltration (NF) 449
  and organic load rate 266
  pesticides 449
  and pH 261–262
  pharmaceuticals 16, 240, 265
  PPCPs 210–211, 220–221, 260
  of reverse osmosis (RO) 225–227, 449
  and sludge retention time (SRT) 263–265
  surfactants and personal care products 27
  and temperature 266–268
  of ultrafiltration (UF) 218, 219, 220–225
  of ultrasound (US) treatment 325–330

of UV/H_2O_2 oxidation 303–307, 311, 312
volatile organic compounds (VOC) 311
in wastewater treatment plants (WWTPs) 16, 240, 263–264, 265, 266, 267–268
wetland treatment 273–274, 452
see also individual compounds
removal mechanisms 254–259
  abiotic degradation 257–258
  biodegradation 258–259
  enhanced coagulation 206
  estrogens 255–258, 269
  pharmaceuticals 255
  photolysis 257–258
  sorption 254–257
  volatization 258
reverse osmosis (RO) 129–131, 225–227, 453
  fouling 227
  hormone removal 227
  micropollutant removal 226–227, 228–233
  organic compound removal 150
  removal efficiency 225–227, 449
  removal mechanisms 225
riverwater
  biodegradation of pharmaceuticals 19
  endocrine disrupting chemicals (EDCs) 253–254
  estrogenicity 253–254
  perfluorinated compounds (PFCs) 32
  pharmaceuticals 16, 19
RO *see* reverse osmosis (RO)
roxithromycin, removal efficiency 278

S
S-ethyl dipropylthiocarbamate (EPTC) 451
sample extraction 67–69
SAPs (sustainable analytical procedures) 60
scaling of filtration membranes 139, 140, 148
scavenging effect 331
screen printed electrodes 106
screen printed sensors 106

SCW (supercritical water) 323
SDI (silt density index) 141
seawater
  carbamazepine (CBZ) in 19
  desalination 148–149
  lithium removal 180–181
  pharmaceuticals in 16
seawater reverse osmosis (SWRO) 148–149
sediments 19, 24
sensors and biosensors 93–114
  biosensors 96–100
  conducting polymers 112–113
  electrochemical 95
  for endocrine disrupting chemicals (EDCs) 94–105
  future developments 114
  immunosensors 100–104
  molecular imprinted polymers (MIPs) 108–111
  nanotechnology 106–108
  screen printed 106
silt density index (SDI) 141
simazine 9–10, 434, 452
  nanofiltration 145
  removal efficiency 449
single-walled carbon nanotubes (SWCNTs) 98, 99, 147
SITE (Superfund Innovative Technology Evaluation) 311
slow sand filtration 146–147
sludge
  analysis 250–252
  biological treatment 278
  sorption onto 254–255
sludge retention time (SRT) 220, 221, 270–271
  and biodegradation 264–265
  in membrane bioreactors (MBRs) 274–275
  and removal efficiency 263–265
softening process *see* water softening
soils 18
sol-gel process 185–186, 192
solar photocatalytic process 377, 381
solid phase extraction (SPE) 67–68, 78, 252

solid phase microextraction (SPME) 68–69, 252
solute rejection 216–217
sonocatalytic degradation 383–388
sonochemical degradation 337–348
　of dyes 340–343
　of pharmaceuticals 336–340 *see also* ultrasound (US) treatment
sonoelectro-Fenton (SEF) process 393–394
sonophotocatalysis 385–388
sorption 254–257
　in activated sludge systems 256–257, 262
　of estrogens 208, 255–257, 262
　kinetics 209
　mechanisms 178–180, 206
　onto sludge 254–255
　and pH 262
　of pharmaceuticals 255
　on sediments 256
sorption sites, ferrocyanides 178
sotalol, removal efficiency 273
source separation sanitation *see* urine source separation
SPE *see* solid phase extraction (SPE)
Sphingobacterium sp. 280
Sphingomonas sp. 278–279
SPME *see* solid phase microextraction (SPME)
SPR (plasmon resonance) 98
SRT *see* sludge retention time (SRT)
steroid hormones 21–25, 227, 255
　excretion rates 21, 242
　see also estrogens
Stockholm Convention 427
stripping 310, 312
substituted ureas 10–13
subsurface flow constructed wetlands (SSFCWs) 273, 451–452
sugar industry, Great Barrier Reef (GBR) 439, 440, 442–443
sulfamethoxazole
　photodegradation 303
　removal efficiency 221, 278
supercritical theory of cavitation 323
supercritical water (SCW) 323

Superfund Innovative Technology Evaluation (SITE) 311
surface water
　personal care products in 27
　pharmaceuticals in 16, 257
　surfactants in 27
surfactants 25–30
　alkylphenol ethoxylates (APEs) 25
　in the aquatic environment 25–30
　bioaccumulation 29
　for bioavailability stimulation 261
　in effluent wastewater 27
　fate 27–29
　legislation 30
　nanofiltration (NF) 149
　nonylphenol ethoxylates (NPEs) 25
　nonylphenol (NP) 25
　removal from wastewater 27
　in surface water 27
　toxicity 29
sustainable analytical procedures (SAPs) 60
SWCNT *see* single-walled carbon nanotubes (SWCNTs)
SWRO (seawater reverse osmosis) 148–149

T

TCE *see* trichloroethylene (TCE)
TCS *see* triclosan (TCS)
tebuthiuron 434
temperature
　and adsorption 268
　and biodegradation 266–267
　and pharmaceuticals removal 267
　and removal efficiency 266–268
　and ultrasound (US) treatment 327
　in wastewater treatment plants (WWTPs) 266–267
terephthalic acid (TPA) 374
tetrachlorethene, removal efficiency 219
tetrachloroethylene (PCE)
　in groundwater 308, 313, 314
　removal efficiency 312
tetracyclines, removal efficiency 265
textile wastewater treatment
　nanofiltration (NF) 150

photocatalytic treatment 378
ultrasound (US) treatment 340–343
thin film composite (TFC) membranes
     132, 137–138, 149
time-of-flight (TOF) mass spectrometry
     73, 74, 251
titanium dioxide 374, 383–385
  nanocatalysts 412
  nanotubes 412–413
  photocatalysis 377–379, 389
TMP see trimethoprim (TMP)
TNP (2,4,6-trinitrophenol) 455
tonalide, removal efficiency 210, 221, 278
toxaphene 430, 434
toxicity
  to aquatic organisms 19–20
  of atrazine 446
  of diclofenac (DCF) 340
  of diuron 11–12, 445–446
  of dyes 341–342
  ecotoxicological assessment 79–80
  of herbicides 11–12, 445–446
  of nonylphenol (NP) 29
  of organochlorines 5–6
  of organophosphates 8
  of perfluorinated compounds (PFCs)
     32–33
  of pharmaceuticals 19–20
  of substituted ureas 11–12, 445–446
  of surfactants and personal care
     products 29
  of triazines 10
  of triclosan (TCS) 29
transformation products 71–78, 82–83
  biological assessment 78–82
  detection and quantification 71–78
transient cavitation 327
transport of persistent organic pollutants
     (POPs) 426
treatability studies 250–251, 315
triadimefon, removal efficiency 225
triazine herbicides 8–10, 434
  persistence 9
  toxicity 10
trichlorethene, removal efficiency 219
trichloroethylene (TCE)
  removal efficiency 311, 312

ultrasonic degradation 326
UV/H_2O_2 oxidation 301
2,4,6-trichlorophenol 302
trickling filters 277
triclosan (TCS) 26, 81
  photodegradation 28, 378–379
  removal efficiency 27, 226
  toxicity 29
trihalomethan 308
trimethoprim (TMP)
  biodegradation 272
  metabolites 14
  removal efficiency 221
2,4,6-trinitrophenol (TNP) 455

U
U-PLS (unfolded partial least squares) 65
UF see ultrafiltration (UF)
ultra-performance liquid chromatography
     (UPLC) 69, 72, 74
ultrafiltration (UF) 129–131, 135
  adsorption 218–219
  fate of micropollutants 218–219
  of ibuprofen 218, 220
  and membrane bioreactors (MBRs)
     220–221
  membrane processes 218–221
  modified fouling index (MFI) 141
  and powdered activated carbon (PAC)
     219
  removal efficiency 218, 219, 220–225
ultrasound (US) treatment 321–353
  acoustic power 326
  additives 327–329
  and adsorption (US/A) 333–334
  and advanced catalytic oxidation
     383–388
  chlorpyrifos 347
  and contaminant properties 326–327
  costs 348, 350–352
  degradation mechanism 323–325
  for dye removal 340–343
  and electrooxidation (US/EO) 334–335
  energy efficiency 329–330, 336, 351,
     353
  and enzymatic catalysis 397
  equipment 329–330

of estrogens 337–338
frequency 325–326, 334–335
hybrid techniques 330–336, 336–348
and hydrogen peroxide (US/H_2O_2) 330–331
industrial scale 348, 350, 352
micropollutant degradation 336–353
and microwave irradiation (US/MW) 335–336
of organochlorines 343, 347
of organophosphates 343
and oxidants 330–332
and ozone (US/O_3) 331–332, 339–340
of pesticides 343–348
pharmaceuticals 336–340
removal efficiency 325–330, 347–348
sonochemical degradation efficiency 325–330
and temperature 327
theory 322–330
and ultraviolet irradiation (US/UV) 333
see also cavitation
ultraviolet (UV) irradiation
with ozone (UV/O_3) 300, 308–309
and ultrasound (US/UV) 333
UV-Vis spectroscopy 71, 76, 251
see also UV/H_2O_2 oxidation
unfolded partial least squares (U-PLS) 65
upflow anaerobic sludge blanket (UASB) 276
UPLC see ultra-performance liquid chromatography (UPLC)
urine
  estrogens in 21
  parabens in 29
  pharmaceuticals in 14, 243, 244–246
  storage 246–247
  treatment 227, 246–248, 249
urine source separation 243–247
  advantages 245–246
  biodegradation 247–249
  fate of micropollutants 247–249
  nutrient recycling 246
  treatment systems 246–248, 249
  urine storage 246–247
US see ultrasound (US) treatment
UV-Visible spectroscopy 76, 251

UV/H_2O_2 oxidation 295–316
  alternative radiation sources 309–310
  commercial systems 311–313
  cost and performance 313–316
  degradation rates 300–303
  disinfection by-product (DBP) removal 308–309
  drinking water treatment 303–305
  endocrine disrupting chemicals (EDCs) 306–307
  estrogens 306–307
  groundwater treatment 300–303, 308
  iron, addition of 307–308
  laboratory scale experiments 300–307
  limitations 310
  mechanisms 299–300
  modifications 307–310
  municipal wastewater treatment 305–307
  paper and pulp industry 307
  pesticides 302
  pharmaceuticals 305–306
  photolysis 298–299
  practical issues 310–313
  removal efficiency 303–307, 311, 312
  stripping 310, 312
  theory 296–300

V
vertical flow constructed wetlands 273
virus removal 146–148
volatile organic compounds (VOCs)
  in groundwater 311–312, 315
  removal efficiency 311
volatization 258

W
WAO see wet air oxidation (WAO)
wastewater
  analysis 250–252
  endocrine disrupting chemicals (EDCs) 252–254
  pH of 212
  pharmaceuticals in 16
wastewater treatment plants (WWTPs)
  effluent micropollutants 242–243, 253–254
  estrogens 253–254, 266

ibuprofen removal 254, 255
influent micropollutants 240–243, 245–246
pharmaceuticals 240–243, 255
PPCP removal mechanisms 254
removal efficiency 16, 240, 263–264, 265, 266, 267–268
temperature variations 266–267
see also WWTP effluent
Water Framework Directive (WFD) 295–296, 452
water purification 172, 174
water quality 143, 151
water softening 131
  groundwater treatment 142–143
  ion exchange resins 174
wet air oxidation (WAO) 397–399
  see also catalytic wet air oxidation (CWAO)
wetland processes
  biological treatment 272–274
  herbicide/pesticide removal 451–452
  ibuprofen removal 273
  removal efficiency 273–274, 452
WFD *see* Water Framework Directive (WFD)
whole-cell biosensors 98–99
WWTP effluent
  estrogen transport 22
  organic matter 305–306
  perfluorinated compounds (PFCs) 32
  personal care products (PCPs) 27
  pharmaceuticals 16
  surfactants 27

Y
yeast estrogen screen (YES) 252–253

Z
zeolites 168, 172–173
  ion exchange 173
  membranes 133–134, 144
zero-valent iron (ZVI) 400, 411

CPSIA information can be obtained at www.ICGtesting.com
Printed in the USA
LVOW070443230212

270055LV00004B/10/P